Filaggrin

Jacob P. Thyssen
Howard I. Maibach
Editors

Filaggrin

Basic Science, Epidemiology, Clinical Aspects and Management

Springer

Editors
Jacob P. Thyssen, MD, PhD, DmSci
Department of Dermato-Allergology
National Allergy Research Centre
Gentofte Hospital
University of Copenhagen
Hellerup
Denmark

Howard I. Maibach, MD
Department of Dermatology
University of California
San Francisco, CA
USA

ISBN 978-3-662-51117-6 ISBN 978-3-642-54379-1 (eBook)
DOI 10.1007/978-3-642-54379-1
Springer Heidelberg New York Dordrecht London

Softcover reprint of the hardcover 1st edition 2014

Printed on acid-free paper

Springer is part of Springer Science+Business Media (www.springer.com)

Preface

Filaggrin. Why a book on filaggrin?

As described below, there are indeed many good reasons to be interested in this particular molecule and its effects on human homeostasis.

In 1977, Beverly Dale identified a "stratum corneum basic protein" (SCBP) that was rich in histidine and worked as an interfilamentous matrix substance. The term "filaggrin" (filament aggregating protein) was later coined when the investigators showed that the SCBP condensed and aligned keratin intermediate filaments in vitro. Epidermal decrease or absence of the precursor molecule pro-filaggrin was then linked to ichthyosis vulgaris, and in 2002, genetic linkage analyses mapped the filaggrin gene (*FLG*) to the epidermal differentiation complex on chromosome 1q21. Ever since 2006, when successful genotyping of the *FLG* was performed for the first time, novel groundbreaking discoveries fundamentally changed the way we understand the atopic disorders.

Our overall understanding of skin barrier functions has also benefited dramatically from filaggrin research being relevant to non-atopic cutaneous diseases. Genotyping of different populations has even suggested a possible evolutionary advantage of *FLG* mutations. While filaggrin is definitely a very important skin protein, filaggrin is also expressed in other tissues. We predict that we are merely at the beginning of research focusing on the broader perspectives of filaggrin. It is therefore of interest not only to dermatologists and allergists but also to clinicians from other specialties, toxicologists, and regulators. Moreover, pharmaceutical companies already show a strong interest in developing therapeutics to help these patients and prevent morbidity.

This first edition provides a broad overview of filaggrin research and should ideally be used as an encyclopedia. We are incredibly thankful to the authors who have so kindly committed to this project and who have carefully reviewed difficult topics and made the information easily accessible to the readers. Without such commitment, obviously no book could have been developed. We wish to thank Springer and especially Dr. Sverre Klemp for facilitating and supporting this project. Diane Lamsback and Suganya Selvaraj are thanked for their skilled, patient, and careful organization of the manuscripts and for communicating extensively with us and the authors. Professor Peter Elias is thanked for sharing his deep insight and knowledge

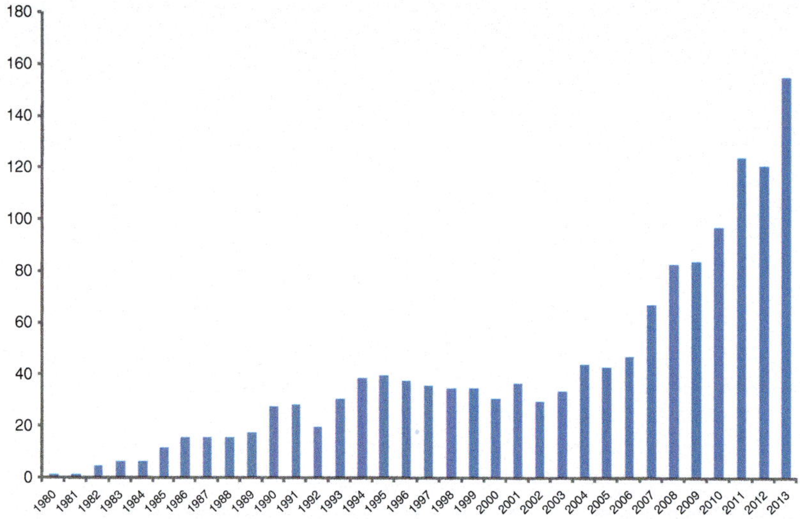

Fig. 1 The number of publications per year retrieved in PubMed using the search term "filaggrin"

on filaggrin and the skin barrier during Jacob Thyssen's sabbatical in San Francisco. Finally, we wish to thank Professor Torkil Menné for the idea of making a book on filaggrin, for being such a strong inspiration to his colleagues, and for always being enthusiastic.

Hellerup, Denmark — Jacob P. Thyssen
San Francisco, CA, USA — Howard I. Maibach

Contents

Contributors

Katrina J. Allen, MD, PhD Population Health, Murdoch Childrens Research Institute, Melbourne, VIC, Australia

Department of Allergy and Immunology, Royal Children's Hospital, Melbourne, VIC, Australia

Department of Paediatrics, The University of Melbourne, Melbourne, VIC, Australia

Charlotte M. Bonefeld, PhD Department of International Health, Immunology and Microbiology, University of Copenhagen, Copenhagen, Denmark

Maria Bradley, MD, PhD Department of Dermatology, Institution of Medicine, Karolinska University Hospital, Stockholm, Sweden

Richard Brans, MD Department of Dermatology, Environmental Medicine, Health Theory, University of Osnabrueck, Osnabrueck, Germany

David Bregnbak, MD Department of Dermato-Allergology, Gentofte Hospital, Copenhagen University, Hellerup, Denmark

Sophie C.S. Cai, BA (Hons), MS Duke-National University of Singapore Graduate Medical School, Singapore, Singapore

Berit Christina Carlsen, MD, PhD Department of Dermato-Allergology, National Allergy Research Centre, Gentofte Hospital, University of Copenhagen, Hellerup, Denmark

Ruhong Cheng, MD Department of Dermatology, Xinhua Hospital, Shanghai Jiao Tong University, Shanghai, China

Dieter L. Deforce, PhD, PharmD Laboratory for Pharmaceutical Biotechnology, University Ghent, Ghent, Belgium

Mette S. Deleuran, MD, DMSc Department of Dermatology, Aarhus University Hospital, Aarhus C, Denmark

Lennart Emtestam, MD, PhD Department of Dermatology, Karolinska Institutet, Stockholm, Sweden

Simon W. Fage, MD Department of Dermato-Allergology, National Allergy Research Centre, Gentofte Hospital, University of Copenhagen, Hellerup, Denmark

Carsten Flohr, MD, MA, MPhil, MSc, PhD Department of Paediatric Dermatology and Children's Allergies, St. John's Institute of Dermatology, Guy's and St. Thomas' Hospitals NHS Foundation Trust and King's College London, London, UK

Emily Y. Gan, MBBS(Hons), MRCP(UK), MMed(IntMed) Department of Dermatology, National Skin Centre, Singapore, Singapore

Carsten Geisler, MD, PhD, DMSc Department of International Health, Immunology and Microbiology, University of Copenhagen, Copenhagen, Denmark

Ana M. Gimenez-Arnau, MD, PhD, MBBS Department of Dermatology, Hospital del Mar Universitat Autonoma de Barcelona, Barcelona, Spain

Elena Godoy-Gijón, MD Department of Dermatology, Hospital De Cabueñes, Secretaría del Servicio de Dermatología, Policlínicas de la Casa del Mar, Gijón, Spain

Francisco J. González, PhD Center for the Investigation and Application of Science and Technology, Autonomous University of San Luis Potosi, San Luis Potosi, Mexico

Robert Gruber, MD Department of Dermatology and Venereology, Innsbruck Medical University, Innsbruck, Austria

Lena Hagströmer, MD, PhD Department of Dermatology and Venereology, South General Hospital, Stockholm, Sweden

Britta Hansmann, PhD Department of Dermatology, University Hospital of Schleswig-Holstein, Schleswig-Holstein, Germany

Steffen Heegaard, PhD Department of Neuroscience and Pharmacology, Eye Pathology Institute, University of Copenhagen, Copenhagen, Denmark

Department of Ophthalmology, Glostrup Hospital, University of Copenhagen, Copenhagen, Denmark

Jennifer R. Heimall, MD Division of Allergy and Immunology, Department of Pediatrics, The Children's Hospital of Philadelphia, Philadelphia, PA, USA

Thomas Iftner, Dr rer nat (PhD) Division of Experimental Virology, Institute for Medical Virology and Epidemiology of Viral Diseases, University Hospital, Eberhard-Karls-Universität Tübingen, Tuebingen, Germany

Peter Jensen, MD, PhD Department of Dermato-Allergology, Gentofte Hospital, Copenhagen University, Hellerup, Denmark

Ulla N. Joensen, MD, PhD Department of Growth and Reproduction, Copenhagen University Hospital (Rigshospitalet), Copenhagen, Denmark

Swen Malte John, MD, PhD Department of Dermatology, Environmental Medicine, Health Theory, University of Osnabrueck, Osnabrueck, Germany

Jakob Mutanu Jungersted, MD, PhD Department of Dermatology, Roskilde Hospital, Roskilde, Denmark

Sanja Kezic, PhD Coronel Institute of Occupational Health, Academic Medical Center, Amsterdam, The Netherlands

Susanne K. Kjaer, MD, DMSc Unit of Virus, Lifestyle and Genes, Gynecologic Clinic, Juliane Marie Centre, Rigshospitalet, and Danish Cancer Research Center, Copenhagen, Denmark

Yasuo Kubota, MD, PhD Department of Dermatology, Kagawa University, Kita-gun, Kagawa, Japan

Sarah Küchler, PhD Department of Pharmaceutical Sciences, Pharmacology/Toxicology, Freie Universität Berlin, Berlin, Germany

Ming Li, MD Department of Dermatology, Xinhua Hospital, Shanghai Jiao Tong University, Shanghai, China

Anne Sofie Lund, MD Department of Neuroscience and Pharmacology, Eye Pathology Institute, University of Copenhagen, Copenhagen, Denmark

Howard I. Maibach, MD Department of Dermatology, University of California, San Francisco, San Francisco, CA, USA

Michael Meldgaard, MS, PhD Department of Clinical Biochemistry, Gentofte Hospital, University of Copenhagen, Hellerup, Denmark

Torkil Menné, MD Department of Dermatology, KA-1502, Gentofte Hospital, University of Copenhagen, Hellerup, Denmark

Kozo Nakai, MD, PhD Department of Dermatology, Kagawa University, Kita-gun, Kagawa, Japan

Morten M. Nielsen, MS Department of International Health, Immunology and Microbiology, University of Copenhagen, Copenhagen, Denmark

Amy S. Paller, MD Department of Dermatology, Feinberg School of Medicine, Northwestern University, Chicago, IL, USA

Anne Marie Lynge Pedersen, PhD, DDS Section of Oral Medicine, Clinical Oral Physiology, Oral Pathology and Anatomy, Department of Odontology, Faculty of Health and Medical Sciences, University of Copenhagen, Copenhagen, Denmark

Jesper Reibel, PhD Department of Odontology, Faculty of Health and Medical Sciences, University of Copenhagen, Copenhagen, Denmark

Karin Sartorius, MD, PhD Department of Dermatology, Södersjukhuset, Stockholm, Sweden

Section of Dermatology, Department of Clinical Science and Education, Karolinska Institutet, Södersjukhuset, Stockholm, Sweden

Jens-Michael Schröder, PhD Department of Dermatology, University Hospital of Schleswig-Holstein, Schleswig-Holstein, Germany

Marie-Louise A. Schuttelaar, MD, PhD Department of Dermatology, University Medical Centre Groningen, Groningen, The Netherlands

Lone Skov, MD, PhD Department of Dermato-Allergology, Gentofte Hospital, Copenhagen University, Hellerup, Denmark

Jonathan M. Spergel, MD, PhD Allergy Section, Division of Allergy and Immunology, The Children's Hospital of Philadelphia, Philadelphia, PA, USA

Katleen F.E. Van Steendam, PhD, PharmD Laboratory for Pharmaceutical Biotechnology, University Ghent, Ghent, Belgium

David L. Stenoien, PhD Biological Sciences Division, Pacific Northwest National Laboratory, Richland, WA, USA

Gunnaporn Suriyaphol, DVM, MSc, Dr. rer. nat. Biochemistry Unit, Department of Veterinary Physiology, Faculty of Veterinary Science, Chulalongkorn University, Pathumwan, Thailand

Prapat Suriyaphol, BSc, PhD Bioinformatics and Data Management for Research Unit, Office for Research and Development, Faculty of Medicine, Siriraj Hospital, Mahidol University, Bangkoknoi, Thailand

Pal B. Szecsi, MD, DMSci Department of Clinical Biochemistry, Gentofte Hospital, University of Copenhagen, Hellerup, Denmark

Sreya Talasila, BS Department of Dermatology, Feinberg School of Medicine, Northwestern University, Chicago, IL, USA

Hern Tze Tina Tan, PhD School of Medical Sciences, Health Campus, Universiti Sains Malaysia, Kota Bharu, Malaysia

Mark B.Y. Tang, MBBS, MMED, FRCP (UK), FAMS Department of Dermatology, National Skin Centre, Singapore, Singapore

Sirin Theerawatanasirikul, DVM, PhD Department of Anatomy, Faculty of Veterinary Medicine, Kasetsart University, Chatuchak, Thailand

Simon F. Thomsen, MD, PhD Department of Dermatology, Bispebjerg Hospital, Copenhagen, Denmark

Andreas P. Thyssen, BS, MD Internal Medicine/Pediatrics, Maine Medical Center, Portland, ME, USA

Erik P. Thyssen, MD Specialists in Gastroenterology, Olde Cabin Road, St. Louis, MO, USA

Jacob P. Thyssen, MD, PhD, DmSci Department of Dermato-Allergology, National Allergy Research Centre, Gentofte Hospital, University of Copenhagen, Hellerup, Denmark

Teresa Tsakok, MA, BM, BCh (Oxon) Core Medical Trainee/NIHR Academic Clinical Fellow, St. John's Institute of Dermatology, Guy's and St. Thomas' Hospitals NHS Foundation Trust, London, UK

Wolfgang J.C. Uter, MD Department of Medical Informatics, Biometry and Epidemiology, University of Erlangen/Nürnberg, Erlangen, Germany

Christian Vestergaard, MD, PhD, DMSc Department of Dermatology, Aarhus University Hospital, Aarhus C, Denmark

Maaike J. Visser, BS Coronel Institute of Occupational Health, Academic Medical Centre, Amsterdam, The Netherlands

Tatiana A. Vogel, MD Department of Dermatology, University Medical Centre Groningen, Groningen, The Netherlands

Marten C.G. Winge, MD, PhD Department of Dermatology, Program in Epithelial Biology, Stanford University, Stanford, CA, USA

Zhirong Yao, MD Department of Dermatology, Xinhua Hospital, Shanghai Jiao Tong University, Shanghai, China

Kozo Yoneda, MD, PhD Department of Dermatology, Kagawa University, Kita-gun, Kagawa, Japan

Claus O.C. Zachariae, MD, DMSci Department of Dermato-Allergology, Gentofte Hospital, Copenhagen University, Hellerup, Denmark

Hui Zhang, MD Department of Dermatology, Xinhua Hospital, Shanghai Jiao Tong University, Shanghai, China

Part I

Basic Science

Function of Filaggrin and Its Metabolites

1

Sanja Kezic

Contents

S. Kezic, PhD
Coronel Institute of Occupational Health, Academic Medical Center, Meibergdreef 15, Amsterdam 1105 AZ, The Netherlands
e-mail: s.kezic@amc.uva.nl

1.1 Background

The skin is a highly specialized, complex, and efficient barrier that prevents loss of water from the body and ingress of exogenous compounds and pathogens into the body. The skin barrier function is primarily conferred by the outermost layer of the epidermis, stratum corneum (SC), in particular by its lipid component. SC is composed of flattened, terminally differentiated keratinocytes (corneocytes), which are embedded in a lipid matrix that creates the highly ordered lipid lamellae. Filaggrin is present in different compartments of the SC. The precursor of filaggrin, pro-filaggrin, is present in the stratum granulosum, whereas filaggrin is expressed in the cornified envelope and within the corneocytes.

A major emphasis for filaggrin has historically been placed on its role in the aggregation of intermediate keratin filaments within the corneocytes, ensuring mechanical integrity of the SC and as a main source of hygroscopic components, comprising the "natural moisturizing factors" (NMFs) [1]. The breakthrough discovery of the loss-of-function mutations in the filaggrin gene (*FLG*) and their importance for the development of ichthyosis vulgaris (IV) and atopic dermatitis (AD) initiated a cascade of studies aiming to elucidate the underlying mechanisms and functional consequences of decreased filaggrin expression for the skin barrier and immune aberrations in AD. Novel molecular biology technologies and in vitro and animal models of filaggrin deficiency provided a body of evidence that filaggrin itself,

J.P. Thyssen, H. Maibach (eds.), *Filaggrin*,
DOI 10.1007/978-3-642-54379-1_1, © Springer-Verlag Berlin Heidelberg 2014

but also its precursor pro-filaggrin, and degradation products interact with relevant structures and biological processes crucial for the homeostasis of the SC. In this chapter, recent knowledge on the main functions of filaggrin will be highlighted, addressing the key players in the processing and degradation of this unique protein as well as discussing their relevance to biological processes crucial for the maintenance of SC homeostasis.

1.2 Pro-Filaggrin and Pro-Filaggrin Processing

Pro-filaggrin is expressed as a giant, highly phosphorylated polyprotein that is present in the keratohyalin granules localized in the outer nucleated layers of the epidermis, the stratum granulosum [1–5]. Carriers of the *FLG* mutations show complete absence of the keratohyalin granules [2, 3]. Pro-filaggrin contains 10–12 filaggrin repeats flanked by an S100-type calcium-binding domain, two domains at the N-terminus and C-terminal domain [2, 3]. The S100-type domain contains two distinct calcium-binding sites and is suggested to play a role in the regulation of calcium-dependent events during epidermal differentiation [1–5]. The N-terminal domain has been shown to translocate to the nucleus, where it might have a signaling function during terminal epidermal differentiation [1]. The function of the C-terminal domain is unknown; however, its presence is important for pro-filaggrin stability and proteolytic processing of pro-filaggrin to functional filaggrin monomers [1].

During terminal differentiation, pro-filaggrin is rapidly dephosphorylated and cleaved to yield functional filaggrin monomers, which bind to and assemble keratin intermediate filaments [1–5]. Dissociation of filaggrin from keratin filaments mediated by peptidyl arginine deiminase is an essential step for further degradation of filaggrin into free amino acids. Cleavage of deiminated filaggrin monomers is catalyzed by several proteases, including caspase 14 [6], calpain 1, and bleomycin hydrolase [7]. As shown by in vivo Raman spectroscopy and biochemical analysis, the levels of filaggrin degradation products are significantly reduced in caspase 14-deficient mice compared with wild-type mice, although pro-filaggrin was fully processed to filaggrin [8]. The importance of filaggrin cleavage is underscored by the remarkably short half-life of filaggrin, specifically, only 6 h before full proteolysis [4, 9].

1.3 Filaggrin and Skin Barrier Function

Filaggrin monomers aggregate and align keratin bundles, contributing to the mechanical strength and integrity of the SC [1]. This has been supported by recent studies in *FLG*-KO mice (Flg-/Flg-) showing immature bundles and an aberrant network of keratin filaments in upper parts of the SG and loss of the keratin pattern in the lower SC [10]. However, as the barrier function largely depends on the composition and structural organization of the lipid extracellular matrix of the SC, the question arose whether and by which mechanisms filaggrin deficiency affects the barrier function. Initial studies in a filaggrin-deficient flaky tail (*ft*/*ft*) mice model demonstrated abnormalities in barrier function in absence of filaggrin [11, 12]. Penetration of water-soluble tracers was enhanced through the SC of *ft*/*ft* mice as compared to wild-type mice and occurred through the extracellular lipid matrix. As supported by ultrastructural analysis, the penetration-enhancing effect was assigned by a partial failure of lamellar body secretion. Importantly, the barrier abnormality correlated with reduced inflammatory thresholds to topical irritants and haptens [12].

Kawasaki et al. [10] recently identified premature detachment of corneocytes in the *FLG*-KO (*Flg*-/*Flg*-) mice that was accompanied with enhanced penetration of liposome-encapsulated calcein through the SC. *Flg*-/*Flg*- mice showed aberrant lipid composition. Interestingly, free fatty acid levels were normal; however, the levels of ceramide and cholesterol were increased, which is in contrast with reduced lamellar body secretion found in *ft*/*ft* mice. Similar to *ft*/*ft* mice [12], *Flg*-/*Flg*- mice showed increased inflammatory response after exposure to a skin irritant and allergen [10]. Notably, the same research group [13] showed in *Flg-/Flg-* mice that filaggrin deficiency affected specifically the lower layer of the SC, allowing specific metal ions to permeate viable layers. Reduced barrier function has also been demonstrated in two in vitro penetration assays in filaggrin-deficient skin models [14, 15].

In contrast with animal and in vitro models of filaggrin deficiency, in vivo data on skin permeability in humans are less consistent, although alterations in the structure of various SC components have been reported in several studies. Angelova-Fischer et al. [16] reported impairment in SC integrity and cohesion in AD patients who were *FLG* mutation carriers, which is consistent with disorganized keratin filaments, impaired lamellar body loading, and abnormal architecture of the lamellar bilayer found in the patients with ichthyosis vulgaris [17]. In the study of Jakasa et al. [18], non-lesional skin of AD patients showed higher permeability for PEG molecule of 370 Da in comparison to healthy skin; however, this was irrespective of their *FLG* genotype. No significant differences in the skin barrier between filaggrin AD and non-filaggrin AD were confirmed also in several studies measuring transepidermal water loss (TEWL) [16, 19–21], and so far only one study [22] reported lower TEWL in AD patients who were *FLG* mutation carriers as compared to *FLG* wild-type AD patients.

As breakdown products of filaggrin contribute to the acidic pH of the SC [17, 19, 23], which in turn regulates activity of various enzymes involved in the synthesis of SC lipids [24], several studies focused on the SC lipids in relation to *FLG* mutations [16, 19, 20]. However, recent studies, which have subdivided AD patients according to their FLG genotype, found no significant impact of FLG mutations on the composition or organization of SC lipids in non-lesional skin [16, 19, 20]. Though, in a study of Janssens et al. [20], the composition and organization of SC lipids was associated with the levels of filaggrin breakdown products although there was no association with the *FLG* genotype. A possible explanation for the absence of the effect of *FLG* mutations on the SC lipids can be explained by other factors than *FLG* mutations which can influence the filaggrin expression such as *FLG* copy numbers [25] and inflammation [26, 27]. To get more insight into the interplay between inflammation, filaggrin expression, and skin barrier function, further investigations in well-defined subgroups of patients and controls are needed.

While most studies focus on the SC and its intercellular lipid layers as a principal barrier of the skin, recent studies emphasize the role of tight junctions (TJs), which play a role in sealing epidermal cell-to-cell integrity. Recent studies have also shown that deficiency of filaggrin influences TJ proteins, suggesting that there is an interplay of SC and TJ barriers [17, 28].

1.4 Filaggrin Degradation Products

Filaggrin is a predominant source of free amino acids and their derivatives, contributing more than 50 % to total content of NMF and a complex mixture of hygroscopic compounds that contain filaggrin degradation products as well as chloride and sodium ions, lactate, and urea [5, 9, 29]. NMF primarily exists within the corneocytes and serves as a humectant, contributing to a proper hydration level in the skin. By regulating the water content in the SC, NMF has an important function in regulating the activity of enzymes crucial for maintaining epidermal barrier homeostasis. Furthermore, components of the NMFs contribute to the plasticity of the skin through interaction with keratin and keratin filaments [1, 30].

The most abundant amino acid in filaggrin is histidine, which is further enzymatically deaminated by histidase into *trans*-urocanic acid (*t-UCA*). *T*-UCA is a major absorber of UV radiation (UV-R) in the SC. Upon exposure to UV, *t*-UCA is isomerized into a *cis*-urocanic acid (*cis*-UCA), which has been shown to be (photo) immunosuppressive [31–34]. Caspase $14^{-/-}$ mice, which are deficient in UCA, had increased susceptibility to UV [8, 35]. An interesting hypothesis has recently been put forward by Thyssen et al. [36] that the deficiency of UCA due to *FLG* mutations may lead to increased penetration of UV-B across the epidermis, resulting in higher levels of 25-OH vitamin D. In addition to its protective role against UV-R, UCA has also been suggested to largely contribute to the “acid mantle” of the skin, although this view has been challenged based on studies with histidase- and filaggrin-deficient mice showing that UCA is not essential for the acidification of the SC pH [37, 38]. However, individuals with *FLG* mutations showed elevated skin surface pH in a dose-dependent fashion, and furthermore, pH values were inversely correlated with the NMF levels [17, 19, 23]. Elevated pH might lead to altered activity of proteases and enzymes within the SC that are involved in biosynthesis of lipids, desquamation, and cleavage of the precursors of IL-1 cytokines into their active form [24]. NMF levels were inversely correlated with IL-1 cytokine concentrations in the uninvolved skin of patients with moderate-to-severe AD [23]. These findings were supported by upregulated expression of IL-1b and IL-RA in filaggrin-deficient mice (flaky tail mice; *ft/ft*), suggesting that reduced filaggrin levels might cause a preexisting or enhanced proinflammatory status in the skin, although causal relationship between pH and IL-1 cytokines still has to be elucidated [23]. The “acid mantle” of the SC is important for antimicrobial defense, and there is evidence that filaggrin breakdown products at physiological concentrations exert an inhibitory effect on the growth of *Staphylococcus aureus* [39]. Thus, filaggrin deficiency may at least partly explain frequent colonization with *S. aureus* in AD patients, although the exact mechanisms still have to be elucidated.

Another important product of filaggrin degradation is pyrrolidone-5-carboxylic acid (PCA), which is derived from glutamine and glutamic acid. PCA is a highly hygroscopic compound and the most abundant single constituent of NMF (contributing for more than 10 %) [1, 9, 29].

In vivo studies based on Raman confocal microscopy showed that the carriers of *FLG* null mutations have reduced NMF levels as compared to noncarriers [21, 40]. This was confirmed also by biochemical analysis of PCA and UCA in the SC samples collected by adhesive tape in children with and without *FLG* mutations [26]. In a large cohort of AD patients, O’Regan et al. showed that the number of *FLG* mutations was sufficiently strong to discriminate three endophenotypes within AD [21]. In addition to *FLG* loss-of-function mutations, Brown et al. [25] showed recently that also intragenic copy number variations (CNVs) in the *FLG* are related to the UCA levels. The frequencies of CNV have been identified in Irish children as a risk factor for AD in a dose-dependent fashion. In addition to genetic variations, reduced filaggrin expression can be acquired likely due to the T_H2 cytokine milieu in AD [27]. Filaggrin expression was lower in patients with AD regardless of the *FLG* genotype, and furthermore, T_H2 cytokines downregulated *FLG* expression in differentiated keratinocytes [27]. This is consistent with the reduced levels of NMFs found in AD patients who are wild-type for *FLG* mutations as compared to wild-type healthy individuals [20, 26, 41, 42]. Severity of the disease was shown to significantly influence NMF levels in the SC, although the filaggrin genotype remains a major determinant of NMFs in the SC [26, 41, 42]. Interestingly, it has been shown recently that the levels of NMF in the SC are associated with altered composition and structure of SC lipids [20, 42], although a relationship between lipid changes and the filaggrin genotype could not be demonstrated. This points to the importance of acquired deficiency of filaggrin on the skin barrier.

Conclusion

In summary, several lines of evidence show that filaggrin, its precursor pro-filaggrin, and degradation products interfere with key processes and structures in the SC. Some of these effects are due to the filaggrin molecule itself (e.g., deficiency of filaggrin affects alignment of keratin bundles). In addition, deficiency of filaggrin leads to a variety of downstream effects that can be ascribed to reduced levels of its degradation products. Filaggrin breakdown products regulate skin hydration and, furthermore, contribute to the acidic pH of the skin, which in turn is crucial for the activity of various SC proteases involved in desquamation and lipid synthesis. A clearer understanding of the mechanisms by which filaggrin deficiency leads to skin barrier failure can help in identifying environmental risk factors for AD and improve therapeutic possibilities in combating this common inflammatory disease.

References

1. Harding CR, Aho S, Bosko CA. Filaggrin-revised. Int J Cosmet Sci. 2013;35(5):412–23.
2. Sandilands A, Sutherland C, Irvine AD, McLean WH. Filaggrin in the frontline: role in skin barrier function and disease. J Cell Sci. 2009;122(Pt 9):1285–94.
3. Brown SJ, McLean WHI. One remarkable molecule: filaggrin. J Invest Dermatol. 2012;132:751–62.
4. O'Regan GM, Sandilands A, McLean WH, Irvine AD. Filaggrin in atopic dermatitis. J Allergy Clin Immunol. 2009;124(3 Suppl 2):R2–6.
5. McAleer MA, Irvine AD. The multifunctional role of filaggrin in allergic skin disease. J Allergy Clin Immunol. 2013;131:280–91.
6. Denecker G, Hoste E, Gilbert B, Hochepied T, Ovaere P, Lippens S, et al. Caspase-14 protects against epidermal UVB photodamage and water loss. Nat Cell Biol. 2007;9(6):666–74.
7. Kamata Y, Taniguchi A, Yamamoto M, Nomura J, Ishihara K, Takahara H, et al. Neutral cysteine protease bleomycin hydrolase is essential for the breakdown of deiminated filaggrin into amino acids. J Biol Chem. 2009;284(19):12829–36.
8. Hoste E, Kemperman P, Devos M, Denecker G, Kezic S, Yau N, et al. Caspase-14 is required for filaggrin degradation to natural moisturizing factors in the skin. J Invest Dermatol. 2011;131(11):2233–41.
9. Rawlings AV, Harding CR. Moisturization and skin barrier function. Dermatol Ther. 2004;17 Suppl 1: 743–8.
10. Kawasaki H, Nagao K, Kubo A, Hata T, Shimizu A, Mizuno H, et al. Altered stratum corneum barrier and enhanced percutaneous immune responses in filaggrin-null mice. J Allergy Clin Immunol. 2012;129: 1538–46.e6.
11. Presland RB, Boggess D, Lewis SP, Hull C, Fleckman P, Sundberg JP. Loss of normal pro-filaggrin and filaggrin in flaky tail (ft/ft) mice: an animal model for the filaggrin-deficient skin disease ichthyosis vulgaris. J Invest Dermatol. 2000;115:1072–81.
12. Scharschmidt TC, Man MQ, Hatano Y, Crumrine D, Gunathilake R, Sundberg JP, et al. Filaggrin deficiency confers a paracellular barrier abnormality that reduces inflammatory thresholds to irritants and haptens. J Allergy Clin Immunol. 2009;124(3): 496–506.
13. Kubo A, Ishizaki I, Kubo A, Kawasaki H, Nagao K, Ohashi Y, et al. The stratum corneum comprises three layers with distinct metal-ion barrier properties. Sci Rep. 2013;3:1731.
14. Mildner M, Jin J, Eckhart L, Kezic S, Gruber F, Barresi C, et al. Knockdown of filaggrin impairs diffusion barrier function and increases UV sensitivity in a human skin model. J Invest Dermatol. 2010;130(9):2286–94.
15. Küchler S, Henkes D, Eckl KM, Ackermann K, Plendl J, Korting HC, et al. Hallmarks of atopic skin mimicked in vitro by means of a skin disease model based on FLG knock-down. Altern Lab Anim. 2011;39:471–80.
16. Angelova-Fischer I, Mannheimer AC, Hinder A, Ruether A, Franke A, Neubert RH, et al. Distinct barrier integrity phenotypes in filaggrin-related atopic eczema following sequential tape stripping and lipid profiling. Exp Dermatol. 2011;20:351–6.
17. Gruber R, Elias PM, Crumrine D, Lin TK, Brandner JM, Hachem JP, Presland RB, et al. Filaggrin genotype in ichthyosis vulgaris predicts abnormalities in epidermal structure and function. Am J Pathol. 2011;178(5):2252–63.
18. Jakasa I, Koster ES, Calkoen F, McLean WHI, Campbell LE, Bos JD, et al. Skin barrier function in healthy subjects and patients with atopic dermatitis in relation to filaggrin loss-of-function mutations. J Invest Dermatol. 2011;131:540–2.
19. Jungersted JM, Scheer H, Mempel M, Baurecht H, Cifuentes L, Høgh JK, et al. Stratum corneum lipids, skin barrier function and filaggrin mutations in patients with atopic eczema. Allergy. 2010;65:911–8.
20. Janssens M, Van Smeden J, Gooris GS, Bras W, Portale G, Caspers PJ, et al. Increase in short-chain ceramides correlates with an altered lipid organization and decreased barrier function in atopic eczema patients. J Lipid Res. 2012;53:2755–66.
21. O'Regan GM, Kemperman PMJH, Sandilands A, Chen H, Campbell LE, Kroboth K, et al. Raman profiles of the stratum corneum define 3 filaggrin genotype-determined atopic dermatitis endophenotypes. J Allergy Clin Immunol. 2010;126:574–80.
22. Nemoto-Hasebe I, Akiyama M, Nomura T, Sandilands A, McLean WHI, Shimizu H. Clinical severity correlates with impaired barrier in filaggrin-related eczema. J Invest Dermatol. 2009;129:682–9.
23. Kezic S, O'Regan GM, Lutter R, Jakasa I, Koster ES, Saunders S, et al. Filaggrin loss-of-function mutations are associated with enhanced expression of IL-1 cytokines in the stratum corneum of patients with atopic dermatitis and in a murine model of filaggrin deficiency. J Allergy Clin Immunol. 2012;129:1031–9.
24. Elias PM, Schmuth M. Abnormal skin barrier in the etiopathogenesis of atopic dermatitis. Curr Allergy Asthma Rep. 2009;9(4):265–72.
25. Brown SJ, Kroboth K, Sandilands A, Campbell LE, Pohler E, Kezic S, et al. Intragenic copy number variation within filaggrin contributes to the risk of atopic dermatitis with a dose-dependent effect. J Invest Dermatol. 2012;132:98–104.
26. Kezic S, O'Regan GM, Yau N, Sandilands A, Chen H, Campbell LE, et al. Levels of filaggrin degradation products are influenced by both filaggrin genotype and atopic dermatitis severity. Allergy. 2011;66: 934–40.

27. Howell MD, Kim BE, Gao P, Grant AV, Boguniewicz M, DeBenedetto A, et al. Cytokine modulation of atopic dermatitis filaggrin skin expression. J Allergy Clin Immunol. 2009;124:R7–12.
28. Nakai K, Yoneda K, Hosokawa Y, Moriue T, Presland RB, Fallon PG, et al. Reduced expression of epidermal growth factor receptor, E-cadherin, and occludin in the skin of flaky tail mice is due to filaggrin and loricrin deficiencies. Am J Pathol. 2012;181:969–77.
29. Rawlings AV, Scott IR, Harding CR, Bowser PA. Stratum corneum moisturization at the molecular level. J Invest Dermatol. 1994;103:731–40.
30. Yates Y. Investigations onto the controlled drying of sheepskins. J Am Leather Chem Assoc. 1996;61:235–400.
31. Kammeyer A, Teunissen MBM, Pavel S, de Rie MA, Bos JD. Photoisomerization spectrum of urocanic acid in human skin and in vitro: effects of simulated solar and artificial ultraviolet radiation. Br J Dermatol. 1995;132:884–91.
32. Gibbs NK, Tye J, Norval M. Recent advances in urocanic acid photochemistry, photobiology and photoimmunology. Photochem Photobiol Sci. 2008;7:655–67.
33. Norval M, Gibbs NK, Gilmour J. The role of urocanic acid in UV-induced immunosuppression: recent advances (1992–1994). Photochem Photobiol. 1989;50: 267–75.
34. Walterscheid JP, Ullrich SE, Nghiem DX. Platelet-activating factor, a molecular sensor for cellular damage, activates systemic immune suppression. J Exp Med. 2002;195:171–9.
35. Devos M, Prawitt J, Staumont-Salle D, Hoste E, Fleury S, Bouchaert E, et al. Filaggrin degradation by caspase-14 is required for UVB photoprotection but does not influence allergic sensitization in a mouse model of atopic dermatitis. J Invest Dermatol. 2012; 132:2857–60.
36. Thyssen JP, Thuesen B, Huth C, Standl M, Carson CG, Heinrich J, et al. Skin barrier abnormality caused by filaggrin (FLG) mutations is associated with increased serum 25-hydroxyvitamin D concentrations. J Allergy Clin Immunol. 2012;130(5):1204–7.
37. Fluhr JW, Elias PM, Man MQ, Hupe M, Selden C, Sundberg JP, et al. Is the filaggrin-histidine-urocanic acid pathway essential for stratum corneum acidification? J Invest Dermatol. 2010;130(8):2141–4.
38. Fluhr JW, Behne MJ, Brown BE, Moskowitz DG, Selden C, Man MQ, et al. Stratum corneum acidification in neonatal skin: secretory phospholipase A2 and the sodium/hydrogen antiporter-1 acidify neonatal rat stratum corneum. J Invest Dermatol. 2004;122(2):320–9.
39. Miajlovic H, Fallon PG, Irvine AD, Foster TJ. Effect of filaggrin breakdown products on growth of and protein expression by Staphylococcus aureus. J Allergy Clin Immunol. 2010;126(6):1184–90.
40. Kezic S, Kemperman PM, Koster ES, de Jongh CM, Thio HB, Campbell LE, et al. Loss-of-function mutations in the filaggrin gene lead to reduced level of natural moisturizing factor in the stratum corneum. J Invest Dermatol. 2008;128:2117–9.
41. Mlitz V, Latreille J, Gardinier S, Jdid R, Drouault Y, Hufnagl P, et al. Impact of filaggrin mutations on Raman spectra and biophysical properties of the stratum corneum in mild to moderate atopic dermatitis. J Eur Acad Dermatol Venereol. 2012;26:983–90.
42. van Smeden J, Janssens M, Lavrijsen AP, Kezic S, Vreeken RJ, Bouwstra JA. Skin barrier dysfunction in non-lesional atopic skin: the role of stratum corneum lipids. Eur J Dermatol. 2013; (early view. 9 April 2013).

Morphology of Filaggrin-Depleted Epidermis

2

Robert Gruber

Contents

R. Gruber, MD
Department of Dermatology and Venereology, Innsbruck Medical University, Anichstraße 35, Innsbruck 6020, Austria
e-mail: r.gruber@i-med.ac.at

In filaggrin-depleted epidermis, the amount of the protein filaggrin is more or less reduced, which leads to multiple morphological changes mainly in the stratum granulosum (SG) and stratum corneum (SC) and consequently to abnormalities in epidermal permeability barrier function [1]. The correlation between filaggrin deficiency and barrier abnormality can primarily be investigated in ichthyosis vulgaris (IV), the most prevalent disorder of cornification in humans, which is characterized by the postnatal onset of a generalized fine scaling phenotype, and associated with palmoplantar hyperlinearity, keratosis pilaris, and atopic dermatitis (AD) [2–6], because this skin disease is caused by loss-of-function mutations in the filaggrin gene (*FLG*) [7]. Mutations result in a truncated pro-filaggrin protein, which is not further processed into functional filaggrin monomers [8]. In IV, which is inherited in a semidominant way, phenotype severity appears to be subject to a dose effect, wherein heterozygous patients show a milder phenotype with reduced filaggrin, whereas homozygous or compound heterozygous *FLG* mutation carriers typically lack filaggrin and exhibit a more severe scaling phenotype with a greater predisposition for early development and more severe AD [7–11]. As approximately 25 % of individuals with AD also display *FLG* mutations [12, 13], the contribution of filaggrin deficiency to an abnormal barrier function was also assessed in AD [14]; however, a bias due to Th2-dominant inflammation, which secondarily compromises barrier function by multiple mechanisms [15], including an acquired reduction in filaggrin, cannot be excluded [16–18]. Furthermore,

J.P. Thyssen, H. Maibach (eds.), *Filaggrin*,
DOI 10.1007/978-3-642-54379-1_2,

for investigating the morphology of filaggrin-depleted epidermis, two mouse models, the filaggrin null mouse, in which *flg* is knocked out [19], and the flaky tail mouse, which bears a distal frameshift mutation in the profilaggrin gene lacking processed murine *flg* [20], however, the latter mouse in addition carries the *Tmem79/Matt* gene causing spontaneous dermatitis and atopy [21].

During keratinocyte differentiation, pro-filaggrin, which is the main component of F-type keratohyalin granules defining the SG, is dephosphorylated and cleaved by proteases (matriptase) into C-terminal filaggrin monomers, which are aggregating keratin filaments in the cytoplasm of early corneocytes, resulting in a flattening of the cells [22]. In parallel, beginning in the SG, the cornified envelope (CE) is formed at the inner plasma membrane of the keratinocytes as involucrin, filaggrin monomers, loricrin, trichohyalin, elafin, small proline-rich proteins, plakins, and additional structural proteins become cross-linked by calcium-dependent glutaminases [23–25]. The outcome of this is a rigid structure surrounding the corneocytes that provides mechanical resistance against keratolytics, organic solvents, and microorganisms [23]. Attached to the external site of the CE is the corneocyte lipid envelope, which is an omega-hydroxyceramide-rich monolayer serving as an initial scaffold for the lipids secreted by lamellar bodies (LBs) and assisting the organization of the extracellular lamellar bilayers [26, 27]. The hydrophobic extracellular matrix, together with the cohesive properties of specialized intercellular junctions in SC and SG (i.e., corneodesmosomes [CDs] and tight junctions [TJs]), provides the epidermis with a formidable barrier to the outward loss of water and electrolytes, while also blocking transcutaneous entry of exogenous xenobiotics [16, 28, 29].

2.1 Sparse Inflammatory Infiltrates, Decreased Stratum Corneum Hydration, and Increased Skin Surface pH in FLG-Depleted Epidermis

Filaggrin proteolysis regulates SC hydration [30] and likely also contributes to the acidification of normal SC [31, 32]. Accordingly, SC hydration declines by about 30 % in *FLG* double-allele mutant individuals in comparison to wild-type controls, but reductions in hydration levels do not achieve statistical significance in single-allele mutant subjects [1, 33]. By steepening the gradient of water loss across the SC, decreased SC hydration alone likely also places further stress on the permeability barrier in IV. Decreased SC hydration correlated with the development of a sparse inflammatory infiltrate in double-allele mutant subjects, which is enriched in mast cells [1]. This confirms prior studies showing that prolonged exposure of normal skin to reduced environmental humidity stimulates mast cell infiltration [34]. Furthermore, filaggrin-depleted epidermis displays an increase in skin surface pH as a consequence of a reduction in filaggrin metabolites such as polycarboxylic acids, including *trans*-urocanic acid and pyrrolidone carboxylic acid, which impart most of the hygroscopic properties that underlie SC hydration (see Chap. 1) [1, 20, 35–37]. In IV patients, an inverse relationship between filaggrin dose and skin surface pH is verifiable, with double-allele mutant subjects displaying the greatest elevations in skin surface pH [1]. However, in a recent study, skin pH was not significantly different in IV patients versus normal controls, but this is most likely due to power limitations [33].

2.2 Distinctive Corneocyte Abnormalities and Altered Epidermal Homeostasis in FLG-Depleted Epidermis

In the pre-genotype era, a reduced or absent SG, which can be detected by histology, immunohistochemistry (IHC), and electron microscopy, has already been requested as a main diagnostic criteria for IV [38–41]; however, today further graduation is feasible as reductions in filaggrin are allele-dose-dependent [1]. Loss of filaggrin causes an absence to near-absence of F-type keratohyalin granules in patients with double-allele mutations, with residual granules and filaggrin expression in single-allele mutation carriers (Fig. 2.1a). However, in rare cases, even individuals with two mutated alleles can

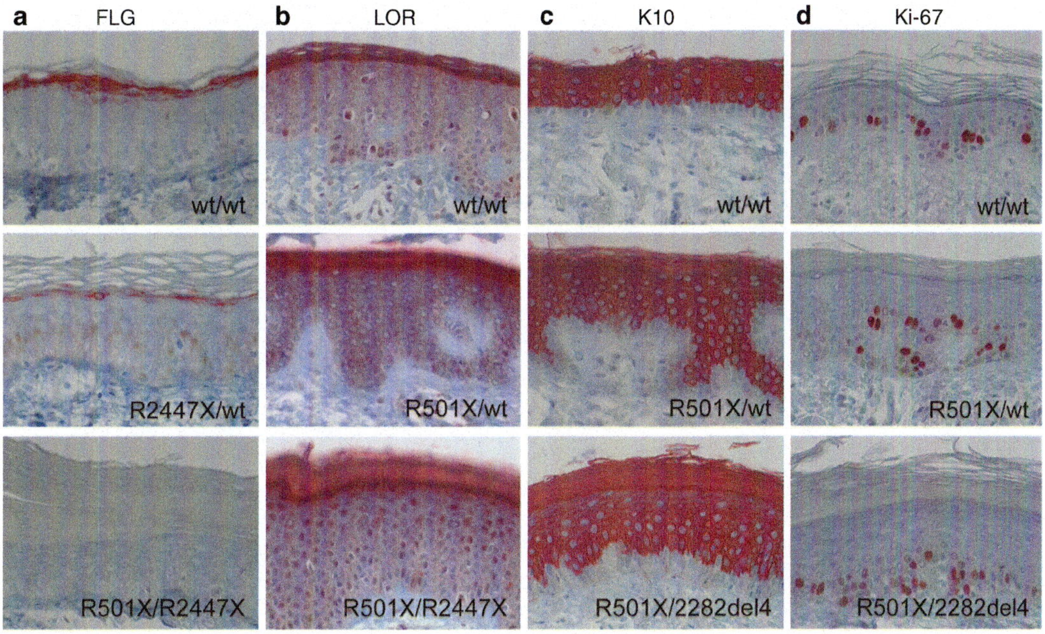

Fig. 2.1 Allele-dose-dependent reductions in filaggrin and epidermal hyperplasia in filaggrin-depleted epidermis. (**a**) Whereas in wild types (wt/wt) filaggrin staining in the SG is normal, in heterozygote individuals it is reduced, and in double-allele mutant IV subjects it is missing. (**b**) No differences in loricrin (*LOR*) expression between wt/wt, filaggrin gene (*FLG*) heterozygotes, and double-allele mutant IV. (**c**) No differences in keratin 10 (*K10*) expression between all three groups. (**d**) While overall Ki-67 staining is not increased, Ki-67-positive cells extend into suprabasal layers in both single- and double-allele mutant IV subjects. Immunohistochemistry. Mag=400×

exhibit remaining keratohyalin granules [10]. On electron microscopy, residual F-type keratohyalin granules appear crumbly [41, 42]. In addition, keratinocytes of the SG of filaggrin-depleted epidermis display a distinctive cytoskeletal abnormality (i.e., perinuclear halos resulting from retraction of keratin intermediate filaments) (Fig. 2.2) [1], but throughout the remaining corneocyte, cytosol keratin intermediate filament organization appears normal. This suggests that low amounts of filaggrin are sufficient for normal keratin aggregation or that other filaggrin-like proteins can replace the keratin binding function of filaggrin. Most likely because of a compensatory repair mechanism, reduction or loss of filaggrin correlates with epidermal hyperplasia. An increased number of SC cell layers, which histologically becomes apparent in an orthohyperkeratotic or basket-weave-like SC (see Fig. 2.1a) [43], and a higher number of suprabasal, Ki-67-positive cells are seen in double-allele mutant IV (see Fig. 2.1d); however, the proliferation rate is much lower compared to lamellar ichthyosis or psoriasis. With the exception of reduced filaggrin, other differentiation markers, such as loricrin and keratin 10, do not differ in filaggrin-deficient versus control epidermis (see Fig. 2.1b, c). In summary, *FLG* mutations result in a characteristic cytoskeletal abnormality, as well as altered epidermal homeostasis.

2.3 Defective Corneocyte Integrity in Filaggrin-Depleted Stratum Corneum

Ultrastructurally, the corneocyte cytosol, the structure of the CE, and the morphology of the corneocyte lipid envelope appear normal in filaggrin-depleted epidermis, although filaggrin is a component of the CE. Yet, CE dimensions, measured in coded, randomized electron micrographs, are moderately decreased in both single- and double-allele filaggrin-deficient versus wild-type

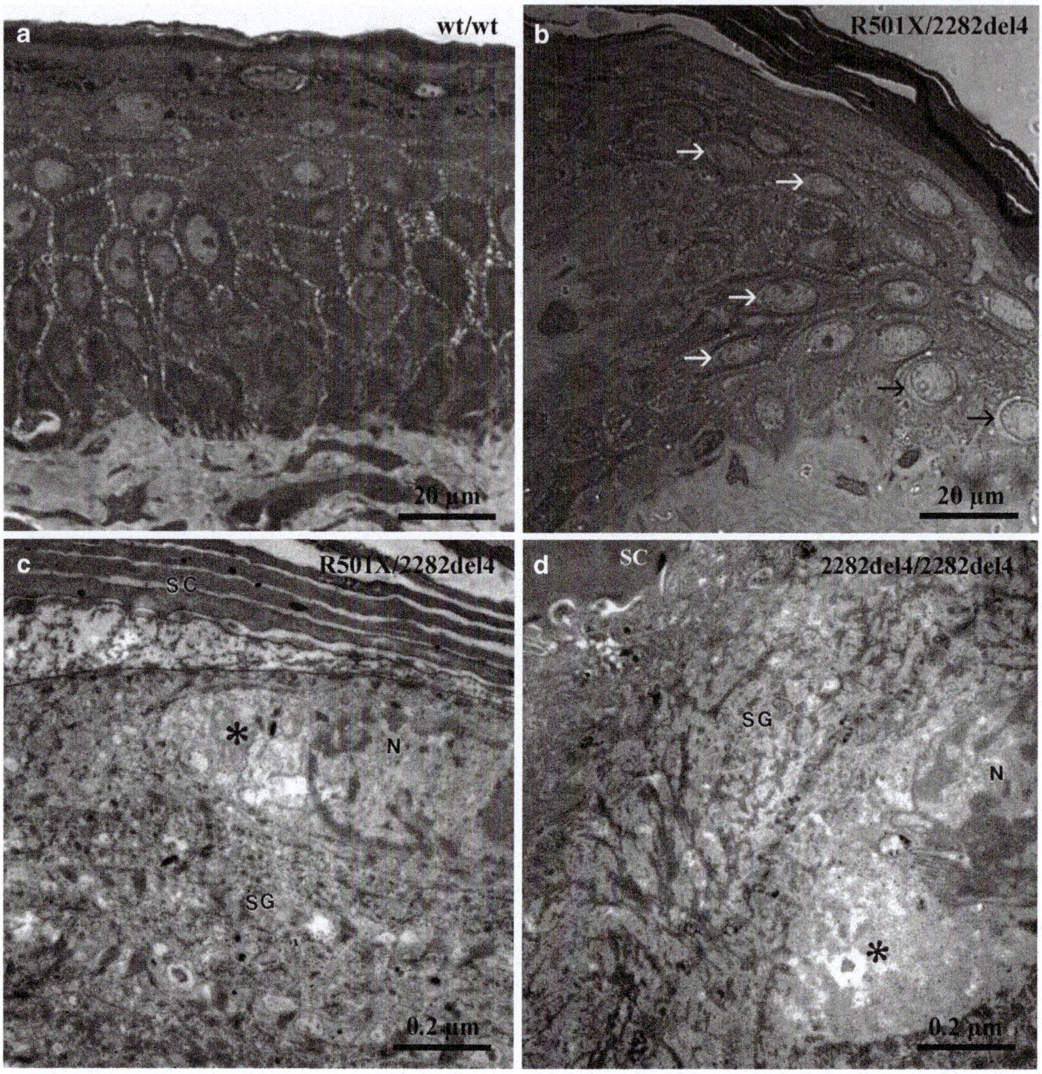

Fig. 2.2 Cytoskeletal abnormalities in filaggrin-depleted epidermis. As compared to control subjects (wt/wt) (**a**), intermediate filament retractions and reductions in F-type keratohyalin are visible in keratinocytes of filaggrin-deficient epidermis (**b–d**). *Arrows* and *asterisks* depict perinuclear keratin intermediate filament retraction. *N* nucleus. (**a, b**) Toluidine blue staining. Mag bar = 20 µm. (**c, d**) Osmium tetroxide post-fixation. Mag bars = 0.2 µm

corneocytes [1]. Thinning of CEs in IV corneocytes correlates with a moderate, gene dose-dependent impairment in corneocyte integrity, which is more subtle than the abnormality previously described for TGM1-deficient, autosomal recessive lamellar ichthyosis [44]. Specifically, in contrast to control corneocytes, which are not disrupted (Fig. 2.3a), a subgroup of corneocytes from IV subjects displays ragged, fragile outlines after detergent and heat (2 % SDS, 80 °C) treatment (see Fig. 2.3b, c). Higher percentages of fragile corneocytes are most obvious in individuals with double-allele mutations (see Fig. 2.3c) [1]. Moreover, the time to corneocyte disruption upon ultrasound treatment is significantly decreased in double-allele mutant IV as compared to wild type; however, differences are not significant between single-allele mutant IV and controls [1]. Thus, despite normal-appearing corneocytes, filaggrin-deficient corneocytes

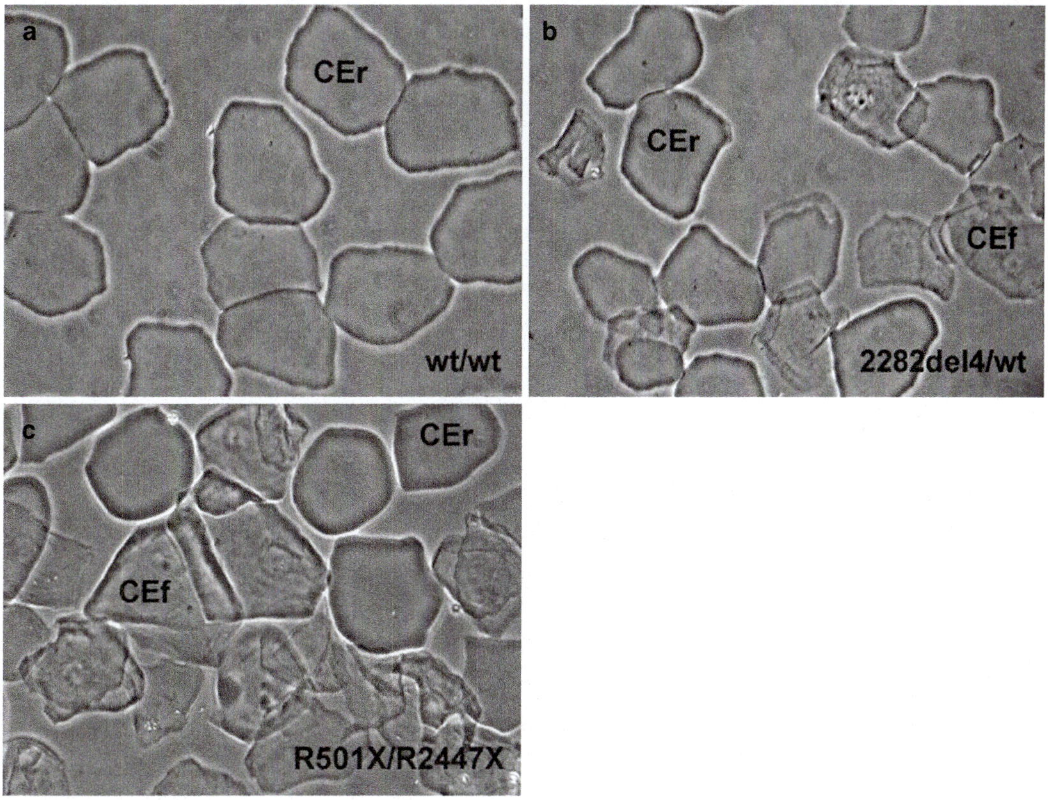

Fig. 2.3 Impaired corneocyte integrity in filaggrin-depleted epidermis. Whereas after detergent and heat treatment (2 % SDS, 80 °C) corneocytes show a normal rigid outline (*CEr*) in wild-type (wt/wt) individuals (**a**), higher percentages of ragged, fragile corneocytes (*CEf*) are verifiable in single-allele (**b**) and double-allele mutant individuals (**c**). Phase microscopy. Mag = 100×

display gene dose-dependent alterations in CE structure that result in reduced corneocyte integrity.

2.4 Impaired Epidermal Permeability Barrier Function Due to Increased Paracellular Permeability in Filaggrin-Depleted Epidermis

Basal permeability barrier function is impaired in filaggrin-depleted epidermis compared to normal skin, as there is an increased transepidermal water loss (TEWL) and a decreased SC hydration; however, differences are only significant in double-allele mutant IV compared to nonmutation carriers, but not in single-allele mutant IV [1, 33, 45]. Moreover, barrier recovery after acute barrier disruption by tape stripping is significantly delayed in both single- and double-allele mutant subjects [1]. This shows that filaggrin deficiency alone without coexistent AD suffices to alter permeability barrier function.

In all of the inherited disorders of corneocyte proteins to date, increased permeability occurs via a paracellular rather than transcellular pathway, due to abnormalities in either corneocyte scaffold function or cytoskeletal disruption of LB secretion [46]. Equally, filaggrin-depleted epidermis shows an extracellular permeability impairment, which can be demonstrated by the perfusion pathway of lanthanum nitrate, a water-soluble, electron-dense tracer. Lanthanum tracer normally is excluded from both the corneocyte cytosol and the SC extracellular matrix in non-mutation carriers (Fig. 2.4a). In contrast, despite

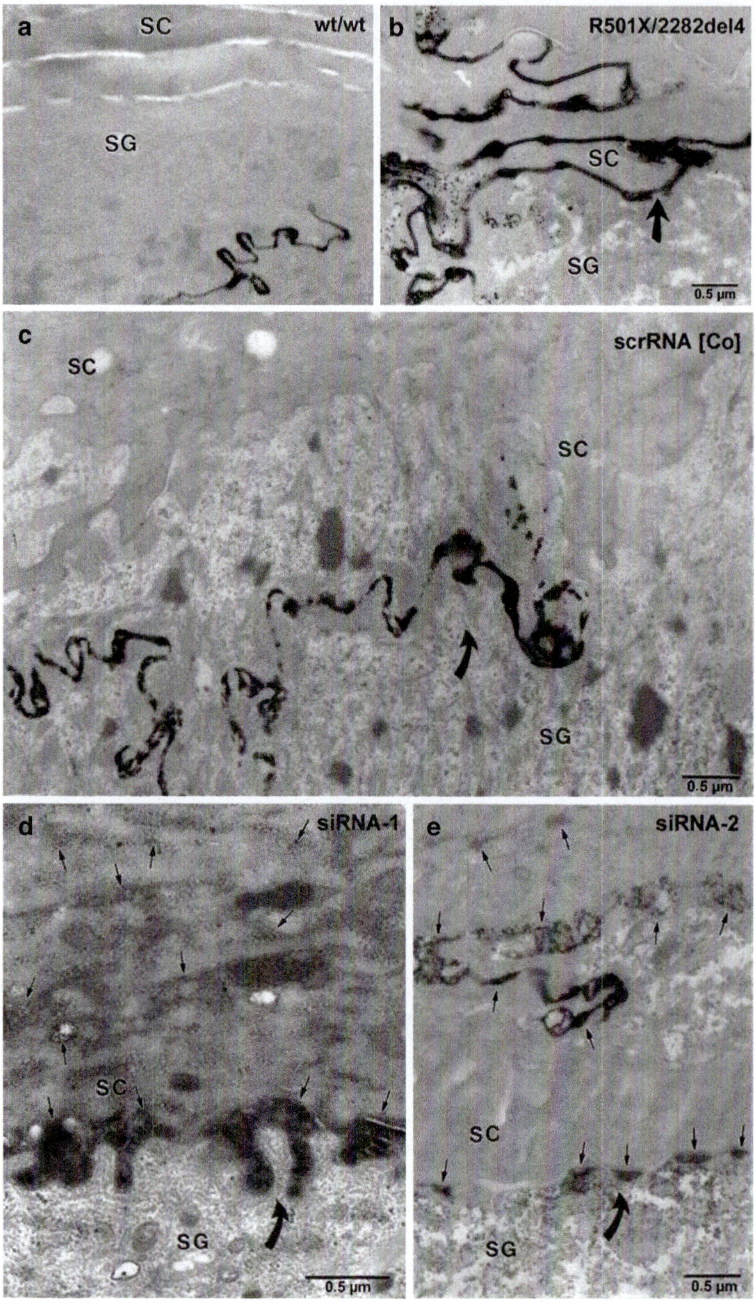

Fig. 2.4 Paracellular barrier defect in filaggrin-depleted epidermis. While lanthanum nitrate tracer is excluded from both the corneocyte cytosol and the extracellular matrix of the SC in control subjects (wt/wt) (**a**), tracer breaches SG-SC interface in both single- (not shown) and double-allele mutant IV patients and crosses SC by a paracellular route (**b**). (**c**) Stop of tracer perfusion at the SG-SC interface in organotypic skin cultures transfected with a scrambled (scr)RNA. (**d**, **e**) Filaggrin-deficient cultures (two different siRNAs) display lanthanum movement into and across the SC via the paracellular pathway. *Curved arrows* indicate (outward) direction of tracer movement. *Small arrows* mark colloidal lanthanum nitrate in the extracellular space. Osmium tetroxide post-fixation. Mag bars = 0.5 µm

the abnormalities in CE structure and integrity described above, tracer breaches SG-SC interface and permeates into and through the SC in both single- and double-allele mutant IV subjects, with passage entirely restricted to extracellular domains (see Fig. 2.4b) [1]. Furthermore, in filaggrin-deficient organotypic skin cultures, generated from keratinocytes transfected with two different small interfering ribonucleic acids (siRNAs) [47], as well as in organotypic human cultures, prepared with keratinocytes from single- and double-allele mutant IV, lanthanum tracer again displays movement into and across the SC solely via the extracellular pathway (see

Fig. 2.4d, e), whereas in scrambled ribonucleic acid and nonmutation human control cultures, tracer is excluded from the SC (see Fig. 2.4c) [1], confirming that filaggrin deficiency alone suffices to provoke a barrier abnormality.

Flaky tail mice display an increased bidirectional paracellular permeability of water-soluble xenobiotics due to impaired secretion of LBs and altered SC extracellular membranes and decreased thresholds to the development of an AD-like dermatosis following exposure to haptens or allergens [48, 49], which could explain the enhanced risk of atopic diathesis in IV [50, 51].

2.5 Abnormal Extracellular Lamellar Bilayer Architecture Due to Impaired Loading of Secretory Cargo Within Lamellar Bodies and Nonuniform Extracellular Dispersion of Secreted Contents in FLG-Depleted Stratum Corneum

Detailed analysis on electron microscopy using reduced osmium and ruthenium tetroxide postfixation reveals impairment of the supramolecular organization of lamellar bilayers as causal for the altered paracellular permeability in IV [1]. Whereas in normal SC mature processed lamellar bilayers replace newly secreted membrane structures one cell layer above the SG-SC interface (Fig. 2.5a), in filaggrin-depleted SC lamellar bilayer organization is disrupted by foci of nonlamellar, amorphous material in the mid-to-upper SC (see Fig. 2.5b); however, the overall quantities of lamellar bilayers are reduced nonsignificantly in control subjects versus IV patients. Moreover, in double-allele mutant IV subjects and the majority of heterozygous IV patients, there is delayed maturation of extracellular lamellar bilayers as incompletely processed immature lamellar material persists several cell layers above the SG-SC interface (see Fig. 2.5c, d) [1]. Although the density of LBs appears almost normal in filaggrin-depleted epidermis, many organelles display microvesicular or amorphous internal contents (Fig. 2.6c, d), suggesting a defective loading of LBs. In parallel, foci of non-lamellar, vesicular contents replace normal lamellar bilayers at the SG-SC interface (see Fig. 2.6a, b). Furthermore, filaggrin-depleted epidermis exhibits abnormalities in the extracellular dispersion of secreted LB contents, assessed as distribution of acid lipase, an ultrastructural marker of LB contents/secretion [1, 52]. Compared to normal epidermis, where acid lipase is concentrated initially within LBs, and then secreted in toto, followed by an uniform dispersion within the SC interstices (Fig. 2.7a), acid lipase is distributed nonuniformly within the extracellular spaces in IV subjects (see Fig. 2.7b); however, in the latter, both total contents and secretion of enzyme activity appear near-normal. Yet, quantitative measurements only show a reduction in LB secretion in *FLG* double-allele mutants, but not in single-allele IV [1]. The abovementioned increase in skin surface pH in IV could reduce the activity of ceramide-generating hydrolases (in particular, beta-glucocerebrosidase and acid sphingomyelinase require acidic pH optima) [53, 54], accounting for the observed delay in maturation of extracellular lamellar bilayers in FLG-depleted skin. In summary, the abnormalities in extracellular lamellar bilayer structure in IV likely reflect impaired loading of secretory cargo within LBs, followed by faulty, post-secretory dispersion of organelle contents.

2.6 Defective Corneocyte Cohesion Based on Decreased Corneodesmosome Density and Length and Abnormalities in Tight Junction Protein Expression in FLG-Depleted Epidermis

Impairment of the epidermal permeability barrier is not only caused by abnormal extracellular lamellar bilayer architecture but also a result of altered cohesive properties of the intercellular junctions in the SC and SG. SC cohesion, which

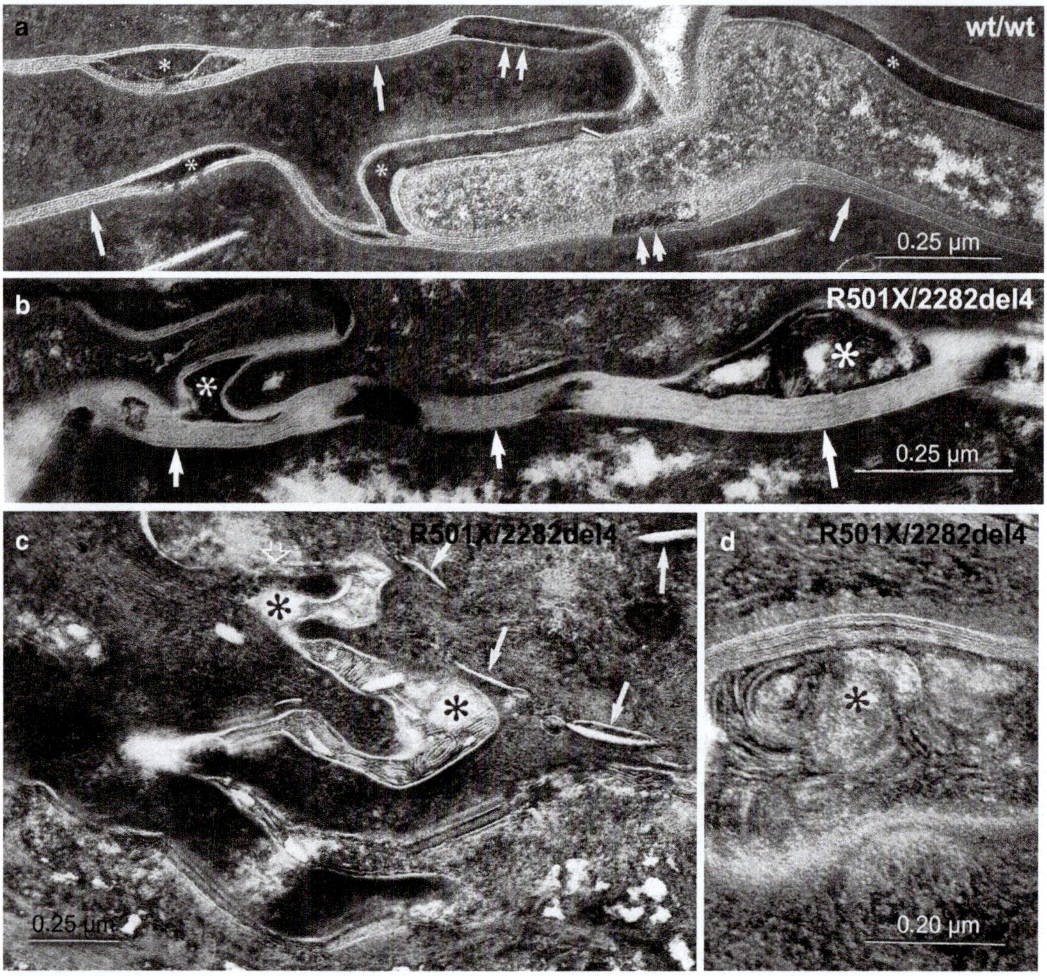

Fig. 2.5 Abnormal extracellular lamellar bilayer organization and delayed maturation in filaggrin-depleted epidermis. (**a**) Normal lamellar bilayers (*arrows*), corneodesmosomes (CDs) (*double-arrows*), and CD-derived lacunae (*asterisks*) in wild-type controls (wt/wt). (**b**) Normal numbers of bilayers (*arrows*), but disruption of membrane arrays by non-lamellar domains (*asterisks*) in double-allele mutant IV subjects. (**c, d**) Delayed transformation of secreted lamellar body contents into bilayers above the SG-SC interface (*asterisks*). Note that CD structure (*double-arrows*) appears normal. Ruthenium tetroxide post-fixation. Mag bars = 0.25 μm (**a–c**); 0.2 μm (**d**)

is commonly measured by the amount of protein removed per strip by sequential tape stripping, is decreased in IV patients in an *FLG* mutation dose-dependent way [1]. Ultrastructurally, this goes together with reduced density of CDs and a significantly diminished CD length in filaggrin-depleted SC compared to nonmutation carriers; however, the structure of individual CDs appears normal [1]. In addition, there is evidence for an impaired TJ formation shown by abnormal expression and localization of TJ proteins in filaggrin-depleted epidermis. While in immunofluorescence stainings of normal epidermis a clear staining at the cell-cell borders of the uppermost living cell layers is observed for both occludin and ZO-1 (Fig. 2.8a, b), staining is reduced with a more cytoplasmic pattern in FLG single-allele mutant (see Fig. 2.8c, d) and almost completely lost in double-allele mutant skin (see Fig. 2.8e, f) [1]. The abovementioned increased skin pH in filaggrin-depleted SC could also impact SC function by altering TJ function and by favoring CD proteolysis, which could further compromise barrier function [55].

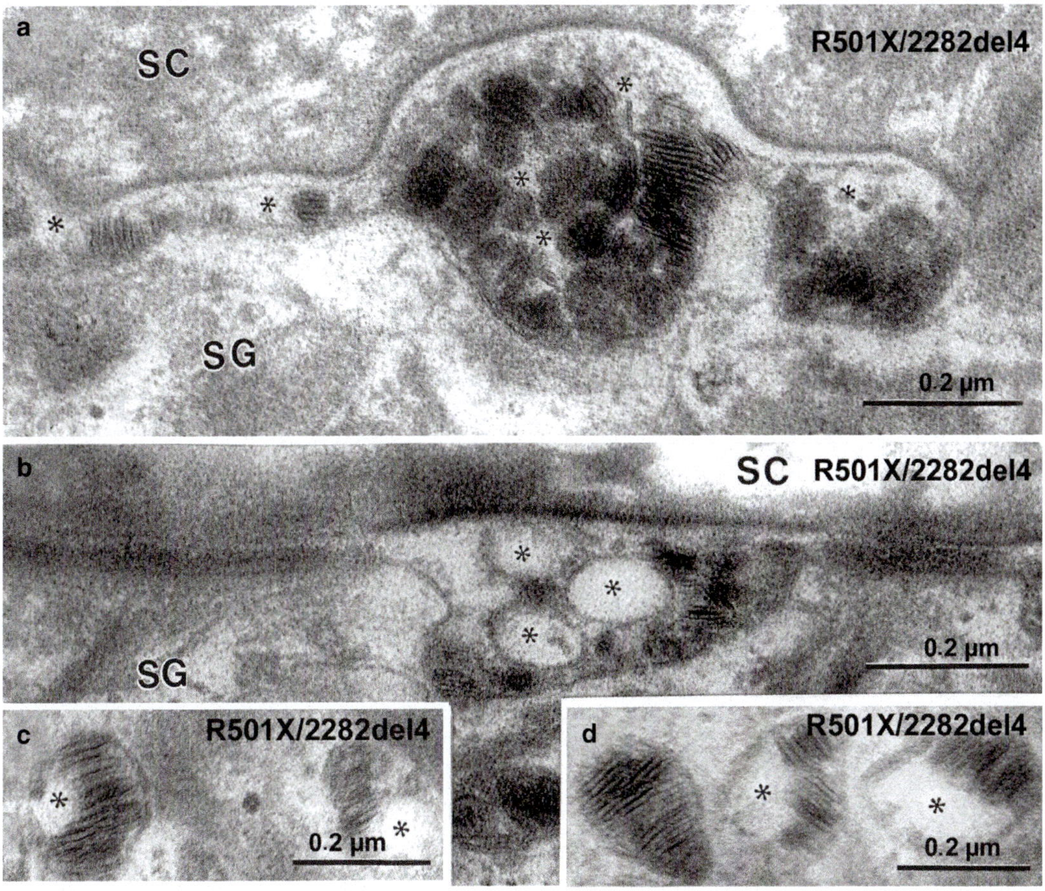

Fig. 2.6 Defective lamellar body (LB) contents in LB secretory system in filaggrin-depleted epidermis. (**a**, **b**) Non-lamellar (vesicular) contents at the SG-SC interface (*asterisks*). (**c**, **d**) Non-lamellar (vesicular) contents (*asterisks*) in single LBs, suggesting defective loading of organelle contents. Osmium tetroxide post-fixation. Mag bars = 0.2 µm

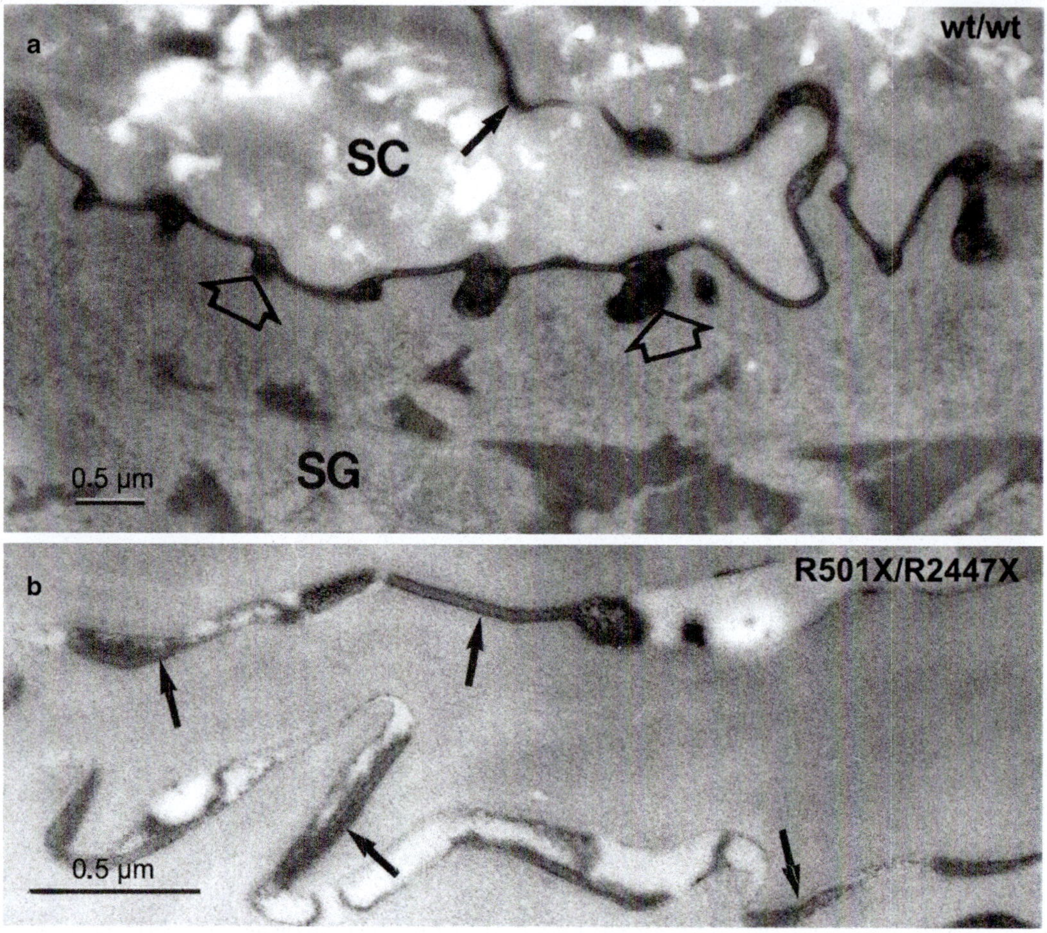

Fig. 2.7 Nonuniform secretion of LB-derived hydrolytic enzymes in filaggrin-depleted epidermis. (**a**) In wild-type controls (wt/wt) acid lipase, a marker of the extent and distribution of lamellar body (LB)-derived enzyme contents, is concentrated initially within LBs (black granular label), and then secreted in toto, followed by a uniform dispersion within the SC interstices (*arrowheads* at the SG-SC interface, *solid arrows* = enzyme activity in SC extracellular domains). (**b**) Although both total contents and secretion of hydrolase activity to the SC interstices appears near-normal in all IV subjects, enzyme activity is distributed nonuniformly within the SC interstices. Osmium tetroxide post fixation. Mag bars = 0.5 µm

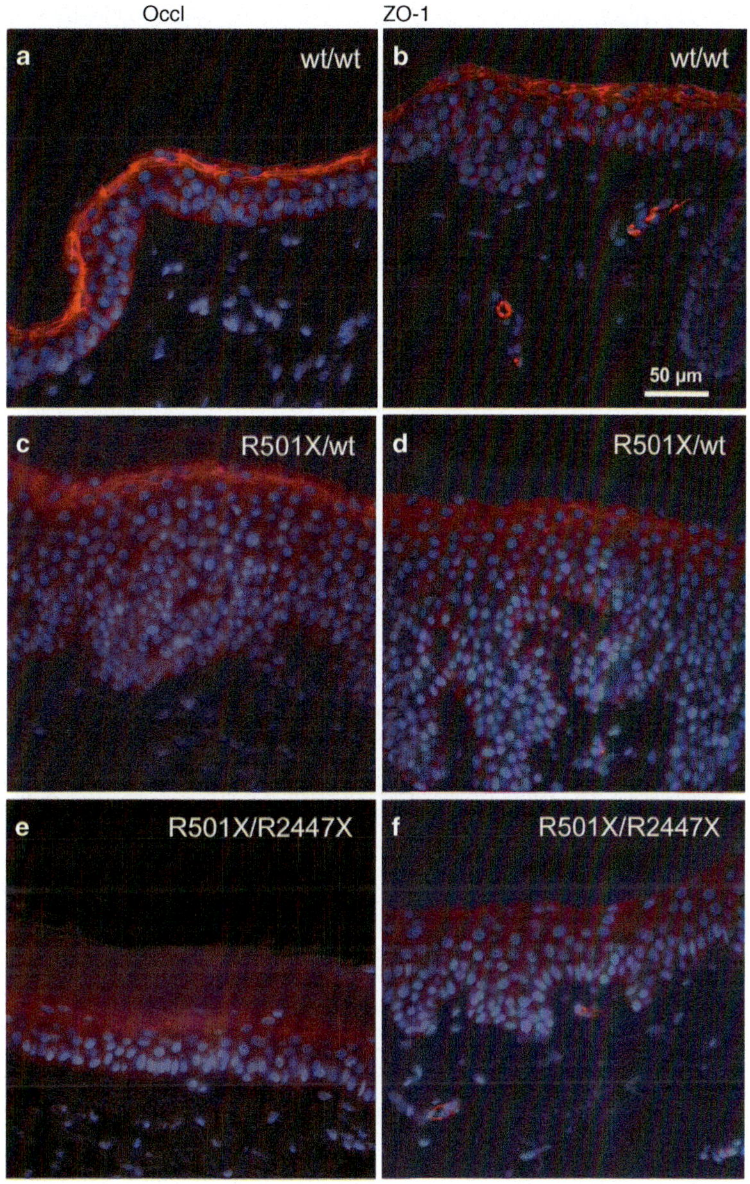

Fig. 2.8 Abnormalities in TJ protein expression in filaggrin-depleted epidermis shown by immunofluorescence staining for occludin (**a**, **c**, **e**) and ZO-1 (**b**, **d**, **f**). (**a**, **b**) In wild-type controls (wt/wt), a clear staining at the cell-cell borders of the uppermost living layers is observed for both occludin and ZO-1 (*red*). (**c**, **d**) Reduced staining with increased cytoplasmic staining in single-allele mutant IV. (**e**, **f**) Staining is almost completely lost in double-allele mutant epidermis. Nuclei are stained with DAPI (*blue*). Overlay of the epifluorescence pictures. Mag bar = 50 μm

References

1. Gruber R, Elias PM, Crumrine D, Lin TK, Brandner JM, Hachem JP, et al. Filaggrin genotype in ichthyosis vulgaris predicts abnormalities in epidermal structure and function. Am J Pathol. 2011;178(5):2252–63.
2. Traupe H, Happle R. Clinical features and genetics of the ichthyosis vulgaris group. Fortschr Med. 1980; 98(46):1809–15.
3. Uehara M, Hayashi S. Hyperlinear palms: association with ichthyosis and atopic dermatitis. Arch Dermatol. 1981;117(8):490–1.
4. Oji V, Tadini G, Akiyama M, Blanchet Bardon C, Bodemer C, Bourrat E, et al. Revised nomenclature and classification of inherited ichthyoses: results of the First Ichthyosis Consensus Conference in Soreze 2009. J Am Acad Dermatol. 2010;63(4):607–41.
5. Schmuth M, Martinz V, Janecke AR, Fauth C, Schossig A, Zschocke J, et al. Inherited ichthyoses/generalized Mendelian disorders of cornification. Eur J Hum Genet. 2013;21(2):123–33.
6. Brown SJ, Relton CL, Liao H, Zhao Y, Sandilands A, Wilson IJ, et al. Filaggrin null mutations and childhood atopic eczema: a population-based case-control study. J Allergy Clin Immunol. 2008;121(4):940–46.e3.
7. Smith FJ, Irvine AD, Terron-Kwiatkowski A, Sandilands A, Campbell LE, Zhao Y, et al. Loss-of-function mutations in the gene encoding filaggrin cause ichthyosis vulgaris. Nat Genet. 2006;38(3):337–42.
8. Sandilands A, Terron-Kwiatkowski A, Hull PR, O'Regan GM, Clayton TH, Watson RM, et al. Comprehensive analysis of the gene encoding filaggrin uncovers prevalent and rare mutations in ichthyosis vulgaris and atopic eczema. Nat Genet. 2007; 39(5):650–4.
9. Sandilands A, O'Regan GM, Liao H, Zhao Y, Terron-Kwiatkowski A, Watson RM, et al. Prevalent and rare mutations in the gene encoding filaggrin cause ichthyosis vulgaris and predispose individuals to atopic dermatitis. J Invest Dermatol. 2006;126(8):1770–5.
10. Gruber R, Janecke AR, Fauth C, Utermann G, Fritsch PO, Schmuth M. Filaggrin mutations p.R501X and c.2282del4 in ichthyosis vulgaris. Eur J Hum Genet. 2007;15(2):179–84.
11. Thyssen JP, Linneberg A, Johansen JD, Carlsen BC, Zachariae C, Meldgaard M, et al. Atopic diseases by filaggrin mutations and birth year. Allergy. 2012;67(5):705–8.
12. Palmer CN, Irvine AD, Terron-Kwiatkowski A, Zhao Y, Liao H, Lee SP, et al. Common loss-of-function variants of the epidermal barrier protein filaggrin are a major predisposing factor for atopic dermatitis. Nat Genet. 2006;38(4):441–6.
13. Irvine AD, McLean WH, Leung DY. Filaggrin mutations associated with skin and allergic diseases. N Engl J Med. 2011;365(14):1315–27.
14. Angelova-Fischer I, Mannheimer AC, Hinder A, Ruether A, Franke A, Neubert RH, et al. Distinct barrier integrity phenotypes in filaggrin-related atopic eczema following sequential tape stripping and lipid profiling. Exp Dermatol. 2011;20(4):351–6.
15. Elias PM, Hatano Y, Williams ML. Basis for the barrier abnormality in atopic dermatitis: outside-inside-outside pathogenic mechanisms. J Allergy Clin Immunol. 2008;121(6):1337–43.
16. Proksch E, Brandner JM, Jensen JM. The skin: an indispensable barrier. Exp Dermatol. 2008;17(12):1063–72.
17. Howell MD, Kim BE, Gao P, Grant AV, Boguniewicz M, Debenedetto A, et al. Cytokine modulation of atopic dermatitis filaggrin skin expression. J Allergy Clin Immunol. 2007;120(1):150–5.
18. Kuo IH, Yoshida T, De Benedetto A, Beck LA. The cutaneous innate immune response in patients with atopic dermatitis. J Allergy Clin Immunol. 2013; 131(2):266–78.
19. Kawasaki H, Nagao K, Kubo A, Hata T, Shimizu A, Mizuno H, et al. Altered stratum corneum barrier and enhanced percutaneous immune responses in filaggrin-null mice. J Allergy Clin Immunol. 2012; 129(6):1538–46.e6.
20. Presland RB, Boggess D, Lewis SP, Hull C, Fleckman P, Sundberg JP. Loss of normal pro-filaggrin and filaggrin in flaky tail (ft/ft) mice: an animal model for the filaggrin-deficient skin disease ichthyosis vulgaris. J Invest Dermatol. 2000;115(6):1072–81.
21. Saunders SP, et al. Tmem79/Matt is the matted mouse gene and is a predisposing gene for atopic dermatitis in human subjects. J Allergy Clin Immunol. 2013;132(5): 1121–9. (PMID 24084074).
22. Brown SJ, McLean WH. One remarkable molecule: filaggrin. J Invest Dermatol. 2012;132(3 Pt 2):751–62.
23. Candi E, Schmidt R, Melino G. The cornified envelope: a model of cell death in the skin. Nat Rev Mol Cell Biol. 2005;6(4):328–40.
24. Steinert PM, Marekov LN. The proteins elafin, filaggrin, keratin intermediate filaments, loricrin, and small proline-rich proteins 1 and 2 are isodipeptide cross-linked components of the human epidermal cornified cell envelope. J Biol Chem. 1995;270(30):17702–11.
25. Simon M, Haftek M, Sebbag M, Montezin M, Girbal-Neuhauser E, Schmitt D, et al. Evidence that filaggrin is a component of cornified cell envelopes in human plantar epidermis. Biochem J. 1996;317(Pt 1):173–7.
26. Elias PM, et al. Formation and functions of the corneocyte lipid envelope (CLE). Biochim Biophys Acta. 2014;1841(3):314–18. (PMID 24076475).
27. Zheng Y, Yin H, Boeglin WE, Elias PM, Crumrine D, Beier DR, et al. Lipoxygenases mediate the effect of essential fatty acid in skin barrier formation: a proposed role in releasing omega-hydroxyceramide for construction of the corneocyte lipid envelope. J Biol Chem. 2011;286(27):24046–56.
28. Feingold KR, Schmuth M, Elias PM. The regulation of permeability barrier homeostasis. J Invest Dermatol. 2007;127(7):1574–6.
29. Elias PM. Stratum corneum defensive functions: an integrated view. J Invest Dermatol. 2005;125(2):183–200.
30. Scott IR, Harding CR. Filaggrin breakdown to water binding compounds during development of the rat stratum corneum is controlled by the water activity of the environment. Dev Biol. 1986;115(1):84–92.

31. Krien PM, Kermici M. Evidence for the existence of a self-regulated enzymatic process within the human stratum corneum -an unexpected role for urocanic acid. J Invest Dermatol. 2000;115(3):414–20.
32. Fluhr JW, Elias PM, Man MQ, Hupe M, Selden C, Sundberg JP, et al. Is the filaggrin-histidine-urocanic acid pathway essential for stratum corneum acidification? J Invest Dermatol. 2010;130(8):2141–4.
33. Perusquia-Ortiz AM, Oji V, Sauerland MC, Tarinski T, Zaraeva I, Seller N, et al. Complete filaggrin deficiency in ichthyosis vulgaris is associated with only moderate changes in epidermal permeability barrier function profile. J Eur Acad Dermatol Venereol. 2013;27(12):1552–8.
34. Denda M, Sato J, Tsuchiya T, Elias PM, Feingold KR. Low humidity stimulates epidermal DNA synthesis and amplifies the hyperproliferative response to barrier disruption: implication for seasonal exacerbations of inflammatory dermatoses. J Invest Dermatol. 1998; 111(5):873–8.
35. Ohman H, Vahlquist A. The pH gradient over the stratum corneum differs in X-linked recessive and autosomal dominant ichthyosis: a clue to the molecular origin of the “acid skin mantle”? J Invest Dermatol. 1998;111(4):674–7.
36. Nachat R, Mechin MC, Takahara H, Chavanas S, Charveron M, Serre G, et al. Peptidylarginine deiminase isoforms 1–3 are expressed in the epidermis and involved in the deimination of K1 and filaggrin. J Invest Dermatol. 2005;124(2):384–93.
37. Scott IR, Harding CR, Barrett JG. Histidine-rich protein of the keratohyalin granules. Source of the free amino acids, urocanic acid and pyrrolidone carboxylic acid in the stratum corneum. Biochim Biophys Acta. 1982;719(1):110–7.
38. Erickson L, Kahn G. The granular layer thickness in atopy and ichthyosis vulgaris. J Invest Dermatol. 1970;54(1):11–2.
39. Sybert VP, Dale BA, Holbrook KA. Ichthyosis vulgaris: identification of a defect in synthesis of filaggrin correlated with an absence of keratohyaline granules. J Invest Dermatol. 1985;84(3):191–4.
40. Fleckman P, Holbrook KA, Dale BA, Sybert VP. Keratinocytes cultured from subjects with ichthyosis vulgaris are phenotypically abnormal. J Invest Dermatol. 1987;88(5):640–5.
41. Anton-Lamprecht I, Hofbauer M. Ultrastructural distinction of autosomal dominant ichthyosis vulgaris and X-linked recessive ichthyosis. Humangenetik. 1972;15(3):261–4.
42. Fleckman P, Brumbaugh S. Absence of the granular layer and keratohyalin define a morphologically distinct subset of individuals with ichthyosis vulgaris. Exp Dermatol. 2002;11(4):327–36.
43. Ziprkowski L, Feinstein A. A survey of ichthyosis vulgaris in Israel. Br J Dermatol. 1972;86(1):1–8.
44. Rice RH, Crumrine D, Uchida Y, Gruber R, Elias PM. Structural changes in epidermal scale and appendages as indicators of defective TGM1 activity. Arch Dermatol Res. 2005;297(3):127–33.
45. Winge MC, Hoppe T, Berne B, Vahlquist A, Nordenskjold M, Bradley M, et al. Filaggrin genotype determines functional and molecular alterations in skin of patients with atopic dermatitis and ichthyosis vulgaris. PLoS One. 2011;6(12):e28254.
46. Schmuth M, Gruber R, Elias PM, Williams ML. Ichthyosis update: towards a function-driven model of pathogenesis of the disorders of cornification and the role of corneocyte proteins in these disorders. Adv Dermatol. 2007;23:231–56.
47. Mildner M, Jin J, Eckhart L, Kezic S, Gruber F, Barresi C, et al. Knockdown of filaggrin impairs diffusion barrier function and increases UV sensitivity in a human skin model. J Invest Dermatol. 2010;130(9): 2286–94.
48. Fallon PG, Sasaki T, Sandilands A, Campbell LE, Saunders SP, Mangan NE, et al. A homozygous frameshift mutation in the mouse Flg gene facilitates enhanced percutaneous allergen priming. Nat Genet. 2009;41(5):602–8.
49. Scharschmidt TC, Man MQ, Hatano Y, Crumrine D, Gunathilake R, Sundberg JP, et al. Filaggrin deficiency confers a paracellular barrier abnormality that reduces inflammatory thresholds to irritants and haptens. J Allergy Clin Immunol. 2009;124(3):496–506, e1–6.
50. Weidinger S, Illig T, Baurecht H, Irvine AD, Rodriguez E, Diaz-Lacava A, et al. Loss-of-function variations within the filaggrin gene predispose for atopic dermatitis with allergic sensitizations. J Allergy Clin Immunol. 2006;118(1):214–9.
51. Thyssen JP, Johansen JD, Linneberg A, Menné T, Nielsen NH, Meldgaard M, et al. The association between null mutations in the filaggrin gene and contact sensitization to nickel and other chemicals in the general population. Br J Dermatol. 2010;162(6):1278–85.
52. Menon GK, Price LF, Bommannan B, Elias PM, Feingold KR. Selective obliteration of the epidermal calcium gradient leads to enhanced lamellar body secretion. J Invest Dermatol. 1994;102(5):789–95.
53. Hachem JP, Man MQ, Crumrine D, Uchida Y, Brown BE, Rogiers V, et al. Sustained serine proteases activity by prolonged increase in pH leads to degradation of lipid processing enzymes and profound alterations of barrier function and stratum corneum integrity. J Invest Dermatol. 2005;125(3):510–20.
54. Hachem JP, Wagberg F, Schmuth M, Crumrine D, Lissens W, Jayakumar A, et al. Serine protease activity and residual LEKTI expression determine phenotype in Netherton syndrome. J Invest Dermatol. 2006;126(7):1609–21.
55. Cork MJ, Danby SG, Vasilopoulos Y, Hadgraft J, Lane ME, Moustafa M, et al. Epidermal barrier dysfunction in atopic dermatitis. J Invest Dermatol. 2009;129(8):1892–908.

Stratum Corneum Lipids and Filaggrin

3

Jakob Mutanu Jungersted

Contents

J.M. Jungersted, MD, PhD
Department of Dermatology, Roskilde Hospital, Køgevej 7-13, Roskilde 4000, Denmark
e-mail: jungersted@gmail.com

3.1 Introduction

Stratum corneum, the most upper layer of the skin, is crucial to skin barrier functions.

The major components of the stratum corneum are corneocytes and lipids. The lipids surround the corneocytes and are arranged in lipid bilayers, historically described as a brick wall, with the bricks being the annucleated corneocytes and the mortar being the lipids [1]. Accordingly, no substance can pass through the skin without first passing through the stratum corneum lipids [1, 2]. This chapter focuses on the stratum corneum lipids, in particular on the influence of filaggrin on the lipids and its possible role in lipid synthesis through influencing pH.

3.2 The Lipids: What They Are Influenced by and Their Regulation

The lipids are mainly composed of three different classes: ceramides, free fatty acids, and cholesterol. The ceramides are further subdivided into 12 subclasses (ceramides 1–12) and, more recently, also according to chain length [3–5]. The synthesis of the lipids takes place in the stratum granulosum, and through lamellar body exocytosis, the lipids are delivered to the stratum corneum [6]. Regulation of the synthesis is linked in a symphony of enzymes, calcium gradient, and pH, which is not fully understood, as yet. Lipid levels in the stratum corneum are related to

J.P. Thyssen, H. Maibach (eds.), *Filaggrin*,
DOI 10.1007/978-3-642-54379-1_3, © Springer-Verlag Berlin Heidelberg 2014

anatomical location and are influenced by environmental factors, such as season and UV exposure. For example, in winter, decreasing levels of ceramides are found on the face compared with more protected areas of the skin [7].

Topical skin treatments also influence stratum corneum lipid levels. Moisturisers, topical corticosteroids, and topical calcineurin inhibitors have been investigated regarding their effect on up- or downregulation of the lipids. Although data are not uniform, it has been shown for moisturisers that the mRNA expression of enzymes involved in lipid synthesis is up- or downregulated according to the ingredients [8, 9]. However, since the moisturisers tested all contained multiple ingredients, the precise influence of each ingredient is unknown. For topical corticosteroids, one group has shown that their use leads to inconsistency in the lipid bilayers [10], which is interpreted as a negative influence. Nevertheless, the use of topical corticosteroids, rather than a moisturiser, has been shown to increase the ceramide/cholesterol ratio, probably benefiting the skin barrier [11]. For topical calcineurin inhibitors, data also point to an increase in the ceramide/cholesterol ratio, compared with no treatment [12].

3.3 Atopic Dermatitis from a Lipid and Filaggrin Perspective

Atopic dermatitis (AD) is the disease that highlighted the importance of ceramides. Compared with those with healthy skin, people with AD have lower levels of ceramides 1 and 3 in their stratum corneum, even in non-lesional skin [13–17]. Seemingly, this partly explains the inferior skin barrier function in those with AD, also because ceramide 1 plays a central role in the structural arrangement of the lipid bilayers [18].

After it became possible to group people with AD according to their filaggrin gene (*FLG*) mutation status, the question arose as to whether having an *FLG* mutation would change the composition of the ceramides. It is likely that a person's *FLG* mutation status influences the ceramides because breakdown products from filaggrin contribute to the acidic pH of the stratum corneum and thereby influence the enzymes that regulate ceramide synthesis [19–23].

Two research groups have analysed the stratum corneum lipids in humans grouped according to their *FLG* mutation status; participants included those with and those without AD.

Jungersted et al.'s study on non-lesional skin showed statistically significant lower levels of ceramide 4 (EOH) in the group with an *FLG* mutation and AD compared with those with healthy skin. This is the first time ceramide 4 has been shown to be statistically lower in AD compared with healthy skin. However, since ceramide 4 was significantly lower only when the AD *FLG* mutation group was compared with healthy skin, it was investigated whether the difference could be ascribed to the *FLG* mutation. Comparisons were made between the two groups with *FLG* mutations (both AD and control skin) and the two groups without FLG mutations (both AD and control skin), and no statistically significant difference was found. The two AD groups were not compared with the two healthy groups because both groups were selected to include as many *FLG* mutation carriers as possible and were therefore biased. On this basis, Jungersted et al. concluded that low ceramide 4 in AD is due to reasons other than FLG. However, since the difference was statistically significant only for the AD *FLG* mutation group, and the study did not correct for multiple testing, more research on larger populations is required before drawing any final conclusion. It was, therefore, highly interesting that the group led by Angelova-Fischer compared the ceramides on the lesional skin of people with AD divided according to *FLG* status. They, too, found lower levels of ceramide 4 (EOH) in the AD *FLG* mutation group than in the AD *FLG* wild-type group.

Taken together, these two studies point towards ceramide 4 being influenced by *FLG* mutations. This is interesting because ceramide 4 has not previously been a ceramide receiving much attention [19, 24].

3.4 pH and Filaggrin

The enzymes involved in lipid synthesis all have slight differences in their optimum pH and are, therefore, influenced by changes in the skin's pH [25]. The pH ranges from neutral in the deeper layers of the skin to acidic on the skin surface.

People with AD have a higher skin surface pH than healthy individuals do, even in non-lesional skin [23]. Filaggrin breakdown products help maintain the acidic environment on the stratum corneum surface assisted by the free fatty acids [26]. In the aforementioned study by Jungersted et al., the skin surface pH was measured and compared between AD and healthy groups according to *FLG* status. The findings showed a statistically significant higher skin surface pH for those carrying *FLG* mutations than for controls. Even participants without AD who were carrying an *FLG* mutation had a higher pH than those with AD without *FLG* mutations. This difference in pH could be expected to influence the synthesis of stratum corneum lipids; nonetheless, the data do not support this expectation. This might be because the pH was measured on the surface of stratum corneum, and the enzymatic processes and lipid synthesis take place deeper in the skin at the stratum corneum-stratum granulosum interface, and hence may not be influenced by the skin surface pH.

3.5 Where to Go from Here

Recent studies in stratum corneum lipids have focused on ceramide chain length rather than the 12 subgroups. They find that although chain length influences barrier properties, the ceramide chain length seem not be influenced by *FLG* mutation status [27]. Clearly, further studies are needed focusing on subgroups of patients according to filaggrin mutation, treatment regime, and skin barrier parameters.

References

1. Michaels AS, Chandrasekaran SK, Shaw JE. Drug penetration through human skin. Theory and in vitro experimental measurements. AIChE J. 1975;21:985–96.
2. Elias PM. Epidermal lipids, barrier function and desquamation. J Invest Dermatol. 1983;80:44–9.
3. van Smeden J, Hoppel L, van der Heijden R, Hankemeier T, Vreeken RJ, Bouwstra JA. LC-MS analysis of stratum corneum lipids: ceramide profiling and discovery. J Invest Dermatol. 2011;52:1211–21.
4. Masukawa Y, Narita H, Sato H, Naoe A, Kondo N, Sugai Y, et al. Comprehensive quantification of ceramide species in human stratum corneum. J Lipid Res. 2009;50:1708–19.
5. Hinder A, Schmelzer CDH, Rawlings AV, Neubert RHH. Investigation of the molecular structure of the human stratum corneum ceramides [NP] and [EOS] by mass spectrometry. Skin Pharmacol Physiol. 2011;24:127–35.
6. Elias PM, Menon GK. Structural and lipid biochemical correlates of the epidermal permeability barrier. Adv Lipid Res. 1991;24:1–26.
7. Rogers J, Harding C, Mayo A, Banks J, Rawlings A. Stratum corneum lipids: the effect of aging and the seasons. Arch Dermatol Res. 1996;228:765–70.
8. Buraczewska I, Berne B, Lindberg M, Lodén M. Törmä. Moisturizers change the mRNA expression of enzymes synthesizing skin barrier lipids. Arch Dermatol Res. 2009;3001:587–94.
9. Buraczewska I, Berne B, Lindberg M, Lodén M. Törmä. Long-term treatment with moisturizers affects the mRNA levels of genes involved in keratinocyte differentiation and desquamation. Arch Dermatol Res. 2009;301:175–81.
10. Jensen JM, Pfeiffer S, Witt M, Bräutigam M, Neumann C, Weichenthal M, et al. Different effects of pimecrolimus and betamethasone on the skin barrier in patients with atopic dermatitis. J Allergy Clin Immunol. 2009;123:1124–33.
11. Jensen JM, Scherer A, Wanke C, Bräutigam M, Bongiovanni S, Letzkus M, et al. Gene expression is differently affected by pimecrolimus and betamethasone in lesional skin of atopic dermatitis. Allergy. 2012;67:413–23.
12. Jungersted JM, Høgh JK, Hellgren LI, Jemec GBE, Agner T. Effect of topical corticosteroid and tacrolimus on ceramides and irritancy to sodium lauryl sulphate in healthy skin. Acta Derm Venereol. 2011;91:290–4.
13. Imokawa G, Abe A, Jin K, Higaki Y, Kawashima M, Hidano A. Decreased level of ceramides in stratum corneum of atopic dermatitis: an etiologic factor in atopic dry skin? J Invest Dermatol. 1991;96:523–6.
14. Yamamoto A, Serizawa S, Ito M, Sato Y. Stratum corneum lipid abnormalities in atopic dermatitis. Arch Dermatol Res. 1991;283:219–23.
15. Di Nardo A, Wertz P, Giannetti A, Seidenari S. Ceramide and cholesterol composition of the skin of patients with atopic dermatitis. Acta Derm Venereol. 1998;78:27–30.
16. Bleck O, Abeck D, Ring J, Hoppe U, Vietzke JP, Wolber R, et al. Two ceramide subfractions detectable in Cer(AS) position by HPTLC in skin surface lipids on non-lesional skin of atopic eczema. J Invest Dermatol. 1999;113:894–900.

17. Matsumoto M, Umemoto N, Sugiura H, Uehara M. Difference in ceramide composition between "dry" and "normal" skin in patients with atopic dermatitis. Acta Derm Venereol. 1998;79:246–7.
18. Bouwstra JA, Gorris GS, Dubbelaar FER, Weerheim AM, Ijzerman AP. Role of ceramide 1 in the molecular organization of the stratum corneum lipids. J Lipid Res. 1998;39:186–96.
19. Jungersted JM, Scheer H, Mempel M, Baurecht H, Cifuentes L, Høgh JK, et al. Stratum corneum lipids, skin barrier function and filaggrin mutations in patients with atopic eczema. Allergy. 2010;5: 911–8.
20. Imokawa G. A possible mechanism underlying the ceramide deficiency in atopic dermatitis: expression of a deacylase enzyme that cleaves the N-acyl linkage of sphingomyelin and glucosylceramide. J Dermatol Sci. 2009;55:1–9.
21. Higuchi K, Hara J, Okamoto R, Kawashima M, Imokawa G. The skin and atopic dermatitis patients contains a novel enzyme, glucosylceramide sphingomyelin deacylase, which cleaves the N-acyl linkage of sphingomyelin and glucosylceramide. Biochem J. 2000;350:747–56.
22. Redoules D, Tarroux R, Périé J. Epidermal enzymes: their role in homeostasis and their relationships with dermatoses. Skin Pharmacol Appl Skin Physiol. 1998; 11:183–92.
23. Cork MJ, Danby SG, Vasilopoulos Y, Hadgraft J, Majella EL, Moustafa M, et al. Epidermal barrier dysfunction in atopic dermatitis. J Invest Dermatol. 2009;129:1892–908.
24. Angelova-Fischer I, Mannheimer AC, Hinder A, Ruether A, Franke A, Neubert RH, et al. Distinct barrier integrity phenotypes in filaggrin-related atopic eczema following sequential tape stripping and lipid profiling. Exp Dermatol. 2011;20:351–6.
25. Jungersted JM, Hellgren LI, Jemec GBE, Agner T. Lipids and skin barrier function – a clinical perspective. Contact Dermatitis. 2008;58:255–62.
26. Chen X, Kwak S, Lafleur M, Bloom M, Kitson N, Thewalt J. Fatty acids influence "solid" phase formation in models of stratum corneum intercellular membranes. Langmuir. 2007;23:5548–56.
27. Janssen M, Smeden JS, Gooris GS, Bras W, Portale G, Caspers PJ, et al. Lamellar lipid organization and ceramide composition in the stratum corneum of patients with atopic eczema. J Invest Dermatol. 2011;131:2136–8.

Inflammatory-Driven Depletion of Filaggrin Proteins

4

Christian Vestergaard and Mette S. Deleuran

Contents

C. Vestergaard, MD, PhD, DMSc (✉)
M.S. Deleuran, MD, DMSc
Department of Dermatology,
Aarhus University Hospital,
PP Oerumsgade 11, Aarhus C 8000, Denmark
e-mail: chrivest@rm.dk, chr-vest@post9.tele.dk;
mettdele@rm.dk

4.1 Introduction

Since the milestone publication by Palmer et al. in 2006 [1], in which it was shown that mutations in the filaggrin gene (*FLG*), which codes for filaggrin "filament aggregating protein," significantly increase the risk of atopic dermatitis (AD), filaggrin and barrier dysfunction have been the epicenter of AD research and other skin diseases with decreased skin barrier function. Filaggrin is important for the aggregation of keratin filaments forming the impenetrable keratin layer in the stratum corneum of the epidermis and also for the hydration of the skin through the "natural moisturizing factors" (NMFs) [2]. The decreased barrier function caused by the lack of filaggrin expression may increase the risk of allergic sensitization and thus offer a molecular background for the so-called atopic march, in which patients suffering from early childhood AD have a higher risk of developing asthma, rhinitis, and food allergies [3]. An interesting fact is that 44–85.8 % of AD patients do not carry an *FLG* mutation, but still all patients have a decreased skin barrier function [4], and that among patients carrying an *FLG* mutation, 40 % never show any signs of decreased skin barrier function [5].

These results indicate that regulatory mechanisms other than null mutations in the *FLG* are responsible for the decreased skin barrier function in AD patients. Mutations in other genes involved in the skin barrier function have been linked to the development of AD such as SPINK5 (LEKTI), a serine protease inhibitor, which inhibits the

J.P. Thyssen, H. Maibach (eds.), *Filaggrin*,
DOI 10.1007/978-3-642-54379-1_4, © Springer-Verlag Berlin Heidelberg 2014

kallikrein's breakdown of the corneodesmosomes [6] and very likely also filaggrin (KLK5) [7]. The inflammatory reaction taking place in the skin can modulate the expression of filaggrin on transcriptional, translational, and posttranslational levels, thus inducing a functional lack of filaggrin in the absence of an *FLG* null mutation [8, 9]. In this chapter, we will focus on the effect of inflammation on expression of filaggrin and the subsequent effect on skin barrier functions.

4.2 Production and Regulation of Filaggrin

Filaggrin has the property of aligning keratin intermediary filaments into a tight matrix in the keratinocytes in the upper layer of the epidermis leading to a collapse of the intracellular structure and the formation of the spindle-shaped and tightly packed cells of the stratum corneum. Filaggrin is thus one of the most important structural proteins in the establishment of an intact skin barrier [10–12].

Filaggrin is produced as the pre-protein profilaggrin containing 10–12 filaggrin repeats flanked by a C- and N-terminal region [13, 14]. The N-terminal region may act as a Ca^{2+}-binding region and play a role for the terminal keratinocyte denucleation, whereas the exact role of the C-terminal region is unknown, although mice lacking this region are unable to process profilaggrin into filaggrin [15]. The gene for profilaggrin, *FLG*, is located on chromosome 1q21 in the epidermal differentiation complex that also contains other genes transcribed in the maturation process of the keratinocyte [14].

The interest in filaggrin was spurred by the finding that homozygous null mutations in the *FLG* lead to varying degrees of ichthyosis vulgaris [16] and that individuals heterozygous for the mutation had a higher risk of AD [1]. Until today, more than 40 different null mutations in the *FLG* have been identified, but the frequency of each mutation differs depending on the specific ancestral group [17].

The maturation process of pro-filaggrin is a very complex multistep process, and only an overview, aimed at describing possible points for regulation by inflammation, will be given here.

The expression of the *FLG* could be under control of the transcription factor AP-1 (fos and/or jun) as AP-1 responsive elements have been described in the promoter region of the *FLG* [18]. Furthermore, oct1, oct6, and skn1a/i bind in vitro to two responsive elements in the promotor region of the *FLG* [19]. Certain isoforms of p63 inhibit the transcription of *FLG* [20], and the *FLG* promotor region also contains glucocorticoid- and retinoic acid-responsive elements that suppress promotor activity [21].

Once pro-filaggrin is transcribed, it is heavily phosphorylated to prevent premature degradation to filaggrin [22]; however, when the keratinocytes reach the granular layer, filaggrin is dephosphorylated as one of the initial steps in the formation of the keratin matrix. The next step involves proteolytic cleavage of the pro-filaggrin molecule, primarily in the C- and N-terminal flanking regions, followed by cleavage of the linker region between the filaggrin repeats. The proteases involved in this process include matriptase [23] and prostasin [24]. In prostasin knockout mice, two- and three-domain filaggrin intermediates accumulate in the stratum corneum, but no filaggrin monomers, whereas profilaggrin accumulates in the stratum corneum in matriptase knockout mice [23–25]. LEKTI (lymphoepithelial Kazal-type-related inhibitor) is a serine protease inhibitor encoded by the SPINK5 gene. Mutations in this gene in humans are responsible for "Netherton's syndrome," in which the skin is very dry with abnormal maturation of the keratinocytes and a defect in skin barrier [26]. Under normal circumstances LEKTI inhibits the proteases that cleaves pro-filaggrin into filaggrin monomers, and in SPINK5 knockout mice, there is premature pro-filaggrin cleavage, but whether this is due to lack of inhibition of matriptase and prostasin is unknown [27, 28]. In mice lacking 12R-lipooxygenase, which catalyze dioxygenation of fatty acids, there is an aberrant profilaggrin cleavage, although the background for this is unclear [29].

The filaggrin monomers are broken down to hygroscopic amino acids that constitute the

NMFs by caspase 14 [2]. Knockout mice lacking caspase 14 have dry skin and an increased susceptibility to ultraviolet (UV) B light. They do cleave filaggrin, but still there is an abnormal maturation of the skin and a decreased production of NMFs [30].

From this short overview of the maturation of pro-filaggrin into filaggrin, it is clear that there are several steps apart from null mutations in which the expression of filaggrin can be regulated.

4.3 Inflammation in the Skin and the Effect of Cytokines

The inflammatory reactions in the skin seen in inflammatory skin diseases are due to a complex interaction between the cells of the immune system, adaptive and innate; the keratinocytes; the endothelial cells; and the connective tissue. An offensive agent, such as bacteria, antigens, or local tissue damage, induces a cascade of events that attract the cells of the immune system such as polymorphnuclear leukocytes and lymphocytes. These cells are activated and differentiated in a manner specific for each inflammatory skin disease [31]. In AD, the dominating lymphocytes are of the Th_2 subtype; in psoriasis, they are Th_1; and in contact dermatitis, the lymphocytes are CD8+ rather than CD4+, although treatment may alter this dramatically [32]. The dermal dendritic cells also have a specific profile of surface receptors and cytokine production profile depending on the inflammatory skin disease [33]. The production and expression of cytokines in inflammatory skin diseases often take place in a specific spatial and timely sequence for each disease [34].

All the cells in the skin produce inflammatory mediators that attract inflammatory cells, and among these mediators are the cytokines that are small, biologically potent glycoproteins that regulate growth, migration, and differentiation of cells. The cytokines may be subgrouped in different, often overlapping, ways depending on function or molecular structure. The interleukins (ILs) are the cytokines initially described as produced by lymphocytes, monocytes, or macrophages, and the chemokines are a subgroup of these, with a highly conserved motif of cysteine residues in the N-terminal. For example, IL-8, which was originally termed so because it was produced by macrophages, was also found to have the structure of a CXC-chemokine; therefore, in accordance with the chemokine nomenclature subcommittee, it was renamed CXCL8 [35].

Other mediators of inflammation, besides the cytokines, are vasoactive amines such as histamine, 5-hydroxytryptamine, and serotonin; the reactive oxygen intermediates and nitrogen oxide; and the arachidonic acid products such as leukotrienes, prostaglandins, and thromboxanes.

4.4 Inflammatory Cytokines and Filaggrin Expression

4.4.1 Interleukin-4 and Interleukin-13

Patients suffering from AD have a decreased skin barrier function, but this is not confined to the group of patients carrying the *FLG* mutation. To address this question, the filaggrin expression on both mRNA and protein levels in the skin of 30 healthy patients was compared to the expression of lesional as well as non-lesional skin in 39 patients suffering from AD. In addition, a subgroup of patients was screened for the *FLG* null mutations 2282del4 and R501X, and 2 of 25 of the healthy controls had mutations, whereas 3 of 17 of the AD patients carried the 2282del4 mutation. None of the patients carried the R501X mutation. The results showed that filaggrin expression was significantly decreased in lesional as well as non-lesional skin of the AD patients not carrying the *FLG* mutation compared to the groups of healthy patients with or without *FLG* null mutation. The filaggrin expression in the skin of lesional AD in patients with *FLG* null mutation was significantly lower than that of AD patients without mutations. In the same study, it was shown that the typical Th_2 cytokines IL-4 and IL-13, which are abundant in AD skin, could decrease the expression of filaggrin in primary keratinocyte cultures in which differentiation had been induced by increased calcium concentrations

[8, 36]. From these studies it was clear that inflammation of the skin could induce a "functional filaggrin defect" in the skin and, thereby, a decreased skin barrier function.

Not only do IL-4 and IL-13 decrease filaggrin expression; they also decrease the expression of caspase 14 [9], which breaks down filaggrin to NMFs and contributes to the integrity of the skin barrier [2].

4.4.2 Interleukin-22 (IL-22)

IL-22 is a cytokine that was first described to be produced by Th_{17} cells, but later as a cytokine produced by a distinct subgroup of skin-homing T lymphocytes expressing the skin-homing chemokine receptor profile CCR4, CCR6, and CCR10, which do not express IL-17 [37]. This subgroup of lymphocytes, known as Th_{22} lymphocytes, are found in inflammatory skin disorders where they induce transcription of genes related to the innate and adaptive immune pathways in keratinocytes [38]. Dermal dendritic cells and Langerhans cells from the skin induce IL-22 production in T lymphocytes stressing the concept of the skin specificity of these cells [39]. Furthermore, exotoxins from *Staphylococcus aureus* that colonize the skin, especially in AD, induce IL-22 expression in CD4+ T lymphocytes [40]. In HaCaT cells (an immortalized keratinocytic cell line) cultured with high calcium concentrations to induce filaggrin expression, IL-22 downregulates pro-filaggrin and filaggrin expression on both mRNA and protein level [41].

4.4.3 Interleukin-17A (IL-17A) and Interleukin-17E (IL-17E)

Interleukin-17 is produced by Th_{17} lymphocytes after stimulation with IL-23. IL-17 was primarily associated with inflammatory diseases in the central nervous system and in the joints, because it was found to be the mediator of the IL-23-dependent models of collagen-induced arthritis and experimental autoimmune encephalitis. Later, important roles for IL-17 were described in autoimmune diseases, such as rheumatoid arthritis, inflammatory bowel disease, multiple sclerosis, and in psoriasis. Interleukin-17 belongs to a family of at least five members – IL-17A–E. IL-17A is often described as a Th_1-like cytokine, and it is the IL-17 member associated with psoriasis [42, 43]. IL-17E, also known as IL-25, is the opposite and is often described as a Th_2-associated cytokine. IL-17E binds to the IL-17B-receptor, whereas IL-17A binds to the IL-17A receptor [44]. IL-17A has been shown to be able to downregulate not only pro-filaggrin and filaggrin expression in HaCaT cells cultured in high calcium concentrations but also proteins associated with cell adhesion between the keratinocytes, thereby affecting the skin barrier on at least two different levels [45]. The role of filaggrin downregulation in psoriasis in which IL-17A levels are high is not clear.

IL-17E (IL-25), on the other hand, is found in very high concentrations in Th_2-mediated diseases such as asthma and AD. Il-17E is important in the response to helminthic infections in murine models [46], but also for the initiation and prolongation of the Th_2 response. In asthma patients, Th_2 cells, mast cells, and lung epithelial cells produce IL-17E [47], whereas in the skin of AD patients, dermal dendritic cells produce IL-25 [48]. Surprisingly, this production is downregulated by TSLP. The pro-filaggrin expression in primary keratinocytes cultured to 80 % confluency is reduced when co-cultured with IL-17E [48], a finding that could be reproduced in keratinocytes cultured with 1.3 mM calcium [36]. These results indicate that IL-17E is a key mediator in AD, as it has the ability both to induce the production of Th_2 cytokines such as IL-4 and IL-13 and to downregulate filaggrin, thereby decreasing the skin barrier function.

4.4.4 Tumor Necrosis Factor-α (TNF-α)

Tumor necrosis factor-α (TNF-α) is a central mediator of inflammation. It inhibits viral replication and tumorigenesis, and in the context of skin diseases, it is a very central mediator of the

Table 4.1 Regulatory stimuli on filaggrin expression

Cytokine/stimuli	Effect on filaggrin	Produced by	Reference
IL-4 + IL-13	Downregulation (direct effect and mediated through downregulation of caspase 14	T cells	Howell et al. [8] Pellerin et al. [36] Hvid et al. [9]
IL-22	Downregulation	Th_{22} cells, *S. aureus* stimulated T cells	Gutowska-Owsiak et al. [41]
IL-17A	Downregulation	Th_{17} cells	Gutowska-Owsiak et al. [45]
IL-25 (IL-17E)	Downregulation	Dermal dendritic cells	Hvid et al. [48] Pellerin et al. [36]
TNF-α	Downregulation	T cells	Kim et al. [51]
IL-31	Downregulation (*in part mediated through IL-20 and IL-24*)	Th_2 cells, dermal dendritic cells, mast cells, monocytes	Cornelissen et al. [55]
Topical corticosteroid	Upregulation		Jensen et al. [56]
Retinoic acid	Downregulation		Asselineau et al. [57] Kautsky et al. [58]
Coal tar	Upregulation		Van den Bogaard et al. [59]

inflammation in psoriasis, as proven by the high efficacy of TNF-α inhibitors in the treatment of psoriasis [49]. TNF-α is also expressed in AD, but since TNF-α inhibitors have almost no effect in the treatment of this disease, its role is very unclear [50]. In psoriasis a skin barrier defect also exists, although its role in the pathogenesis of psoriasis is not as clear as in AD. TNF-α acts on the filaggrin synthesis by downregulating the expression in primary keratinocytes cultured in 0.06 % calcium [51].

4.4.5 Interleukin 31 (IL-31)

IL-31 was originally described as a cytokine produced by Th_2 lymphocytes, but other cells such as mast cells, dendritic cells, and monocytes may also produce the cytokine [52, 53]. It is increased in the skin and serum of AD patients, but also in a murine model of AD, the NC/Nga mouse, in which it induces pruritus through binding to IL-31 receptors in the dorsal ganglia. Blocking of IL-31 with monoclonal antibodies in this model decreases the itching behavior of the mice [54]. In a 3D model of the skin based on HaCaT cells, in which a doxycycline-inducible IL-31 receptor was inserted, IL-31 could repress the expression of pro-filaggrin. This effect was in part mediated through an IL-31-dependent induction of IL-20 and IL-24 production in the HaCaT cells, and these cytokines in turn downregulated pro-filaggrin [55] (Table 4.1).

4.5 The Effect of Anti-inflammatory Treatment on Filaggrin Expression in the Skin

4.5.1 Topical Steroids and Calcineurin Inhibitors

The most widely used therapy for inflammatory skin diseases of mild to moderate severity is topical corticosteroids, and in the case of AD also the topical calcineurin inhibitors tacrolimus and pimecrolimus. Topical corticosteroids exert their action partly through binding to steroid receptors in the cytosol of the cell, which then translocate into the nucleus, where it inhibits the transcription of inflammatory genes (transrepression) and induces expression of anti-inflammatory genes (transactivation), but also affects the synthesis of structural proteins [60]. Calcineurin inhibitors inhibit calcium-dependent dephosphorylation of the transcription factor nuclear factor of activated T cells (NFATs) that is required for the transcription of inflammatory cytokines such as IL-2 [61]. A study of genes expressed in the skin of AD patients treated with either betamethasone valerate or pimecrolimus showed that both reduced inflammation, betamethasone valerate more than pimecrolimus; both restored filaggrin and loricrin expression to normal levels; but that betamethasone valerate reduced expression of lipid synthesis rate-limiting enzymes and involucrin, which may be part of the explanation for the lack of effect of corticosteroids on the restoration of the skin barrier function [56].

4.5.2 Retinoic Acid

Retinoic acid is used in inflammatory skin diseases and has an anti-proliferative effect on the keratinocytes and also an anti-inflammatory effect; both effects are exerted through the RAR (retinoid acid receptors) and the RXR (retinoid X receptors). Filaggrin production is downregulated by retinoic acid as seen in keratinocytes cultured in multilayer on a collagen matrix at the air-liquid surface [57] and in oral keratinocytes in a lifted culture system [58]. The effect may be due to the presence of a retinoic acid-responsive element in the promoter region of the pro-*FLG* [21].

Chronic hand eczema may be treated with the retinoid alitretinoin (9-*cis*-retinoid) with good effect [62]. A side effect of the treatment is dry skin, and in this light, AD treated with alitretinoin (9-*cis*-retinoid) surprisingly shows good effect [63]. Larger studies, however, are needed.

4.5.3 Coal Tar

Coal tar is a treatment used for the treatment of skin diseases for more than two millennia. A recent publication has shown that the aryl hydrocarbon receptor (AHR) is involved not only in normalization of the Th_2-dominated response by interfering with the STAT6-mediated response but also through an induction of filaggrin production along with other markers of epidermal differentiation in primary keratinocyte cell cultures and in an organotypic skin model of human keratinocytes both with and without filaggrin mutations, a finding that could be reproduced in skin biopsies from AD patients [59]. Thus, the AHR seems to be a very interesting and promising candidate for new filaggrin-targeted therapies for AD, as it both inhibits the inflammatory reaction and increases the differentiation of keratinocytes.

4.6 Discussion and Conclusion

The finding of the correlation between *FLG* null mutations, and AD and ichthyosis vulgaris spurred the interest in skin barrier defects, especially in AD but in also other eczematous diseases. The fact that AD patients have a skin barrier defect has long been recognized and has led to the discussion of the "inside-outside" hypothesis versus the "outside-inside" hypothesis. The "inside-outside" hypothesis favors the notion that the inflammatory reaction, initiated by allergens, infections, or autoimmune reactions, in the skin leads to skin barrier breakdown, whereas the "outside-inside" hypothesis favors the notion of the skin barrier defect as the initial event followed by an inflammatory/allergic reaction on the basis of noxious stimuli entering the body through the open barrier [64]. In this review, inflammatory stimuli leading to reduced filaggrin expression have been reviewed, and this could be read as a support of the "inside-outside" hypothesis. However, other studies have shown that a lack of filaggrin itself can enhance production of inflammatory stimuli such as IL-1 [65] and TSLP [66] (thymic stromal lymphopoietin). Thus, the two subelements of AD pathogenesis, inflammation and skin barrier defect, seem to be closely interconnected.

The downregulation of filaggrin expression by inflammation may also be of great importance in other inflammatory skin diseases such as contact dermatitis, in which inflammation leads to a decreased skin barrier, and this in turn increases the inflammatory reaction in a perpetual vicious circle, in which the risk of acquiring new allergies through the broken skin barrier is increased, as is the case in patients homozygous for the *FLG* mutations R501X and 2282del4 who have an odds Ration of 5.71 for sensitization to other allergens than nickel compared to healthy individuals [67].

Until now, the main focus on research in filaggrin and inflammation has been on the effect of various inflammatory mediators on the production of filaggrin in keratinocytes. However, inflammation also affects the posttranslational processing of both filaggrin and pro-filaggrin, as exemplified by the downregulation of caspase 14 by inflammatory cytokines [2, 9]. However, the effect of inflammation on posttranslational processes as well as transcriptional regulation is not well examined and should constitute a major focus area for research in skin barrier function in the next decade.

It is clear from the research so far that treatment of skin inflammation restores the skin barrier function with regard to filaggrin. The experiments performed so far indicate, for a large part, that this is due to the decreased production of inflammatory mediators by the inflammatory cells in the skin. However, the promoter region of pro-filaggrin contains retinoid as well as glucocorticoid responsive elements [21], and the effect of these immunosuppressants themselves on filaggrin production still remains to be clarified.

In conclusion, inflammation of all types, including Th_1-, Th_2-, Th_{17}-, and Th_{22}-dominated responses, induces skin barrier defects, and the current results suggest that this may, partly, be due to a downregulation of filaggrin.

References

1. Palmer CN, Irvine AD, Terron-Kwiatkowski A, Zhao Y, Liao H, Lee SP, et al. Common loss-of-function variants of the epidermal barrier protein filaggrin are a major predisposing factor for atopic dermatitis. Nat Genet. 2006;38(4):441–6. PubMed PMID: 16550169.
2. Hoste E, Kemperman P, Devos M, Denecker G, Kezic S, Yau N, et al. Caspase-14 is required for filaggrin degradation to natural moisturizing factors in the skin. J Invest Dermatol. 2011;131(11):2233–41. PubMed PMID: 21654840.
3. Boguniewicz M, Leung DY. Atopic dermatitis: a disease of altered skin barrier and immune dysregulation. Immunol Rev. 2011;242(1):233–46. PubMed PMID: 21682749. Pubmed Central PMCID: PMC3122139. Epub 2011/06/21. eng.
4. Irvine AD. Fleshing out filaggrin phenotypes. J Invest Dermatol. 2007;127(3):504–7. PubMed PMID: 17299430.
5. O'Regan GM, Sandilands A, McLean WH, Irvine AD. Filaggrin in atopic dermatitis. J Allergy Clin Immunol. 2008;122(4):689–93. PubMed PMID: 18774165.
6. Weidinger S, Baurecht H, Wagenpfeil S, Henderson J, Novak N, Sandilands A, et al. Analysis of the individual and aggregate genetic contributions of previously identified serine peptidase inhibitor Kazal type 5 (SPINK5), kallikrein-related peptidase 7 (KLK7), and filaggrin (FLG) polymorphisms to eczema risk. J Allergy Clin Immunol. 2008;122(3):560–8 e4. PubMed PMID: 18774391.
7. Sakabe JI, Yamamoto M, Hirakawa S, Motoyama A, Ohta I, Tatsuno K, et al. Kallikrein-related peptidase 5 functions in proteolytic processing of pro-filaggrin in cultured human keratinocytes. J Biol Chem. 2013;288(24):17179–89. PubMed PMID: 23629652.
8. Howell MD, Kim BE, Gao P, Grant AV, Boguniewicz M, Debenedetto A, et al. Cytokine modulation of atopic dermatitis filaggrin skin expression. J Allergy Clin Immunol. 2007;120(1):150–5. PubMed PMID: 17512043. Pubmed Central PMCID: PMC2669594. Epub 2007/05/22. eng.
9. Hvid M, Johansen C, Deleuran B, Kemp K, Deleuran M, Vestergaard C. Regulation of caspase 14 expression in keratinocytes by inflammatory cytokines – a possible link between reduced skin barrier function and inflammation? Exp Dermatol. 2011;20(8):633–6. PubMed PMID: 21539619. Epub 2011/05/05. eng.
10. Barker JN, Palmer CN, Zhao Y, Liao H, Hull PR, Lee SP, et al. Null mutations in the filaggrin gene (FLG) determine major susceptibility to early-onset atopic dermatitis that persists into adulthood. J Invest Dermatol. 2007;127(3):564–7. PubMed PMID: 16990802.
11. Candi E, Schmidt R, Melino G. The cornified envelope: a model of cell death in the skin. Nat Rev Mol Cell Biol. 2005;6(4):328–40. PubMed PMID: 15803139.
12. McGrath JA, Uitto J. The filaggrin story: novel insights into skin-barrier function and disease. Trends Mol Med. 2008;14(1):20–7. PubMed PMID: 18068483.
13. Gan SQ, McBride OW, Idler WW, Markova N, Steinert PM. Organization, structure, and polymorphisms of the human pro-filaggrin gene. Biochemistry. 1990;29(40):9432–40. PubMed PMID: 2248957.
14. McKinley-Grant LJ, Idler WW, Bernstein IA, Parry DA, Cannizzaro L, Croce CM, et al. Characterization of a cDNA clone encoding human filaggrin and localization of the gene to chromosome region 1q21. Proc Natl Acad Sci U S A. 1989;86(13):4848–52. PubMed PMID: 2740331. Pubmed Central PMCID: 297512.
15. Sandilands A, Terron-Kwiatkowski A, Hull PR, O'Regan GM, Clayton TH, Watson RM, et al. Comprehensive analysis of the gene encoding filaggrin uncovers prevalent and rare mutations in ichthyosis

vulgaris and atopic eczema. Nat Genet. 2007;39(5):650–4. PubMed PMID: 17417636.
16. Smith FJ, Irvine AD, Terron-Kwiatkowski A, Sandilands A, Campbell LE, Zhao Y, et al. Loss-of-function mutations in the gene encoding filaggrin cause ichthyosis vulgaris. Nat Genet. 2006;38(3):337–42. PubMed PMID: 16444271.
17. Akiyama M. FLG mutations in ichthyosis vulgaris and atopic eczema: spectrum of mutations and population genetics. Br J Dermatol. 2010;162(3):472–7. PubMed PMID: 19958351.
18. Jang SI, Steinert PM, Markova NG. Activator protein 1 activity is involved in the regulation of the cell type-specific expression from the proximal promoter of the human pro-filaggrin gene. J Biol Chem. 1996;271(39):24105–14. PubMed PMID: 8798649.
19. Jang SI, Karaman-Jurukovska N, Morasso MI, Steinert PM, Markova NG. Complex interactions between epidermal POU domain and activator protein 1 transcription factors regulate the expression of the pro-filaggrin gene in normal human epidermal keratinocytes. J Biol Chem. 2000;275(20):15295–304. PubMed PMID: 10809764.
20. King KE, Ponnamperuma RM, Gerdes MJ, Tokino T, Yamashita T, Baker CC, et al. Unique domain functions of p63 isotypes that differentially regulate distinct aspects of epidermal homeostasis. Carcinogenesis. 2006;27(1):53–63. PubMed PMID: 16081516.
21. Presland RB, Tomic-Canic M, Lewis SP, Dale BA. Regulation of human pro-filaggrin promoter activity in cultured epithelial cells by retinoic acid and glucocorticoids. J Dermatol Sci. 2001;27(3):192–205. PubMed PMID: 11641059.
22. Resing KA, Johnson RS, Walsh KA. Characterization of protease processing sites during conversion of rat pro-filaggrin to filaggrin. Biochemistry. 1993;32(38):10036–45. PubMed PMID: 8399131.
23. List K, Szabo R, Wertz PW, Segre J, Haudenschild CC, Kim SY, et al. Loss of proteolytically processed filaggrin caused by epidermal deletion of Matriptase/MT-SP1. J Cell Biol. 2003;163(4):901–10. PubMed PMID: 14638864. Pubmed Central PMCID: 2173680.
24. Leyvraz C, Charles RP, Rubera I, Guitard M, Rotman S, Breiden B, et al. The epidermal barrier function is dependent on the serine protease CAP1/Prss8. J Cell Biol. 2005;170(3):487–96. PubMed PMID: 16061697. Pubmed Central PMCID: 2171460.
25. Netzel-Arnett S, Currie BM, Szabo R, Lin CY, Chen LM, Chai KX, et al. Evidence for a matriptase-prostasin proteolytic cascade regulating terminal epidermal differentiation. J Biol Chem. 2006;281(44):32941–5. PubMed PMID: 16980306.
26. Chavanas S, Bodemer C, Rochat A, Hamel-Teillac D, Ali M, Irvine AD, et al. Mutations in SPINK5, encoding a serine protease inhibitor, cause Netherton syndrome. Nat Genet. 2000;25(2):141–2. PubMed PMID: 10835624.
27. Descargues P, Deraison C, Bonnart C, Kreft M, Kishibe M, Ishida-Yamamoto A, et al. Spink5-deficient mice mimic Netherton syndrome through degradation of desmoglein 1 by epidermal protease hyperactivity. Nat Genet. 2005;37(1):56–65. PubMed PMID: 15619623.
28. Hewett DR, Simons AL, Mangan NE, Jolin HE, Green SM, Fallon PG, et al. Lethal, neonatal ichthyosis with increased proteolytic processing of filaggrin in a mouse model of Netherton syndrome. Hum Mol Genet. 2005;14(2):335–46. PubMed PMID: 15590704.
29. Epp N, Furstenberger G, Muller K, de Juanes S, Leitges M, Hausser I, et al. 12R-lipoxygenase deficiency disrupts epidermal barrier function. J Cell Biol. 2007;177(1):173–82. PubMed PMID: 17403930. Pubmed Central PMCID: 2064121.
30. Denecker G, Hoste E, Gilbert B, Hochepied T, Ovaere P, Lippens S, et al. Caspase-14 protects against epidermal UVB photodamage and water loss. Nat Cell Biol. 2007;9(6):666–74. PubMed PMID: 17515931.
31. Bieber T. Atopic dermatitis. N Engl J Med. 2008;358(14):1483–94. PubMed PMID: 18385500.
32. Vestergaard C, Deleuran M, Kragballe K. Two cases of atopic dermatitis-like conditions induced in psoriasis patients treated with infliximab. J Eur Acad Dermatol Venereol. 2007;21(9):1272–4. PubMed PMID: 17894730.
33. Zaba LC, Krueger JG, Lowes MA. Resident and "inflammatory" dendritic cells in human skin. J Invest Dermatol. 2009;129(2):302–8. PubMed PMID: 18685620. Pubmed Central PMCID: 2746703.
34. Nestle FO, Kaplan DH, Barker J. Psoriasis. N Engl J Med. 2009;361(5):496–509. PubMed PMID: 19641206.
35. Homey B, Meller S. Chemokines and other mediators as therapeutic targets in psoriasis vulgaris. Clin Dermatol. 2008;26(5):539–45. PubMed PMID: 18755373.
36. Pellerin L, Henry J, Hsu CY, Balica S, Jean-Decoster C, Mechin MC, et al. Defects of filaggrin-like proteins in both lesional and nonlesional atopic skin. J Allergy Clin Immunol. 2013;131(4):1094–102. PubMed PMID: 23403047. Epub 2013/02/14. Eng.
37. Duhen T, Geiger R, Jarrossay D, Lanzavecchia A, Sallusto F. Production of interleukin 22 but not interleukin 17 by a subset of human skin-homing memory T cells. Nat Immunol. 2009;10(8):857–63. PubMed PMID: 19578369.
38. Eyerich S, Eyerich K, Pennino D, Carbone T, Nasorri F, Pallotta S, et al. Th22 cells represent a distinct human T cell subset involved in epidermal immunity and remodeling. J Clin Invest. 2009;119(12):3573–85. PubMed PMID: 19920355. Pubmed Central PMCID: 2786807.
39. Fujita H, Nograles KE, Kikuchi T, Gonzalez J, Carucci JA, Krueger JG. Human Langerhans cells induce distinct IL-22-producing CD4+ T cells lacking IL-17 production. Proc Natl Acad Sci U S A. 2009;106(51):21795–800. PubMed PMID: 19996179. Pubmed Central PMCID: 2799849.
40. Niebuhr M, Scharonow H, Gathmann M, Mamerow D, Werfel T. Staphylococcal exotoxins are strong

inducers of IL-22: a potential role in atopic dermatitis. J Allergy Clin Immunol. 2010;126(6):1176–83 e4. PubMed PMID: 20864149.
41. Gutowska-Owsiak D, Schaupp AL, Salimi M, Taylor S, Ogg GS. Interleukin-22 downregulates filaggrin expression and affects expression of pro-filaggrin processing enzymes. Br J Dermatol. 2011;165(3):492–8. PubMed PMID: 21564072. Epub 2011/05/14. eng.
42. Asarch A, Barak O, Loo DS, Gottlieb AB. Th17 cells: a new paradigm for cutaneous inflammation. J Dermatolog Treat. 2008;19(5):259–66. PubMed PMID: 18629676.
43. McKenzie BS, Kastelein RA, Cua DJ. Understanding the IL-23-IL-17 immune pathway. Trends Immunol. 2006;27(1):17–23. PubMed PMID: 16290228.
44. Lee J, Ho WH, Maruoka M, Corpuz RT, Baldwin DT, Foster JS, et al. IL-17E, a novel proinflammatory ligand for the IL-17 receptor homolog IL-17Rh1. J Biol Chem. 2001;276(2):1660–4. PubMed PMID: 11058597.
45. Gutowska-Owsiak D, Schaupp AL, Salimi M, Selvakumar TA, McPherson T, Taylor S, et al. IL-17 downregulates filaggrin and affects keratinocyte expression of genes associated with cellular adhesion. Exp Dermatol. 2012;21(2):104–10. PubMed PMID: 22229441.
46. Fallon PG, Ballantyne SJ, Mangan NE, Barlow JL, Dasvarma A, Hewett DR, et al. Identification of an interleukin (IL)-25-dependent cell population that provides IL-4, IL-5, and IL-13 at the onset of helminth expulsion. J Exp Med. 2006;203(4):1105–16. PubMed PMID: 16606668. Pubmed Central PMCID: 2118283.
47. Wang YH, Angkasekwinai P, Lu N, Voo KS, Arima K, Hanabuchi S, et al. IL-25 augments type 2 immune responses by enhancing the expansion and functions of TSLP-DC-activated Th2 memory cells. J Exp Med. 2007;204(8):1837–47. PubMed PMID: 17635955. Pubmed Central PMCID: 2118667.
48. Hvid M, Vestergaard C, Kemp K, Christensen GB, Deleuran B, Deleuran M. IL-25 in atopic dermatitis: a possible link between inflammation and skin barrier dysfunction? J Invest Dermatol. 2011;131(1):150–7. PubMed PMID: 20861853. Epub 2010/09/24. eng.
49. Richardson SK, Gelfand JM. Update on the natural history and systemic treatment of psoriasis. Adv Dermatol. 2008;24:171–96. PubMed PMID: 19256309. Pubmed Central PMCID: 2634854.
50. Deleuran MS, Vestergaard C. Therapy of severe atopic dermatitis in adults. J Dtsch Dermatol Ges. 2012;10(6):399–406. PubMed PMID: 22606964.
51. Kim BE, Howell MD, Guttman-Yassky E, Gilleaudeau PM, Cardinale IR, Boguniewicz M, et al. TNF-alpha downregulates filaggrin and loricrin through c-Jun N-terminal kinase: role for TNF-alpha antagonists to improve skin barrier. J Invest Dermatol. 2011;131(6):1272–9. PubMed PMID: 21346775. Epub 2011/02/25. eng.
52. Dillon SR, Sprecher C, Hammond A, Bilsborough J, Rosenfeld-Franklin M, Presnell SR, et al. Interleukin 31, a cytokine produced by activated T cells, induces dermatitis in mice. Nat Immunol. 2004;5(7):752–60. PubMed PMID: 15184896.
53. Niyonsaba F, Ushio H, Hara M, Yokoi H, Tominaga M, Takamori K, et al. Antimicrobial peptides human beta-defensins and cathelicidin LL-37 induce the secretion of a pruritogenic cytokine IL-31 by human mast cells. J Immunol. 2010;184(7):3526–34. PubMed PMID: 20190140.
54. Grimstad O, Sawanobori Y, Vestergaard C, Bilsborough J, Olsen UB, Gronhoj-Larsen C, et al. Anti-interleukin-31-antibodies ameliorate scratching behaviour in NC/Nga mice: a model of atopic dermatitis. Exp Dermatol. 2009;18(1):35–43. PubMed PMID: 19054054.
55. Cornelissen C, Marquardt Y, Czaja K, Wenzel J, Frank J, Luscher-Firzlaff J, et al. IL-31 regulates differentiation and filaggrin expression in human organotypic skin models. J Allergy Clin Immunol. 2012;129(2):426–33, 33 e1–8. PubMed PMID: 22177328. Epub 2011/12/20. eng.
56. Jensen JM, Scherer A, Wanke C, Brautigam M, Bongiovanni S, Letzkus M, et al. Gene expression is differently affected by pimecrolimus and betamethasone in lesional skin of atopic dermatitis. Allergy. 2012;67(3):413–23. PubMed PMID: 22142306. Epub 2011/12/07. eng.
57. Asselineau D, Dale BA, Bernard BA. Filaggrin production by cultured human epidermal keratinocytes and its regulation by retinoic acid. Differentiation. 1990;45(3):221–9. PubMed PMID: 2090523.
58. Kautsky MB, Fleckman P, Dale BA. Retinoic acid regulates oral epithelial differentiation by two mechanisms. J Invest Dermatol. 1995;104(4):546–53. PubMed PMID: 7706775.
59. van den Bogaard EH, Bergboer JG, Vonk-Bergers M, van Vlijmen-Willems IM, Hato SV, van der Valk PG, et al. Coal tar induces AHR-dependent skin barrier repair in atopic dermatitis. J Clin Invest. 2013;123(2):917–27. PubMed PMID: 23348739. Pubmed Central PMCID: PMC3561798. Epub 2013/01/26. Eng.
60. Barnes PJ. Corticosteroids: the drugs to beat. Eur J Pharmacol. 2006;533(1–3):2–14. PubMed PMID: 16436275.
61. Liu J, Farmer Jr JD, Lane WS, Friedman J, Weissman I, Schreiber SL. Calcineurin is a common target of cyclophilin-cyclosporin A and FKBP-FK506 complexes. Cell. 1991;66(4):807–15. PubMed PMID: 1715244.
62. Ruzicka T, Lynde CW, Jemec GB, Diepgen T, Berth-Jones J, Coenraads PJ, et al. Efficacy and safety of oral alitretinoin (9-cis retinoic acid) in patients with severe chronic hand eczema refractory to topical corticosteroids: results of a randomized, double-blind, placebo-controlled, multicentre trial. Br J Dermatol. 2008;158(4):808–17. PubMed PMID: 18294310.
63. Grahovac M, Molin S, Prinz JC, Ruzicka T, Wollenberg A. Treatment of atopic eczema with oral alitretinoin. Br J Dermatol. 2010;162(1):217–8. PubMed PMID: 19886882.

64. Cork MJ, Danby SG, Vasilopoulos Y, Hadgraft J, Lane ME, Moustafa M, et al. Epidermal barrier dysfunction in atopic dermatitis. J Invest Dermatol. 2009;129(8):1892–908. PubMed PMID: 19494826.
65. Kezic S, O'Regan GM, Lutter R, Jakasa I, Koster ES, Saunders S, et al. Filaggrin loss-of-function mutations are associated with enhanced expression of IL-1 cytokines in the stratum corneum of patients with atopic dermatitis and in a murine model of filaggrin deficiency. J Allergy Clin Immunol. 2012;129(4):1031–9 e1. PubMed PMID: 22322004. Epub 2012/02/11. eng.
66. Lee KH, Cho KA, Kim JY, Kim JY, Baek JH, Woo SY, et al. Filaggrin knockdown and toll-like receptor 3 (TLR3) stimulation enhanced the production of thymic stromal lymphopoietin (TSLP) from epidermal layers. Exp Dermatol. 2011;20(2):149–51. PubMed PMID: 21255094. Epub 2011/01/25. eng.
67. Thyssen JP, Linneberg A, Ross-Hansen K, Carlsen BC, Meldgaard M, Szecsi PB, et al. Filaggrin mutations are strongly associated with contact sensitization in individuals with dermatitis. Contact Dermatitis. 2013;68(5):273–6. PubMed PMID: 23343419.

Immune Activity and Vitamin D

5

Charlotte M. Bonefeld, Morten M. Nielsen, and Carsten Geisler

Contents

C.M. Bonefeld, PhD (✉) • M.M. Nielsen, MS
C. Geisler, MD, PhD, DMSc
Department of International Health, Immunology and Microbiology, University of Copenhagen, Blegdamsvej 3, Copenhagen 2200, Denmark
e-mail: cmenne@sund.ku.dk; mmilek@sund.ku.dk; cge@sund.ku.dk

5.1 Introduction

Atopic diseases are characterized by elevated levels of IgE and classically described as Th2 mediated. It is believed that following entry of an allergen into the skin, it is processed by local dendritic cells (DCs) that then migrate to the draining lymph node. Here, the DC presents the allergen for naïve $CD4^+$ T cells, which subsequently differentiate into Th2 cells characterized by their production of IL-4, IL-5, and IL-13 (Fig. 5.1). The Th2 cells activate allergen-specific B cells and promote isotype switch to IgE. The allergen-specific IgE will bind and prime mast cells, which then can be activated following exposure to the allergen (see Fig. 5.1). During the last years it has become clear that several new players are involved in allergen-specific immune responses, among these the epithelial cells, innate lymphoid cells (ILCs), Th17 cells, and vitamin D. This chapter focuses on how filaggrin (or lack of) affects the immune response and vitamin D synthesis in the skin. Both human and mice studies are discussed as much of the knowledge about the effects of filaggrin on the immune system comes from studies using the Flaky tail mice (Flg^{ft} mice) lacking filaggrin [1–3] (Fig. 5.2).

5.2 The T-Cell Response

Atopic dermatitis (AD) is a complex disease dependent on both genetic and environmental factors that induce a complex immune response.

J.P. Thyssen, H. Maibach (eds.), *Filaggrin*,
DOI 10.1007/978-3-642-54379-1_5,

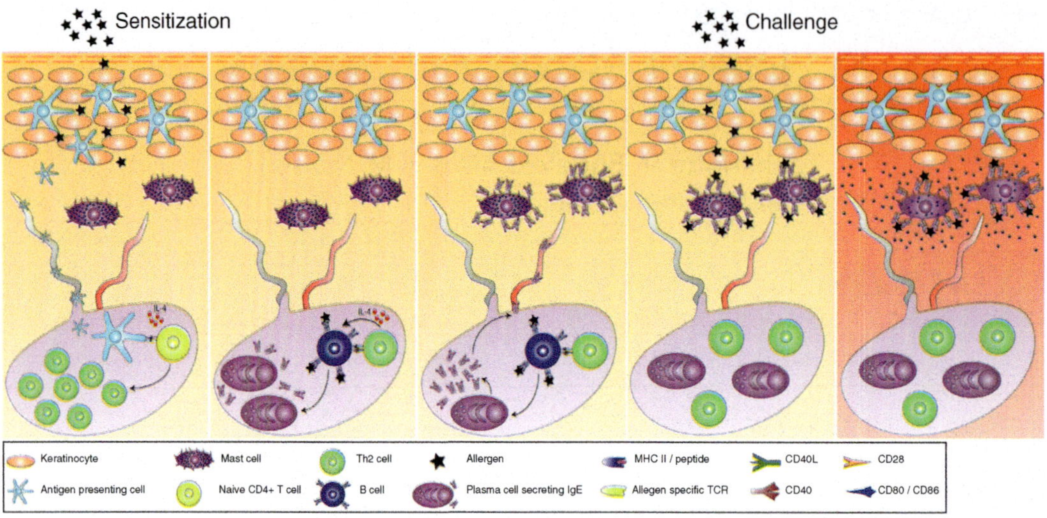

Fig. 5.1 Simple model for immune responses leading to AD. Skin exposure to allergens leads to activation of skin DC that migrates to the draining lymph nodes, where they present allergen for allergen-specific naïve CD4$^+$ T cells. Due to the presence of IL-4, allergen-specific CD4$^+$ T cells differentiate into Th2 cells. These subsequently activate allergen-specific B cells that differentiate into IgE-producing plasma cells. IgE bind to FcεR1 receptors on mast cells in the skin. Upon subsequent exposure to the allergen, IgE on the mast cells bind allergen and induce mast cell activation and thereby skin inflammation

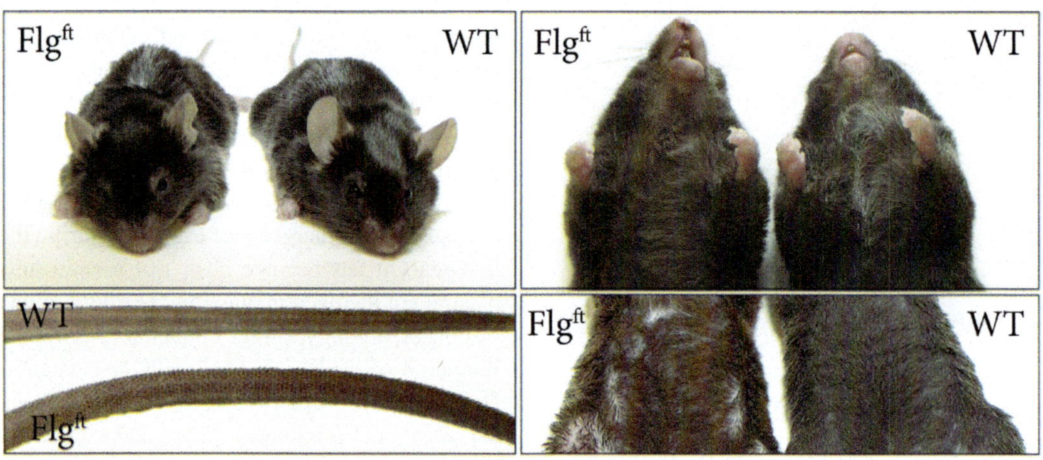

Fig. 5.2 Appearance of age-matched Flg^{ft} and WT mice (C57bl/6)

It is well known that T cells play a central role in the pathogenesis of AD [4, 5]. An AD mouse model showed that αβ T cells were required for skin inflammation, whereas γδ T cells and B cells were not required [5]. In addition, IL-4 expression was upregulated in inflamed skin and found to be produced by αβ T cells [5]. Interestingly, skin inflammation could be induced in mice lacking either B cells or CD40L indicating that IgE is not required for the development of skin inflammation [5].

CD4$^+$ T-cell differentiation is classically divided into Th1 and Th2 responses dominated by IFNγ, IL-2 and IL-4, IL-5 and IL-13, respectively (Fig. 5.3) [6, 7]. Recently, a new CD4$^+$ T-cell subtype has been identified, namely, the

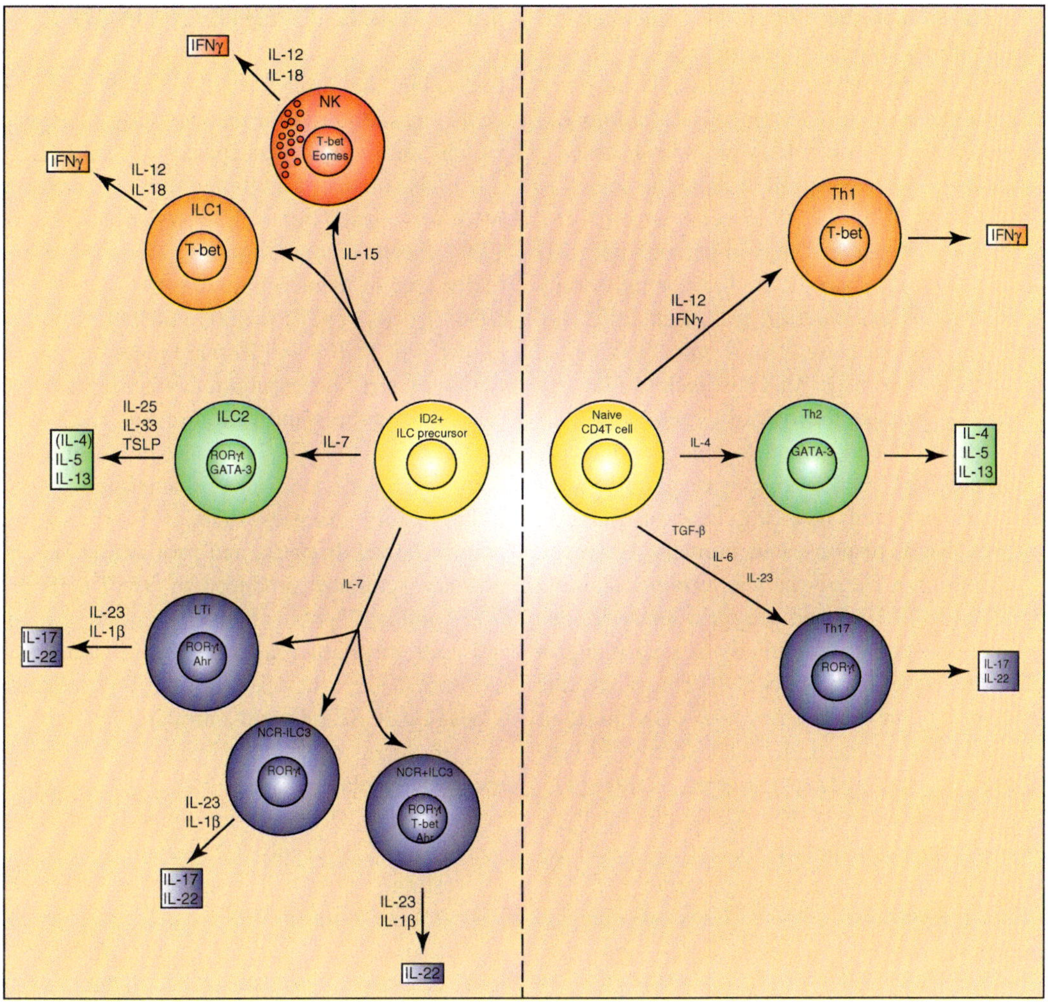

Fig. 5.3 Schematic representation of differentiation of ILC and naïve CD4+ T cells showing cytokines and transcription factor involved in the differentiation as well as effector cytokines produced by the cells

Th17 cells (see below). A mouse model for AD-like skin inflammation was used to further investigate the role of Th1 and Th2 cells in AD. Impaired eosinophil recruitment to the skin was seen in mice lacking either IL-4 or IL-5, whereas IFNγ did not seem to be involved in eosinophil recruitment [8]. However, IFNγ seemed to be involved in the response by other mechanisms as reduced skin inflammation was seen in mice lacking either IL-5 or IFNγ [8]. In accordance with this, IL-4, IL-5, IL-13, IFNγ, and IL-12 are upregulated in the skin from AD patients compared to healthy skin [9, 10]. Interestingly, these cytokines appear to be involved at different stages of the disease. The initial inflammation seems to be driven by IL-4 and IL-13, as these are the dominating cytokines in acute lesional skin [9, 10]. In contrast, chronic inflammation seems to be maintained by IL-5, IL-12, and IFNγ [9, 10]. Increased eosinophil infiltration was seen in chronic compared to acute lesions, which correlates with the increased expression of IL-5 in chronic lesions [9]. Thus, the acute inflammation appears to be mediated by Th2 cells and their production of IL-4 and IL-13, whereas the chronic inflammation seems to involve both Th1 and Th2 cells.

Exposure of the skin to protein allergens (ovalbumin (OVA), *Dermatophagoides pteronyssinus* (Derp1)), contact allergens, and irritants induces a more vigorous inflammatory response in Flg^{ft} mice than in control mice [1, 3]. The response to protein antigens seems to involve Th1, Th2, and Th17 cells, which correlates well with the findings in patients with AD (see below [1, 3, 9, 10]). Interestingly, only minor cytokine induction was seen following allergen exposure in control mice having an intact skin barrier [1, 3]. Thus, lack of filaggrin seems to increase the risk of developing allergen-specific T-cell responses mediated by Th1, Th2, and Th17 cells. To date very few studies have investigated how the lack of filaggrin affects T-cell responses in humans. However, by combining Derp1-specific tetramer staining together with IL-4 and filaggrin analysis, one study showed that individuals lacking filaggrin had an increased number of Derp1-specific IL-4-producing $CD4^+$ positive cells in their blood compared to individual with wild-type filaggrin gene (*FLG*) mutation status [11]. Taken together, even though AD classically is described as mediated by Th2 cells, other effector $CD4^+$ T cells seem to be important at different stages of the disease. Lack of filaggrin seems to increase the risk of developing allergen-specific T-cell responses, probably due to increased skin penetration of allergens that elicit a basic inflammatory response and thereby provides a reduced activation threshold for the T cells. In addition, a recent study shows that AD patients with *FLG* loss-of-function mutations had increased levels of IL-1α and IL-1β in the skin compared to both healthy controls and AD patients without *FLG* mutations [12]. Interestingly, the increased IL-1α and IL-1β correlated inversely with the "natural moisturizing factors" (NMFs) that correlated inversely with skin pH [12]. As IL-1 is a pro-inflammatory cytokine known to be involved in the initiation of the immune response in general (e.g., by inducing maturation and migration of DC), increased IL-1 levels are likely to lower the immune activation threshold within the skin of patients lacking filaggrin [13].

5.3 The IL-23/TH17 Axis in AD

Following the discovery that IL-23, and not IL-12, was required for the induction of experimental autoimmune encephalomyelitis (EAE), the mouse model of multiple sclerosis, intensive work was carried out to characterize the effector $CD4^+$ T cells responding to IL-23, which eventually lead to identification of IL-17-producing $CD4^+$ T cells (Th17) in 2005 [14–16]. It is now known that TGF-β and IL-6 are required for initiation of Th17 differentiation, whereas IL-23 is required for the stabilization of the Th17 cells (see Fig. 5.3) [7]. Th17 cells are involved in the pathogenesis of a variety of autoimmune and inflammatory diseases such as EAE, inflammatory bowel diseases, and psoriasis [17]. The primary cytokines produced by Th17 cells are IL-17 and IL-22, both of which stimulate epithelial cells to produce a variety of inflammatory cytokines (e.g., IL-1β, IL-6, TNFα), chemokines (e.g., CXCL8), and antimicrobial peptides (e.g., β-defensin, cathelicidin) [17]. Th17 cells most likely play a role in AD as an increased percentage of Th17 cells are found in the blood and lesional skin of AD patients [18, 19]. Interestingly, Th17 cells seem to serve as an initial cytokine source as they are more prevalent in acute than chronic lesions [18, 19]. Furthermore, Th17 cells are likely associated with the severity of AD as a direct correlation between severity of the inflammation and the percentage of Th17 cells in the blood has been found [18]. The role of Th17 cells in the immune response to protein allergen has been further investigated by using an OVA sensitization mouse model. Here mice were exposed to OVA either epicutaneously (EC) or intraperitoneally (IP) [20]. It was shown that EC OVA sensitization induced both a local and a systemic Th17 response, whereas IP OVA sensitization did not [20]. In contrast, the production of IL-4 and IFNγ following OVA sensitization seemed to be independent on the sensitization route [20]. The Th17 response also appears to drive airway inflammation as neutrophil influx and bronchial hyperactivity induced by OVA inhalation in EC-sensitized mice could be reversed by IL-17 blockade [20].

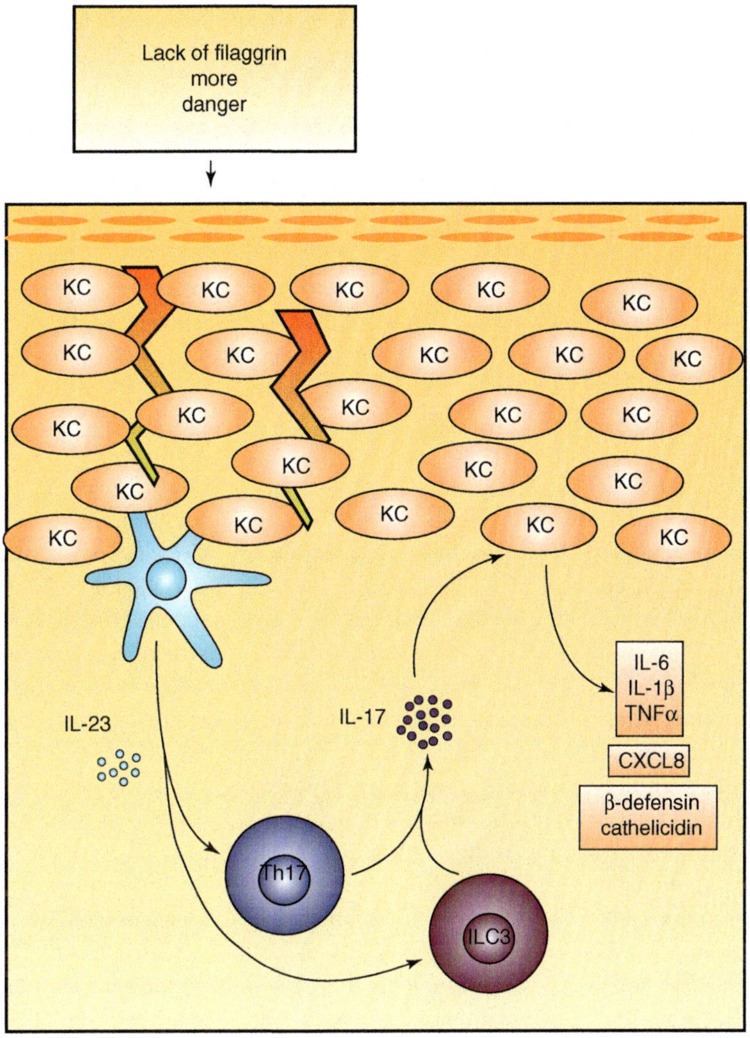

Fig. 5.4 Model for the role of IL-17 in initiation of AD

The reason why EC sensitization, in contrast to IP sensitization, leads to Th17 responses might be explained by the ability of skin-derived DC to produce IL-23, a feature that is not observed in splenic DC [20]. In this model, mice were tape-stripped before OVA exposure of the skin, a procedure that is known to induce disruption of the skin barrier. As EC sensitization with allergens seems to play an important role in allergen sensitization of patients with AD, and as patients lacking filaggrin have an increased risk of developing asthma [21, 22], it was suggested that allergen exposure of skin lacking filaggrin leads to a Th17 response, which upon later allergen exposure of the airways induces a Th17-dependent airway inflammation [20]. In agreement with this, three studies on Flg^{ft} mice have shown an increased IL-17 production in Flg^{ft} mice compared to control mice [1–3]. The increased IL-17 production was found both in the skin at steady state and in OVA-specific $CD4^{+}$ T cells after EC OVA sensitization [1, 3]. Interestingly, a similar Th17 response was seen in Flg^{ft} mice and control mice following IP sensitization with OVA [3]. Taken together, allergen exposure of the skin seems to favor a Th17 response, and lack of filaggrin increases the risk of developing allergen-specific Th17 responses that again increases the risk of developing severe AD and asthma (Fig. 5.4).

Even though Th17 cells were first described as the IL-17-producing cells, it is now clear that several other types of cells can produce IL-17, i.e., CD8$^+$ T cells, γδ T cells, and ILC [3, 17, 23]. During the last years much focus has been on ILC. The ILC are characterized by lack of expression of markers associated with T cells, B cells, DCs, macrophages, and granulocytes [23]. Interestingly, it seems that ILCs can be subdivided based on transcription factors and cytokine production in a way similar to the CD4$^+$ T effector cells (see Fig. 5.3). Thus, ILC1 are the innate analogs to Th1 cells, ILC2 are the innate analogs to Th2 cells, and, finally, LTi and ILC3 are the innate analogs to Th17 cells [7, 23]. A recent study indicated that cells other than Th17 cells might be responsible of the increased level of IL-17 found in the skin of Flg^{ft} mice at steady state [24]. In this study, Flg^{ft} mice were crossed to RAG2-deficient mice lacking both T and B cells [24]. Lesional skin inflammation characterized by fur loss, erythematous scaly skin, and periocular swelling was seen in 88 % of Flg^{ft} mice after 32 weeks. In contrast, no sign of skin inflammation was seen in mice lacking both filaggrin and RAG2, indicating that the adaptive immune response is required for the skin inflammation in Flg^{ft} mice [24]. However, increased levels of both IL-17A and IL-22 were found in the skin of RAG2$^{-/-}$ Flg^{ft} mice compared to mice only lacking RAG2, indicating that LTi and/or ILC3 might be involved in the increased level of IL-17A found in Flg^{ft} mice at steady state [24]. Thus, it is likely that ILC subtypes are involved in the inflammatory response induced by the lack of filaggrin; however, this needs further investigations (see Fig. 5.4).

Taken together, IL-17 is most likely produced by both Th17 cells and ILC in the inflammatory response observed in AD. Lack of filaggrin leads to an impaired barrier function and thereby probably to danger signals that activate skin DC to produce IL-23. IL-23 subsequently stimulates IL-17 production from ILC and Th17 cells. IL-17 in the skin stimulates keratinocytes to produce pro-inflammatory cytokines and chemokines that eventually lead to increased skin inflammation (see Fig. 5.4).

5.4 The Role of TSLP in the Response

Keratinocytes constitute the majority of the cells in the epidermis and were originally described mechanistically as the cells that form the physical barrier between the environment and the body. However, it has become clear that keratinocytes also play an important immune-modulating role due to their ability to produce a variety of cytokines (e.g., IL-1β, IL-23, TNFα, and IL-10 in response to pathogens, stress, and other environmental triggers) [25–28]. Cytokines produced by keratinocytes can modify the activation and differentiation of skin DC. Thymic stromal lymphopoietin (TSLP) mainly produced by keratinocytes, fibroblasts, and stromal cells can induce a Th2 response [29]. TSLP is highly expressed in the epidermis of patients with AD [30]. Stimulation of CD11c$^+$ DC with TLSP in vitro leads to activation and differentiation of DC that promote Th2 differentiation (Fig. 5.5) [30]. By use of a transgenic mouse model, where TSLP specifically can be induced in the keratinocytes, it was shown that mice developed spontaneous AD characterized by skin inflammation, increased number of skin-homing Th2 cells, and elevated levels of serum IgE 2–3 weeks after TSLP induction [31]. Interestingly, induction of TSLP in TCRβKO mice lacking all CD4$^+$ and CD8$^+$ αβ T cells still lead to skin inflammation, suggesting that T cells are not necessary for the induction of the allergic response [31]. It was suggested that the response could be induced by TSLP acting directly on activated macrophages, eosinophils, mast cells, and other myeloid effector cells [31]. However, ILC2 cells could also be involved. TSLP can induce cytokine production by ILC2 in the skin independent of IL-33 and IL-25 [32]. An increased frequency of ILC2 was found in lesional skin from patients with AD [32]. Furthermore, AD-like skin inflammation could be significantly reduced either by depleting ILC or by using TSLP receptor KO mice [32]. This indicates that TSLP might play an important role in the induction of AD by stimulating ILC2 to produce IL-5 and IL-13. In agreement with studies in human

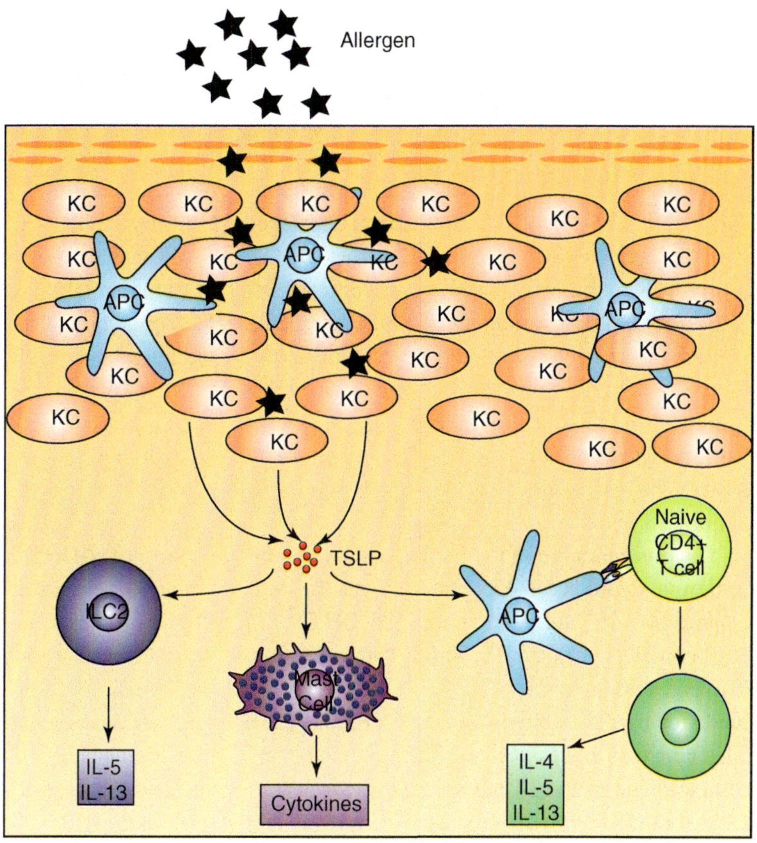

Fig. 5.5 Model for the involvement of TSLP in the immune response during AD

AD patients, TSLP was found to be more expressed in skin from Flg^{ft} mice than in control mice [33]. Furthermore, it was found that the expression and activity of the endogenous proteases kallikrein 5, 7, and 14, which activate TSLP production in keratinocytes, were higher in skin from Flg^{ft} mice compared to control mice in steady state [33]. It can, therefore, be suggested that the increased activity of the endogenous proteases caused by the lack of filaggrin leads to increased production of TSLP via the protease-activated receptor-2 in keratinocytes and that this plays an important role in the induction of both ILC2 and Th2 cells.

5.5 Vitamin D, Filaggrin, and Immune Responses

Several studies have demonstrated that vitamin D regulates keratinocyte growth and differentiation and affects immune responses [34–39]. The major source of vitamin D for most humans is 7-dehydrocholesterol (7-DHC) in the plasma membrane of keratinocyte [40, 41]. The first stage of vitamin D synthesis depends on the UVB (280–320 nm)-mediated photoconversion of 7-DHC to previtamin D_3 in the skin. Once formed, previtamin D_3 is rapidly converted to vitamin D_3 that diffuses to the blood circulation, where it is bound to the vitamin D-binding protein (DBP). Vitamin D_3 is subsequently metabolized in the liver to 25-hydroxyvitamin D_3 ($25(OH)D_3$) and then in the kidney to its biologically active form 1,25-dihydroxyvitamin D_3 ($1,25(OH)_2D_3$) [40, 42, 43]. $1,25(OH)_2D_3$ is classically considered to function as a endocrine regulator of calcium homeostasis. However, the understanding of vitamin D metabolism and physiological function has evolved dramatically in recent years. Vitamin D is now recognized as a pleiotropic regulator of human physiology with emerging roles in several tissues including the immune system and the skin [44].

The biological actions of $1,25(OH)_2D_3$ are mediated by the vitamin D receptor (VDR) that belongs to the nuclear hormone receptor superfamily [45, 46]. Interaction of $1,25(OH)_2D_3$ with VDR induces heterodimerization with the retinoid X receptor (RXR) and translocation of $1,25(OH)_2D_3$-VDR/RXR complexes into the nucleus [44, 47, 48]. The $1,25(OH)_2D_3$-VDR/RXR complexes bind to specific DNA sequences called vitamin D response elements (VDREs) in target genes, and dependent on the recruited co-regulators either augment or inhibit transcription of the target gene [48–50]. Both keratinocytes and various cells of the immune system express VDR, especially after their activation [38, 51–54].

The normal range of the $1,25(OH)_2D_3$ concentration in serum is 50–175 pM, whereas the concentration of the precursor $25(OH)D_3$ is approximately 1,000-fold higher (50–160 nM). The conversion of $25(OH)D_3$ to $1,25(OH)_2D_3$ is mediated by the 1-α hydroxylase CYP27B1 [55]. This conversion was at first believed exclusively to take place in the kidneys; however, it is now clear that CYP27B1 is expressed in various cell types including keratinocytes, macrophages, and activated T cells [56–58], and evidence is rapidly accumulating that local CYP27B1-catalyzed production of $1,25(OH)_2D_3$ is critical for its physiological actions [37, 59]. In this context, the keratinocytes are the only cell type where the complete enzymatic machinery for the synthesis of $1,25(OH)_2D_3$ from 7-DHC has been shown [41, 60–62]. Thus, it can be assumed that UVB-induced production of vitamin D_3 in the skin might result in formation of substantial amounts of local $1,25(OH)_2D_3$, which regulate keratinocyte growth and differentiation and affect the local immune response.

The keratohyalin granules in the stratum granulosum of the epidermis consist primarily of profilaggrin polymers [63] that are proteolytically cleaved into filaggrin monomers. Monomeric filaggrin binds to keratin to form tight bundles facilitating the collapse and flattening of the cells in the stratum corneum [64]. Subsequently, filaggrin is fully degraded to its constituents amino acids dominated by glutamine, arginine, and histidine [65]. Histidine is a substrate for histidase that is highly expressed in the stratum granulosum [66]. Histidase converts histidine to urocanic acid (UCA) in the upper layers of the epidermis. UCA has been suggested to be an important UV photoprotectant as it has a high extinction coefficient in the wavelength range from 260 to 310 nm [67, 68], and it was for several years used as a component of commercial sunscreens [69]. All the prevalent *FLG* mutations are either nonsense or frameshift mutations that result in loss of filaggrin production in the epidermis [70]. Because of the lower levels of filaggrin, individuals with *FLG* mutations have reduced levels of epidermal UCA and thereby reduced UCA-mediated absorption of UVB. This should, in theory, lead to a higher photoconversion of 7-DHC to previtamin D_3 and thereby higher levels of $25(OH)D_3$ and $1,25(OH)_2D_3$. This hypothesis is supported by in vitro experiments demonstrating that knockdown of filaggrin increased UVB sensitivity [71], by in vivo experiments demonstrating that mice with a mutated histidase gene have reduced levels of UCA in the skin and show increased sensitivity to UVB radiation [66], and finally by five general population studies that showed that *FLG* mutation carriers have 10 % higher mean serum $25(OH)D_3$ levels than controls [72]. How *FLG* mutations influence the local concentration of $1,25(OH)_2D_3$ is not known, but it could well be assumed to augment the concentration and thereby have an impact on immune responses. $1,25(OH)_2D_3$ also stimulates keratinocytes and macrophages to increased production of the antimicrobial peptide cathelicidin, which might be of benefit for AD patients [73, 74].

References

1. Fallon PG, Sasaki T, Sandilands A, Campbell LE, Saunders SP, Mangan NEJ, et al. A homozygous frameshift mutation in the mouse Flg gene facilitates enhanced percutaneous allergen priming. Nat Genet. 2009;41:602–8.
2. Moniaga CS, Egawa G, Kawasaki H, Hara-Chikuma M, Honda T, Tanizaki H, et al. Flaky tail mouse denotes human atopic dermatitis in the steady state and by topical application with Dermatophagoides pteronyssinus extract. Am J Pathol. 2010;176: 2385–93.

3. Oyoshi MK, Murphy GF, Geha RS. Filaggrin-deficient mice exhibit TH17-dominated skin inflammation and permissiveness to epicutaneous sensitization with protein antigen. J Allergy Clin Immunol. 2009;124(3):485–93, 493.e1.
4. Mihm Jr MC, Soter NA, Dvorak HF, Austen KF. The structure of normal skin and the morphology of atopic eczema. J Invest Dermatol. 1976;67:305–12.
5. Woodward AL, Spergel JM, Alenius H, Mizoguchi E, Bhan AK, Castigli E, et al. An obligate role for T-cell receptor alphabeta+ T cells but not T-cell receptor gammadelta+ T cells, B cells, or CD40/CD40L interactions in a mouse model of atopic dermatitis. J Allergy Clin Immunol. 2001;107:359–66.
6. Mosmann TR, Cherwinski H, Bond MW, Giedlin MA, Coffman RL. Two types of murine helper T cell clone. I. Definition according to profiles of lymphokine activities and secreted proteins. J Immunol. 1986;136:2348–57.
7. Zhou L, Chong MM, Littman DR. Plasticity of CD4+ T cell lineage differentiation. Immunity. 2009;30: 646–55.
8. Spergel JM, Mizoguchi E, Oettgen H, Bhan AK, Geha RS. Roles of TH1 and TH2 cytokines in a murine model of allergic dermatitis. J Clin Invest. 1999;103:1103–11.
9. Hamid Q, Boguniewicz M, Leung DY. Differential in situ cytokine gene expression in acute versus chronic atopic dermatitis. J Clin Invest. 1994;94:870–6.
10. Hamid Q, Naseer T, Minshall EM, Song YL, Boguniewicz M, Leung DY. In vivo expression of IL-12 and IL-13 in atopic dermatitis. J Allergy Clin Immunol. 1996;98:225–31.
11. McPherson T, Sherman VJ, Aslam A, Crack L, Chan H, Lloyd-Lavery A, et al. Filaggrin null mutations associate with increased frequencies of allergen-specific CD4+ T-helper 2 cells in patients with atopic eczema. Br J Dermatol. 2010;163:544–9.
12. Kezic S, O'Regan GM, Lutter R, Jakasa I, Koster ES, Saunders S, et al. Filaggrin loss-of-function mutations are associated with enhanced expression of IL-1 cytokines in the stratum corneum of patients with atopic dermatitis and in a murine model of filaggrin deficiency. J Allergy Clin Immunol. 2012;129:1031–9.
13. Shornick LP, Bisarya AK, Chaplin DD. IL-1beta is essential for langerhans cell activation and antigen delivery to the lymph nodes during contact sensitization: evidence for a dermal source of IL-1beta. Cell Immunol. 2001;211:105–12.
14. Cua DJ, Sherlock J, Chen Y, Murphy CA, Joyce B, Seymour B, et al. Interleukin-23 rather than interleukin-12 is the critical cytokine for autoimmune inflammation of the brain. Nature. 2003;421:744–8.
15. Harrington LE, Hatton RD, Mangan PR, Turner H, Murphy TL, Murphy KM, et al. Interleukin 17-producing CD4+ effector T cells develop via a lineage distinct from the T helper type 1 and 2 lineages. Nat Immunol. 2005;6:1123–32.
16. Park H, Li Z, Yang XO, Chang SH, Nurieva R, Wang YH, et al. A distinct lineage of CD4 T cells regulates tissue inflammation by producing interleukin 17. Nat Immunol. 2005;6:1133–41.
17. Korn T, Bettelli E, Oukka M, Kuchroo VK. IL-17 and Th17 cells. Annu Rev Immunol. 2009;27:485–517.
18. Koga C, Kabashima K, Shiraishi N, Kobayashi M, Tokura Y. Possible pathogenic role of Th17 cells for atopic dermatitis. J Invest Dermatol. 2008;128:2625–30.
19. Toda M, Leung DY, Molet S, Boguniewicz M, Taha R, Christodoulopoulos P, et al. Polarized in vivo expression of IL-11 and IL-17 between acute and chronic skin lesions. J Allergy Clin Immunol. 2003;111:875–81.
20. He R, Oyoshi MK, Jin H, Geha RS. Epicutaneous antigen exposure induces a Th17 response that drives airway inflammation after inhalation challenge. Proc Natl Acad Sci U S A. 2007;104:15817–22.
21. O'Regan GM, Sandilands A, McLean WH, Irvine AD. Filaggrin in atopic dermatitis. J Allergy Clin Immunol. 2009;124:R2–6.
22. Palmer CN, Irvine AD, Terron-Kwiatkowski A, Zhao Y, Liao H, Lee SP, et al. Common loss-of-function variants of the epidermal barrier protein filaggrin are a major predisposing factor for atopic dermatitis. Nat Genet. 2006;38:441–6.
23. Walker JA, Barlow JL, McKenzie AN. Innate lymphoid cells – how did we miss them? Nat Rev Immunol. 2013;13:75–87.
24. Leisten S, Oyoshi MK, Galand C, Hornick JL, Gurish MF, Geha RS. Development of skin lesions in filaggrin-deficient mice is dependent on adaptive immunity. J Allergy Clin Immunol. 2013;131:1247–50.
25. Enk AH, Katz SI. Identification and induction of keratinocyte-derived IL-10. J Immunol. 1992;149:92–5.
26. Enk AH, Katz SI. Early molecular events in the induction phase of contact sensitivity. Proc Natl Acad Sci U S A. 1992;89:1398–402.
27. Larsen JM, Bonefeld CM, Poulsen SS, Geisler C, Skov L. IL-23 and T(H)17-mediated inflammation in human allergic contact dermatitis. J Allergy Clin Immunol. 2009;123:486–92.
28. Schleimer RP, Kato A, Kern R, Kuperman D, Avila PC. Epithelium: at the interface of innate and adaptive immune responses. J Allergy Clin Immunol. 2007;120: 1279–84.
29. Ziegler SF. Thymic stromal lymphopoietin and allergic disease. J Allergy Clin Immunol. 2012;130: 845–52.
30. Soumelis V, Reche PA, Kanzler H, Yuan W, Edward G, Homey B, Lu YJ, et al. Human epithelial cells trigger dendritic cell mediated allergic inflammation by producing TSLP. Nat Immunol. 2002;3:673–80.
31. Yoo J, Omori M, Gyarmati D, Zhou B, Aye T, Brewer A, et al. Spontaneous atopic dermatitis in mice expressing an inducible thymic stromal lymphopoietin transgene specifically in the skin. J Exp Med. 2005;202:541–9.
32. Kim BS, Siracusa MC, Saenz SA, Noti M, Monticelli LA, Sonnenberg GF, et al. TSLP elicits IL-33-independent innate lymphoid cell responses to promote skin inflammation. Sci Transl Med. 2013;5: 170ra16.

33. Moniaga CS, Jeong SK, Egawa G, Nakajima S, Hara-Chikuma M, Jeon JE, et al. Protease activity enhances production of thymic stromal lymphopoietin and basophil accumulation in flaky tail mice. Am J Pathol. 2013;182:841–51.
34. Cianferotti L, Cox M, Skorija K, Demay MB. Vitamin D receptor is essential for normal keratinocyte stem cell function. Proc Natl Acad Sci U S A. 2007;104: 9428–33.
35. Svendsen ML, Daneels G, Geysen J, Binderup L, Kragballe K. Proliferation and differentiation of cultured human keratinocytes is modulated by 1,25(OH)2D3 and synthetic vitamin D3 analogues in a cell density-, calcium- and serum-dependent manner. Pharmacol Toxicol. 1997;80:49–56.
36. Rid R, Wagner M, Maier CJ, Hundsberger H, Hintner H, Bauer JW, et al. Deciphering the calcitriol-induced transcriptomic response in keratinocytes: presentation of novel target genes. J Mol Endocrinol. 2013;50: 131–49.
37. Hewison M. Vitamin D and immune function: autocrine, paracrine or endocrine? Scand J Clin Lab Invest Suppl. 2012;243:92–102.
38. Nagy L, Szanto A, Szatmari I, Szeles L. Nuclear hormone receptors enable macrophages and dendritic cells to sense their lipid environment and shape their immune response. Physiol Rev. 2012;92:739–89.
39. White JH. Vitamin D metabolism and signaling in the immune system. Rev Endocr Metab Disord. 2012; 13:21–9.
40. Holick MF. Sunlight and vitamin D for bone health and prevention of autoimmune diseases, cancers, and cardiovascular disease. Am J Clin Nutr. 2004;80: 1678S–88.
41. Lehmann B, Meurer M. Vitamin D metabolism. Dermatol Ther. 2010;23:2–12.
42. Holick MF, MacLaughlin JA, Clark MB, Holick SA, Potts Jr JT, Anderson RR, et al. Photosynthesis of previtamin D3 in human skin and the physiologic consequences. Science. 1980;210:203–5.
43. Webb AR, Holick MF. The role of sunlight in the cutaneous production of vitamin D3. Annu Rev Nutr. 1988;8:375–99.
44. Nagpal S, Na S, Rathnachalam R. Noncalcemic actions of vitamin D receptor ligands. Endocr Rev. 2005;26:662–87.
45. Baker AR, McDonnell DP, Hughes M, Crisp TM, Mangelsdorf DJ, Haussler MR, et al. Cloning and expression of full-length cDNA encoding human vitamin D receptor. Proc Natl Acad Sci U S A. 1988; 85:3294–8.
46. Moore DD, Kato S, Xie W, Mangelsdorf DJ, Schmidt DR, Xiao R, et al. International Union of Pharmacology. LXII. The NR1H and NR1I receptors: constitutive androstane receptor, pregnene X receptor, farnesoid X receptor alpha, farnesoid X receptor beta, liver X receptor alpha, liver X receptor beta, and vitamin D receptor. Pharmacol Rev. 2006;58:742–59.
47. Prufer K, Racz A, Lin GC, Barsony J. Dimerization with retinoid X receptors promotes nuclear localization and subnuclear targeting of vitamin D receptors. J Biol Chem. 2000;275:41114–23.
48. Haussler MR, Whitfield GK, Kaneko I, Haussler CA, Hsieh D, Hsieh JC, et al. Molecular mechanisms of vitamin D action. Calcif Tissue Int. 2013;92:77–98.
49. Kerner SA, Scott RA, Pike JW. Sequence elements in the human osteocalcin gene confer basal activation and inducible response to hormonal vitamin D3. Proc Natl Acad Sci U S A. 1989;86:4455–9.
50. Pike JW, Meyer MB, Bishop KA. Regulation of target gene expression by the vitamin D receptor - an update on mechanisms. Rev Endocr Metab Disord. 2012;13:45–55.
51. Hong SP, Kim MJ, Jung MY, Jeon H, Goo J, Ahn SK, et al. Biopositive effects of low-dose UVB on epidermis: coordinate upregulation of antimicrobial peptides and permeability barrier reinforcement. J Invest Dermatol. 2008;128:2880–7.
52. Demetriou SK, Ona-Vu K, Teichert AE, Cleaver JE, Bikle DD, Oh DH. Vitamin D receptor mediates DNA repair and is UV inducible in intact epidermis but not in cultured keratinocytes. J Invest Dermatol. 2012; 132:2097–100.
53. von Essen MR, Kongsbak M, Schjerling P, Olgaard K, Odum N, Geisler C. Vitamin D controls T cell antigen receptor signaling and activation of human T cells. Nat Immunol. 2010;11:344–9.
54. Yasmin N, Konradi S, Eisenwort G, Schichl YM, Seyerl M, Bauer T, et al. beta-catenin promotes the differentiation of epidermal langerhans dendritic cells. J Invest Dermatol. 2013;133:1250–9.
55. Prosser DE, Jones G. Enzymes involved in the activation and inactivation of vitamin D. Trends Biochem Sci. 2004;29:664–73.
56. Zehnder D, Bland R, Williams MC, McNinch RW, Howie AJ, Stewart PM, et al. Extrarenal expression of 25-hydroxyvitamin d(3)-1 alpha-hydroxylase. J Clin Endocrinol Metab. 2001;86:888–94.
57. Hewison M, Burke F, Evans KN, Lammas DA, Sansom DM, Liu P, et al. Extra-renal 25-hydroxyvitamin D3-1alpha-hydroxylase in human health and disease. J Steroid Biochem Mol Biol. 2007;103: 316–21.
58. Adams JS, Hewison M. Extrarenal expression of the 25-hydroxyvitamin D-1-hydroxylase. Arch Biochem Biophys. 2012;523:95–102.
59. White JH. Regulation of intracrine production of 1,25-dihydroxyvitamin D and its role in innate immune defense against infection. Arch Biochem Biophys. 2012;523:58–63.
60. Lehmann B, Rudolph T, Pietzsch J, Meurer M. Conversion of vitamin D3 to 1alpha,25-dihydroxyvitamin D3 in human skin equivalents. Exp Dermatol. 2000;9:97–103.
61. Lehmann B, Sauter W, Knuschke P, Dressler S, Meurer M. Demonstration of UVB-induced synthesis of 1 alpha,25-dihydroxyvitamin D3 (calcitriol) in human skin by microdialysis. Arch Dermatol Res. 2003;295:24–8.
62. Lehmann B, Schattiger K, Meurer M. Conversion of vitamin D3 to hormonally active 1alpha,25-dihydroxyvitamin

D3 in cultured keratinocytes: relevance to cell growth and differentiation. J Steroid Biochem Mol Biol. 2010;121: 322–3.
63. Sybert VP, Dale BA, Holbrook KA. Ichthyosis vulgaris: identification of a defect in synthesis of filaggrin correlated with an absence of keratohyaline granules. J Invest Dermatol. 1985;84:191–4.
64. Manabe M, Sanchez M, Sun TT, Dale BA. Interaction of filaggrin with keratin filaments during advanced stages of normal human epidermal differentiation and in ichthyosis vulgaris. Differentiation. 1991;48:43–50.
65. Candi E, Schmidt R, Melino G. The cornified envelope: a model of cell death in the skin. Nat Rev Mol Cell Biol. 2005;6:328–40.
66. Barresi C, Stremnitzer C, Mlitz V, Kezic S, Kammeyer A, Ghannadan M, et al. Increased sensitivity of histidinemic mice to UVB radiation suggests a crucial role of endogenous urocanic acid in photoprotection. J Invest Dermatol. 2011;131:188–94.
67. Zenisek A, Kral JA, Hais IM. Sun-screening effect of urocanic acid. Biochim Biophys Acta. 1955;18: 589–91.
68. Tabachnick J. Urocanic acid, the major acid-soluble, ultraviolet-absorbing compound in guinea pig epidermis. Arch Biochem Biophys. 1957;70:295–8.
69. Gibbs NK, Tye J, Norval M. Recent advances in urocanic acid photochemistry, photobiology and photoimmunology. Photochem Photobiol Sci. 2008;7:655–67.
70. Sandilands A, Terron-Kwiatkowski A, Hull PR, O'Regan GM, Clayton TH, Watson RM, et al. Comprehensive analysis of the gene encoding filaggrin uncovers prevalent and rare mutations in ichthyosis vulgaris and atopic eczema. Nat Genet. 2007; 39:650–4.
71. Mildner M, Jin J, Eckhart L, Kezic S, Gruber F, Barresi C, et al. Knockdown of filaggrin impairs diffusion barrier function and increases UV sensitivity in a human skin model. J Invest Dermatol. 2010;130: 2286–94.
72. Thyssen JP, Thuesen B, Huth C, Standl M, Carson CG, Heinrich J, et al. Skin barrier abnormality caused by filaggrin (FLG) mutations is associated with increased serum 25-hydroxyvitamin D concentrations. J Allergy Clin Immunol. 2012;130:1204–7.
73. Heine G, Hoefer N, Franke A, Nothling U, Schumann RR, Hamann L, et al. Association of vitamin D receptor gene polymorphisms with severe atopic dermatitis in adults. Br J Dermatol. 2013;168:855–8.
74. Schauber J, Dorschner RA, Coda AB, Buchau AS, Liu PT, Kiken D, et al. Injury enhances TLR2 function and antimicrobial peptide expression through a vitamin D-dependent mechanism. J Clin Invest. 2007; 117:803–11.

Filaggrin-2

6

Jens-Michael Schröder and Britta Hansmann

Contents

J.-M. Schröder, PhD • B. Hansmann, PhD (✉)
Department of Dermatology,
University Hospital of Schleswig-Holstein,
Schittenhelmstr. 7, Kiel, Schleswig-Holstein
24105, Germany
e-mail: jschroeder@dermatology.uni-kiel.de;
bhansmann@dermatology.uni-kiel.de

6.1 Filaggrin-2: The Structural Relation

Human filaggrin-2 (also known as ifapsoriasin) is another member of the S100 fused-type protein (SFTP) family [1, 2]. As the other genes of this family, the *filaggrin-2* gene is encoded within the epidermal differentiation complex (EDC) on human chromosome 1q21.3. In the cluster of so far seven known SFTPs, it is flanked by the genes of *pro-filaggrin* (centromeric) and *cornulin* (telomeric). As it is long known for pro-filaggrin, all SFTPs are organized in a very similar way on genomic and protein level (Fig. 6.1). The gene consists of three exons of which the first small one (around 50 bp in humans) is noncoding. Whereas the second exon (in humans around 150–160 bp) carries the start codon and the coding sequence for the S100 domain, the very large third exon (up to several kb in size) carries the EF hand domain and remaining coding sequence. The deduced protein contains an N-terminal potential Ca^{2+}-binding S100/EF hand domain that is homologous to the S100A proteins' N-terminal domain. This domain is "fused" to a multiple tandem repeat region, a structural characteristic attributed to the cornified envelope (CE) precursor proteins involucrin and loricrin and the small proline-rich repeat proteins, the SPRRs [3]. Within this repetitive region, the repeats strongly differ in size and sequence among the proteins. Usually repeats are flanked by an N-terminal spacer and a C-terminal sequence unique to each protein. In case of filaggrin-2, its protein sequence

J.P. Thyssen, H. Maibach (eds.), *Filaggrin*,
DOI 10.1007/978-3-642-54379-1_6,

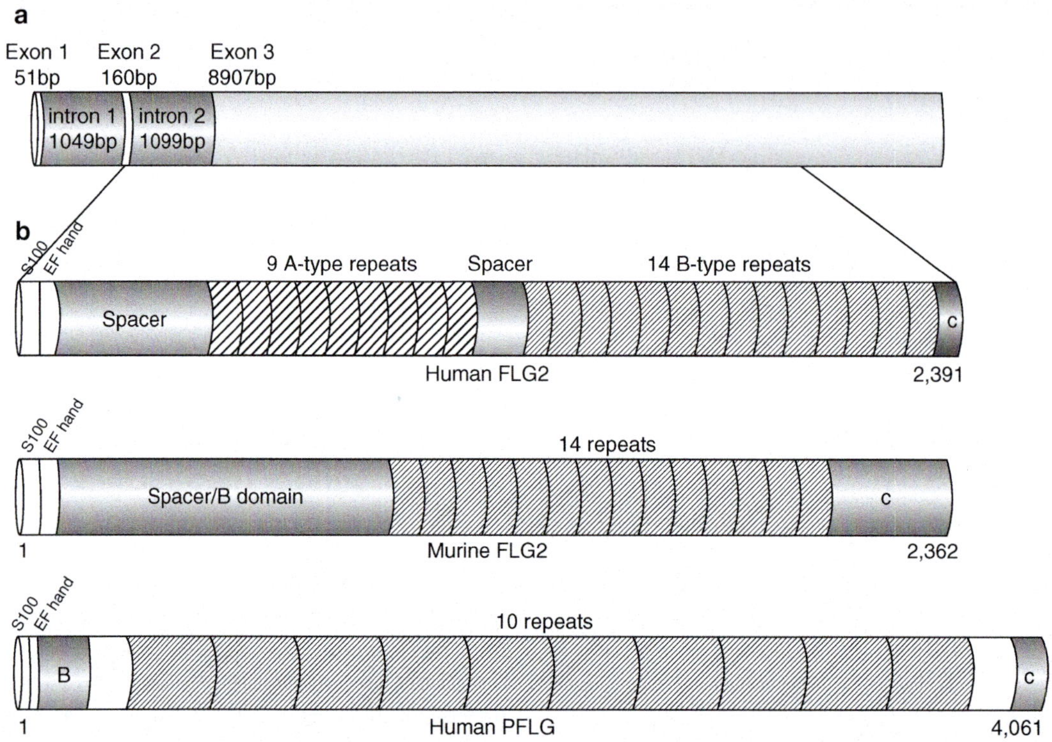

Fig. 6.1 S100 fused-type protein structure on genomic and protein level. (**a**) Conserved genomic structure shared by all SFTPs. The first small exon is noncoding; the slightly larger exon 2 contains the 5′ untranslated region, the ATG start codon, and the coding sequence for the first 47 amino acids; and the very large exon 3 carries the remaining coding sequence and the 3′ untranslated region. The sizes of exons and introns are indicated for human *filaggrin-2*. (**b**) Structure of human and murine filaggrin-2 as well as human pro-filaggrin. The repeats of the respective protein are shown in different shadings. For human pro-filaggrin a 10-repeat variant is shown, where the complete filaggrin units are flanked N- and C-terminal by truncated repeats. The amino acids length of the proteins is indicated underneath each column

is 2,391 amino acids long, which correspond to a molecular mass of 248 kDa and a pI of 8.45 (see Fig. 6.1b). With almost 67 % similarity and 86 % identity and 60 % similarity and 83 % identity, respectively, the filaggrin-2 N-terminal S100/EF hand domain is most similar to that of hornerin and pro-filaggrin. The N-terminal domain is then followed by a 385-amino-acid-spanning spacer region (corresponding to the B-domain in pro-filaggrin) of unknown function. Whereas pro-filaggrin has only one type of repeat, filaggrin-2 is composed of two types of repeats, the so-called A- and B-type repeats. Both types are similar in length (75 or 77 amino acids) but vary in sequence. They are rich in serine (~32 %) and glycine (~28 and 17 %, respectively). Even though the repeats are not rich in lysine, they contain a high amount of glutamine (~15 and 8 %, respectively). Those amino acids are candidates to function as glutamine acceptor site for the formation of protein-protein cross-links during keratinocyte differentiation [4, 5]. The nine A-repeats show homology to hornerin repeats, while the B-repeats are rather homologous to a part of the 324- to 325-amino-acid-spanning filaggrin repeats [2]. The C-terminus, 88 amino acids in length, shows sequence similarities to the pro-filaggrin C-terminus. In summary, filaggrin-2 can be considered as a "fused-type" protein in two aspects: it shares sequence similarity with both hornerin within the N-terminal and A-type repeat region and filaggrin in the B-type repeat and C-terminal region, therefore not only "fusing" structural characteristics of S100 and CE precursor protein structures but also features of two other SFTP family members.

The sequences of human, mouse, and rat pro-filaggrin contain at least one nuclear localization signal (NLS) that shows the typical sequence for bipartite NLS [6, 7]. For the pro-filaggrin N-terminal/B-domain, a translocation into the nucleus and the functionality of NLS were clearly shown [6, 8–10]. The pro-filaggrin N-terminal domain may play a substantial role in the change in gene expression and nuclear loss during the terminal differentiation process and probably even in wound healing. When analyzing the human filaggrin-2 sequence for an NLS with the subcellular localization prediction program PSORT II [11], no bipartite NLS is detected, but three monopartite-like sequences in the N-terminal spacer sequence (aa100–103, overlapping with aa101–104) as well as in the B1 repeat (aa1266–1269). The cleavage of the N-terminal domain is an essential step during the processing of pro-filaggrin to active filaggrin subunits and in consequence for the translocation of the N-terminal domain(s) into the nucleus [8, 12, 13]. For human pro-filaggrin it was shown to be cleaved in vitro by furin [6], and both mouse and rat pro-filaggrin B-domain sequences also possess canonical furin cleavage sites. Except for the rat sequence, where two sites are located within the B-domain, the cleavage site in both the human and mouse pro-filaggrin sequences is situated at the very end of their respective B-domain [12]. A prediction tool developed for propeptide cleavage sites as by the subtilisin-like proprotein convertases furin (ProP1.0, http://www.cbs.dtu.dk/services/ProP, [14]) highlighted within the 10-repeat variant of human pro-filaggrin 22 further potential furin cleavage sites additional to the already known at R_{293}, among those four more of the canonical type (R X R/K R↓, [15], at R_{789}, R_{2886}, R_{3210}, R_{3858}). For human filaggrin-2 the prediction tool predicts three potential furin cleavage sites after amino acids R_{183} within the N-terminal spacer sequence, R_{1269} located in the B1 repeat, and R_{2324} in the C-terminus. The cleavage site at R_{1269} shows the highest score and is the only one out of these three sites to possess a furin consensus sequence. Nothing is known so far about the functionality of these predicted cleavage sites of filaggrin-2.

The mouse *filaggrin-2* gene is located on mouse chromosome 3F2.1, which is syntenic to the region of the human EDC [16]. Its genomic and protein organization is similar to that of the other human and murine SFTPs. With 2,362 amino acids, the murine protein is slightly smaller than its potential human ortholog and has a predicted molecular mass of approximately 250 kDa and a pI of 7.66 (see Fig. 6.1b). Its N-terminal domain sequence shares almost 89 % identity and 96 % similarity with the human N-terminal filaggrin-2 domain [17]. The repeat organization of murine filaggrin-2 differs from that of the human; it contains 14 repeats similar to the B-repeats of the human filaggrin-2 (see Fig. 6.1b). The repeats are 73–80 amino acids in length and also are rich in serine, glycine, glutamine, proline, arginine, and histidine, as known for the human filaggrin-2 and filaggrin [2, 17]. The numerous glutamine and various lysine residues, which are exclusively located outside the repeat units, can possibly serve as donor or acceptor sides for cross-linking [4, 5]. Instead of A-repeats, the corresponding region in the murine ortholog is not organized in repetitive units, yet similar in sequence composition (see Fig. 6.1b). The C-terminus of mouse filaggrin-2 spans 325 amino acids, exceeding the length of the human C-terminus by far, and is only partially similar to the human sequence. As in the human filaggrin-2 sequence, a bipartite NLS cannot be detected in the murine sequence. Nevertheless, a total of five monopartite sequences of which all are located within the spacer/B-domain and some also overlapping according to PSORT II (aa100–103, 101–104, 151–154, 174–177, 149–155) are identified. The furin cleavage site prediction tool was able to find four sites in mouse filaggrin-2, of which two are located within the N-terminal spacer/B-domain (cleavage after R_{511} with the highest score and R_{679}) and two within the third (R_{1122}) and seventh repeats (R_{1442}), respectively. Just as little as is known about the role of the human filaggrin-2 N-terminal domains, the same is true for the murine N-terminal domains and they therefore remain subject of investigation.

6.2 Filaggrin-2: The Functional Relation

The members of the S100 fused-type proteins are mainly expressed and analyzed in skin, skin appendages, and other keratinized epithelial tissues. Most of them are regulated in skin during terminal differentiation at the transition of keratinocytes to corneocytes. Differentiation is a Ca^{2+}-dependent process; therefore, most of the SFTPs (and other differentiation markers as loricrin and involucrin) are expressed in a Ca^{2+}-dependent manner in keratinocytes, and their expression is inducible by an elevated Ca^{2+} level in cultivated keratinocytes [1, 17–27]. Similar to pro-filaggrin, a long-known marker of keratinocyte differentiation, human *filaggrin-2* was found to be highly upregulated during keratinocyte differentiation [1], suggesting an involvement during the process of terminal differentiation. The differentiation-dependent expression of filaggrin-2 was confirmed and shown on mRNA and/or protein level to be expressed in skin in different locations, as well as on mRNA level in the thymus, esophagus, tonsils, stomach, testis, and placenta, but not kidneys, pancreas, mammary gland, bladder, thyroid, salivary gland, trachea, and pharynx [2]. Immunohistochemistry revealed strong staining in distal parts of the hair but only scattered staining in central and proximal parts of the outer root sheath as well as in sections of the esophagus, testis, and tonsils for filaggrin-2 protein [2]. Furthermore, its mRNA is upregulated in cultivated keratinocytes by an elevated Ca^{2+} level pointing to an involvement during differentiation. The protein seems to co-localize—at least partially—with pro-filaggrin in keratohyalin granules in the stratum granulosum and stratum corneum [2, 28]. An identical expression pattern is seen for the potential mouse ortholog. It is expressed in skin and related tissues (nose, tongue, paw pad, esophagus, forestomach) and is also upregulated upon Ca^{2+} stimulation in cultured mouse keratinocytes [17]. In both human and murine skin, filaggrin-2 shows a comparable intense staining in the upper stratum granulosum and the lower stratum corneum (Fig. 6.2) [2, 29]. The disappearance of immune reactivity in the upper parts of the stratum corneum might be due to extensive posttranslational modifications as cross-linking, formation of disulfide bridges, deimination, or degradation, as already well known for pro-filaggrin.

The N-terminal S100/EF hand domains, which are potential Ca^{2+}-binding domains, have been shown to be functional and indeed are able to bind Ca^{2+} ions for human and mouse pro-filaggrin, pig trichohyalin, human cornulin, human repetin, and mouse filaggrin-2 [22, 23, 30–32]. Due to the high similarity of the N-terminal S100/EF hand domain among the SFTP members, it seems to be likely that these are functional in human filaggrin-2 and also have the ability of binding Ca^{2+} ions.

In contrast to pro-filaggrin, the information about the function of filaggrin-2 is rather sparse. The large protein is most likely stored in keratohyalin granules [2] and subsequently processed and modified as known for pro-filaggrin [2, 29]. During its extensive processing that is highly important for the correct function of its subunits, pro-filaggrin is dephosphorylated, citrullinated, and cleaved by a variety of enzymes (reviewed in [33, 34]).

For the dephosphorylated processed filaggrin monomers, it is known that these are responsible for the bundling of keratin filaments [35, 36] that might also be responsible for a collapse of the intermediate filament network and change in the cell shape of keratinocytes in the upper stratum granulosum [37]. Indeed a murine filaggrin-2 repetitive unit is able to aggregate isolated bovine keratin filaments in vitro [38], one of the main functions attributed to the filaggrin subunits.

Upon Western blot analyses, immunoreactive filaggrin-2 bands of lower molecular mass than the full-length protein indicate a processing of the large precursor protein [2, 29]. Calpain 1 was identified to be one protease being responsible for the cleavage of filaggrin-2 in vitro and in vivo [29], which is also known to cleave citrullinated filaggrin into smaller fragments [29, 39]. Citrullination is a modification of arginine residues that become deiminated. The responsible enzymes are peptidylarginine deiminases (PADs). In this context, it was shown that filaggrin-2 was utilized as a substrate by PADs 1 and 3, similarly to filaggrin, and calpain 1 cleavage was even

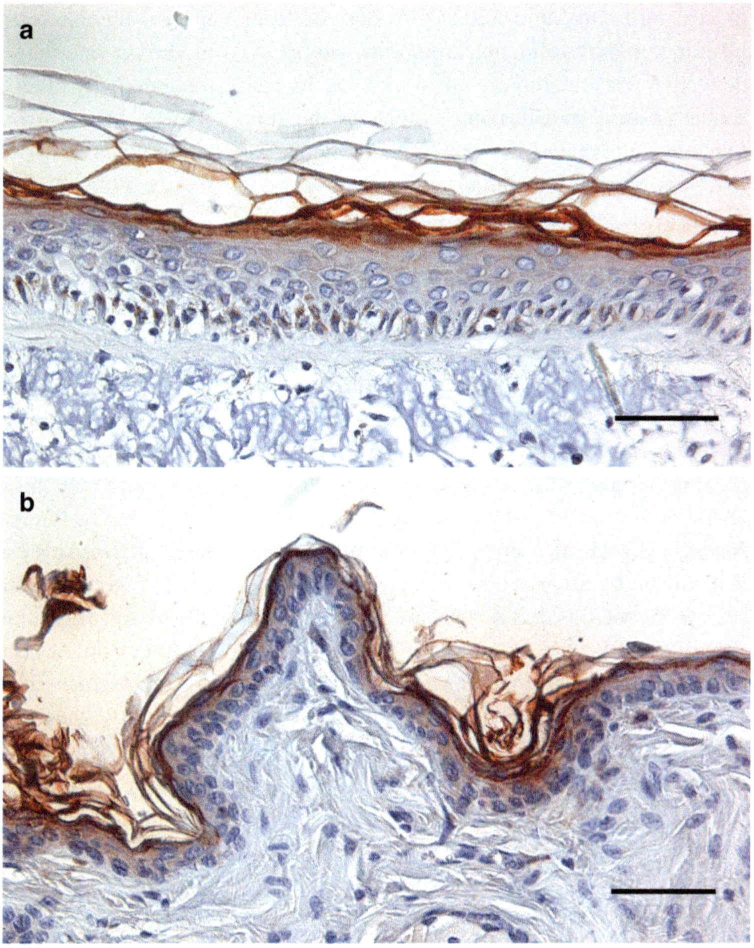

Fig. 6.2 Immunohistochemical staining of human and murine skin. (**a**) Section of human cheek skin stained with an antibody purified against the human N-terminal spacer region (amino acids 96–235). (**b**) Section of murine trunk skin stained with an antibody purified against the murine N-terminal spacer region (amino acids 96–231). Scale bar—50 μm

more efficient when filaggrin-2 was citrullinated [29]. Not only genetic mutations of the *pro-filaggrin* gene lead to truncated and incompletely translated pro-filaggrin that account for a reduced availability or lack of full-length protein and correctly processed subunits. Additionally to these mutations, also disturbed or defective transcriptional, posttranscriptional, or posttranslational control and mutations or dysfunctions in processing enzymes can account to a disease with insufficiently processed pro-filaggrin. For kallikrein 5, it has been shown very recently that it is able to cleave pro-filaggrin within the linker sequence and might therefore be involved in the processing from pro-filaggrin to filaggrin [40]. Earlier it has been reported that pro-filaggrin processing in SPINK5 knockout mice is enhanced, and in consequence the filaggrin monomers accumulate in the epidermis [41, 42]. This might be the case for other enzymes involved in pro-filaggrin processing and could also be taken into account for filaggrin-2 and its processing.

Human filaggrin-2 peptides of the N-terminal spacer region have been shown to be a component of the CE, and some of its cysteines are involved in disulfide bond formation following oxidation [43]. This confirms the observation that filaggrin-2 can only be detected in skin extracts treated with either urea, SDS, DTT, trypsin, or a combination of these and favors the hypothesis that it is poorly water soluble and—at least in part—covalently bound to other proteins in the CE [2, 29].

In the layers of the upper stratum corneum, the citrullinated filaggrin monomers are degraded into single amino acids which then—after further processing to urocanic acid (UCA, derived from histidine) and pyrrolidone carboxylic acid (PCA, derived from glutamine)—also serve as hygroscopic "natural moisturizing factors" of the skin to retain water and as UV protection (reviewed in [44, 45]). Due to the similar amino acid composition of filaggrin-2 B-repeats and filaggrin, it is possible that its function is partially corresponding to that of filaggrin subunits. Possibly having a similar fate of proteolytically processing and modification, they might be able to reinforce or support the function of filaggrin subunits and also contribute to the generation of natural moisturizing factors and UV protection of the skin.

There is also some evidence from proteomic analyses for posttranslational modifications of filaggrin-2 such as acetylation, ubiquitinylation, and phosphorylation in other cell types than keratinocytes, but this has to be analyzed in more detail [46–48].

The epidermis as the outermost barrier of the skin is in permanent contact with external factors. It is important for the barrier function of the skin to immediately react to environmental and internal challenges such as xenobiotics or toxic substances by sustaining or improving the differentiation process of keratinocytes to ensure the integrity of the barrier. The skin as a barrier organ is constantly in contact with oxygen and therefore exposed to high levels of ROS (reactive oxygen species); therefore, an efficient ROS quenching mechanism is required [43]. Besides the SPRRs, also filaggrin-2, loricrin, and keratinocyte proline-rich protein have been found to be involved in the ROS quenching ability of the CE. These proteins are able to reduce ROS via the formation of disulfide bonds of cysteines and further oxidation [43]. So filaggrin-2 can also function as a protection protein against hazardous substances, either external or internal.

Gene expression of *filaggrin-2* has been found to be regulated upon dioxin treatment [49, 50]. Presumably dioxins are involved in the development of inflammatory and allergic diseases, including dermatitis and chloracne [51]. In dioxin-treated mouse hepatoma cells, *filaggrin-2* expression is downregulated [50], whereas it has been found to be upregulated in dioxin-treated normal human keratinocytes [49]. The response to dioxin or other xenobiotics is known to be mediated by the aryl hydrocarbon receptor (AhR). Upon ligand binding (e.g., dioxin-like compounds, polycyclic aromatic hydrocarbons), the cytosolic AhR translocates into the nucleus and binds to a nuclear transcription factor, ARNT (AhR nuclear translocator). This AhR/ARNT complex then is able to interact with DNA regions, so-called XREs (xenobiotic response elements). The binding to XREs then modulates the expression of genes that carry these elements in their regulatory DNA sequence. Some of the SFTP genes, as *filaggrin-2*, are known to carry XREs and respond to AhR/ARNT-mediated signals [49, 52]. Interestingly, mice with a constitutively active AhR exhibit skin lesions as typically seen in atopic dermatitis (AD) [51], indicating a disordered regulation of the skin barrier proteins.

Just as little as is understood about the function and involvement of filaggrin-2 in cellular processes is currently known about contribution in or to disease. Since filaggrin-2, like several other members of the SFTP family, is a marker of terminal differentiation in keratinocytes, it is supposed to be affected in skin-related diseases as AD or psoriasis. Mutations in the *pro-filaggrin* gene are apparently not associated with psoriasis [53] but with AD, ichthyosis vulgaris, and some types of allergies, which probably are at least partially caused by a defective skin barrier function [54–59].

To date, no association between mutations in *filaggrin-2* and AD or psoriasis has been documented. Although the *filaggrin-2* genomic sequence contains almost four hundred annotated SNPs (single nucleotide polymorphism; UCSC genome browser, http://genome.ucsc.edu, accessed in April 2013), so far none of them was confirmed to be associated with a certain disease. One study considered *filaggrin-2* as candidate gene to be involved in early-onset psoriasis, but no significant correlation was found [60]. Even though mutations in the *pro-filaggrin* gene represent the strongest currently known predisposition factors of both ichthyosis vulgaris and AD [55, 59, 61], correlations of SNPs in *filaggrin-2* to AD have not been found [62–64]. Nevertheless, a regulated expression of filaggrin-2 on both mRNA and protein levels has been observed in AD, in psoriasis, or under similar disease-like conditions [2, 64–67].

Filaggrin-2 expression is even more downregulated than that of *pro-filaggrin* in lipid raft-disrupted keratinocytes, a situation that is described as resemblance of AD concerning its transcriptional profile [68]. At the protein level, a downregulation has been observed in nonlesional and even more in lesional skin samples of AD patients [64]. Although mutations in the *filaggrin-2* gene seem not to be associated directly with AD, other factors affect the expression of *filaggrin-2* on transcriptional level: it was shown on both mRNA and protein levels that filaggrin-2 was strongly downregulated by the T_H2-associated cytokines IL-4, IL-13, and IL-25, but not influenced by IL-22 [64, 69]. A regulation via the T_H2 cytokines IL-4, IL-13, and IL-25 has been shown before for filaggrin [70, 71] and has been confirmed together with a similar regulation of hornerin [64]. In early- and acute-phase AD, those T_H2 cytokines are known to play an important role in the development and manifestation of the disease and now have been shown to modulate the expression of SFTPs, suggesting an involvement for them in disease development and progression.

Several factors can account for a disturbed barrier. Besides skin diseases that are characterized by a disturbed skin barrier function, the epidermis can come in contact with potentially noxious substances or is subject to mechanical stress that also results in a disturbed barrier. Upon tape stripping, a downregulation of filaggrin-2 mRNA and protein is seen in healthy skin as well as in nonlesional AD and psoriatic skin [72]. The downregulation is similar for other SFTP members *pro-filaggrin* and *repetin*, but hornerin showed a converse regulation on both mRNA and protein levels. The same converse regulation could be observed for SDS-treated healthy skin. Interestingly, the relative change of hornerin, pro-filaggrin, and filaggrin-2 protein levels is similar in healthy skin compared to uninvolved AD and psoriasis skin upon barrier disruption treatment [72], although these SFTPs are all less expressed in nonlesional and lesional AD skin compared to healthy skin [64]. Somehow the regulation upon a mechanical- or detergent-induced barrier disruption seems to differ from a disease-associated disturbed barrier for the different SFTPs.

In hairless SKH-1 mice, neither the murine *filaggrin-2* nor *pro-filaggrin* expression was significantly changed upon an acetone-induced barrier disruption, whereas SKH-1 mice fed an essential fatty acid-deficient (EFAD) diet exhibited an opposed regulation of *filaggrin-2* and *pro-filaggrin*, with a significant downregulation for *filaggrin-2* but upregulation for *pro-filaggrin* [17]. This reduction for filaggrin-2 is also clearly seen at protein level in immunohistochemical staining of EFAD mice skin. The EFAD mice develop a metabolically induced barrier dysfunction due to the lack of essential fatty acids. That leads to an altered lipid organization where linoleic acid is replaced by oleic acid in ceramide 1 [73, 74]. This results in an eczema-like skin phenotype displaying an elevated transepidermal water loss and epidermal hyperplasia that goes along with a raised DNA synthesis rate [75–77]. Under this barrier-disrupted conditions, protein levels of the basal keratins 5 and 14 are decreased, while increased for the suprabasal keratins 1 and 10 in EFAD mice [78]. Obviously the keratinocytes are shifted toward differentiating cells, which is causing the increased levels of differentiation markers such as keratins 5 and 14 and *pro-filaggrin*, but interestingly not filaggrin-2.

Preliminary results in our group suggest a regulation of *filaggrin-2* in other skin-related diseases; it was found to be upregulated in basal cell carcinoma, squamous cell carcinoma, and precancerous lesions and downregulated in psoriasis, acute allergic contact dermatitis, and inflamed skin. A similar regulation is seen in those samples for other SFTPs including *pro-filaggrin* [65].

In solid tumor lysates of skin melanoma in a melanoma mouse model, several filaggrin-2 peptides (mainly located within the repetitive region) were found [79], and a downregulation of filaggrin-2 has been observed in another melanocyte-involved disease, the giant congenital melanocytic nevus [80]. In an approach to search for markers of atherosclerosis, a screening of atherosclerotic plaques by subtractive phage display identified filaggrin-2 among 21 other proteins [81]. Even analyses of cellular proteins from human immunodeficiency virus (HIV) type 1 isolated from infected monocyte-derived macrophages revealed the presence of filaggrin-2 peptides [82].

A more complex functional relevance becomes evident considering that human filaggrin-2 fragments have been identified as potential interaction partners of a variety of different molecules, such as Rho kinase 1 (ROCK1, [83]), an epidermal growth factor receptor (EGFR) antagonist [84], human cytomegalovirus DNA polymerase subunit UL44 [85], proliferating cell nuclear antigen (PCNA)-associated factor (PAF15, [86]), cyclin D1 [87], and the integrin β1 complex [88].

ROCK1 (Ras homolog family member A (RhoA) kinase 1) is a kinase that only is active when bound to the GTP-bound RhoA. ROCKs are known to influence various cellular processes, among them ordering of the cytoskeleton, involvement during assembling of stratified epithelia [89], and keratinocyte differentiation [90]. One of the most important functions of filaggrin subunits is to assemble keratin filaments into bundles [37], but it also affects other cytoskeletal filaments as actin and is capable of disrupting the arrangement of desmosomal proteins [91]. Taking these together, filaggrin-2 might also be able to influence the formation of a proper barrier during keratinocyte differentiation in various ways.

The treatment of cultured keratinocytes with epidermal growth factor (EGF) decreased the *filaggrin-2* gene expression 20-fold [49]. EGF as growth factor stimulates the proliferation of keratinocytes and vice versa inhibits differentiation [92–94]. Filaggrin-2 expression is rather localized to the upper suprabasal layers of the epidermis, the stratum granulosum and the stratum corneum, indicating that proliferating cells express no or barely any filaggrin-2. So either keratinocytes are in the state of proliferation providing new cells to differentiate in the stratum basale or the cells are differentiating in the suprabasal layers and therefore changing their gene expression profile accordingly. When then treated with EGF (or EGF receptor is activated), the cells get a proliferation signal and stop their differentiation program by either downregulating or abolishing gene expression for differentiation markers [93, 95]. This is also true for the SFTPs *filaggrin*, *repetin*, *hornerin*, and *filaggrin-2* [49]. Filaggrin-2 peptides have been identified to interact with the EGF receptor antagonist AX14596. This is an immobilized derivate of gefitinib, a small EGF receptor inhibitor and tumor growth suppressor. Considering this, it might be possible that filaggrin-2 can also interfere with the signal transduction of EGF or EGF receptor by blocking proliferation signals in keratinocytes that then lead to differentiation.

In human cytomegalovirus-infected foreskin fibroblasts, filaggrin-2 peptides were identified as interaction partner of the UL44 accessory subunit of the human cytomegalovirus DNA polymerase [85]. This subunit interacts with both DNA and the polymerase catalytic subunit. UL44 clamps the DNA as a homodimer and shows a high structural similarity to the eukaryotic processivity factor of DNA polymerases δ and ε, PCNA (proliferating cell nuclear antigen) [96–99]. PCNA as a processivity factor is involved not only in DNA replication but also in DNA repair and cell cycle control [100, 101]. Among others as cyclin D1 and the cyclin-dependent kinase inhibitor p21 (reviewed in [101]), one of the PCNA interacting factors is the PCNA-associated factor, PAF15. PAF15 is able to bind to PCNA via a conserved motif and is primarily located in the nucleus [102]; its expression peaks during G2/M phase and seems to be involved in cell cycle regulation [86, 103]. Filaggrin-2 has been detected to interact with both the viral DNA polymerase subunit UL44 and PAF15 [85, 86]. Since the two proteins are mainly located in the nucleus, it might be pointing to a nuclear localization and also function for filaggrin-2 fragments, as already known for the pro-filaggrin N-terminal domains [6, 8–10].

An involvement of filaggrin-2 in cell cycle regulation is furthermore supported by interaction with cyclin D1 [104]. Cyclins are proteins that are playing an important role during cell cycle progression. The expression of most cyclins peaks in a certain phase of the cell cycle; the expression of others—as cyclin D—is present throughout the complete cycle but in different concentrations.

They can only act together with a corresponding kinase, the cyclin-dependent kinase. The type D cyclins are responsible for progression of the G1 phase. Rat epidermal keratinocytes, transfected with a tetracycline-inducible overexpression plasmid containing the coding sequence of a filaggrin subunit without linker sequence, displayed a disruption of the cytoskeleton, altered distribution of desmosomal components, and cell cycle arrest in S/G2 phase [91]. The association of filaggrin-2 with cell cycle regulation points to an involvement of the SFTPs maybe in cell cycle exit at a very early step of differentiation as postulated for filaggrin [91].

In precipitates of β1 integrin complexes, filaggrin-2 peptides were discovered in Jurkat cells and platelets. The β1 integrin is present on epidermal stem cells in high amounts in the basal layer of the epidermis and is therefore considered as a stem cell marker in keratinocytes [105]. This integrin is necessary for the basal keratinocyte to attach to the basement membrane [106], and this adhesion in turn prohibits terminal differentiation. In consequence, only cells detached from the basement membrane will be able to differentiate. Nevertheless, β1 integrins seem not to be essential for the survival or the differentiation of keratinocytes [107, 108]. The β1 integrin is also able to interact with the EGF receptor [104], and upon a conditional knockdown of β1 integrin in chondrocytes, the cyclin D1 amount was diminished, leading to impaired G1/S transition [109]. Filaggrin-2 peptides have been detected as possible interacting partners for both an EGF receptor inhibitor and cyclin D1, as well as for integrin β1 [84, 87, 88]; it can be speculated that filaggrin-2 might be involved directly and indirectly via interaction partners in regulation of cell cycle and differentiation.

Filaggrin-2 was also detected as an interaction partner within autophagy networks [110]. The process of keratinocyte differentiation with the formation of corneocytes is a unique type of programmed cell death that differs from canonical apoptosis [111–113]. The normal turnover time of a keratinocyte to differentiate is about 4 weeks from *stratum basale* to desquamation in the *stratum corneum*, making it a rather slow process [105] compared to apoptosis, which is a relatively fast process occurring within several hours. During apoptosis the chromatin and the cytoplasm condense, the DNA is fragmented in a very characteristic way, the cytoskeleton assembly is broken down, and also organelles are affected. In this context, autophagy ("self-eating") sometimes goes hand in hand with apoptosis, and sometimes it is able to prevent apoptosis [114]. During differentiation the corneocytes in the *stratum corneum* are characterized by the lack of the nucleus and organelles. So possibly during differentiation, an autophagy-like process might be involved. Filaggrin-2 as a protein related to the differentiation process could also be able to affect apoptotic-like programmed cell death processes during terminal differentiation.

Filaggrin-2 peptides also were found in cervicovaginal fluids [115] and in aqueous humor (a liquid of the anterior and posterior chamber of the eye) after cataract surgery [116]. This points to filaggrin-2 playing a role in other barrier organs and not only in skin, in either structural or other functions.

Taking all these findings together, filaggrin-2 not only seems to be expressed as a consequence of differentiation of keratinocytes as a structural component but might be an element in networks as interactor and regulator of a variety of cellular processes related to differentiation, for example, also somehow affecting cell cycle regulation and proliferation. It contributes to maintain the barrier integrity of the skin and might also be a susceptibility factor for certain diseases or could compensate in those diseases characterized by a constrained or lost function of filaggrin.

Acknowledgment This work was supported by grants from the Deutsche Forschungsgemeinschaft to J.-M. Schröder (Schr 305⁄6–1).

References

1. Toulza E, Mattiuzzo NR, Galliano MF, Jonca N, Dossat C, Jacob D, et al. Large-scale identification of human genes implicated in epidermal barrier function. Genome Biol. 2007;8(6):R107.
2. Wu Z, Hansmann B, Meyer-Hoffert U, Gläser R, Schröder JM. Molecular identification and expression analysis of filaggrin-2, a member of the S100 fused-type protein family. PLoS ONE. 2009;4(4):e5227.

3. Gibbs S, Fijneman R, Wiegant J, van Kessel AG, van De Putte P, Backendorf C. Molecular characterization and evolution of the SPRR family of keratinocyte differentiation markers encoding small proline-rich proteins. Genomics. 1993;16(3):630–7.
4. Rice RH, Green H. The cornified envelope of terminally differentiated human epidermal keratinocytes consists of cross-linked protein. Cell. 1977;11(2):417–22.
5. Greenberg CS, Birckbichler PJ, Rice RH. Transglutaminases: multifunctional cross-linking enzymes that stabilize tissues. FASEB J. 1991;5(15): 3071–7.
6. Pearton DJ, Dale BA, Presland RB. Functional analysis of the pro-filaggrin N-terminal peptide: identification of domains that regulate nuclear and cytoplasmic distribution. J Invest Dermatol. 2002;119(3):661–9.
7. Fontes MR, Teh T, Kobe B. Structural basis of recognition of monopartite and bipartite nuclear localization sequences by mammalian importin-alpha. J Mol Biol. 2000;297(5):1183–94.
8. Zhang D, Karunaratne S, Kessler M, Mahony D, Rothnagel JA. Characterization of mouse pro-filaggrin: evidence for nuclear engulfment and translocation of the pro-filaggrin B-domain during epidermal differentiation. J Invest Dermatol. 2002;119(4):905–12.
9. Ishida-Yamamoto A, Takahashi H, Presland RB, Dale BA, Iizuka H. Translocation of pro-filaggrin N-terminal domain into keratinocyte nuclei with fragmented DNA in normal human skin and loricrin keratoderma. Lab Invest. 1998;78(10):1245–53.
10. Aho S, Harding CR, Lee JM, Meldrum H, Bosko CA. Regulatory role for the pro-filaggrin N-terminal domain in epidermal homeostasis. J Invest Dermatol. 2012;132(10):2376–85.
11. Horton P, Nakai K. Better prediction of protein cellular localization sites with the k nearest neighbors classifier. Proc Int Conf Intell Syst Mol Biol. 1997;5: 147–52.
12. Pearton DJ, Nirunsuksiri W, Rehemtulla A, Lewis SP, Presland RB, Dale BA. Proprotein convertase expression and localization in epidermis: evidence for multiple roles and substrates. Exp Dermatol. 2001;10(3):193–203.
13. Presland RB, Kimball JR, Kautsky MB, Lewis SP, Lo CY, Dale BA. Evidence for specific proteolytic cleavage of the N-terminal domain of human pro-filaggrin during epidermal differentiation. J Invest Dermatol. 1997;108(2):170–8.
14. Duckert P, Brunak S, Blom N. Prediction of proprotein convertase cleavage sites. Protein Eng Des Sel. 2004;17(1):107–12.
15. Nakayama K. Furin: a mammalian subtilisin/Kex2p-like endoprotease involved in processing of a wide variety of precursor proteins. Biochem J. 1997;327(Pt 3):625–35.
16. Mathur P, Murray B, Crowell T, Gardner H, Allaire N, Hsu YM, et al. Murine peptidoglycan recognition proteins PglyrpIalpha and PglyrpIbeta are encoded in the epidermal differentiation complex and are expressed in epidermal and hematopoietic tissues. Genomics. 2004;83(6):1151–63.
17. Hansmann B, Ahrens K, Wu Z, Proksch E, Meyer-Hoffert U, Schröder JM. Murine filaggrin-2 is involved in epithelial barrier function and down-regulated in metabolically induced skin barrier dysfunction. Exp Dermatol. 2012;21(4):271–6.
18. Yuspa SH, Kilkenny AE, Steinert PM, Roop DR. Expression of murine epidermal differentiation markers is tightly regulated by restricted extracellular calcium concentrations in vitro. J Cell Biol. 1989;109(3): 1207–17.
19. Makino T, Takaishi M, Morohashi M, Huh NH. Hornerin, a novel pro-filaggrin-like protein and differentiation-specific marker isolated from mouse skin. J Biol Chem. 2001;276(50):47445–52.
20. Wu Z, Meyer-Hoffert U, Reithmayer K, Paus R, Hansmann B, He Y, et al. Highly complex peptide aggregates of the S100 fused-type protein hornerin are present in human skin. J Invest Dermatol. 2009; 129(6):1446–58.
21. Krieg P, Schuppler M, Koesters R, Mincheva A, Lichter P, Marks F. Repetin (Rptn), a new member of the "fused gene" subgroup within the S100 gene family encoding a murine epidermal differentiation protein. Genomics. 1997;43(3):339–48.
22. Huber M, Siegenthaler G, Mirancea N, Marenholz I, Nizetic D, Breitkreutz D, et al. Isolation and characterization of human repetin, a member of the fused gene family of the epidermal differentiation complex. J Invest Dermatol. 2005;124(5):998–1007.
23. Contzler R, Favre B, Huber M, Hohl D. Cornulin, a new member of the "fused gene" family, is expressed during epidermal differentiation. J Invest Dermatol. 2005;124(5):990–7.
24. Hamilton EH, Payne Jr RE, O'Keefe EJ. Trichohyalin: presence in the granular layer and stratum corneum of normal human epidermis. J Invest Dermatol. 1991; 96(5):666–72.
25. Dale BA, Scofield JA, Hennings H, Stanley JR, Yuspa SH. Identification of filaggrin in cultured mouse keratinocytes and its regulation by calcium. J Invest Dermatol. 1983;81(1Suppl):90s–595.
26. Deucher A, Efimova T, Eckert RL. Calcium-dependent involucrin expression is inversely regulated by protein kinase C (PKC)alpha and PKCdelta. J Biol Chem. 2002;277(19):17032–40.
27. Hohl D, Mehrel T, Lichti U, Turner ML, Roop DR, Steinert PM. Characterization of human loricrin. Structure and function of a new class of epidermal cell envelope proteins. J Biol Chem. 1991;266(10): 6626–36.
28. Mattiuzzo NR, Toulza E, Jonca N, Serre G, Guerrin M. A large-scale multi-technique approach identifies forty-nine new players of keratinocyte terminal differentiation in human epidermis. Exp Dermatol. 2011;20(2):113–8.
29. Hsu CY, Henry J, Raymond AA, Mechin MC, Pendaries V, Nassar D, et al. Deimination of human filaggrin-2 promotes its proteolysis by calpain 1. J Biol Chem. 2011;286(26):23222–33.

30. Lee SC, Kim IG, Marekov LN, O'Keefe EJ, Parry DA, Steinert PM. The structure of human trichohyalin. Potential multiple roles as a functional EF-hand-like calcium-binding protein, a cornified cell envelope precursor, and an intermediate filament-associated (crosslinking) protein. J Biol Chem. 1993;268(16):12164–76.
31. Presland RB, Bassuk JA, Kimball JR, Dale BA. Characterization of two distinct calcium-binding sites in the amino-terminus of human pro-filaggrin. J Invest Dermatol. 1995;104(2):218–23.
32. Presland RB, Rothnagel JA, Lawrence OT. Profilaggrin and the fused S100 family of calcium-binding proteins. In: Elias PM, Feingold KR, editors. Skin barrier. New York: Taylor & Francis; 2006. p. 111–40.
33. Sandilands A, Sutherland C, Irvine AD, McLean WH. Filaggrin in the frontline: role in skin barrier function and disease. J Cell Sci. 2009;122(Pt 9):1285–94.
34. Matsui T, Miyamoto K, Kubo A, Kawasaki H, Ebihara T, Hata K, et al. SASPase regulates stratum corneum hydration through pro-filaggrin-to-filaggrin processing. EMBO Mol Med. 2011;3(6):320–33.
35. Lonsdale-Eccles JD, Teller DC, Dale BA. Characterization of a phosphorylated form of the intermediate filament-aggregating protein filaggrin. Biochemistry. 1982;21(23):5940–8.
36. Steinert PM, Cantieri JS, Teller DC, Lonsdale-Eccles JD, Dale BA. Characterization of a class of cationic proteins that specifically interact with intermediate filaments. Proc Natl Acad Sci U S A. 1981;78(7): 4097–101.
37. Dale BA, Presland RB, Lewis SP, Underwood RA, Fleckman P. Transient expression of epidermal filaggrin in cultured cells causes collapse of intermediate filament networks with alteration of cell shape and nuclear integrity. J Invest Dermatol. 1997;108(2):179–87.
38. Listwan P, Rothnagel JA. Keratin bundling proteins. Methods Cell Biol. 2004;78:817–27.
39. Yamazaki M, Ishidoh K, Suga Y, Saido TC, Kawashima S, Suzuki K, et al. Cytoplasmic processing of human pro-filaggrin by active mu-calpain. Biochem Biophys Res Commun. 1997;235(3):652–6.
40. Sakabe JI, Yamamoto M, Hirakawa S, Motoyama A, Ohta I, Tatsuno K, et al. Kallikrein-related peptidase 5 functions in proteolytic processing of pro-filaggrin in cultured human keratinocytes. J Biol Chem. 2013; 288(24):17179–89.
41. Descargues P, Deraison C, Bonnart C, Kreft M, Kishibe M, Ishida-Yamamoto A, et al. Spink5-deficient mice mimic Netherton syndrome through degradation of desmoglein 1 by epidermal protease hyperactivity. Nat Genet. 2005;37(1):56–65.
42. Hewett DR, Simons AL, Mangan NE, Jolin HE, Green SM, Fallon PG, et al. Lethal, neonatal ichthyosis with increased proteolytic processing of filaggrin in a mouse model of Netherton syndrome. Hum Mol Genet. 2005;14(2):335–46.
43. Vermeij WP, Alia A, Backendorf C. ROS quenching potential of the epidermal cornified cell envelope. J Invest Dermatol. 2011;131(7):1435–41.
44. Rawlings AV, Harding CR. Moisturization and skin barrier function. Dermatol Ther. 2004;17 Suppl 1:43–8.
45. Gibbs NK, Tye J, Norval M. Recent advances in urocanic acid photochemistry, photobiology and photoimmunology. Photochem Photobiol Sci. 2008;7(6):655–67.
46. Chuang C, Lin SH, Huang F, Pan J, Josic D, Yu-Lee LY. Acetylation of RNA processing proteins and cell cycle proteins in mitosis. J Proteome Res. 2010;9(9): 4554–64.
47. Danielsen JM, Sylvestersen KB, Bekker-Jensen S, Szklarczyk D, Poulsen JW, Horn H, et al. Mass spectrometric analysis of lysine ubiquitylation reveals promiscuity at site level. Mol Cell Proteomics. 2011;10(3):M110.
48. Huttlin EL, Jedrychowski MP, Elias JE, Goswami T, Rad R, Beausoleil SA, et al. A tissue-specific atlas of mouse protein phosphorylation and expression. Cell. 2010;143(7):1174–89.
49. Sutter CH, Bodreddigari S, Campion C, Wible RS, Sutter TR. 2,3,7,8-Tetrachlorodibenzo-p-dioxin increases the expression of genes in the human epidermal differentiation complex and accelerates epidermal barrier formation. Toxicol Sci. 2011;124(1):128–37.
50. Thornley JA, Trask HW, Ridley CJ, Korc M, Gui J, Ringelberg CS, et al. Differential regulation of polysome mRNA levels in mouse Hepa-1C1C7 cells exposed to dioxin. Toxicol In Vitro. 2011;25(7): 1457–67.
51. Tauchi M, Hida A, Negishi T, Katsuoka F, Noda S, Mimura J, et al. Constitutive expression of aryl hydrocarbon receptor in keratinocytes causes inflammatory skin lesions. Mol Cell Biol. 2005;25(21):9360–8.
52. van den Bogaard EH, Bergboer JG, Vonk-Bergers M, Vlijmen-Willems IM, Hato SV, van der Valk PG, et al. Coal tar induces AHR-dependent skin barrier repair in atopic dermatitis. J Clin Invest. 2013;123(2):917–27.
53. Zhao Y, Terron-Kwiatkowski A, Liao H, Lee SP, Allen MH, Hull PR, et al. Filaggrin null alleles are not associated with psoriasis. J Invest Dermatol. 2007; 127(8):1878–82.
54. Irvine AD, McLean WH. Breaking the (un)sound barrier: filaggrin is a major gene for atopic dermatitis. J Invest Dermatol. 2006;126(6):1200–2.
55. Palmer CN, Irvine AD, Terron-Kwiatkowski A, Zhao Y, Liao H, Lee SP, et al. Common loss-of-function variants of the epidermal barrier protein filaggrin are a major predisposing factor for atopic dermatitis. Nat Genet. 2006;38(4):441–6.
56. Palmer CN, Ismail T, Lee SP, Terron-Kwiatkowski A, Zhao Y, Liao H, et al. Filaggrin null mutations are associated with increased asthma severity in children and young adults. J Allergy Clin Immunol. 2007; 120(1):64–8.
57. Weidinger S, O'Sullivan M, Illig T, Baurecht H, Depner M, Rodriguez E, et al. Filaggrin mutations, atopic eczema, hay fever, and asthma in children. J Allergy Clin Immunol. 2008;121(5):1203–9.

58. Sandilands A, O'Regan GM, Liao H, Zhao Y, Terron-Kwiatkowski A, Watson RM, et al. Prevalent and rare mutations in the gene encoding filaggrin cause ichthyosis vulgaris and predispose individuals to atopic dermatitis. J Invest Dermatol. 2006;126(8):1770–5.
59. Smith FJ, Irvine AD, Terron-Kwiatkowski A, Sandilands A, Campbell LE, Zhao Y, et al. Loss-of-function mutations in the gene encoding filaggrin cause ichthyosis vulgaris. Nat Genet. 2006;38(3):337–42.
60. Chen H, Toh TK, Szeverenyi I, Ong RT, Theng CT, McLean WH, et al. Association of skin barrier genes within the PSORS4 locus is enriched in Singaporean Chinese with early-onset psoriasis. J Invest Dermatol. 2009;129(3):606–14.
61. Marenholz I, Nickel R, Ruschendorf F, Schulz F, Esparza-Gordillo J, Kerscher T, et al. Filaggrin loss-of-function mutations predispose to phenotypes involved in the atopic march. J Allergy Clin Immunol. 2006;118(4):866–71.
62. Stemmler S, Nothnagel M, Parwez Q, Petrasch-Parwez E, Epplen JT, Hoffjan S. Variation in genes of the epidermal differentiation complex in German atopic dermatitis patients. Int J Immunogenet. 2009; 36(4):217–22.
63. Marenholz I, Rivera VA, Esparza-Gordillo J, Bauerfeind A, Lee-Kirsch MA, Ciechanowicz A, et al. Association screening in the Epidermal Differentiation Complex (EDC) identifies an SPRR3 repeat number variant as a risk factor for eczema. J Invest Dermatol. 2011;131(8):1644–9.
64. Pellerin L, Henry J, Hsu CY, Balica S, Jean-Decoster C, Mechin MC, et al. Defects of filaggrin-like proteins in both lesional and nonlesional atopic skin. J Allergy Clin Immunol. 2013;131(4):1094–102.
65. Wu Z. Analysis of two human gene clusters involved in innate immunity. Thesis, Christian-Albrechts-University of Kiel; 2005.
66. Broccardo CJ, Mahaffey S, Schwarz J, Wruck L, David G, Schlievert PM, et al. Comparative proteomic profiling of patients with atopic dermatitis based on history of eczema herpeticum infection and Staphylococcus aureus colonization. J Allergy Clin Immunol. 2011;127(1):186–93, 193.e1–11.
67. Suarez-Farinas M, Tintle SJ, Shemer A, Chiricozzi A, Nograles K, Cardinale I, et al. Nonlesional atopic dermatitis skin is characterized by broad terminal differentiation defects and variable immune abnormalities. J Allergy Clin Immunol. 2011;127(4):954–64.
68. Mathay C, Pierre M, Pittelkow MR, Depiereux E, Nikkels AF, Colige A, et al. Transcriptional profiling after lipid raft disruption in keratinocytes identifies critical mediators of atopic dermatitis pathways. J Invest Dermatol. 2011;131(1):46–58.
69. Gutowska-Owsiak D, Schaupp AL, Salimi M, Taylor S, Ogg GS. Interleukin-22 downregulates filaggrin expression and affects expression of pro-filaggrin processing enzymes. Br J Dermatol. 2011;165(3):492–8.
70. Howell MD, Kim BE, Gao P, Grant AV, Boguniewicz M, Debenedetto A, et al. Cytokine modulation of atopic dermatitis filaggrin skin expression. J Allergy Clin Immunol. 2007;120(1):150–5.
71. Hvid M, Vestergaard C, Kemp K, Christensen GB, Deleuran B, Deleuran M. IL-25 in atopic dermatitis: a possible link between inflammation and skin barrier dysfunction? J Invest Dermatol. 2011;131(1):150–7.
72. de Koning HD, van den Bogaard EH, Bergboer JG, Kamsteeg M, Vlijmen-Willems IM, Hitomi K, et al. Expression profile of cornified envelope structural proteins and keratinocyte differentiation-regulating proteins during skin barrier repair. Br J Dermatol. 2012;166(6):1245–54.
73. Hou SY, Mitra AK, White SH, Menon GK, Ghadially R, Elias PM. Membrane structures in normal and essential fatty acid-deficient stratum corneum: characterization by ruthenium tetroxide staining and x-ray diffraction. J Invest Dermatol. 1991;96(2):215–23.
74. Wertz PW, Cho ES, Downing DT. Effect of essential fatty acid deficiency on the epidermal sphingolipids of the rat. Biochim Biophys Acta. 1983;753(3): 350–5.
75. Bibel DJ, Miller SJ, Brown BE, Pandey BB, Elias PM, Shinefield HR, et al. Antimicrobial activity of stratum corneum lipids from normal and essential fatty acid-deficient mice. J Invest Dermatol. 1989;92(4):632–8.
76. Proksch E, Feingold KR, Man MQ, Elias PM. Barrier function regulates epidermal DNA synthesis. J Clin Invest. 1991;87(5):1668–73.
77. Proksch E, Feingold KR, Elias PM. Epidermal HMG CoA reductase activity in essential fatty acid deficiency: barrier requirements rather than eicosanoid generation regulate cholesterol synthesis. J Invest Dermatol. 1992;99(2):216–20.
78. Ekanayake-Mudiyanselage S, Aschauer H, Schmook FP, Jensen JM, Meingassner JG, Proksch E. Expression of epidermal keratins and the cornified envelope protein involucrin is influenced by permeability barrier disruption. J Invest Dermatol. 1998; 111(3):517–23.
79. Zanivan S, Gnad F, Wickstrom SA, Geiger T, Macek B, Cox J, et al. Solid tumor proteome and phosphoproteome analysis by high resolution mass spectrometry. J Proteome Res. 2008;7(12):5314–26.
80. Kim HK, Kim YK, Song IS, Lee SR, Jeong SH, Kim MH, et al. Human giant congenital melanocytic nevus exhibits potential proteomic alterations leading to melanotumorigenesis. Proteome Sci. 2012; 10(1):50.
81. Cooksley-Decasper S, Reiser H, Thommen DS, Biedermann B, Neidhart M, Gawinecka J, et al. Antibody phage display assisted identification of junction plakoglobin as a potential biomarker for atherosclerosis. PLoS ONE. 2012;7(10):e47985.
82. Chertova E, Chertov O, Coren LV, Roser JD, Trubey CM, Bess Jr JW, et al. Proteomic and biochemical analysis of purified human immunodeficiency virus type 1 produced from infected monocyte-derived macrophages. J Virol. 2006;80(18):9039–52.
83. Ongusaha PP, Qi HH, Raj L, Kim YB, Aaronson SA, Davis RJ, et al. Identification of ROCK1 as an upstream activator of the JIP-3 to JNK signaling axis in response to UVB damage. Sci Signal. 2008;1(47):ra14.

84. Sharma K, Weber C, Bairlein M, Greff Z, Keri G, Cox J, et al. Proteomics strategy for quantitative protein interaction profiling in cell extracts. Nat Methods. 2009;6(10):741–4.
85. Strang BL, Boulant S, Coen DM. Nucleolin associates with the human cytomegalovirus DNA polymerase accessory subunit UL44 and is necessary for efficient viral replication. J Virol. 2010;84(4):1771–84.
86. Emanuele MJ, Ciccia A, Elia AE, Elledge SJ. Proliferating cell nuclear antigen (PCNA)-associated KIAA0101/PAF15 protein is a cell cycle-regulated anaphase-promoting complex/cyclosome substrate. Proc Natl Acad Sci U S A. 2011;108(24):9845–50.
87. Jirawatnotai S, Hu Y, Michowski W, Elias JE, Becks L, Bienvenu F, et al. A function for cyclin D1 in DNA repair uncovered by protein interactome analyses in human cancers. Nature. 2011;474(7350):230–4.
88. Klockenbusch C, Kast J. Optimization of formaldehyde cross-linking for protein interaction analysis of non-tagged integrin beta1. J Biomed Biotechnol. 2010;2010:927585.
89. Vaezi A, Bauer C, Vasioukhin V, Fuchs E. Actin cable dynamics and Rho/Rock orchestrate a polarized cytoskeletal architecture in the early steps of assembling a stratified epithelium. Dev Cell. 2002;3(3):367–81.
90. McMullan R, Lax S, Robertson VH, Radford DJ, Broad S, Watt FM, et al. Keratinocyte differentiation is regulated by the Rho and ROCK signaling pathway. Curr Biol. 2003;13(24):2185–9.
91. Presland RB, Kuechle MK, Lewis SP, Fleckman P, Dale BA. Regulated expression of human filaggrin in keratinocytes results in cytoskeletal disruption, loss of cell-cell adhesion, and cell cycle arrest. Exp Cell Res. 2001;270(2):199–213.
92. Rheinwald JG, Green H. Epidermal growth factor and the multiplication of cultured human epidermal keratinocytes. Nature. 1977;265(5593):421–4.
93. Monzon RI, McWilliams N, Hudson LG. Suppression of cornified envelope formation and type 1 transglutaminase by epidermal growth factor in neoplastic keratinocytes. Endocrinology. 1996;137(5):1727–34.
94. Chen CS, Lavker RM, Rodeck U, Risse B, Jensen PJ. Use of a serum-free epidermal culture model to show deleterious effects of epidermal growth factor on morphogenesis and differentiation. J Invest Dermatol. 1995;104(1):107–12.
95. Boisnic S, Ouhayoun JP, Branchet MC, Frances C, Beranger JY, Le Charpentier Y, et al. Alteration of cytokeratin expression in oral lichen planus. Oral Surg Oral Med Oral Pathol Oral Radiol Endod. 1995;79(2):207–15.
96. Appleton BA, Loregian A, Filman DJ, Coen DM, Hogle JM. The cytomegalovirus DNA polymerase subunit UL44 forms a C clamp-shaped dimer. Mol Cell. 2004;15(2):233–44.
97. Appleton BA, Brooks J, Loregian A, Filman DJ, Coen DM, Hogle JM. Crystal structure of the cytomegalovirus DNA polymerase subunit UL44 in complex with the C terminus from the catalytic subunit, Differences in structure and function relative to unliganded UL44. J Biol Chem. 2006;281(8):5224–32.
98. Krishna TS, Kong XP, Gary S, Burgers PM, Kuriyan J. Crystal structure of the eukaryotic DNA polymerase processivity factor PCNA. Cell. 1994;79(7):1233–43.
99. Komazin-Meredith G, Petrella RJ, Santos WL, Filman DJ, Hogle JM, Verdine GL, et al. The human cytomegalovirus UL44 C clamp wraps around DNA. Structure. 2008;16(8):1214–25.
100. Moldovan GL, Pfander B, Jentsch S. PCNA, the maestro of the replication fork. Cell. 2007;129(4):665–79.
101. Strzalka W, Ziemienowicz A. Proliferating cell nuclear antigen (PCNA): a key factor in DNA replication and cell cycle regulation. Ann Bot. 2011;107(7):1127–40.
102. Yu P, Huang B, Shen M, Lau C, Chan E, Michel J, et al. p15(PAF), a novel PCNA associated factor with increased expression in tumor tissues. Oncogene. 2001;20(4):484–9.
103. Simpson F, Lammerts VB, Butterfield N, Bennetts JS, Bowles J, Adolphe C, et al. The PCNA-associated factor KIAA0101/p15(PAF) binds the potential tumor suppressor product p33ING1b. Exp Cell Res. 2006;312(1):73–85.
104. Moro L, Dolce L, Cabodi S, Bergatto E, Boeri EE, Smeriglio M, et al. Integrin-induced epidermal growth factor (EGF) receptor activation requires c-Src and p130Cas and leads to phosphorylation of specific EGF receptor tyrosines. J Biol Chem. 2002;277(11):9405–14.
105. Houben E, De Paepe K, Rogiers V. A keratinocyte's course of life. Skin Pharmacol Physiol. 2007;20(3):122–32.
106. Brakebusch C, Fassler R. beta 1 integrin function in vivo: adhesion, migration and more. Cancer Metastasis Rev. 2005;24(3):403–11.
107. Brakebusch C, Grose R, Quondamatteo F, Ramirez A, Jorcano JL, Pirro A, et al. Skin and hair follicle integrity is crucially dependent on beta 1 integrin expression on keratinocytes. EMBO J. 2000;19(15):3990–4003.
108. Raghavan S, Bauer C, Mundschau G, Li Q, Fuchs E. Conditional ablation of beta1 integrin in skin, Severe defects in epidermal proliferation, basement membrane formation, and hair follicle invagination. J Cell Biol. 2000;150(5):1149–60.
109. Aszodi A, Hunziker EB, Brakebusch C, Fassler R. Beta1 integrins regulate chondrocyte rotation, G1 progression, and cytokinesis. Genes Dev. 2003;17(19):2465–79.
110. Behrends C, Sowa ME, Gygi SP, Harper JW. Network organization of the human autophagy system. Nature. 2010;466(7302):68–76.
111. Lippens S, Denecker G, Ovaere P, Vandenabeele P, Declercq W. Death penalty for keratinocytes: apoptosis versus cornification. Cell Death Differ. 2005;12 Suppl 2:1497–508.
112. Polakowska RR, Haake AR. Apoptosis: the skin from a new perspective. Cell Death Differ. 1994;1(1):19–31.
113. Takahashi H, Aoki N, Nakamura S, Asano K, Ishida-Yamamoto A, Iizuka H. Cornified cell envelope formation is distinct from apoptosis in epidermal keratinocytes. J Dermatol Sci. 2000;23(3):161–9.

114. Maiuri MC, Zalckvar E, Kimchi A, Kroemer G. Self-eating and self-killing: crosstalk between autophagy and apoptosis. Nat Rev Mol Cell Biol. 2007;8(9):741–52.
115. Shaw JL, Smith CR, Diamandis EP. Proteomic analysis of human cervico-vaginal fluid. J Proteome Res. 2007;6(7):2859–65.
116. Stastna M, Behrens A, McDonnell PJ, Van Eyk JE. Analysis of protein composition of rabbit aqueous humor following two different cataract surgery incision procedures using 2-DE and LC-MS/MS. Proteome Sci. 2011;9(1):8.

Part II

Identification and Study Methods

Animal Models

7

Kozo Nakai, Kozo Yoneda, and Yasuo Kubota

Contents

7.1 Introduction

The roles of human filaggrin in health and disease have been modeled with experimental animal models, mostly mice (Table 7.1). Mouse models have provided important information relevant to the roles of filaggrin, but it must be always considered that mice are not humans and that the role of filaggrin in skin pathology and/or physiology still remains unclear. Most altered

K. Nakai, MD, PhD (✉) • K. Yoneda, MD, PhD
Y. Kubota, MD, PhD
Department of Dermatology, Kagawa University, 1750-1 Ikenobe, miki-cho, Kita-gun, Kagawa 761-0793, Japan
e-mail: kozo@kms.ac.jp; kyoneda@kms.ac.jp; kubotay@kms.ac.jp

J.P. Thyssen, H. Maibach (eds.), *Filaggrin*,
DOI 10.1007/978-3-642-54379-1_7,

Table 7.1 Animal models for filaggrin research

Filaggrin gene-mutated mouse and filaggrin-null mouse
Flaky tail mouse
Filaggrin-null mouse
Mutant mice with abnormal filaggrin expression
Epilation mutant mouse
Pupoid fetus mutant mouse
SENCAR mouse
Flaky skin mouse
Epidermal-specific glycosylphosphatidylinositol-anchor-deficient mouse
Mouse lacking the C-terminal region of connexin43
Peroxisome-proliferator-activated receptor-γ-deficient mouse
25OHD 1α-hydroxylase knockout mouse
Spermidine/spermine N1-acetyltransferase transgenic mouse
Nude mouse
Spink5 (R820X/R820X) mouse
CAP1/Prss8-deficient skin mouse
Heterozygous Inv-Cldn6 mouse
DHCR24–/– mouse
Arnt-null skin mouse
12R-LOX-deficient mouse
Caspase-14-deficient mouse
PPARβ/δ knockout mouse
Glucocorticoid receptor knockout mouse
AP-2γ-deficient mouse
High-chimeric Hai-1/Spint1(–/–) mouse
St14 transgenic mouse
ELA2 overexpressing transgenic mouse
UBC-Taps mouse
Abca12-disrupted mouse
BMP-receptor-IB transgenic mouse
CB1R and CB2R knockout mouse
Chemicals or UV exposure that alters the expression of filaggrin in mice
Retinoic acid
Cantharidin
12-O-tetradecanoylphorbol-13-acetate
UVB
Filaggrin in other animals
Porcine
Guinea pig
Pup
Dog
Snake
Bird
Gecko lizard
Crocodile

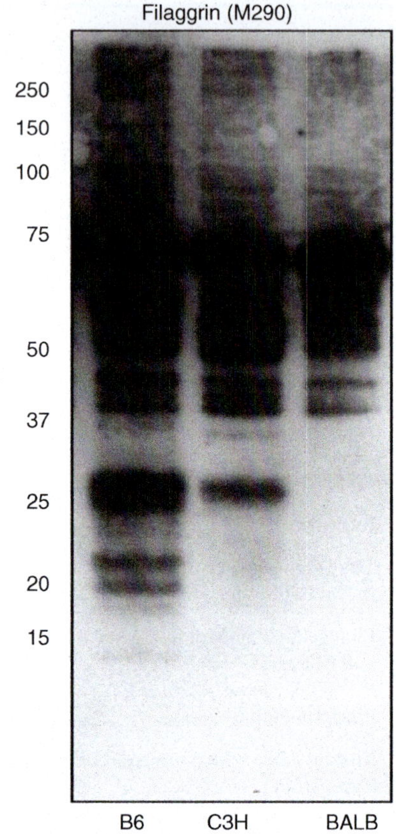

Fig. 7.1 Western blot analysis of filaggrin expression in the skin of C57BL/6 (B6), C3H, and BALB mouse

expression levels or pattern of filaggrin could appear as a result of abnormal keratinization and may not be substantial. Moreover, the processing appears to depend on the strain of mice. Although mature filaggrin is detected in the skin of C57BL/6 mouse, it is reduced in the skin of C3H mouse and BALB mouse (Fig. 7.1). Therefore, careful attention is necessary for interpretation of experiments utilizing animal models.

7.2 The Discovery of Filaggrin in Animal Models (Mouse and Rat)

In 1981, Steinert et al. isolated a class of cationic structural proteins from the stratum corneum of mouse epidermis [1]. The protein associated specifically with intermediate filaments but not with other types of cytoskeletal proteins. They named the protein "filaggrin." The presence of a human filaggrin was confirmed 2 years later [2]. In 1987, Fisher et al. demonstrated the localization of pro-filaggrin mRNA in newborn rat skin by in situ hybridization [3], and Rothnagel et al. characterized the gene for mouse epidermal filaggrin precursor [4]. In 1988, filaggrin was confirmed as a component of the cell envelope of the newborn rat [5]. Rothnagel et al. studied the structure of the mouse filaggrin gene (*FLG*) [6]. Mouse and rat filaggrins are very similar at both the nucleic acid and amino acid levels [4]; however, in contrast, mouse and human filaggrins have almost no sequence homology [7]. Although human filaggrin repeats show heterogeneity in their composition, the filaggrin repeats of rat and mouse show more than 90 % identity in amino acid sequence due to their inbred strains [6]. There are two types of filaggrin repeats in the mouse gene: 255 amino acid long and 250 amino acid long [6, 8]. The number of filaggrin repeats depends on the strain of mice. In 2002, Pearton et al. demonstrated that the mouse pro-filaggrin N-terminus undergoes proteolytic processing in two steps, first releasing an N-terminal peptide containing some filaggrin sequence and finally the free N-terminus of 28–30 kDa [9].

7.3 Filaggrin Gene-Mutated Mouse and Filaggrin-Null Mouse

There are two animal models, in which the direct functions of filaggrin have been studied.

7.3.1 Flaky Tail Mouse

The flaky tail mouse has been utilized to investigate the direct functions of filaggrin (Fig. 7.2, Table 7.2). The mouse arose spontaneously in 1958 at The Jackson Laboratory, and two distinct mutations on chromosome 3, flaky tail (*ft*) and matted hair (*ma*) allele, were reported in 1972 [10]. These two mutations are tightly linked and maintained as a compound, but their phenotypes do not overlap. Presland et al. showed the loss of normal pro-filaggrin and filaggrin in the flaky tail mouse. The flaky tail mouse expresses a lower-molecular-weight form of pro-filaggrin (220 kDa) instead of normal molecular pro-filaggrin

Fig. 7.2 Flaky tail mouse

Table 7.2 Summary of flaky tail mouse

Genetic abnormality
bp deletion (5303delA) in murine *Flg* gene
ma mutation
Epidermal abnormality
Structure
A lower-molecular-weight form of pro-filaggrin (220 kDa) instead of normal molecular pro-filaggrin (500 kDa)
Abnormal filaggrin processing
Abnormal keratohyalin F granules formation
No cornified layer
Reduced expression of loricrin, epidermal growth factor receptor, E-cadherin, and occludin
Function
Elevated transepidermal water loss
Elevated outside-to-inside skin penetration of foreign materials
Immune abnormality
Steady state
Increased levels of IgE in the serum and CD4/CD8 cells in the skin-draining axillary and inguinal lymph nodes
Allergen (ovalbumin) exposure
Exacerbated skin inflammation and cytokine productions: IL-4, IL-5, IL-13, INF-γ, IL-10, and IL-17

(500 kDa). The abnormal pro-filaggrin does not form normal keratohyalin F granules and is not proteolytically processed to filaggrin. Hence, filaggrin is absent from the cornified layers in the epidermis. Due to these observations, they suggested the mouse as an animal model for the filaggrin-deficient skin disease ichthyosis vulgaris [11]. A 1-bp deletion (5303delA) in murine *Flg* gene is detected in flaky tail mouse [12]. By backcrossing on the C57BL/6 background, it is possible to remove the *ma* allele in flaky tail mouse. Even in the backcrossed flaky tail mouse, the disturbed barrier function and subsequent enhanced percutaneous allergen priming could be observed in the mouse, suggesting that the mouse is a model of atopic dermatitis with barrier dysfunction [13]. In fact, spontaneous dermatitis in flaky tail mouse under specific pathogen-free conditions is clinically observed after 5 weeks. Transepidermal water loss and outside-to-inside skin penetration of foreign materials are elevated in the mouse, indicating a skin barrier abnormality. The barrier dysfunction is also supported as the reduced expression of loricrin, epidermal growth factor receptor, E-cadherin, and occludin is reduced in the skin of flaky tail mice [14]. The immune status of flaky tail mouse in a steady state shows the increased levels of IgE in the serum and CD4/CD8 cells in the skin-draining axillary and inguinal lymph nodes [13]. Allergen (ovalbumin) exposure exacerbates skin inflammation and cytokine productions: IL-4, IL-5, IL-13, INF-γ, IL-10, and IL-17 [12].

7.3.2 Filaggrin-Null Mouse

The flaky tail mouse spontaneously has itchy skin lesions, providing an attractive model for atopic dermatitis research. However, the mouse does not completely lack filaggrin, meaning that it was difficult to show the precise roles of filaggrin. Thereafter, filaggrin-null mouse was a pure filaggrin-deficient mouse [15]. The mouse exhibits dry scaly skin. Although the levels of natural moisturizing factor (NMF) were decreased, hydration and transepidermal water loss are normal in the filaggrin-null mice. Premature shedding of stratum corneum layers, increased desquamation under mechanical stress, and loss of keratin patterns were reported in the filaggrin-null mice. The repeated allergen (ovalbumin) exposure increases the serum levels of IgG_1 and IgE.

7.4 Mutant Mice with Abnormal Filaggrin Expression

Numerous mutant mice show unique epidermal keratinization. The expression of filaggrin is indirectly deranged in some of these mice.

7.4.1 Epilation Mutant Mouse

Repeated epilation (Er) is a radiation-induced, autosomal, incomplete dominant mutation in a mouse. The epidermis in Er/Er mouse is adhesive, has a hyperplastic granular layer with prominent keratohyalin granules, and fails to undergo terminal differentiation. The skin of Er/Er mouse does not synthesize filaggrin proteins but produces abnormal keratins [16].

7.4.2 Pupoid Fetus Mutant Mouse

The epidermis of the pupoid fetus mutant mouse is invaded by cells from the dermis. These invading cells establish a network of cells including fibroblasts, endothelial cells, and nerve fibers, throughout the epidermis. Subsequent to these events, filaggrin is reduced and keratinization is inhibited [17].

7.4.3 SENCAR Mouse

SENCAR mice, the name derived from SENsitivity to CARcinogenesis, are sensitive to skin tumor initiators. The distribution of filaggrin is deranged in the initiated papillomas and its expression is reduced in the invasive carcinomas [18].

7.4.4 Flaky Skin Mouse

Flaky skin (fsn) mouse is a model of psoriasis. Keratohyalin abnormalities can be observed by using immunocytochemical staining for profilaggrin [19].

7.4.5 Epidermal-Specific Glycosylphosphatidylinositol-Anchor-Deficient Mouse

The epidermal-specific glycosylphosphatidylinositol (GPI)-anchor-deficient mouse exhibits wrinkled and dry skin with hyperkeratosis and abnormal differentiation. Pro-filaggrin to its monomeric form is impaired in the mice [20].

7.4.6 Mouse Lacking the C-Terminal Region of Connexin43

Mouse in which the C-terminal region of connexin43 (Cx43) is removed (designated as Cx43K258stop) shows the thinness and fragility of corneocytes and the defects of epidermal barrier [21]. A broader expression of filaggrin can be observed in the epidermis of the mouse. Full filaggrin processing is delayed in the skin of the mouse.

7.4.7 Peroxisome-Proliferator-Activated Receptor-γ-Deficient Mouse

PPAR-γ activators increase the expression of filaggrin in the skin of mice. The increase in filaggrin expression is suppressed in the skin of PPAR-γ-deficient mouse [22].

7.4.8 25ohd 1α-Hydroxylase Knockout Mouse

Keratinocytes express high levels of 25-hydroxy vitamin D (25OHD) 1α-hydroxylase. The expression levels of filaggrin are reduced in the epidermis of 25OHD 1α-hydroxylase knockout mouse [23].

7.4.9 Spermidine/Spermine N^1-Acetyltransferase Transgenic Mouse

Overexpression of the rate-limiting enzyme in polyamine catabolism spermidine/spermine N^1-acetyltransferase in transgenic mouse leads to accumulation of putrescine in the skin and permanent hair loss [24]. The mouse is characterized by delayed onset of epidermal differentiation with reduced expression of filaggrin.

7.4.10 Nude Mouse

Nude mouse has a mutation in the transcription factor *Foxn1^{nu}*. The expression of hair keratins is downregulated [25]. Sulfur concentrations are reduced in the nude mouse nail plates. Keratin 1 protein expression is suppressed, and filaggrin expression is deranged in the nail plate of the mouse.

7.4.11 Spink5 (R820X/R820X) Mouse

Spink5 (R820X/R820X) mouse is a model of Netherton syndrome [26]. The mouse develops a lethal, severe ichthyosis with a loss of skin barrier function and dehydration. The spink5 (R820X/R820X) mouse shows a substantial increase in the proteolytic processing of profilaggrin into its constituent filaggrin monomers.

7.4.12 CAP1/Prss8-Deficient Skin Mouse

Severe malformation of the stratum corneum and lethal hydration caused by skin barrier dysfunction are observed in mice lacking the membrane-anchored channel-activating serine protease (CAP) 1 (protease serine S1 family member 8

[*Prss8*]) in skin [27]. The skin of mice shows aberrant pro-filaggrin to filaggrin processing.

7.4.13 Heterozygous Inv-Cldn6 Mouse

Claudin-6 is a tight junction protein. Homozygous mouse overexpressing Claudin-6 (Cldn6) dies within 48 h of birth. Heterozygous Inv-Cldn6 mouse exhibits a distinct coat phenotype and mild epidermal hyperkeratosis. The processing of pro-filaggrin to filaggrin is altered in the mouse [28].

7.4.14 DHCR24–/– Mouse

Desmosterolosis is an autosomal recessive disorder due to mutations in the *3β-hydroxysterol-Δ24 reductase* (*DHCR24*) gene that encodes an enzyme catalyzing the conversion of desmosterol to cholesterol in skin. *DHCR24–/–* mouse has thickened epidermis with few and smaller keratohyalin granules [29]. Filaggrin expression is altered in the epidermis of *DHCR24–/–* mouse.

7.4.15 Arnt-Null Skin Mouse

The aryl hydrocarbon receptor nuclear translocator (Arnt) is a transcription factor that mediates biological functions including sensing of light, reactive oxygen species. Targeted K14-driven deletion of *Arnt* in the mouse epidermis resulted in early death, associated with a failure of epidermal barrier function. The expression of filaggrin is reduced in Arnt-null epidermis [30].

7.4.16 12R-LOX-Deficient Mouse

12R-lipoxygenase (12R-LOX) represents a key enzyme of an eicosanoid pathway in the skin. 12R-LOX/epidermal LOX-3 (eLOX-3) is involved in terminal differentiation in skin. Inactivating 12R-LOX/eLOX-3 is linked to the development of autosomal recessive congenital ichthyosis. Lipid metabolism and processing of pro-filaggrin to filaggrin are impaired in 12R-LOX-deficient mouse [31].

7.4.17 Caspase-14-Deficient Mouse

Caspase-14 is an aspartate-specific proteinase which is involved in terminal keratinocyte differentiation and cornification. The skin of *caspase-14-deficient* mouse is shiny and lichenified [32]. *Caspase-14*-deficient epidermis contains significantly more alveolar keratohyalin F-granules, the pro-filaggrin stores. The pro-filaggrin processing is altered in the epidermis of *caspase-14-deficient* mouse.

7.4.18 PPARβ/δ Knockout Mouse

Peroxisome-proliferator-activated receptors (PPAR) are nuclear receptor proteins of lipid metabolites that function as transcription factors. PPAR*β*/*δ* is involved in differentiation and proliferation of epidermal keratinocytes. The expression of filaggrin is increased in PPAR*β*/*δ* knockout mouse [33].

7.4.19 Glucocorticoid Receptor Knockout Mouse

Glucocorticoid receptor (GR) is required for skin barrier competence. Incomplete epidermal stratification with dramatically abnormal differentiation is observed in GR knockout mouse [34]. Terminal differentiation is absent due to an impaired activation of caspase-14, and filaggrin expression is suppressed.

7.4.20 AP-2γ-Deficient Mouse

AP-2γ is a transcription factor that regulates both proliferation and differentiation of epidermal keratinocytes. Skin development is delayed in AP-2γ-deficient mouse, and filaggrin expression pattern in epidermis is altered [35].

7.4.21 High-Chimeric Hai-1/Spint1(−/−) Mouse

Hai-1/Spint1 is a serine proteinase inhibitor that inhibits matriptase and prostasin; subsequently epidermal differentiation is deranged. High-chimeric *Hai-1/Spint1*$^{-/-}$ mouse develops scaly skin with hyperkeratinization and abnormal hair shafts [36]. Proteolytic processing of pro-filaggrin is altered in the mouse.

7.4.22 St14 Transgenic Mouse

Suppressor of tumorigenicity 14 (*St14*) encodes matriptase, a serine protease, which regulates processing of pro-filaggrin to filaggrin. Transgenic mouse with 1 % of wild-type *St14* levels displays aberrant processing of pro-filaggrin. Scaling of the skin with epidermal acanthosis and orthohyperkeratosis is observed in the mouse [37].

7.4.23 ELA2 Overexpressing Transgenic Mouse

Elastase 2 (ELA2) is localized to keratohyalin granules and participates in pro-filaggrin processing. ELA2 is hyperactive in skin from human Netherton syndrome patients. Transgenic mouse overexpressing ELA2 in the epidermis shows abnormal pro-filaggrin processing and impaired lipid lamellae structure [38].

7.4.24 UBC-Taps Mouse

AP-1-dependent target gene encodes a retroviral-like aspartic proteinase (Taps/Asprv1) in 12-O-tetradecanoylphorbol-13-acetate (TPA)-treated mouse back skin. Taps expression was detected almost exclusively in stratified epithelia of mouse embryos and adult tissues, and enhanced protein levels were present in several nonneoplastic human skin disorders. A mouse model in which Taps transgene expression is under the control of the human ubiquitin C promoter (UBC-Taps) shows a significant delay in cutaneous wound closure [39]. A hypergranulosum-like phenotype with increased numbers of filaggrin-positive keratinocytes was also observed in UBC-Taps mouse after administration of TPA.

7.4.25 Abca12-Disrupted Mouse

ABca12 is a lipid transporter in the keratinocytes. Loss-of-function mutations in the keratinocyte ABCA12 causes Harlequin ichthyosis. Abca12-disrupted neonatal epidermis lacks pro-filaggrin/filaggrin [40].

7.4.26 BMP-Receptor-IB Transgenic Mouse

Bone morphogenetic protein (BMP) plays an important role during embryonic organ development. BMP is involved in the regulation of filaggrin expression in the epidermis. Mouse expressing a constitutively active form of BMP-receptor-IB in the epidermis exhibits a phenotype resembling human ichthyosis vulgaris, and filaggrin expression is downregulated in the mouse [41].

7.4.27 CB1R and CB2R Knockout Mouse

Cannabinoids are amides, esters, and ethers of long-chain polyunsaturated fatty acids. Filaggrin expression is decreased in CB1R and CB2R knockout mouse [42].

7.5 Mouse Models in Which Filaggrin Expression Is Altered by Chemicals or UV Exposure

The expression levels and/or pattern of filaggrin in the skin of mice can be altered by topical application of some chemicals such as retinoic acid [43], cantharidin, and 12-O-tetradecanoylphorbol-13-acetate [44] or UVB exposure [45].

7.6 Filaggrin in Other Animals

The expression of filaggrin has been reported in porcine [46], guinea pig [47], pup, and canine skin. Human c-*myc* transgenic pup develops a hyperkeratotic phenotype. Inhibition of terminal differentiation and aberrant expression of filaggrin are observed due to c-Myc overexpression [48]. Glucocorticoids accelerate fetal stratum corneum maturation and barrier formation. The expression of filaggrin and skin barrier maturation is reduced in glucocorticoid-deficient murine pup [49]. There is a canine model of atopic dermatitis. Decreased epidermal filaggrin expression is present in the dogs, as well as increased transepidermal water loss [50]. Filaggrin-like protein is observed in the epidermis of snakes [51], birds [52], gecko lizards [53], and crocodiles [54].

References

1. Steinert PM, Cantieri JS, Teller DC, Lonsdale-Eccles JD, Dale BA. Characterization of a class of cationic proteins that specifically interact with intermediate filaments. Proc Natl Acad Sci U S A. 1981;78(7):4097–101.
2. Lynley AM, Dale BA. The characterization of human epidermal filaggrin. A histidine-rich, keratin filament-aggregating protein. Biochim Biophys Acta. 1983;744(1):28–35.
3. Fisher C, Haydock PV, Dale BA. Localization of profilaggrin mRNA in newborn rat skin by in situ hybridization. J Invest Dermatol. 1987;88(5):661–4.
4. Rothnagel JA, Mehrel T, Idler WW, Roop DR, Steinert PM. The gene for mouse epidermal filaggrin precursor. Its partial characterization, expression, and sequence of a repeating filaggrin unit. J Biol Chem. 1987;262(32):15643–8.
5. Richards S, Scott IR, Harding CR, Liddell JE, Powell GM, Curtis CG. Evidence for filaggrin as a component of the cell envelope of the newborn rat. Biochem J. 1988;253(1):153–60.
6. Rothnagel JA, Steinert PM. The structure of the gene for mouse filaggrin and a comparison of the repeating units. J Biol Chem. 1990;265(4):1862–5.
7. McKinley-Grant LJ, Idler WW, Bernstein IA, Parry DA, Cannizzaro L, Croce CM, et al. Characterization of a cDNA clone encoding human filaggrin and localization of the gene to chromosome region 1q21. Proc Natl Acad Sci U S A. 1989;86(13):4848–52.
8. Zhang D, Karunaratne S, Kessler M, Mahony D, Rothnagel JA. Characterization of mouse pro-filaggrin: evidence for nuclear engulfment and translocation of the pro-filaggrin B-domain during epidermal differentiation. J Invest Dermatol. 2002;119(4):905–12.
9. Pearton DJ, Dale BA, Presland RB. Functional analysis of the pro-filaggrin N-terminal peptide: identification of domains that regulate nuclear and cytoplasmic distribution. J Invest Dermatol. 2002;119(3):661–9.
10. Lane PW. Two new mutations in linkage group XVI of the house mouse. Flaky tail and varitint-waddler-J. J Hered. 1972;63:135–40.
11. Presland RB, Boggess D, Lewis SP, Hull C, Fleckman P, Sundberg JP. Loss of normal pro-filaggrin and filaggrin in flaky tail (ft/ft) mice: an animal model for the filaggrin-deficient skin disease ichthyosis vulgaris. J Invest Dermatol. 2000;115(6):1072–81.
12. Fallon PG, Sasaki T, Sandilands A, Campbell LE, Saunders SP, Mangan NE, et al. A homozygous frameshift mutation in the mouse Flg gene facilitates enhanced percutaneous allergen priming. Nat Genet. 2009;41(5):602–8.
13. Moniaga CS, Egawa G, Kawasaki H, Hara-Chikuma M, Honda T, Tanizaki H, et al. Flaky tail mouse denotes human atopic dermatitis in the steady state and by topical application with Dermatophagoides pteronyssinus extract. Am J Pathol. 2010;176(5):2385–93.
14. Nakai K, Yoneda K, Hosokawa Y, Moriue T, Presland RB, Fallon PG, et al. Reduced expression of epidermal growth factor receptor, E-cadherin, and occludin in the skin of flaky tail mice is due to filaggrin and loricrin deficiencies. Am J Pathol. 2012;181(3):969–77.
15. Kawasaki H, Nagao K, Kubo A, Hata T, Shimizu A, Mizuno H, et al. Altered stratum corneum barrier and enhanced percutaneous immune responses in filaggrin-null mice. J Allergy Clin Immunol. 2012;129(6):1538–46 e6.
16. Holbrook KA, Dale BA, Brown KS. Abnormal epidermal keratinization in the repeated epilation mutant mouse. J Cell Biol. 1982;92(2):387–97.
17. Fisher C, Dale BA, Kollar EJ. Abnormal keratinization in the pupoid fetus (pf/pf) mutant mouse epidermis. Dev Biol. 1984;102(2):290–9.
18. Mamrack MD, Klein-Szanto AJ, Reiners Jr JJ, Slaga TJ. Alteration in the distribution of the epidermal protein filaggrin during two-stage chemical carcinogenesis in the SENCAR mouse skin. Cancer Res. 1984;44(6):2634–41.
19. Morita K, Hogan ME, Nanney LB, King Jr LE, Manabe M, Sun TT, et al. Cutaneous ultrastructural features of the flaky skin (fsn) mouse mutation. J Dermatol. 1995;22(6):385–95.
20. Hara-Chikuma M, Takeda J, Tarutani M, Uchida Y, Holleran WM, Endo Y, et al. Epidermal-specific defect of GPI anchor in Pig-a null mice results in Harlequin ichthyosis-like features. J Invest Dermatol. 2004;123(3):464–9.
21. Maass K, Ghanem A, Kim JS, Saathoff M, Urschel S, Kirfel G, et al. Defective epidermal barrier in neonatal mice lacking the C-terminal region of connexin43. Mol Biol Cell. 2004;15(10):4597–608.

22. Mao-Qiang M, Fowler AJ, Schmuth M, Lau P, Chang S, Brown BE, et al. Peroxisome-proliferator-activated receptor (PPAR)-gamma activation stimulates keratinocyte differentiation. J Invest Dermatol. 2004;123(2):305–12.
23. Bikle DD, Chang S, Crumrine D, Elalieh H, Man MQ, Dardenne O, et al. Mice lacking 25OHD 1alpha-hydroxylase demonstrate decreased epidermal differentiation and barrier function. J Steroid Biochem Mol Biol. 2004;89–90(1–5):347–53.
24. Pietila M, Pirinen E, Keskitalo S, Juutinen S, Pasonen-Seppanen S, Keinanen T, et al. Disturbed keratinocyte differentiation in transgenic mice and organotypic keratinocyte cultures as a result of spermidine/spermine N-acetyltransferase overexpression. J Invest Dermatol. 2005;124(3):596–601.
25. Mecklenburg L, Paus R, Halata Z, Bechtold LS, Fleckman P, Sundberg JP. FOXN1 is critical for onycholemmal terminal differentiation in nude (Foxn1) mice. J Invest Dermatol. 2004;123(6):1001–11.
26. Hewett DR, Simons AL, Mangan NE, Jolin HE, Green SM, Fallon PG, et al. Lethal, neonatal ichthyosis with increased proteolytic processing of filaggrin in a mouse model of Netherton syndrome. Hum Mol Genet. 2005;14(2):335–46.
27. Leyvraz C, Charles RP, Rubera I, Guitard M, Rotman S, Breiden B, et al. The epidermal barrier function is dependent on the serine protease CAP1/Prss8. J Cell Biol. 2005;170(3):487–96.
28. Troy TC, Rahbar R, Arabzadeh A, Cheung RM, Turksen K. Delayed epidermal permeability barrier formation and hair follicle aberrations in Inv-Cldn6 mice. Mech Dev. 2005;122(6):805–19.
29. Mirza R, Hayasaka S, Takagishi Y, Kambe F, Ohmori S, Maki K, et al. DHCR24 gene knockout mice demonstrate lethal dermopathy with differentiation and maturation defects in the epidermis. J Invest Dermatol. 2006;126(3):638–47.
30. Geng S, Mezentsev A, Kalachikov S, Raith K, Roop DR, Panteleyev AA. Targeted ablation of Arnt in mouse epidermis results in profound defects in desquamation and epidermal barrier function. J Cell Sci. 2006;119(Pt 23):4901–12.
31. Epp N, Furstenberger G, Muller K, de Juanes S, Leitges M, Hausser I, et al. 12R-lipoxygenase deficiency disrupts epidermal barrier function. J Cell Biol. 2007;177(1):173–82.
32. Denecker G, Hoste E, Gilbert B, Hochepied T, Ovaere P, Lippens S, et al. Caspase-14 protects against epidermal UVB photodamage and water loss. Nat Cell Biol. 2007;9(6):666–74.
33. Man MQ, Barish GD, Schmuth M, Crumrine D, Barak Y, Chang S, et al. Deficiency of PPARbeta/delta in the epidermis results in defective cutaneous permeability barrier homeostasis and increased inflammation. J Invest Dermatol. 2008;128(2):370–7.
34. Bayo P, Sanchis A, Bravo A, Cascallana JL, Buder K, Tuckermann J, et al. Glucocorticoid receptor is required for skin barrier competence. Endocrinology. 2008;149(3):1377–88.
35. Guttormsen J, Koster MI, Stevens JR, Roop DR, Williams T, Winger QA. Disruption of epidermal specific gene expression and delayed skin development in AP-2 gamma mutant mice. Dev Biol. 2008;317(1):187–95.
36. Nagaike K, Kawaguchi M, Takeda N, Fukushima T, Sawaguchi A, Kohama K, et al. Defect of hepatocyte growth factor activator inhibitor type 1/serine protease inhibitor, Kunitz type 1 (Hai-1/Spint1) leads to ichthyosis-like condition and abnormal hair development in mice. Am J Pathol. 2008;173(5):1464–75.
37. Scharschmidt TC, List K, Grice EA, Szabo R, Renaud G, Lee CC, et al. Matriptase-deficient mice exhibit ichthyotic skin with a selective shift in skin microbiota. J Invest Dermatol. 2009;129(10):2435–42.
38. Bonnart C, Deraison C, Lacroix M, Uchida Y, Besson C, Robin A, et al. Elastase 2 is expressed in human and mouse epidermis and impairs skin barrier function in Netherton syndrome through filaggrin and lipid misprocessing. J Clin Invest. 2010;120(3):871–82.
39. Hildenbrand M, Rhiemeier V, Hartenstein B, Lahrmann B, Grabe N, Angel P, et al. Impaired skin regeneration and remodeling after cutaneous injury and chemically induced hyperplasia in taps-transgenic mice. J Invest Dermatol. 2010;130(7):1922–30.
40. Yanagi T, Akiyama M, Nishihara H, Ishikawa J, Sakai K, Miyamura Y, et al. Self-improvement of keratinocyte differentiation defects during skin maturation in ABCA12-deficient harlequin ichthyosis model mice. Am J Pathol. 2010;177(1):106–18.
41. Yu X, Espinoza-Lewis RA, Sun C, Lin L, He F, Xiong W, et al. Overexpression of constitutively active BMP-receptor-IB in mouse skin causes an ichthyosis-vulgaris-like disease. Cell Tissue Res. 2010;342(3):401–10.
42. Roelandt T, Heughebaert C, Bredif S, Giddelo C, Baudouin C, Msika P, et al. Cannabinoid receptors 1 and 2 oppositely regulate epidermal permeability barrier status and differentiation. Exp Dermatol. 2012;21(9):688–93.
43. Lutzow-Holm C, Heyden A, Huitfeldt HS, Brandtzaeg P, Clausen OP. Differential effects of topical retinoic acid application on keratin K1 and filaggrin expression in mouse epidermis. Differentiation. 1994;57(3):179–85.
44. Heyden A, Lutzow-Holm C, Clausen OP, Thrane EV, Brandtzaeg P, Roop DR, et al. Application of cantharidin or 12-O-tetradecanoylphorbol-13-acetate on mouse epidermis induces a cell population shift that causes altered keratin distribution. Differentiation. 1994;57(3):187–93.
45. Hong SP, Kim MJ, Jung MY, Jeon H, Goo J, Ahn SK, et al. Biopositive effects of low-dose UVB on epidermis: coordinate upregulation of antimicrobial peptides and permeability barrier reinforcement. J Invest Dermatol. 2008;128(12):2880–7.
46. Wollina U, Berger U, Mahrle G. Immunohistochemistry of porcine skin. Acta Histochem. 1991;90(1):87–91.

47. Scott IR. Alterations in the metabolism of filaggrin in the skin after chemical- and ultraviolet-induced erythema. J Invest Dermatol. 1986;87(4):460–5.
48. Waikel RL, Wang XJ, Roop DR. Targeted expression of c-Myc in the epidermis alters normal proliferation, differentiation and UV-B induced apoptosis. Oncogene. 1999;18(34):4870–8.
49. Hanley K, Feingold KR, Komuves LG, Elias PM, Muglia LJ, Majzoub JA, et al. Glucocorticoid deficiency delays stratum corneum maturation in the fetal mouse. J Invest Dermatol. 1998;111(3):440–4.
50. Marsella R, Girolomoni G. Canine models of atopic dermatitis: a useful tool with untapped potential. J Invest Dermatol. 2009;129(10):2351–7.
51. Alibardi L, Toni M. Immunolocalization and characterization of cornification proteins in snake epidermis. Anat Rec A Discov Mol Cell Evol Biol. 2005;282(2):138–46.
52. Alibardi L, Toni M. Localization and characterization of specific cornification proteins in avian epidermis. Cells Tissues Organs. 2004;178(4):204–15.
53. Alibardi L, Toni M. Distribution and characterization of proteins associated with cornification in the epidermis of gecko lizard. Tissue Cell. 2005;37(6):423–33.
54. Alibardi L, Toni M. Characterization of keratins and associated proteins involved in the cornification of crocodilian epidermis. Tissue Cell. 2007;39(5):311–23.

In Vitro Models of Filaggrin-Associated Diseases

8

Sarah Küchler

Contents

S. Küchler, PhD
Department of Pharmaceutical Sciences, Pharmacology/Toxicology, Freie Universität Berlin, Königin-Luise-Str. 2-4, Berlin 14195, Germany
e-mail: sarah.kuechler@fu-berlin.de

8.1 Introduction

Filaggrin gene (*FLG*) loss-of-function mutations contribute to several dermatological disorders such as ichthyosis vulgaris (IV) and atopic dermatitis (AD) [1, 2]. IV is the most prevalent cornification disorder and is often associated with AD [3, 4]. In both, IV and AD mutations in the *FLG* were discovered that are localized in the epidermal differentiation complex on chromosome 1q21 [1, 3]. Mutations in the *FLG* are the most widely occurring genetic risk factors for the development of AD known to date and have been identified in up to 20–40 % of the patients [5, 6]. Such mutations provoke an intercellular barrier abnormality that reduces the skin's inflammatory threshold to topical irritants and antigens, triggering inflammatory processes [7–9]. The most common mutations occurring in approximately 9 % of the European population are R501X and 2282del4 [1, 10]. Various other common and rare mutations are also known but with lower prevalence [11, 12], suggesting that the real mutation frequency is even higher. Although *FLG* mutations are the most common mutations in prominent skin disorders such as IV and AD, the overall impact of the filaggrin deficiency on the pathogenesis is not yet completely understood yet.

To gain more insight and to understand the fundamental processes resulting from the mutations, filaggrin-deficient skin models can be useful research tools. In most animal studies, the flaky tail (*ft*) mouse serves as a model system for filaggrin-deficient skin diseases because

J.P. Thyssen, H. Maibach (eds.), *Filaggrin*,
DOI 10.1007/978-3-642-54379-1_8, © Springer-Verlag Berlin Heidelberg 2014

these mice exhibit a barrier dysfunction similar to AD patients [13]. *ft* mice are characterized by a dry and flaky skin, an impaired barrier function, and higher susceptibility to skin irritation caused by a pro-filaggrin, which cannot be further processed to filaggrin monomers and fails to form keratohyalin granules [7, 8, 14]. Considering the findings of Harding and Scott in 1983, filaggrin data obtained from animal studies should be interpreted cautiously due to high species-related variability in the behavior of filaggrin [15]. Additionally, more evidence emerges that animal models do not provide reliable and predictive data for humans, especially when the immune system is involved, as it is the case in AD [16]. To overcome this obstacle, reconstructed skin models emerged as a suitable tool [17]. These skin models are composed of a collagen-based dermal equivalent containing human fibroblasts and a viable epidermis with a stratum corneum that is generated by adding keratinocytes on top of the dermal equivalent and lifting the construct to the air-liquid interface to induce the skin differentiation [18, 19]. In addition to full-thickness skin models, epidermal skin models without a dermal equivalent also exist [17, 20], but they are of minor relevance for fundamental studies on skin physiology. In reconstructed skin models, the gene(s) of interest can be knocked down specifically in order to study the physiological function of the gene product in the tissue [18, 21]. The major advantage of the skin models is that only the gene of interest can be modulated, making it possible to study the role of the single gene, while the impact of confounding factors such as cytokines, which alter the skin barrier function, too, can be excluded. This is of uttermost importance, especially for studying the role of filaggrin in the pathogenesis of AD. Although it is known that filaggrin mutations are the most predisposing factor for the development of AD, the mechanism leading from the genetic defect to clinical manifestation and subclinical consequences are not yet fully understood [1, 2].

This chapter aims to highlight the research efforts studying the role of filaggrin for the skin physiology and the pathogenesis of IV and AD by the means of reconstructed skin models. As this type of research is still at its infancy, only a limited number of publications are currently available.

8.2 Filaggrin-Deficient Reconstructed Skin Models

Filaggrin-deficient skin models are based on a *FLG* knockdown using RNA interference. To our best knowledge, no other approach resulting in filaggrin deficiency in vitro has been published so far. Prior to the cultivation of the reconstructed skin, the human (primary) keratinocytes are transfected with the siRNA being specific for the target gene, in this case for *FLG*. After 24 h, the transfected cells are seeded onto the dermal equivalent [18, 21, 22]. Subsequently, the constructs are lifted to the air-liquid interface, where they are further cultivated for several days (Fig. 8.1) [18, 21]. Depending on the protocol, the submerged cultivation time prior to the airlift may vary.

The length of the cultivation period at the air-liquid interface has major impact on the outcome and should be chosen carefully depending on the aim of the study. For example, in constructs that have been only cultivated for 7 days, no histological differences have been observed between healthy and filaggrin-deficient skin models [23]. In contrast, our group observed major differences in terms of skin maturation and differentiation between healthy and *FLG* knockdown constructs as shown by histological evaluations [22]. Aside from histological differences, the skin barrier function also seems to change over time in filaggrin-deficient constructs. Mildner et al. described increased skin absorption of the dye Lucifer yellow in filaggrin-deficient skin after 7 days of cultivation [23]. Interestingly, in our group we did only see differences in the skin absorption of the OECD standard compounds caffeine and testosterone after 14 days [22], whereas no differences were detected after 7 days [24]. Obviously, the cultivation period at the air-liquid interface has major impact on the skin barrier function as filaggrin-associated influences become particularly detectable after longer cultivation times and maturation of the reconstructed skin.

Fig. 8.1 Common setup for in vitro cultivation of skin constructs in six-well plates

One important aspect that needs to be considered when performing a gene knockdown is the knockdown efficiency. Using commercially available siRNA, sound knockdown rates varying between 80 and 90 % can be achieved easily [18, 21, 22]. Most importantly, the transfection has to remain stable until the end of the experiment, which should be monitored frequently, especially while establishing the model.

Mildner et al. were the first to publish data on *FLG* knockdown models, and they presented very interesting results. As expected, keratohyalin granules were missing in filaggrin-deficient skin models and abnormalities in the lamellar body formation have been observed, too [23]. This is well in accordance with clinical observations in IV patients [3, 25] and characteristics in the *ft* mouse model [14]. Interestingly, the keratin aggregation in the filaggrin-deficient skin models was not disturbed, which is in contrast to previous assumptions [23]. Earlier studies suggested that one major function of filaggrin is the aggregation of keratin intermediates into aligned keratin bundles [26]. In fact, no keratin aggregation defect was found in the filaggrin-deficient skin models [23]. Furthermore, although the lamellar body formation was disturbed in the knockdown models, shown by smaller structures and the lack of the typical lamellae, no alterations in the lipid profile were detected, which is in contrast to many clinical reports describing significant differences of the lipid profile and particularly the ceramides between *FLG* mutation carriers and unaffected individuals [27–31]. Just recently, we also reported major changes in the stratum corneum lipid organization and composition of filaggrin deficient skin constructs. We detected significantly increased levels of free fatty acids in the knock down models correlating well with increased activity of the secretory phospholipase (sPLA) IIA [24]. Additionally, we also saw no significant differences in the ceramide profile. Furthermore, we investigated the skin surface pH of the skin models and detected no difference between normal and filaggrin knock down models giving direct evidence that filaggrin is not crucial for the maintenance of the skin surface pH [24]. Most interestingly, we measured values of pH 5.5 which is the physiological pH of human skin. This finding clearly underlines the quality of in skin constructs. Additionally, for the first time we discovered a feedback mechanism maintaining the skin surface pH when a lack of filaggrin occurs [24]. However, contradictory reports also exist in which no differences were detected in the lipid and ceramide profile of AD patients [32]. Obviously, this aspect requires further clarification.

The interdependence of skin lipid composition and organization and its importance for the skin absorption of xenobiotics is well known. The fact that changes in the skin lipid organization correlate with increased skin permeability was already described more than 20 years ago [33]. Different studies demonstrate that also in AD patients the skin lipid organization is disturbed, suggesting differences in the skin absorption of xenobiotics. Changes of the lamellar lipid organization in AD patients were observed [34], as well as an increased hexagonal lipid organization in both AD and IV patients [35, 36]. However, it has to be kept in mind that these studies did not correlate the data with the presence of *FLG* mutations. In contrast, *FLG* mutation carriers have been excluded [34], meaning that studies on the correlation of *FLG* mutations and the skin lipids organization are still required. Although the exact interdependence of filaggrin and the skin barrier function is still ambiguous, evidence emerges that filaggrin deficiency does impair the skin penetration of xenobiotics. The perfusion of lanthanum in healthy and *FLG* knockdown skin has been investigated just recently, showing a lanthanum penetration into and across the stratum corneum exclusively in the knockdown models via the paracellular pathway, whereas in the control models any lanthanum absorption was missing [4]. This observation was verified by cultures of patient-derived samples and in vivo data, clearly demonstrating that the paracellular permeability abnormality that is observed in IV patients is specifically due to the filaggrin deficiency [4]. This study is of particular interest as in vivo and in vitro data were compared directly showing a significant role for filaggrin.

For filaggrin-deficient skin, not only altered skin permeability is discussed but consequentially also an increased susceptibility towards allergens [37]. AD patients showing higher susceptibility to allergens have been reported repeatedly, substantiating the paradigm shift from the historical "inside-outside" view towards the current "outside-inside-outside" view [38–41]. A filaggrin-induced barrier deficiency resulting in enhanced percutaneous allergen priming and increased susceptibility towards skin sensitization has also been observed in *ft* mice [7]. Similar results were obtained in our group using filaggrin-deficient skin constructs: We detected significantly increased levels of the pro-inflammatory cytokines IL-6 and IL-8 as well as a reduced cell viability following the application of the skin irritant sodium dodecyl sulfate [22]. Interestingly, already the application of the control substance, phosphate buffered saline, resulted in enhanced release of the pro-inflammatory cytokines IL-6 and IL-8, which is well in line with the clinical fact that even water can cause skin irritation in AD patients [42].

8.3 Reconstructed Skin Models to Study the Function of Filaggrin

As mentioned before, only a limited number of publications are available describing FLG knockdown skin models. Hopefully, this number will increase over time as fundamental research is still necessary to elucidate the role of this important structural protein for skin physiology. Aside from the skin constructs discussed above, several publications report the use of in vitro skin models for fundamental research associated with filaggrin.

A very interesting publication by Zhang et al. describes the possibility to assess the efficiency of filaggrin modulators by the means of reconstructed skin models [43]. Here, clofibrate and docosahexaenoic acid (DHA), both peroxisome proliferator activator receptor (PPAR) modulators, have been tested in reconstructed skin models for their role in epidermal maturation, differentiation, and growth response. The presence of PPAR receptors in human keratinocyte nuclei of reconstructed skin models had been described previously [44, 45]. When treating skin models with a PPAR modulator, an increased expression of enzymes associated with the lipid metabolism and enhanced lipid deposition were observed [44]. These observations have been confirmed later when using clofibrate [45]. When treating reconstructed skin models with the PPAR modulators clofibrate and DHA, a significant stimulation of the keratinocyte differentiation, especially of the terminal differentiation, has been detected indicated by an increased expression of late differentiation markers such as filaggrin and the ABCA12 transporter, which is involved in the transport of lipids into the lamellar bodies [43]. These observations have not been possible before in animal studies or in 2D cell cultures, highlighting once more the importance of reconstructed skin models and their benefit for fundamental dermatological research. Similar positive effects of PPAR agonists have been found before in a murine hyperproliferative model, demonstrating that PPAR modulators are capable of normalizing the epidermal homeostasis by inhibiting hyperproliferative processes and stimulating keratinocyte differentiation [46]. Hatano et al. verified these results and demonstrated that PPAR ligands are equally effective as glucocorticoids in the treatment of mild and moderate oxazolone-induced AD [47]. In conclusion, in the future, reconstructed skin models may have the potential to serve as test matrices for the preclinical assessment of drug actions in vitro.

Aside from its central role in the pathogenesis of AD, clinical investigations revealed that in mutation carriers the FLG expression is lower in acute lesions compared to uninvolved skin, suggesting that other factors also modulate the FLG expression in vivo [48]. For example, the cytokine IL-31 has been identified as a potential confounding factor [49]. Frequently, increased IL-31 gene expression and serum levels are detected in AD patients correlating with the severity of the phenotype, but without knowing the exact role of the cytokine in AD [49, 50]. Using reconstructed skin models, it was observed that IL-31 considerably interferes with the epidermal differentiation process causing a reduced epidermal thickness, a disturbed epidermal organization, and an altered alignment of the stratum basale. In the course of this study, filaggrin was identified as a relevant downstream target of IL-31, suggesting that the cytokine can further aggravate the clinical condition of AD patients [49].

The impact of exogenous and environmental factors on the skin homeostasis can also be studied using in vitro skin constructs. For example, Yang et al. in 2012 investigated the effect of ionizing radiation on the phosphorylation pattern of the skin. Even very low doses of ionizing radiation resulted in an altered protein phosphorylation pattern including the filaggrin precursor pro-filaggrin and, thus, in a potentially altered protein function [51]. Radiation not only with ionizing but also with UVB light seems to influence the FLG expression: Irradiating reconstructed skin models with UVB light (10–100 mJ/cm^2) resulted in a significantly increased FLG expression without affecting other structural proteins [52]. The importance of filaggrin for UV protection was highlighted before using a *FLG* knockdown model [23]. Filaggrin is the main source of the metabolite urocanic acid, which is involved in the UV protection by acting as a UV-absorbing substance [53]. Irradiating *FLG* knockdown models resulted in an increased formation of cyclobutane pyrimidine dimers and a caspase-3 activation indicating DNA damage and the induction of apoptosis [23]. However, whether filaggrin is relevant for UV protection in vivo still needs to be elucidated, as only a few clinical reports describe an increased photosensitivity in AD or IV patients [54, 55].

In a very recent and highly interesting publication, the role of atopic fibroblasts for the skin homeostasis has been investigated by reconstructing in vitro skin models with fibroblasts and/or keratinocytes obtained from punch biopsies of AD patients [56]. The results of this study were not correlated to FLG mutations in the patients, however. Nevertheless, it becomes clear that fibroblasts play a crucial role in the pathogenesis of AD by modulating factors like the FLG expression in keratinocytes. In constructs containing atopic fibroblasts and healthy keratinocytes, a significantly reduced expression of filaggrin and other structural proteins such as loricrin was detected, accompanied by an increased epidermal thickening. The fibroblast-derived leukemia inhibitory factor (LIF) was identified as a potential key player, as LIF levels were significantly reduced in atopic fibroblasts, a fact that correlated well with the reduced expression of terminal differentiation markers such as filaggrin or involucrin. This study once more shows that there are still many unknown factors contributing to the pathogenesis of AD and IV.

Conclusion

Reconstructed skin models gained more interest during the last decade in the quest for simple and cost-effective test systems to study drug actions and fundamental processes of the skin physiology and the pathogenesis of skin diseases. Filaggrin was identified as one of the key players in the pathogenesis of AD and IV, although the exact mechanism leading from the mutations to inflammatory skin diseases or barrier deficiency are not yet fully understood. Reconstructed skin models in general, and FLG knockdown models in particular, have a huge potential to gain a more detailed understanding about basic processes in the skin. The major advantage of reconstructed skin models is that the function of a single gene can be investigated, while other confounding factors such as immunological reactions can be excluded. In vitro skin models are useful tools and can contribute to a more fundamental and detailed understanding of skin homeostasis in the future.

References

1. Palmer CNA, Irvine AD, Terron-Kwiatkowski A, Zhao YW, Liao HH, Lee SP, et al. Common loss-of-function variants of the epidermal barrier protein filaggrin are a major predisposing factor for atopic dermatitis. Nat Genet. 2006;38(4):441–6.
2. Brown SJ, McLean WH. One remarkable molecule: filaggrin. J Invest Dermatol. 2012;132(3 Pt 2):751–62.
3. Smith FJ, Irvine AD, Terron-Kwiatkowski A, Sandilands A, Campbell LE, Zhao Y, et al. Loss-of-function mutations in the gene encoding filaggrin cause ichthyosis vulgaris. Nat Genet. 2006;38(3):337–42.
4. Gruber R, Elias PM, Crumrine D, Lin TK, Brandner JM, Hachem JP, et al. Filaggrin genotype in ichthyosis vulgaris predicts abnormalities in epidermal structure and function. Am J Pathol. 2011;178(5):2252–63.
5. Jung T, Stingl G. Atopic dermatitis: therapeutic concepts evolving from new pathophysiologic insights. J Allergy Clin Immunol. 2008;122(6):1074–81.
6. O'Regan GM, Sandilands A, McLean WH, Irvine AD. Filaggrin in atopic dermatitis. J Allergy Clin Immunol. 2009;124(3 Suppl 2):R2–6.
7. Fallon PG, Sasaki T, Sandilands A, Campbell LE, Saunders SP, Mangan NE, et al. A homozygous frameshift mutation in the mouse Flg gene facilitates enhanced percutaneous allergen priming. Nat Genet. 2009;41(5):602–8.
8. Scharschmidt TC, Man MQ, Hatano Y, Crumrine D, Gunathilake R, Sundberg JP, et al. Filaggrin deficiency confers a paracellular barrier abnormality that reduces inflammatory thresholds to irritants and haptens. J Allergy Clin Immunol. 2009;124(3):496–506, e1–6.
9. Elias PM. Therapeutic implications of a barrier-based pathogenesis of atopic dermatitis. Ann Dermatol. 2010;22(3):245–54.
10. McGrath JA, Uitto J. The filaggrin story: novel insights into skin-barrier function and disease. Trends Mol Med. 2008;14(1):20–7.
11. Sandilands A, Sutherland C, Irvine AD, McLean WH. Filaggrin in the frontline: role in skin barrier function and disease. J Cell Sci. 2009;122(Pt 9):1285–94.
12. Sandilands A, Terron-Kwiatkowski A, Hull PR, O'Regan GM, Clayton TH, Watson RM, et al. Comprehensive analysis of the gene encoding filaggrin uncovers prevalent and rare mutations in ichthyosis vulgaris and atopic eczema. Nat Genet. 2007;39(5):650–4.
13. Moniaga CS, Egawa G, Kawasaki H, Hara-Chikuma M, Honda T, Tanizaki H, et al. Flaky tail mouse denotes human atopic dermatitis in the steady state and by topical application with Dermatophagoides pteronyssinus extract. Am J Pathol. 2010;176(5):2385–93.
14. Presland RB, Boggess D, Lewis SP, Hull C, Fleckman P, Sundberg JP. Loss of normal pro-filaggrin and filaggrin in flaky tail (ft/ft) mice: an animal model for the filaggrin-deficient skin disease ichthyosis vulgaris. J Invest Dermatol. 2000;115(6):1072–81.
15. Harding CR, Scott IR. Histidine-rich proteins (filaggrins): structural and functional heterogeneity during epidermal differentiation. J Mol Biol. 1983;170(3):651–73.
16. Seok J, Warren HS, Cuenca AG, Mindrinos MN, Baker HV, Xu W, et al. Genomic responses in mouse models poorly mimic human inflammatory diseases. Proc Natl Acad Sci U S A. 2013;110(9):3507–12.
17. Poumay Y, Coquette A. Modelling the human epidermis in vitro: tools for basic and applied research. Arch Dermatol Res. 2007;298(8):361–9.
18. Eckl KM, Alef T, Torres S, Hennies HC. Full-thickness human skin models for congenital ichthyosis and related keratinization disorders. J Invest Dermatol. 2011;131(9):1938–42.
19. Paquet C, Larouche D, Bisson F, Proulx S, Simard-Bisson C, Gaudreault M, et al. Tissue engineering of skin and cornea: development of new models for in vitro studies. Ann N Y Acad Sci. 2010;1197:166–77.
20. Auxenfans C, Fradette J, Lequeux C, Germain L, Kinikoglu B, Bechetoille N, et al. Evolution of three dimensional skin equivalent models reconstructed in vitro by tissue engineering. Eur J Dermatol. 2009; 19(2):107–13.
21. Mildner M, Ballaun C, Stichenwirth M, Bauer R, Gmeiner R, Buchberger M, et al. Gene silencing in a human organotypic skin model. Biochem Biophys Res Commun. 2006;348(1):76–82.
22. Küchler S, Henkes D, Eckl KM, Ackermann K, Plendl J, Korting HC, et al. Hallmarks of atopic skin mimicked in vitro by means of a skin disease model based on FLG knockdown. Altern Lab Anim. 2011; 39(5):471–80.
23. Mildner M, Jin J, Eckhart L, Kezic S, Gruber F, Barresi C, et al. Knockdown of filaggrin impairs diffusion barrier function and increases UV sensitivity in a human skin model. J Invest Dermatol. 2011; 130(9):2286–94.
24. Vávrová K, Henkes D, Strüver K, Sochorová M, Skolová B, Witting MY, et al. Filaggrin deficiency leads to impaired lipid profile and altered acidification pathways in a 3D skin construct. J Invest Dermatol. 2014;134(3):746–53. doi:10.1038/jid.2013.402.
25. Fleckman P, Brumbaugh S. Absence of the granular layer and keratohyalin define a morphologically distinct subset of individuals with ichthyosis vulgaris. Exp Dermatol. 2002;11(4):327–36.

26. Steinert PM, Cantieri JS, Teller DC, Lonsdale-Eccles JD, Dale BA. Characterization of a class of cationic proteins that specifically interact with intermediate filaments. Proc Natl Acad Sci U S A. 1981;78(7):4097–101.
27. Jungersted JM, Scheer H, Mempel M, Baurecht H, Cifuentes L, Hogh JK, et al. Stratum corneum lipids, skin barrier function and filaggrin mutations in patients with atopic eczema. Allergy. 2010;65(7):911–8.
28. Angelova-Fischer I, Mannheimer AC, Hinder A, Ruether A, Franke A, Neubert RH, et al. Distinct barrier integrity phenotypes in filaggrin-related atopic eczema following sequential tape stripping and lipid profiling. Exp Dermatol. 2011;20(4):351–6.
29. Di Nardo A, Wertz P, Giannetti A, Seidenari S. Ceramide and cholesterol composition of the skin of patients with atopic dermatitis. Acta Derm Venereol. 1998;78(1):27–30.
30. Imokawa G, Abe A, Jin K, Higaki Y, Kawashima M, Hidano A. Decreased level of ceramides in stratum corneum of atopic dermatitis: an etiologic factor in atopic dry skin? J Invest Dermatol. 1991;96(4):523–6.
31. Janssens M, van Smeden J, Gooris GS, Bras W, Portale G, Caspers PJ, et al. Increase in short-chain ceramides correlates with an altered lipid organization and decreased barrier function in atopic eczema patients. J Lipid Res. 2012;53(12):2755–66.
32. Farwanah H, Raith K, Neubert RH, Wohlrab J. Ceramide profiles of the uninvolved skin in atopic dermatitis and psoriasis are comparable to those of healthy skin. Arch Dermatol Res. 2005;296(11):514–21.
33. Potts RO, Francoeur ML. Lipid biophysics of water loss through the skin. Proc Natl Acad Sci U S A. 1990;87(10):3871–3.
34. Janssens M, van Smeden J, Gooris GS, Bras W, Portale G, Caspers PJ, et al. Lamellar lipid organization and ceramide composition in the stratum corneum of patients with atopic eczema. J Invest Dermatol. 2011;131(10):2136–8.
35. Janssens M, Mulder AA, van Smeden J, Pilgram GS, Wolterbeek R, Lavrijsen AP, et al. Electron diffraction study of lipids in non-lesional stratum corneum of atopic eczema patients. Biochim Biophys Acta. 2013;1828(8):1814–21.
36. Pilgram GS, Vissers DC, van der Meulen H, Pavel S, Lavrijsen SP, Bouwstra JA, et al. Aberrant lipid organization in stratum corneum of patients with atopic dermatitis and lamellar ichthyosis. J Invest Dermatol. 2001;117(3):710–7.
37. Hudson TJ. Skin barrier function and allergic risk. Nat Genet. 2006;38(4):399–400.
38. Elias PM, Hatano Y, Williams ML. Basis for the barrier abnormality in atopic dermatitis: outside-inside-outside pathogenic mechanisms. J Allergy Clin Immunol. 2008;121(6):1337–43.
39. Elias PM, Steinhoff M. "Outside-to-inside" (and now back to "outside") pathogenic mechanisms in atopic dermatitis. J Invest Dermatol. 2008;128(5):1067–70.
40. Osawa R, Akiyama M, Shimizu H. Filaggrin gene defects and the risk of developing allergic disorders. Allergol Int. 2011;60(1):1–9.
41. Spergel JM. From atopic dermatitis to asthma: the atopic march. Ann Allergy Asthma Immunol. 2010;105(2):99–106; quiz 7–9, 17.
42. Tsai TF, Maibach HI. How irritant is water? An overview. Contact Dermatitis. 1999;41(6):311–4.
43. Zhang C, Gurevich I, Aneskievich BJ. Organotypic modeling of human keratinocyte response to peroxisome proliferators. Cells Tissues Organs. 2012;196(5):431–41.
44. Rivier M, Castiel I, Safonova I, Ailhaud G, Michel S. Peroxisome proliferator-activated receptor-alpha enhances lipid metabolism in a skin equivalent model. J Invest Dermatol. 2000;114(4):681–7.
45. Batheja P, Song Y, Wertz P, Michniak-Kohn B. Effects of growth conditions on the barrier properties of a human skin equivalent. Pharm Res. 2009;26(7):1689–700.
46. Demerjian M, Man MQ, Choi EH, Brown BE, Crumrine D, Chang S, et al. Topical treatment with thiazolidinediones, activators of peroxisome proliferator-activated receptor-gamma, normalizes epidermal homeostasis in a murine hyperproliferative disease model. Exp Dermatol. 2006;15(3):154–60.
47. Hatano Y, Elias PM, Crumrine D, Feingold KR, Katagiri K, Fujiwara S. Efficacy of combined peroxisome proliferator-activated receptor-alpha ligand and glucocorticoid therapy in a murine model of atopic dermatitis. J Invest Dermatol. 2011;131(9):1845–52.
48. Howell MD, Kim BE, Gao P, Grant AV, Boguniewicz M, Debenedetto A, et al. Cytokine modulation of atopic dermatitis filaggrin skin expression. J Allergy Clin Immunol. 2007;120(1):150–5.
49. Cornelissen C, Luscher-Firzlaff J, Baron JM, Luscher B. Signaling by IL-31 and functional consequences. Eur J Cell Biol. 2011;91(6–7):552–66.
50. Raap U, Wichmann K, Bruder M, Stander S, Wedi B, Kapp A, et al. Correlation of IL-31 serum levels with severity of atopic dermatitis. J Allergy Clin Immunol. 2008;122(2):421–3.
51. Yang F, Waters KM, Webb-Robertson BJ, Sowa MB, von Neubeck C, Aldrich JT, et al. Quantitative phosphoproteomics identifies filaggrin and other targets of ionizing radiation in a human skin model. Exp Dermatol. 2012;21(5):352–7.
52. Torii K, Nakamura M, Morita A. NB-UVB irradiation increases filaggrin expression in a three-dimensional human skin model. J Dermatol Sci. 2013;70(2):146–7.
53. Zenisek A, Kral JA, Hais IM. Sun-screening effect of urocanic acid. Biochim Biophys Acta. 1955;18(4):589–91.
54. Tajima T, Ibe M, Matsushita T, Kamide R. A variety of skin responses to ultraviolet irradiation in patients with atopic dermatitis. J Dermatol Sci. 1998;17(2):101–7.
55. Murphy GM. Diseases associated with photosensitivity. J Photochem Photobiol B. 2001;64(2–3):93–8.
56. Berroth A, Kuhnl J, Kurschat N, Schwarz A, Stab F, Schwarz T, et al. Role of fibroblasts in the pathogenesis of atopic dermatitis. J Allergy Clin Immunol. 2013;131:1547–54.

Mapping Posttranslational Regulation of Filaggrin Using Phosphoproteomics

9

David L. Stenoien

Contents

D.L. Stenoien, PhD
Biological Sciences Division, Pacific Northwest National Laboratory, 790 6th St, Richland, WA 99352, USA
e-mail: david.stenoien@pnnl.gov

9.1 Phosphoproteomics Overview

Biological systems are exceedingly more complex than defined by the genome due to differential protein expression and degradation, altered mRNA splicing, and the presence of multiple PTMs including phosphorylation that occur in different cellular contexts. Scientists have made great strides in identifying, and in many cases quantifying, genes, proteins, protein splice variants, and PTMs present in different biological systems before and after various stimuli or over time. However, the technologies used to map these molecular alterations are also demonstrating how extremely complex cell signaling networks are and that many proteins can be regulated by multiple mechanisms to create various biological outcomes. Moreover, it is increasingly evident that proteins function within complex signaling networks, so changes in upstream regulators, downstream effectors, and binding partners can all influence protein behavior within specific cellular environments.

Phosphorylation is an important PTM that plays a critical role in multiple cell signaling pathways including those that potentially regulate filaggrin functions [1–3]. Studies aimed at analyzing signaling pathways require methods that can specifically detect, identify, and quantify phosphoproteins. While traditional methods (e.g., immunohistochemistry) typically allow for characterization of one phosphoprotein (often only one phosphorylation site) at a time, recent advancements in liquid chromatography (LC)

J.P. Thyssen, H. Maibach (eds.), *Filaggrin*,
DOI 10.1007/978-3-642-54379-1_9, © Springer-Verlag Berlin Heidelberg (outside the USA) 2014

and mass spectrometry (MS) technology now enable proteome-wide study of phosphorylation: phosphoproteomics [4–8]. Based on data obtained through global phosphoproteomics and curated in a number of phosphorylation site databases such as PhosphoSite (www.phosphosite.org), Phospho.ELM (www.phospho.elm.eu.org), and Phosida (www.phosida.com), it is estimated that there are at least 200,000 potential phosphorylation sites in mammalian cells. While this fundamentally important PTM typically occurs on serine, threonine, or tyrosine residues, at least six other amino acids can also be phosphorylated [9].

The basic setup for a typical phosphoproteomic analysis is shown in Fig. 9.1. Proteins isolated from cells, tissues, or whole organisms are digested with trypsin to generate mixtures of phosphorylated and non-phosphorylated peptides. To increase sensitivity for detection of low abundant phosphopeptides, a phosphopeptide enrichment step is used prior to separation of peptides by liquid chromatography (LC) and qualitative and quantitative analysis by mass spectrometry (MS). Software programs are then used to identify the parent protein and assign locations of phosphorylation sites. A number of bioinformatics tools can then be employed to identify signal transduction networks and identify kinases responsible for altered phosphorylation in different physiological settings. Specific phosphorylation sites can then be validated using phosphorylation-specific antibodies, and the functional significance of specific phosphorylation sites can be validated through site-directed mutagenesis. A brief overview of these steps will be presented in the following sections, followed by the current state of phosphoproteomics involving filaggrin.

9.2 Phosphopeptide Enrichment and Separation Strategies

Proteins can be isolated from a number of sources, including cell lines, tissues, and whole organisms. Prefractionation of proteins using subcellular fractionation, immunoprecipitation, upstream LC, or other methods can be used to

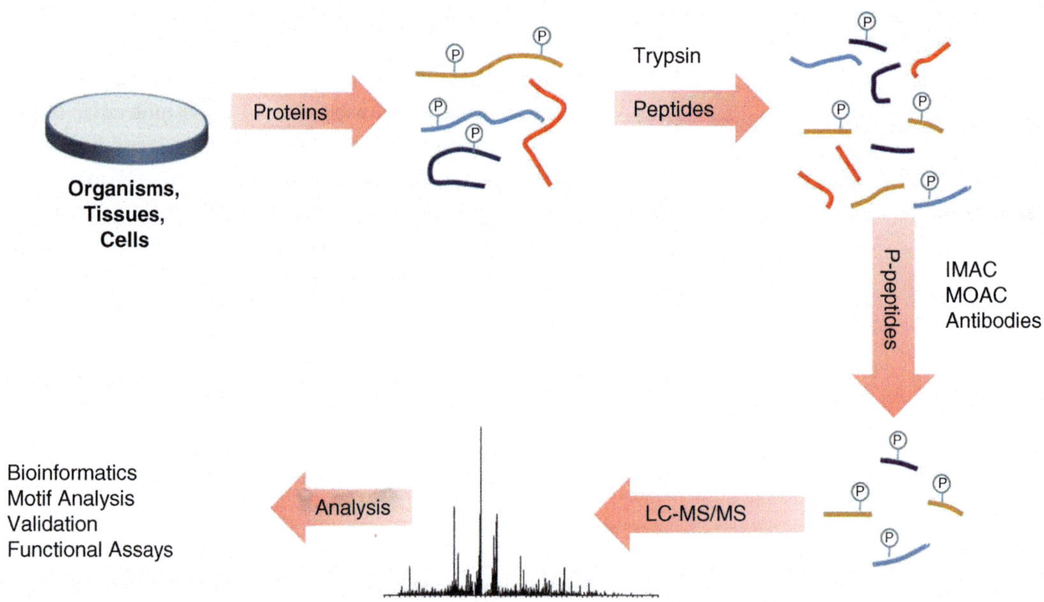

Fig. 9.1 Basic setup for a phosphoproteomic experiment. Proteins isolated from cells, tissues, or organisms are digested with a protease such as trypsin to generate a mixture of phosphorylated and non-phosphorylated peptides. Phosphorylated peptides are enriched and subjected to LC-MS/MS to identify phosphorylation sites, which can then be used in bioinformatics analyses, identification of activated kinases, and assays to assess functions of specific phosphorylation sites

analyze phosphorylation sites in specific cellular compartments and protein complexes or to improve the dynamic range of detection. A highly abundant protein such as filaggrin can be readily detected in a skin tissue lysate, so this could serve as a feasible starting point. However, samples containing more purified filaggrin should provide better sequence coverage and mapping of phosphorylation sites. The isolated proteins are then typically digested with proteases such as trypsin to generate peptides suitable in size for analysis by MS. Since most phosphorylation sites are present at substoichiometric levels and many are present on relatively low abundant proteins that are key regulators of signal transduction pathways, some type of phosphopeptide enrichment strategy is usually employed. Several enrichment techniques have been developed to purify phosphopeptides from complex tryptic digests of proteins including anti-phosphorylation site-specific antibodies [10, 11], immobilized metal affinity chromatography (IMAC) [12, 13], and metal oxide affinity chromatography (MOAC) including titanium dioxide (TiO_2)-based enrichment [14]. Antibody enrichment strategies have been used very successfully to pull out phosphotyrosine-containing peptides using anti-phosphotyrosine-specific antibodies. In the case of IMAC, the phosphate groups on peptides are bound to metal-ligand complexes, typically iminodiacetate and nitrilotriacetate, impurities are removed using a rinsing step, and the phosphopeptides are eluted using a competing ligand to yield a concentrated solution of purified phosphopeptides. IMAC is applicable to a wide range of phosphoproteomic analyses but non-phosphorylated peptides, particularly those containing acidic amino acids, are often present as contaminants. TiO_2-based enrichment has emerged as a popular alternative to IMAC since it appears to be more selective for phosphopeptides. The principle behind this enrichment technique is as follows: at acidic pH, TiO_2 has a positively charged surface that selectively adsorbs phosphopeptides, which are then eluted by raising the pH levels. It should be noted that each enrichment strategy can influence which phosphopeptides are ultimately identified in a particular experiment.

Once enriched, complex mixtures of phosphopeptides must be fractionated using LC methods similar to those employed by conventional proteomic analyses for non-phosphorylated peptides to improve the dynamic range of detection. Phosphopeptides have been successfully fractionated by a variety of chromatographic methods including strong cation exchange [15], anion exchange [16], reverse phase [17], and hydrophilic interaction liquid chromatography (HILIC) [18]. A comparison of the phosphopeptides identified from the same sample using different LC fractionation methods reveals that there is surprisingly little overlap in the phosphopeptides identified [19], suggesting that comprehensive global identification of phosphopeptides requires multiple analyses employing different types of phosphopeptide enrichment and fractionation methods. The sensitivity in phosphopeptide detection can also be greatly increased by performing two-dimensional LC (2D-LC) to fractionate peptides based on multiple properties (charge, hydrophobicity, size, etc.). For example, combining HILIC with reversed-phase chromatography greatly improves both peptide and phosphopeptide identification [20].

9.3 Mass Spectrometry of Phosphopeptides

Once enriched and fractionated, phosphopeptides can be analyzed by a number of different types of mass spectrometers, but analysis of phosphopeptides poses additional challenges compared to unmodified peptides. Typically, MS is used which first provides information on the mass of the phosphopeptide as well as information about its relative abundance. The intact phosphopeptide is then broken apart into small fragment ions and reanalyzed by tandem MS (MS/MS) to yield information about the amino acid sequence and locations of the phosphorylation sites. A number of different fragmentation methods are available to analyze unmodified peptides, and these can have significant impacts for the identification of phosphorylation sites. Collision-induced dissociation (CID) in which the original peptides are physically fragmented by acceleration and collision with neutral ions in the mass spectrometer is typically used for the analysis of unmodified peptides [21]. The masses of these smaller fragments can then be used to determine the amino acid sequences of the precursor peptides. The major problem for using CID for phosphopeptide identification, however, is that the O-phosphate bond in serine- and threonine-phosphorylated peptides is labile resulting in neutral loss of the phosphate group and an inability to identify the phosphorylation site. This problem can be averted in several ways, including the use of alternate dissociation methods such as electron-transfer dissociation (ETD) and electron-capture dissociation (ECD), which preserve labile PTMs, including phosphorylation [22, 23]. Additional strategies such as MS/MS/MS and multistage activation can also be employed using CID for phosphopeptide analysis [24], and as with the enrichment and separation techniques described above, employing multiple fragmentation methods such as ETD/ECD along with CID (with multistage activation) often yields nonoverlapping information. Once the spectra generated by these different fragmentation methods are detected by MS/MS, a number of search engines, typically searching databases of known protein sequences, including Sequest, OMSSA, X!Tandem, or Protein Prospector can be applied to identify the phosphopeptide [25]; however, localization of the phosphorylation site is often nontrivial, especially in cases where the peptides contain multiple potential phosphorylation sites. Additional software routines also must be employed to provide probability scores that determine the degree of confidence in phosphosite localization [26]. Preferably, identified phosphorylation sites should be validated using phosphosite-specific antibodies when available; however, this is not always feasible for global phosphoproteomic studies identifying hundreds to thousands of novel phosphorylation sites.

9.4 Quantitative Analysis of Phosphopeptides

To assess differences in phosphorylation in different biological samples, quantitative methods can be employed in phosphoproteomic analyses. A number of factors, including the sample source (cells vs. tissues), the accuracy required, and the number of samples to be analyzed, all influence the method selection. Several so-called label-free methods, including spectral counting and measurements of MS peak intensities, can be used to generate semiquantitative information on the relative abundance of phosphopeptides in different biological samples [27] and are particularly useful for measuring large-fold changes [28]. Multiple labeling methods that are widely used for peptide level analyses [1, 29, 30] can be used to obtain more reliable quantitative information on the changes in phosphopeptide abundance in two or more samples. In one popular approach, stable isotope labeling of amino acids in culture (SILAC) [30–32] labels proteins by growing cells in media containing light (e.g., ^{14}N or $^{12C6\text{-}Arg}$) or heavy (e.g., ^{15}N or $^{13}C_6$-Arg) amino acids. Cells from two biologically distinct samples are then combined a 1:1 ratio and carried through the phosphoproteomic workflow. The differentially labeled peptides (heavy and light) with the same amino acid sequence are detected in the MS spectra, and their ratio can be used to calculate protein abundance. Since the two samples are combined at an early stage, technical variations that could occur at any stage in the sample processing and phosphopeptide analysis are minimized. However, a significant limitation of this method is that the label incorporation requires growth in defined tissue culture media for an extended period of time, which is not feasible for the analysis of most tissue samples. Peptides can also be labeled after trypsin digestion using the popular isobaric tag for relative and absolute quantitation (iTRAQ) method [33]. A major advantage of the iTRAQ method is the ability to analyze up to eight samples in a single run; therefore, the required number of mass spectrometry (MS) acquisitions is greatly reduced by combining multiple samples labeled with different isobaric tags at the reactive n-terminal amine of peptides and proteins. Because MS acquisitions are not required for each individual sample, more instrument time can be devoted to analyzing increased sample fractionates, which contributes to a larger dynamic range and higher sensitivity for phosphoproteome quantitation.

9.5 Analysis of Intact Phosphoproteins

High-throughput analyses of protein PTMs, predominantly focused on phosphorylation, are demonstrating how extremely complex the array of PTMs on a single protein can be. While peptide level phosphoproteomics can quantitatively identify the PTMs on peptides, it cannot accurately quantify multiple PTMs dispersed along the entire protein length. In many well-documented cases, multisite PTMs resulting from the convergence of multiple signal transduction pathways on critical proteins more accurately reflect the activity status of these proteins and the resulting cellular response than a single-site PTM. Prominent examples of these multiply modified proteins include histone proteins that have specific combinations of acetylations, methylations, and phosphorylations (the histone code) that regulate chromatin functions [34]. While multiple modifications have been found on these and many other proteins including filaggrin, we are only beginning to gain an understanding of the complex combinations of PTMs that regulate protein function within complex signaling networks. When performing peptide level or bottom-up analyses, most phosphorylation sites may be missed due to incomplete sequence coverage especially in complex mixtures. While this problem can be alleviated to some extent by focusing on the modifications present on a purified protein, many phosphopeptides may be missed due to a number of factors, including the length of the peptide and the presence of other unknown PTMs, which alter the mass.

Given the limitations of conventional (bottom-up) proteomics, there has been increasing interest in top-down proteome characterization strategies, where individual proteins are selected for analysis by MS/MS, without the need for prior enzymatic proteolysis. This approach has demonstrated 100 % protein sequence coverage and allowed for the identification of protein isoforms, proteolytic processing events, and PTMs [35–38]. A new variant, so-called middle-down, recently emerged for more effective characterization of multisite PTMs by incorporating LysC or AspN proteolysis prior to MS analysis. This step creates large peptides, due to the relative rarity of potential cleavage sites available to LysC or AspN [39]. Therefore, this approach combines some benefits from both bottom-up and top-down approaches.

Characterization of histone PTMs and their dynamics has been the most notable application of top-down proteomics to date [37, 38, 40–42]. Top-down platforms have also been applied to characterize limited subsets of proteins (~100) from microorganisms [35, 43] and humans [36] and even on proteins with molecular masses above 200 kDa [44], but with limited sensitivity and throughput. Notably, MS/MS of intact proteins enables characterization of coexisting modifications and can potentially reveal information about PTM hierarchy (e.g., which modification occurs first?). Another distinct advantage of the top-down approach is that the relative abundance of different protein isoforms can be determined directly, while it is very challenging to do so using bottom-up strategy due to the extensively observed shared peptides. The top-down platform has been successfully applied for the characterization of various protein PTMs including phosphorylation [36, 37]. However, its application is typically limited to the study of a single purified phosphoprotein [45].

9.6 Downstream Analysis of Phosphoproteomic Data

A number of bioinformatics and other tools are available to analyze phosphoproteomic datasets, and their use depends on whether one wants to perform a global analysis of the biological pathways and processes affected in different biological samples or whether one is interested in a specific protein or protein complex. Global phosphoproteomic datasets can be analyzed as with any other "omic" (transcriptomic, proteomic, metabolomic, etc.) dataset by commercially or publicly available bioinformatics tools such as MetaCore (www.thomsonreuters.com), Ingenuity Pathway Analysis (www.ingenuity.com), or DAVID (david.abcc.ncifcrf.gov). An additional type of analysis available for phosphoproteomic data is to identify potential kinases responsible for altered phosphorylation both in global datasets and on specific proteins of interest. While far from complete, consensus phosphorylation sites have been identified for a number of different kinases, and this data can be used to ascertain which kinases are activated or inhibited in different biological samples or which kinase is responsible for a specific phosphorylation event. Several kinase prediction tools have been developed for this purpose and include ScanSite (scanscite.mit.edu) and NetPhos (www.cbs.dtu.dk/services/NetPhos).

9.7 Validation and Functional Analysis

Optimally, the existence of phosphorylation sites and differential regulation observed in different biological conditions should be validated with phosphorylation-specific antibodies. While hundreds of phosphorylation-specific antibodies are now commercially available, this represents only a small percentage of identified phosphorylation sites. In the case of filaggrin, for example, there are at least 96 putative phosphorylation sites reported, but no phosphorylation-specific antibodies are commercially available or have been reported in the literature. Custom phosphorylation-specific antibodies can be made in the laboratory (or by a number of antibody-generating companies), but extreme care must be used in selecting specific phosphorylation sites to analyze due to the time and costs associated with this endeavor. Functions of identified phosphorylation sites can also be analyzed using site-directed mutagenesis approaches [10] to determine effects on protein structure, stability, localization, protein complex formation, activity, etc. But again, care must be taken to prioritize the large number of identified phosphorylation sites for follow-up studies.

9.8 Phosphoproteomics and Filaggrin

Filaggrin is a protein important for maintaining skin barrier function and hydration and exists as a large 435 kD pro-filaggrin form that gets processed into filaggrin monomers that bind to and cross-link keratin fibers in the stratum corneum of skin and provide the protein degradation products that act as natural moisturizing factors [46]. Pro-filaggrin has long been known to be a heavily phosphorylated protein, making it highly insoluble, and a general decrease in phosphorylation is thought to be required prior to its proteolysis into filaggrin monomers [47]. As mentioned above, global phosphoproteomic studies have identified at least 96 filaggrin phosphorylation sites, and this likely underrepresents the total number of phosphorylation sites. Little is known about how most of these phosphorylation sites affect filaggrin functions and how they are regulated during development and differentiation and in pathophysiological conditions involving filaggrin.

Phosphoproteomic analysis of filaggrin poses some unique challenges compared to other proteins. First off, its expression is largely confined to highly differentiated tissues such as skin, and therefore, tissue culture cells, for which many of the phosphoproteomic analyses have been optimized, cannot be used. Since skin tissue grown in vitro or obtained from animal or patient samples must be used, the SILAC labeling approach described above is not feasible; however, label-free or iTRAQ approaches are available. The large size of pro-filaggrin makes mapping all or most phosphorylation sites a formidable task, and the use of a top-down approach to analyze intact pro-filaggrin would prove to be difficult, although this could be more easily performed on filaggrin monomers. One advantage of working with filaggrin is its high abundance in epidermal fractions, and an upstream purification strategy

[48] could be employed to focus on filaggrin phosphorylation sites.

While to date there have been no phosphoproteomic studies focused on filaggrin, a recent global phosphoproteomic study on a reconstituted skin model identified novel phosphorylation sites on pro-filaggrin that are differentially regulated after exposure to ionizing radiation and could impact pro-filaggrin processing [1]. The iTRAQ method was used to demonstrate that radiation affected a pro-filaggrin phosphopeptide at both a low (10 cGy) and high (2 Gy) doses of radiation (Fig. 9.2a). Alignment of this phosphopeptide indicated that it is present in multiple copies in the repetitive linker regions located between the filaggrin monomers where proteolytic processing of pro-filaggrin occurs [49]. Radiation exposure also affected processing of pro-filaggrin into filaggrin monomers (see Fig. 9.2b), suggesting these linker region phosphorylation sites could regulate pro-filaggrin processing (see Fig. 9.2c). Sequence alignment of the rat, mouse, and human linker regions demonstrate that the amino acids comprising the phosphorylation site identified here are unique to human pro-filaggrin [50] and may present a novel mechanism for humans to regulate pro-filaggrin processing in response to ionizing radiation and possibly other types of skin stress.

9.9 Future Directions

The field of phosphoproteomics has made great progress in identifying global alterations in phosphorylation in multiple biological settings. Improvements in both methods and instrumentation now make it possible to identify greater than 15,000 phosphorylation sites in a single experiment and generate quantitative data on the majority of these [51]. Several important issues, however, remain to be solved. Precise localization

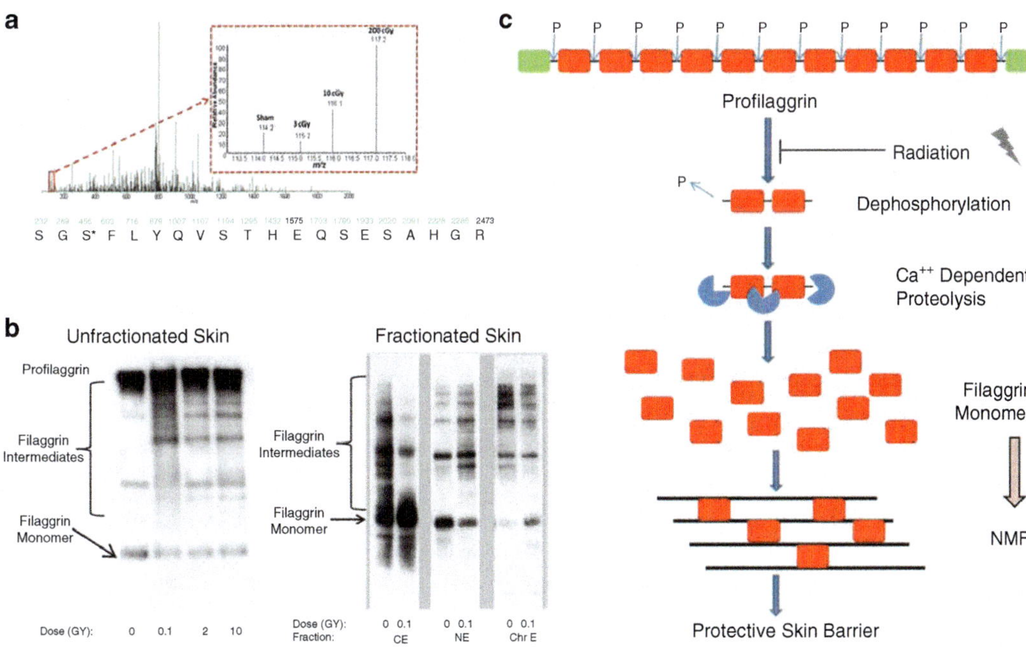

Fig. 9.2 Phosphoproteomic analysis identifies filaggrin as a target of ionizing radiation. (**a**) The spectra from an iTRAQ-labeled phosphopeptide are shown and match the sequence SGpSFLYQVSTHEQSESAHGR. The boxed area shows the relative peptide levels in sham, low-dose (3 and 10 cGy) and high-dose (200 cGy) treated samples. (**b**) Radiation also affects processing of profillagrin into filaggrin monomers, which is evident in both whole skin and subcellular skin compartments including cytoplasmic extracts (*CE*), nuclear extracts (*NE*), and chromatin extracts (*ChrE*). (**c**) From this, a model is proposed in which linker dephosphorylation, a step inhibited by radiation exposure, is required prior to calcium-dependent proteolysis into filaggrin monomers

of phosphorylation sites within a phosphopeptide remains a key issue, and assignment of a phosphate group to the wrong amino acid will impact downstream validation and functional assays relying on phosphorylation site-specific antibodies and site-directed mutagenesis. This issue is often the result of limited MS/MS fragmentation efficiencies that provide missing information on phosphorylation site localization. This can be overcome by applying multiple fragmentation methods (CID/ETD/ECD) and developing new methods, such as ultraviolet photodissociation [52], which can generate more informative MS/MS spectra. In addition, even with the ability to identify thousands of phosphopeptides in a single experiment, most phosphorylation sites are unaccounted for and key regulatory sites may be missed. Experiments focused on a purified protein of interest can provide much greater sequence coverage, and continued development of new technologies such as top-down proteomics will enable protein-wide PTM analysis. Given the number of PTMs and their complexity, validation and functional characterization of phosphorylation sites will remain a daunting challenge.

A phosphoproteomic experiment dedicated to the analysis of purified filaggrin would undoubtedly identify a much larger number of phosphorylation sites than currently reported. Quantitative comparisons of these sites in normal skin and from patients suffering from atopic dermatitis (AD) could provide insight into posttranslational mechanisms that regulate filaggrin in normal skin and during pathogenesis. Functional validation of these sites would be challenging due to filaggrin expression and function in terminally differentiated skin layers. For example, a functional assay for a filaggrin phosphorylation site may require knocking out the endogenous protein in keratinocytes, replacing it with a phosphorylation site mutant, allowing the cells to differentiate, and then performing some type of functional assay such as measuring skin barrier function. The repetitive nature of filaggrin would also further complicate these experiments since a phosphorylation site may be present in multiple copies. Another approach that has not been explored is to perform a comprehensive and quantitative phosphoproteomic analysis of normal and AD skin that could be combined with transcriptomic [53] and proteomic [54] approaches to identify affected signaling pathways. This could result in new targets for therapeutic intervention to help improve skin barrier function and/or alleviate the associated inflammation.

References

1. Yang F, Waters KM, Webb-Robertson BJ, Sowa MB, von Neubeck C, Aldrich JT, et al. Quantitative phosphoproteomics identifies filaggrin and other targets of ionizing radiation in a human skin model. Exp Dermatol. 2012;21(5):352–7. PubMed PMID: 22509832. Pubmed Central PMCID: 3387810.
2. Zanivan S, Gnad F, Wickstrom SA, Geiger T, Macek B, Cox J, et al. Solid tumor proteome and phosphoproteome analysis by high resolution mass spectrometry. J Proteome Res. 2008;7(12):5314–26. PubMed PMID: 19367708.
3. Lander ES. The new genomics: global views of biology. Science. 1996;274(5287):536–9. PubMed PMID: 8928008. Epub 1996/10/25. eng.
4. Oda Y, Nagasu T, Chait BT. Enrichment analysis of phosphorylated proteins as a tool for probing the phosphoproteome. Nat Biotechnol. 2001;19(4):379–82. PubMed PMID: 11283599. Epub 2001/04/03. eng.
5. Shi SDH, Hemling ME, Carr SA, Horn DM, Lindh I, McLafferty FW. Phosphopeptide/phosphoprotein mapping by electron capture dissociation mass spectrometry. Anal Chem. 2001;73(1):19–22. PubMed PMID: WOS:000166262500012.
6. Collins MO, Yu L, Coba MP, Husi H, Campuzano I, Blackstock WP, et al. Proteomic analysis of in vivo phosphorylated synaptic proteins. J Biol Chem. 2005;280(7):5972–82. PubMed PMID: 15572359. Epub 2004/12/02. eng.
7. Delom F, Chevet E. Phosphoprotein analysis: from proteins to proteomes. Proteome Sci. 2006;4:15. PubMed PMID: 16854217. Epub 2006/07/21. eng.
8. Collins MO, Yu L, Choudhary JS. Analysis of protein phosphorylation on a proteome-scale. Proteomics. 2007;7(16):2751–68. PubMed PMID: 17703509. Epub 2007/08/19. eng.
9. Ciesla J, Fraczyk T, Rode W. Phosphorylation of basic amino acid residues in proteins: important but easily missed. Acta Biochim Pol. 2011;58(2):137–48. PubMed PMID: 21623415.
10. Yang F, Camp 2nd DG, Gritsenko MA, Luo Q, Kelly RT, Clauss TR, et al. Identification of a novel mitotic phosphorylation motif associated with protein localization to the mitotic apparatus. J Cell Sci. 2007;120(Pt 22):4060–70. PubMed PMID: 17971412.

11. Zhang Y, Wolf-Yadlin A, Ross PL, Pappin DJ, Rush J, Lauffenburger DA, et al. Time-resolved mass spectrometry of tyrosine phosphorylation sites in the epidermal growth factor receptor signaling network reveals dynamic modules. Mol Cell Proteomics. 2005;4(9):1240–50. PubMed PMID: 15951569.
12. Porath J, Carlsson J, Olsson I, Belfrage G. Metal chelate affinity chromatography, a new approach to protein fractionation. Nature. 1975;258(5536):598–9. PubMed PMID: 1678.
13. Reinders J, Sickmann A. State-of-the-art in phosphoproteomics. Proteomics. 2005;5(16):4052–61. PubMed PMID: 16196093.
14. Dunn JD, Reid GE, Bruening ML. Techniques for phosphopeptide enrichment prior to analysis by mass spectrometry. Mass Spectrom Rev. 2010;29(1):29–54. PubMed PMID: 19263479.
15. Ballif BA, Villen J, Beausoleil SA, Schwartz D, Gygi SP. Phosphoproteomic analysis of the developing mouse brain. Mol Cell Proteomics. 2004;3(11):1093–101. PubMed PMID: 15345747.
16. Han G, Ye M, Zhou H, Jiang X, Feng S, Jiang X, et al. Large-scale phosphoproteome analysis of human liver tissue by enrichment and fractionation of phosphopeptides with strong anion exchange chromatography. Proteomics. 2008;8(7):1346–61. PubMed PMID: 18318008.
17. Ham BM, Yang F, Jayachandran H, Jaitly N, Monroe ME, Gritsenko MA, et al. The influence of sample preparation and replicate analyses on HeLa Cell phosphoproteome coverage. J Proteome Res. 2008;7(6):2215–21. PubMed PMID: 18412383. Pubmed Central PMCID: 2517255.
18. McNulty DE, Annan RS. Hydrophilic interaction chromatography reduces the complexity of the phosphoproteome and improves global phosphopeptide isolation and detection. Mol Cell Proteomics. 2008;7(5):971–80. PubMed PMID: 18212344.
19. Zarei M, Sprenger A, Metzger F, Gretzmeier C, Dengjel J. Comparison of ERLIC-TiO2, HILIC-TiO2, and SCX-TiO2 for global phosphoproteomics approaches. J Proteome Res. 2011;10(8):3474–83. PubMed PMID: 21682340.
20. Di Palma S, Boersema PJ, Heck AJ, Mohammed S. Zwitterionic hydrophilic interaction liquid chromatography (ZIC-HILIC and ZIC-cHILIC) provide high resolution separation and increase sensitivity in proteome analysis. Anal Chem. 2011;83(9):3440–7. PubMed PMID: 21443167.
21. Wells JM, McLuckey SA. Collision-induced dissociation (CID) of peptides and proteins. Methods Enzymol. 2005;402:148–85. PubMed PMID: 16401509.
22. Syka JE, Coon JJ, Schroeder MJ, Shabanowitz J, Hunt DF. Peptide and protein sequence analysis by electron transfer dissociation mass spectrometry. Proc Natl Acad Sci U S A. 2004;101(26):9528–33. PubMed PMID: 15210983. Pubmed Central PMCID: 470779.
23. Zubarev RA, Horn DM, Fridriksson EK, Kelleher NL, Kruger NA, Lewis MA, et al. Electron capture dissociation for structural characterization of multiply charged protein cations. Anal Chem. 2000;72(3):563–73. PubMed PMID: 10695143.
24. Boersema PJ, Mohammed S, Heck AJ. Phosphopeptide fragmentation and analysis by mass spectrometry. J Mass Spectrom. 2009;44(6):861–78. PubMed PMID: 19504542.
25. Bakalarski CE, Haas W, Dephoure NE, Gygi SP. The effects of mass accuracy, data acquisition speed, and search algorithm choice on peptide identification rates in phosphoproteomics. Anal Bioanal Chem. 2007;389(5):1409–19. PubMed PMID: 17874083.
26. Beausoleil SA, Villen J, Gerber SA, Rush J, Gygi SP. A probability-based approach for high-throughput protein phosphorylation analysis and site localization. Nat Biotechnol. 2006;24(10):1285–92. PubMed PMID: 16964243.
27. Yang F, Waters KM, Miller JH, Gritsenko MA, Zhao R, Du X, et al. Phosphoproteomics profiling of human skin fibroblast cells reveals pathways and proteins affected by low doses of ionizing radiation. PLoS One. 2010;5(11):e14152. PubMed PMID: 21152398. Pubmed Central PMCID: 2994767.
28. Old WM, Meyer-Arendt K, Aveline-Wolf L, Pierce KG, Mendoza A, Sevinsky JR, et al. Comparison of label-free methods for quantifying human proteins by shotgun proteomics. Mol Cell Proteomics. 2005;4(10):1487–502. PubMed PMID: 15979981.
29. Yang F, Wu S, Stenoien DL, Zhao R, Monroe ME, Gritsenko MA, et al. Combined pulsed-Q dissociation and electron transfer dissociation for identification and quantification of iTRAQ-labeled phosphopeptides. Anal Chem. 2009;81(10):4137–43. PubMed PMID: 19371082. Pubmed Central PMCID: 2749721.
30. Martinovic S, Veenstra TD, Anderson GA, Pasa-Tolic L, Smith RD. Selective incorporation of isotopically labeled amino acids for identification of intact proteins on a proteome-wide level. J Mass Spectrom. 2002;37(1):99–107. PubMed PMID: 11813317.
31. Ong SE, Blagoev B, Kratchmarova I, Kristensen DB, Steen H, Pandey A, et al. Stable isotope labeling by amino acids in cell culture, SILAC, as a simple and accurate approach to expression proteomics. Mol Cell Proteomics. 2002;1(5):376–86. PubMed PMID: 12118079.
32. Ong SE, Kratchmarova I, Mann M. Properties of 13C-substituted arginine in stable isotope labeling by amino acids in cell culture (SILAC). J Proteome Res. 2003;2(2):173–81. PubMed PMID: 12716131.
33. Ross PL, Huang YLN, Marchese JN, Williamson B, Parker K, Hattan S, et al. Multiplexed protein quantitation in Saccharomyces cerevisiae using amine-reactive isobaric tagging reagents. Mol Cell Proteomics. 2004;3(12):1154–69. PubMed PMID: WOS:000226100700002.
34. Jenuwein T, Allis CD. Translating the histone code. Science. 2001;293(5532):1074–80. PubMed PMID: 11498575. Epub 2001/08/11. eng.
35. Du Y, Parks BA, Sohn S, Kwast KE, Kelleher NL. Top-down approaches for measuring expression ratios of intact yeast proteins using Fourier transform

mass spectrometry. Anal Chem. 2006;78(3):686–94. PubMed PMID: WOS:000235195200022.
36. Roth MJ, Forbes AJ, Boyne MT, Kim YB, Robinson DE, Kelleher NL. Precise and parallel characterization of coding polymorphisms, alternative splicing, and modifications in human proteins by mass spectrometry. Mol Cell Proteomics. 2005;4(7):1002–8. PubMed PMID: WOS:000230334500013.
37. Siuti N, Kelleher NL. Decoding protein modifications using top-down mass spectrometry. Nat Methods. 2007;4:817–21. PubMed PMID: WOS:000249778200014.
38. Thomas CE, Kelleher NL, Mizzen CA. Mass spectrometric characterization of human histone H3: a bird's eye view. J Proteome Res. 2006;5(2):240–7. PubMed PMID: WOS:000235317600003.
39. Young NL, Dimaggio PA, Plazas-Mayorca MD, Baliban RC, Floudas CA, Garcia BA. High-throughput characterization of combinatorial histone codes. Mol Cell Proteomics. 2009;8:2266–84. PubMed PMID: 19654425. Epub 2009/08/06. Eng.
40. Boyne 2nd MT, Pesavento JJ, Mizzen CA, Kelleher NL. Precise characterization of human histones in the H2A gene family by top down mass spectrometry. J Proteome Res. 2006;5(2):248–53. PubMed PMID: 16457589. Epub 2006/02/07. eng.
41. Pesavento JJ, Kim YB, Taylor GK, Kelleher NL. Shotgun annotation of histone modifications: a new approach for streamlined characterization of proteins by top down mass spectrometry. J Am Chem Soc. 2004;126(11):3386–7. PubMed PMID: WOS:000220286400009.
42. Siuti N, Roth MJ, Mizzen CA, Kelleher NL, Pesavento JJ. Gene-specific characterization of human histone H2B by electron capture dissociation. J Proteome Res. 2006;5(2):233–9. PubMed PMID: WOS:000235317600002.
43. Forbes AJ, Patrie SM, Taylor GK, Kim YB, Jiang LH, Kelleher NL. Targeted analysis and discovery of post-translational modifications in proteins from methanogenic archaea by top-down MS. Proc Natl Acad Sci USA. 2004;101(9):2678–83. PubMed PMID: WOS:000220065300008.
44. Han XM, Jin M, Breuker K, McLafferty FW. Extending top-down mass spectrometry to proteins with masses greater than 200 kilodaltons. Science. 2006;314(5796):109–12. PubMed PMID: WOS:000241031200046.
45. Zabrouskov V, Ge Y, Schwartz J, Walker JW. Unraveling molecular complexity of phosphorylated human cardiac troponin I by top down electron capture dissociation/electron transfer dissociation mass spectrometry. Mol Cell Proteomics. 2008;7(10):1838–49. PubMed PMID: 18445579. Epub 2008/05/01. eng.
46. Sandilands A, Sutherland C, Irvine AD, McLean WH. Filaggrin in the frontline: role in skin barrier function and disease. J Cell Sci. 2009;122(Pt 9):1285–94. PubMed PMID: 19386895. Pubmed Central PMCID: 2721001. Epub 2009/04/24. eng.
47. Lonsdale-Eccles JD, Teller DC, Dale BA. Characterization of a phosphorylated form of the intermediate filament-aggregating protein filaggrin. Biochemistry. 1982;21(23):5940–8. PubMed PMID: 6185144. Epub 1982/11/09. eng.
48. Palosuo T, Lukka M, Alenius H, Kalkkinen N, Aho K, Kurki P, et al. Purification of filaggrin from human epidermis and measurement of antifilaggrin autoantibodies in sera from patients with rheumatoid arthritis by an enzyme-linked immunosorbent assay. Int Arch Allergy Immunol. 1998;115(4):294–302. PubMed PMID: 9566352.
49. Resing KA, Walsh KA, Haugen-Scofield J, Dale BA. Identification of proteolytic cleavage sites in the conversion of pro-filaggrin to filaggrin in mammalian epidermis. J Biol Chem. 1989;264(3):1837–45. PubMed PMID: 2912987. Epub 1989/01/25. eng.
50. McKinley-Grant LJ, Idler WW, Bernstein IA, Parry DA, Cannizzaro L, Croce CM, et al. Characterization of a cDNA clone encoding human filaggrin and localization of the gene to chromosome region 1q21. Proc Natl Acad Sci U S A. 1989;86(13):4848–52. PubMed PMID: 2740331. Pubmed Central PMCID: 297512. Epub 1989/07/01. eng.
51. Nagaraj N, D'Souza RC, Cox J, Olsen JV, Mann M. Feasibility of large-scale phosphoproteomics with higher energy collisional dissociation fragmentation. J Proteome Res. 2010;9(12):6786–94. PubMed PMID: 20873877.
52. Shaw JB, Madsen JA, Xu H, Brodbelt JS. Systematic comparison of ultraviolet photodissociation and electron transfer dissociation for peptide anion characterization. J Am Soc Mass Spectrom. 2012;23(10):1707–15. PubMed PMID: 22895858.
53. Choy DF, Hsu DK, Seshasayee D, Fung MA, Modrusan Z, Martin F, et al. Comparative transcriptomic analyses of atopic dermatitis and psoriasis reveal shared neutrophilic inflammation. J Allergy Clin Immunol. 2012;130(6):1335–43 e5. PubMed PMID: 22920495. Pubmed Central PMCID: 3511596.
54. Broccardo CJ, Mahaffey S, Schwarz J, Wruck L, David G, Schlievert PM, et al. Comparative proteomic profiling of patients with atopic dermatitis based on history of eczema herpeticum infection and Staphylococcus aureus colonization. J Allergy Clin Immunol. 2011;127(1):186–93. 93 e1–11. PubMed PMID: 21211653. Pubmed Central PMCID: 3059191.

Noninvasive Detection of Filaggrin Molecules by Raman Spectroscopy

10

Francisco J. González

Contents

F.J. González, PhD
Center for the Investigation and Application of Science and Technology,
Autonomous University of San Luis Potosi,
Av. Sierra Leona 550,
San Luis Potosi 78210, Mexico
e-mail: javier.gonzalez@uaslp.mx

10.1 Raman Spectroscopy

Chandrasekhara Venkata Raman (1888–1970) started working on the scattering of light around 1921, when he published a note in *Nature* entitled "The colour of the sea" [1]. In this note, he explained that the color of the ocean is due to scattering and does not depend on the reflection of the sky. His subsequent work on scattering resulted in the discovery of the Raman effect in 1928, for which he was awarded the Nobel Prize just 2 years later in 1930 [2].

The Raman effect consists of a shift in photon energy due to inelastic collisions of photons with molecules [2]. When light interacts with matter, it gets absorbed or scattered, and the scattered light is most of the time of the same wavelength as the incident radiation (Rayleigh scattering); however, approximately one out of 10^7 photons is scattered at a different wavelength due to the gain or loss of energy because of inelastic collisions with molecules [2]. The amount of shift in wavelength depends on whether the molecule is moving, rotating, or vibrating; therefore, the wavelength shift can be of a rotational, vibrational, or translational nature. These wavelength shifts are unique for each molecule, and they provide a fingerprint of the molecular structure of the sample that can be used to identify the material that is being analyzed [2].

The Raman spectrum shows the scattering intensity as a function of the frequency difference between the incident and the scattered light; this difference is usually known as the "Raman shift." A Raman spectrum of a given sample consists of a

J.P. Thyssen, H. Maibach (eds.), *Filaggrin*,
DOI 10.1007/978-3-642-54379-1_10,

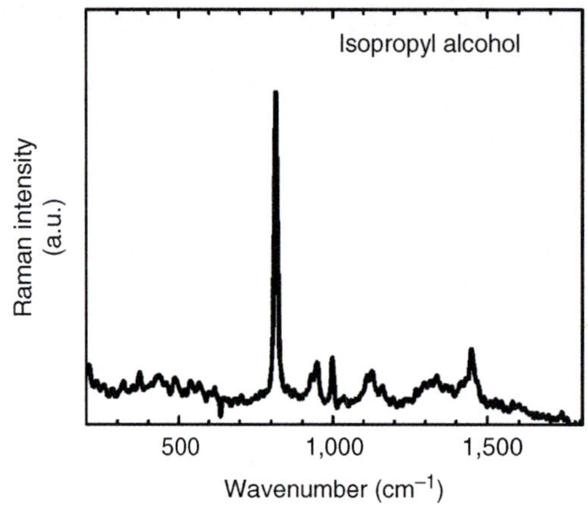

Fig. 10.1 Raman spectrum of isopropyl alcohol

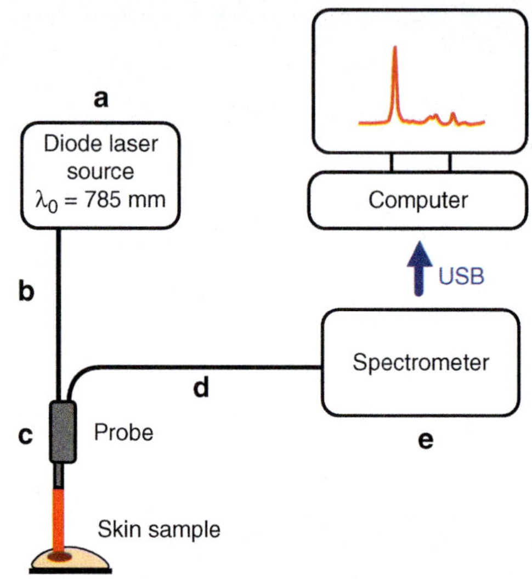

Fig. 10.2 Schematic of a typical Raman system for biological applications, which consists of an (**a**) illumination source, (**b**) fiber-based light delivery, (**c**) Raman probe, (**d**) Raman scattering light collection, and (**e**) detector and computer processing

series of peaks that give detailed information about its chemical composition. It is also worth noting that the intensity of the peaks is proportional to the concentration of the molecule from which this peak arises; therefore, a quantitative analysis and molecular concentration profiles can be obtained using Raman spectroscopy [2]. Figure 10.1 shows the Raman spectrum of isopropyl alcohol.

10.2 Instrumentation and Experimental Techniques

A Raman spectrometer useful in a clinical setting should be an integrated system that can provide real-time spectral acquisition and analysis [2, 3].

A Raman spectrometer for biological applications consists of five major components [3]:

1. An illumination source
2. Light delivery
3. Raman probe
4. Raman signal delivery
5. Detector and computer control/processing

A schematic of the typical arrangement of these components is shown in Fig. 10.2.

Lasers are the ideal illumination sources for Raman spectroscopy; they can provide a highly monochromatic optical source at relatively high powers [2]. The choice of wavelength depends mainly on the application. In the case of biological tissue, a near-infrared (NIR) laser is used since this wavelength can penetrate farther in biological tissue, which is due to the existence of a so-called therapeutic window from 700 to 1,200 nm, where the absorption of the human tissue is the lowest and, therefore, is more transparent to light. Also, NIR lasers produce less fluorescence than lower-wavelength lasers, making them ideal for highly fluorescent samples such as biological tissue [2, 3]. Most of the Raman spectroscopy systems for skin diagnosis use a 785-nm diode laser as the illumination source, since it provides an inexpensive light source that generates low fluorescence and can penetrate deep into human tissue.

The most commonly used method for light delivery in Raman spectrometers for clinical applications is optical fibers. Raman probes vary depending on the clinical application. For endoscopy, the Raman probe is entirely composed of fibers due to size limitations [3]. In the case of taking Raman spectra of skin, the probe consists of a bundle of fibers where usually the middle one is bigger and transports the illumination source and several smaller ones carry the light scattered by the sample. Figure 10.3 shows a schematic of a typical Raman probe used for Raman spectroscopy; in the figure it can be seen how in the same bundle of fibers lie the illumination source and the fibers that collect the scattered light.

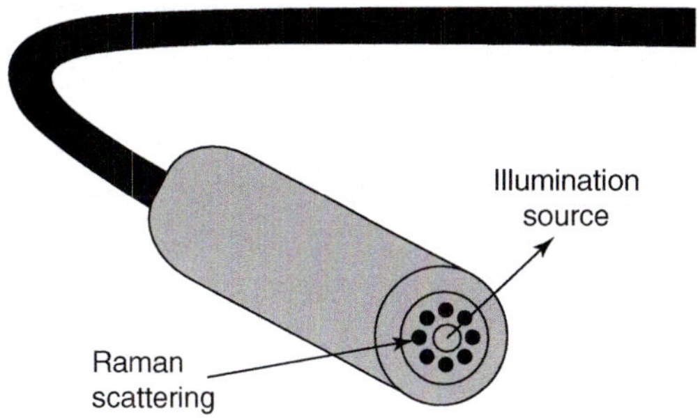

Fig. 10.3 Raman probe illustrating a typical fiber placement used to illuminate and collect scattered light in a typical Raman system

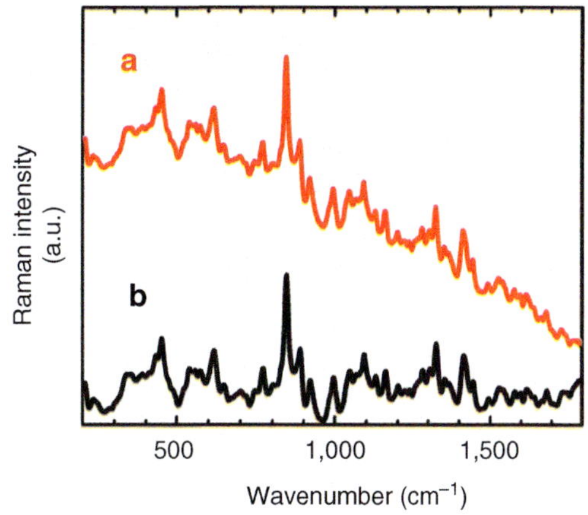

Fig. 10.4 Raman spectra of glutamine (**a**) with fluorescence emission and (**b**) without fluorescence, which has been removed by the Vancouver computer algorithm

10.3 Fluorescence Removal Algorithms

A big issue in Raman spectroscopy applied to biological tissue comes from the fact that the probability of Raman scattering is several orders of magnitude less than the intrinsic fluorescence emission of biological tissues; therefore, fluorescence is an important source of interference for the Raman signal [4].

Fluorescence removal can be done by using instrumentation or computational approaches. The instrumentation methods include shifted excitation and time gating [5–7]. The computational methods include polynomial fitting [8], Fourier transformation [9–11], wavelet transform [12–14], and first- and second-order differentiation [15, 16], among others.

Polynomial curve fitting has a distinct advantage over other fluorescence reduction techniques because of its simplicity and effectiveness [8]. It is faster than other methods and has been widely used for in vivo biomedical Raman applications [12, 13].

Another advantage of polynomial curve fitting resides in the fact that it does not alter the spectral signatures and intensities of the original Raman spectra. The method is based on the principle that fluorescence can be mathematically modeled as a polynomial function. Empirically, fifth-order polynomials provide the best fluorescence approximations for in vivo biomedical applications [17].

The Vancouver Raman algorithm is an iterative automated algorithm for fluorescence removal based on polynomial fitting [8]. The advantage of this method is that it includes a statistical method to account for noise effects and Raman signal contribution. The final polynomial fit is regarded as the fluorescence background. The Raman spectra are derived from the raw spectra by subtracting the final polynomial function. This algorithm was developed by researchers at the BC Cancer Agency (British Columbia, Canada) and is free for noncommercial use and can be downloaded from http://www.flintbox.com/public/project/1956.

Figure 10.4 shows the Raman spectra of glutamine with and without fluorescence removal.

10.4 Raman Spectrum of Filaggrin

In order to quantify the amount of a specific substance in Raman spectra, it is necessary to know in advance the Raman spectrum for that specific substance. In order to measure the Raman spectrum of filaggrin, pure filaggrin human recombinant protein was obtained from GenWay Biotech, Inc. (San Diego, CA).

The pure filaggrin human recombinant protein was first characterized using sodium dodecyl sulfate-polyacrylamide gel electrophoresis (SDS-PAGE). SDS-PAGE is the most commonly used gel electrophoresis technique for proteins; it is particularly useful for monitoring the fractions

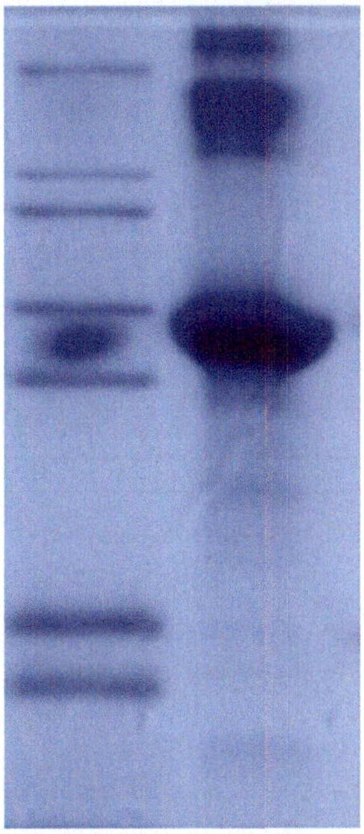

Fig. 10.5 Polyacrylamide gel electrophoresis (PAGE) results. The lanes contain a marker protein (*left*) and the filaggrin (*right*). The band in the second lane (filaggrin) corresponds to 53 kDa

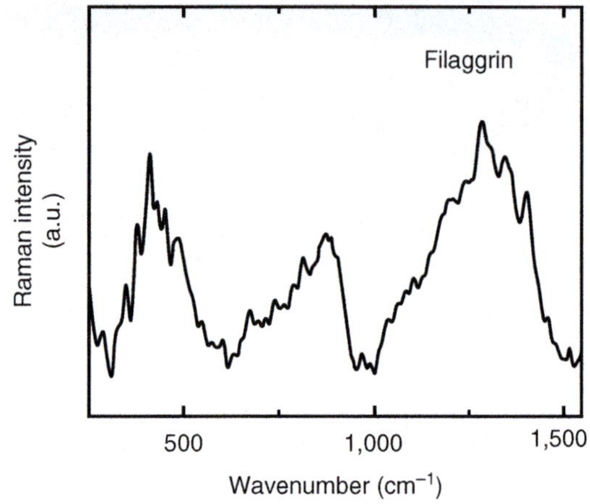

Fig. 10.6 Raman spectrum of pure filaggrin human recombinant protein obtained from GenWay Biotech, Inc. (San Diego, CA)

obtained during chromatographic or other purification procedures [18]. It also allows samples from different sources to be compared for protein content. One of the most important features of PAGE is that it is a simple, reliable method to estimate the molecular weight of proteins.

SDS-PAGE (12 % gel) was performed on a vertical gel electrophoresis system. Samples were subjected to electrophoresis at constant voltage, 70 V for 30 min followed by 120 V for 1 h in a Tris–HCl buffer. The gel was stained for 1 h with 0.25 % Coomassie brilliant blue R-250 in a water/methanol/acetic acid mixture. To ensure accuracy in the results, both the marker protein and the filaggrin were electrophoresed on the same gel under identical separation conditions. Three gels were generated for statistical significance.

The PAGE of filaggrin is shown in Fig. 10.5. In this configuration, the first lane contains the marker protein and the second lane contains the filaggrin. As shown in Fig. 10.5, the band in the second lane (filaggrin) corresponds to 53 kDa which agrees with the technical specifications provided by GenWay Biotech, Inc., regarding the human recombinant protein used.

After the characterization of the Raman spectrum, an in vitro measurement of pure filaggrin was performed in order to have a fixed spectrum that will help analyze the Raman measurements and assess their filaggrin concentration. The measurements were performed at room temperature using a quartz cuvette and a Raman Systems R3000 spectrometer (Ocean Optics, Dunedin, FL, USA) with a 785-nm laser diode and a spectral resolution of 8 cm^{-1} and at a laser power of 300 mW. Figure 10.6 shows the measured Raman spectrum of the pure filaggrin protein.

10.5 Quantification of Filaggrin in Raman Spectra of Human Skin

Several published works have used Raman spectroscopy to analyze the molecular composition of skin and correlate it with history of atopic dermatitis (AD) and filaggrin gene (*FLG*) mutations; however, very few of them have measured directly the filaggrin molecule.

Mlitz et al. [19] used Raman spectroscopy to analyze the skin of 196 French adults, which included 97 with a history of AD. Noninvasive Raman skin measurements were used to analyze the concentrations of water, total natural moisturizing factors (NMFs), alanine, glycine, histidine, lactate, ornithine, proline, urocanic acid (UCA), pyrrolidone-5-carboxylic acid (PCA), urea, ceramide 3, and cholesterol, which are breakdown products of filaggrin that usually accumulate in the stratum corneum. The results showed that stratum corneum concentrations of total NMFs, water, ornithine, and UCA were significantly lower in AD patients than in healthy controls; also, *FLG* mutations were associated with increased stratum corneum levels of lactate. These findings can be used to predict the presence of FLG mutations by noninvasive Raman spectroscopy.

Kezic et al. [20] also measured NMFs noninvasively on the skin of 137 Irish children with a history of moderate to severe AD. Results showed that *FLG* mutations have a more significant effect on NMFs than AD severity, which the same group also proved using noninvasive Raman spectroscopy [21, 22].

The methods to quantify the presence of a specific molecule in a Raman spectrum require that the Raman spectrum of the molecule be known in advance. The most popular methods rely on quantifying the similarities of the reference spectrum with the measured spectra, which can be done by performing a correlation between spectra or by analyzing a set of spectra by transforming them into their principal components using a technique known as principal component analysis (PCA).

10.5.1 Correlation Between the Measured Spectra and a Reference Spectrum

The simplest way of quantifying the presence of a particular substance whose spectrum is known is to calculate the correlation between the measured spectra and the reference spectrum in order to determine the similarities between them. The main advantage of using this method is that it can be done with a single measured spectrum, whereas principal component analysis works better when a large set of spectra is analyzed.

10.5.2 Principal Component Analysis (PCA)

The PCA transforms N spectra into N principal components, which are independent of each other, and it gives an N×N matrix, which contains the coefficients for the transformation between the original data and the principal components, and also provides N eigenvalues describing the importance of the corresponding principal component.

These principal components are given in decreasing order of importance, which means that the first principal component can explain the largest quantity of the variance of the original data, the second one explains more variance than the third, and so on. In the terminology used by the PCA, these variance contributions are known as eigenvalues. One of the advantages of the PCA is that, by evaluating the relative importance of the consecutive principal components, it is possible to reduce the dimension of the data set by finding a smaller collection of spectra that explain a given and acceptable amount of variance [23, 24].

Taking into account the properties of PCA, a method has been developed to analyze Raman spectra in order to find the contribution of a reference spectrum in the measurements [25]. This method consists of including the reference spectrum with the original data set containing the Raman spectra of the measurements and analyzing the results obtained by applying the PCA method.

By plotting the coefficients obtained by the PCA as a function of each other and knowing information about the spectra, one can find which coefficients correspond to certain physical or clinical attributes; therefore, the coefficients can be used to predict that particular physical phenomena in similar spectra analyzed by PCA.

Another method used to evaluate spectra by PCA involves grouping together eigenvalues in order to reduce the dimension of the data and only take into account the first independent principal components and consider the rest as noise contributions. Afterwards, the principal components that are related to the reference spectrum are identified. A new set of reconstructed spectra is generated using only the reference spectrum-related principal components and compared with the original ones. The square mean deviation between the original and the reconstructed spectra is calculated for each spectrum in order to determine the contribution of the reference spectrum-related principal components to the original spectra. A larger square mean deviation will mean that those spectra will have a lower contribution from the reference spectrum, which might indicate a degraded or the absence of the molecule used as reference.

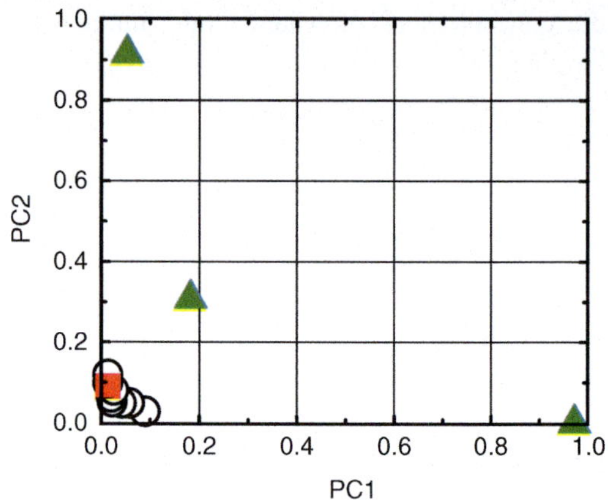

Fig. 10.7 Plot of the two first principal components of the Raman spectra for each newborn and the Raman spectrum of filaggrin. The distance between each point representing a measurement and the filaggrin spectrum indicates which subjects have a higher quantity of filaggrin in their skin

10.6 Clinical Applications

10.6.1 Noninvasive Detection of Filaggrin in Newborns and Its Association with AD

In this work, the presence of the protein filaggrin in the skin of newborns was detected using Raman spectroscopy and PCA as an early detection procedure for filaggrin-related AD [25].

Twelve healthy infants born without complications from healthy mothers participated in this study. Raman scattering measurements were performed on the right thigh for each infant at birth; the measurements were performed at room temperature using a Raman Systems R3000 spectrometer (Ocean Optics, Dunedin, FL, USA) with a 785-nm laser diode, a spectral resolution of 8 cm^{-1}, and a laser power of 90 mW. The irradiance of this laser diode is below the ANSI standard for skin, and none of the participants showed any kind of discomfort when the measurement was performed.

The measurements were performed in the 200–1,800 cm^{-1} spectral range, and the instrument was calibrated using a Teflon standard every day before each round of measurements.

The participants of this study were monitored for 1 year to see whether any of them developed AD. The dermatologist's diagnosis of AD was made according to standard criteria in the presence of a chronic or chronically relapsing pruritic dermatitis with the typical morphology and distribution [26, 27]. From the 12 infants measured and monitored, 3 of them developed AD in the course of 1 year.

The analysis consisted in applying the PCA to the 12 measured spectra and the filaggrin reference spectrum, producing 13 principal components. Using a statistical criterion to group together the principal components [23], it was found that, with a level of confidence of 99 %, the first eight components are statistically independent from each other, and after the eighth component, there is a statistical dependence between them that may be associated with noise [23]. The first eight principal components explain 99.84 % of the total variance of the data.

In order to find the presence of filaggrin in the spectra, the coefficients of the principal components for each spectrum were plotted. The first and second principal components account for 93.86 % of all the variance of the original data. Figure 10.7 shows a graph of these two principal components for each of the spectra measured from the newborns.

The green triangles correspond to those infants who developed AD; the rest of the subjects are grouped together around the location of the filaggrin spectrum (represented as a red square). The geometrical distance of each of the measured spectra to the spectrum of filaggrin in the principal component plane indicates which subjects have a higher content of pure filaggrin and which ones have less amount of filaggrin or which ones have a filaggrin with a different molecular structure than the molecule that was taken as reference spectrum. This result indicates that this approach can be used to identify the persons who are more susceptible to develop AD, making it possible to use this technique as a method for early detection of AD [25].

10.6.2 Correlation of *FLG* Mutations and Skin Filaggrin Content Measured Using Raman Spectroscopy

In this work, the filaggrin content in the skin is assessed using Raman spectroscopy, and the results are compared to their filaggrin genotype in order to validate the use of Raman spectroscopy as a noninvasive tool to detect filaggrin gene mutations [28].

In this study, 19 Mexican-Mestizo patients with major complaints of pruritus and dry skin were analyzed. Raman scattering measurements were performed on the inner area of the forearm of the 19 patients, approximately 10 cm above the wrist. The measurements were made at room temperature using a Raman Systems R3000 spectrometer (Ocean Optics, Dunedin, FL, USA) with a 785-nm laser diode, a spectral resolution of 8 cm^{-1}, and a laser power of 90 mW.

All the measured spectra were preprocessed by subtracting a fifth-grade polynomial to the raw spectra applying the fluorescence removal algorithm proposed by Zhao et al. [8], also known as the Vancouver algorithm, in order to remove the background NIR fluorescence and leave the pure Raman signal.

DNA was obtained from buccal cells using the Gentra Puregene (Qiagen) protocol using a glycogen and isopropanol precipitation with freezing for higher recovery. The FLG was amplified by polymerase chain reaction (PCR) based on sequences published by Smith et al. [29]. For the R501X mutation detection, a 311pb PCR fragment was incubated with 5 units of Nla III restriction enzyme (New England Biolabs), and the products were resolved by a 3 % agarose 1,000 (Invitrogen) or 12 % acrylamide gel electrophoresis; normal sequence produces 204 and 107 pb fragments and the mutant sequence generates the same 107 pb fragment plus additional 128 and 76 pb fragments. The 2284del4 mutation, which creates a new restriction site, was detected incubating an 811 pb filaggrin DNA fragment with the Dra III (New England Biolabs) enzyme; products were resolved on a 2 % regular agarose gel.

The amount of filaggrin on the measured samples was estimated by performing the correlation between the pure filaggrin spectrum and the spectra obtained from the subjects under study; this correlation was performed using the Mathematica (Wolfram Research, Inc., Champaign, IL, USA) computer software. This approach gives an advantage over principal component analysis since it does not need a large data sample for good accuracy and can be performed over just one spectrum.

The genetic analysis showed that 8 out of the 19 patients (42 %) presented a FLG mutation. These 8 patients presented the 2282del4 FLG mutation, 2 of which (10.5 %) were homozygous and 6 (31.5 %) heterozygous, whereas 1 (5.2 %) resulted in a compound heterozygote for the 2282del4 and the R501X mutations.

These genetic results were compared to the filaggrin content estimation; a lower correlation value of the spectra with the pure filaggrin spectrum indicates a lower filaggrin concentration relative to the rest of the correlation results.

Figure 10.8 shows the results of the correlation for the patients with a *FLG* mutation (*FLG* –) and without a *FLG* mutation (*FLG* +). The patients with a *FLG* mutation presented an average correlation of 0.286 (SD = 0.082), while the patients without a *FLG* mutation showed an average correlation of 0.4 (SD = 0.08).

From Fig. 10.8 it can be seen that 7 out of 8 patients with a *FLG* mutation (87.5 %) presented a correlation lower than 0.35 and 8 out of the 11 patients without the R501X and 2284del4 *FLG* mutations (72.7 %) presented a correlation higher than 0.35 (dotted line in Fig. 10.8).

Even though the patients that participated in this study did not undergo the diagnostic criteria for AD, the percentage that presented *FLG* mutations (42 %) is not too far from the percentage of

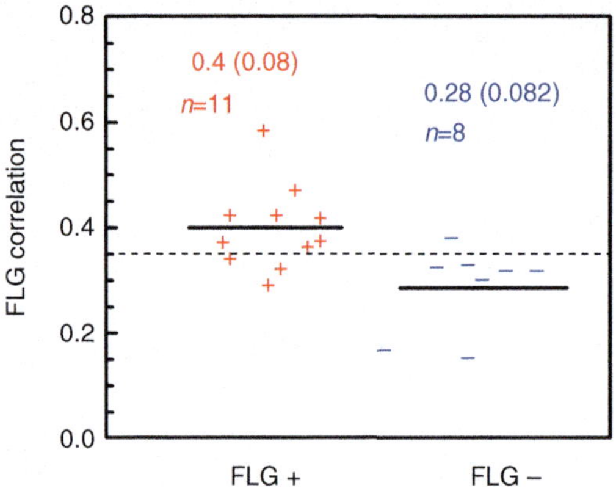

Fig. 10.8 Correlation between the filaggrin Raman spectrum and the skin spectrum of subjects with (*FLG* –) and without (*FLG* +) filaggrin gene mutations

FLG mutations present in the Caucasian population, where loss-of-function mutations are detectable in about one-third of patients with AD [30]. A study with a larger number of Mexican-Mestizo patients who comply with the diagnostic criteria for AD is currently underway in order to compare the incidence of these mutations to the European population.

Conclusion

Raman spectroscopy is one of the techniques that have been actively researched as a possible method to perform "optical biopsies" where optical techniques are used to distinguish healthy from diseased tissue. One of the main advantages of using Raman spectroscopy in optical biopsies is that water is not Raman active, so it does not interfere with the measurements; also, Raman spectroscopy gives a molecular fingerprint of a substance that can be used to noninvasively detect it for clinically relevant information. Among the disadvantages of Raman spectroscopy is that not all the molecules are Raman active, which means that some molecules do not give a strong Raman signal. Another disadvantage is that the small quantities of clinically relevant substances in biological tissue go beyond the detection limit of conventional Raman spectrometers, making them very difficult or impossible to detect; a way around this problem is by increasing the sensitivity of Raman spectroscopy measurements using techniques such as surface-enhanced Raman spectroscopy (SERS). In the particular case of filaggrin, this molecule has been successfully characterized by Raman spectroscopy and can be detected at physiological levels in the skin, and Raman spectroscopy measurements have been successfully correlated to clinical findings. This indicates that Raman spectroscopy can be a useful technique for early diagnosis and monitoring of filaggrin deficiencies and filaggrin-related diseases.

Acknowledgments The author would like to acknowledge the work of Dr. Benjamín Moncada and the Dermatology Department at the "Dr. Ignacio Morones Prieto" Central Hospital (San Luis Potosi, Mexico), where most of the clinical work was performed. Also, the author would like to acknowledge the valuable collaboration of Dr. Victor Saavedra-Alanís, who performed the filaggrin genotyping; Dr. Javier Alda from the University Complutense of Madrid and Dr. Miguel G. Ramírez-Elías, who performed most of the principal component analysis; and Dr. Edgar Briones, who made the figures for this chapter.

References

1. Raman CV. The colour of the sea. Nature. 1921;108(2716):367.
2. Ferraro JR, Nakamoto K, Brown C. Introductory Raman spectroscopy. 2nd ed. San Diego: Academic; 2003.
3. Zhao J, Lui H, McLean DI, Zheng H. Real-time Raman spectroscopy for noninvasive in vivo skin analysis and diagnosis. In: Campolo R, editor. Recent advances in biomedical engineering. Vienna: IN-TECH; 2010. p. 455–74.
4. Ramírez-Elías MG, Alda J, González FJ. Noise and artifact characterization of in-vivo Raman spectroscopy skin measurements. Appl Spectrosc. 2012;66(6):650–5.
5. Shreve P, Cherepy NJ, Mathies RA. Effective rejection of fluorescence interference in Raman spectroscopy using a shifted excitation difference technique. Appl Spectrosc. 1992;46(4):707–11.
6. Baraga JJ, Feld MS, Rava RP. Rapid near-infrared Raman spectroscopy of human tissue with a spectrograph and CCD detector. Appl Spectrosc. 1992;46(2):187–90.
7. Knorr F, Smith ZJ, Wachsmann-Hogiu S. Development of a time-gated system for Raman spectroscopy of biological samples. Opt Exp. 2010;18(19):20049–58.

8. Zhao J, Lui H, McLean D, Zeng H. Automated autofluorescence background subtraction algorithm for biomedical Raman spectroscopy. Appl Spectrosc. 2007;61(11):1225–32.
9. Schulze G, Jirasek A, Yu ML, Lim A, Turner RFB, Blades MW. Investigation of selected baseline removal techniques as candidates for automated implementation. Appl Spectrosc. 2005;59(5):545–74.
10. Villanueva-Luna AE, Castro-Ramos J, Vazquez-Montiel S, Flores-Gil A, Delgado-Atencio JA, Vázquez-Villa A. Raman spectra and optical coherence tomography images skin. Proc SPIE. 2010;7883:788310.
11. Ehrentreich F, Summchen L. Spike removal and denoising of Raman spectra by wavelet transform methods. Anal Chem. 2001;73(1):4364–73.
12. Huang Z, Chen X, Chen Y, Chen J, Dou M, Feng S, et al. Raman spectroscopic characterization and differentiation of seminal plasma. J Biomed Opt. 2011; 16:110501.
13. Wang HH, Huang N, Zhao J, Lui H, Korbelik M, Zeng H. Depth-resolved in vivo micro Raman spectroscopy of a murine skin tumor model reveals cancer-specific spectral biomarkers. J Raman Spectrosc. 2011;42(2):160–1660.
14. Ramos PM, Ruisánchez I. Noise and background removal in Raman spectra of ancient pigments using wavelet transform. J Raman Spectrosc. 2005;36(9):848–56.
15. Zhang DM, Ben-Amotz D. Enhanced chemical classification of Raman images in the presence of strong fluorescence interference. Appl Spectrosc. 2000;54(9): 1379–83.
16. O'Grady A, Dennis AC, Denvir D, McGarvey JJ, Bell SEJ. Quantitative Raman spectroscopy of highly fluorescent samples using pseudosecond derivatives and multivariate analysis. Anal Chem. 2001;73(9):2058–65.
17. Mahadevan-Jansen A, Richards-Kortum R. Raman spectroscopy for the detection of cancers and precancers. J Biomed Opt. 1996;1:31.
18. Garfin DE. Gel electrophoresis of proteins. In: Davey J, Lord M, editors. Essential cell biology, Cell structure, a practical approach, vol. 1. Oxford: Oxford University Press; 2003. p. 197–268.
19. Mlitz V, Latreille J, Gardinier S, Jdid R, Drouault Y, Hufnagl P, et al. Impact of filaggrin mutations on Raman spectra and biophysical properties of the stratum corneum in mild to moderate atopic dermatitis. J Eur Acad Dermatol Venereol. 2012;26(8):983–90.
20. Kezic S, O'Regan GM, Lutter R, Jakasa I, Koster ES, Saunders S, Caspers P, et al. Filaggrin loss-of-function mutations are associated with enhanced expression of IL-1 cytokines in the stratum corneum of patients with atopic dermatitis and in a murine model of filaggrin deficiency. J Allergy Clin Immunol. 2012;129(4): 1031–9.
21. O'Regan GM, Kemperman PM, Sandilands A, Chen H, Campbell LE, Kroboth K, et al. Raman profiles of the stratum corneum define 3 filaggrin genotype-determined atopic dermatitis endophenotypes. J Allergy Clin Immunol. 2010;126(3):574–80.
22. Kezic S, O'Regan GM, Yau N, Sandilands A, Chen H, Campbell LE, et al. Levels of filaggrin degradation products are influenced by both filaggrin genotype and atopic dermatitis severity. Allergy. 2011;66: 934–40.
23. López-Alonso JM, Alda J, Bernabeu E. Principal component characterization of noise for infrared images. Appl Opt. 2002;41:320–31.
24. López-Alonso JM, Alda J. Operational parametrization of the 1/f noise of a sequence of frames by means of the principal component analysis in focal plane arrays. Opt Eng. 2004;42:1915–22.
25. González FJ, Alda J, Moreno-Cruz B, Martínez-Escanamé M, Ramírez-Elías MG, Torres-Álvarez B, et al. Use of Raman spectroscopy in the early detection of filaggrin-related atopic dermatitis. Skin Res Technol. 2011;17(1):45–50.
26. Hannifin JM, Rajka G. Diagnostic features of atopic dermatitis. Acta Derm Venereol. 1980;92:44–7.
27. Williams HC, Jburney PG, Hay RJ, Archer CB, Shipley MJ, Hunter JJ, et al. The U.K. Working Party's diagnostic criteria for atopic dermatitis. I. Derivation of a minimum set of discriminators for atopic dermatitis. Br J Dermatol. 1994;131: 383–96.
28. González FJ, Valdes-Rodríguez R, Ramírez-Elías MG, Castillo-Martínez C, Saavedra-Alanis VM, Moncada B. Noninvasive detection of filaggrin gene mutations using Raman spectroscopy. Biomed Opt Express. 2011;2(12):3363–6.
29. Smith FJD, Irvine AD, Terron-Kwiatkowski A, Sandilands A, Campbell LE, Zhao Y, et al. Loss-of-function mutations in the gene encoding filaggrin cause ichthyosis vulgaris. Nat Genet. 2006;38(3): 337–42.
30. Novak N, Simon D. Atopic dermatitis – from new pathophysiologic insights to individualized therapy. Allergy. 2011;66(7):830–9.

Filaggrin Genotyping

11

Pal B. Szecsi and Michael Meldgaard

Contents

P.B. Szecsi, MD, DMSci (✉)
M. Meldgaard, MS, PhD
Department of Clinical Biochemistry,
Gentofte Hospital, University of Copenhagen,
Niels Andersensvej 65, Hellerup 2900, Denmark
e-mail: pal.szecsi@regionh.dk

11.1 General Genotyping

The human genome features a large number of nucleotide polymorphisms (SNPs), microsatellites, deletions, and insertions. Detection of genetic variations has evolved from the classic labor-intensive polymerase chain reaction (PCR)-mediated amplification coupled with restriction fragment length polymorphism (RFLP) to high-throughput systems using DNA microarrays, (Affymetrix, Illumina), PCR-based TaqMan chemistry [1, 2], or DNA sequencing-based methods such as pyrosequencing (Biotage) or minisequencing (Perkin Elmer) and to mass spectroscopy methods such as iPLEX/MassARRAY (Sequenom). All methods have their pros and cons, and the method of choice depends upon whether known or unknown genetic variants are of interest, whether few or many samples need to be investigated, and available financial resources and instruments.

11.2 Filaggrin

Profilaggrin is a member of the S100 calcium-binding protein family and is one of the largest human proteins (~400 kDa). For the major part, it consists of 10–12 repetitive filaggrin domains, each with 324 amino acid residues and a very similar composition. The filaggrin gene (*FLG*) (GenBank NM_002016.1) is located on human chromosome 1q21 and comprises three exons and two introns [3, 4]. More than 1,300 genetic

J.P. Thyssen, H. Maibach (eds.), *Filaggrin*,
DOI 10.1007/978-3-642-54379-1_11, © Springer-Verlag Berlin Heidelberg 2014

variants in the *FLG* are registered in different databases. About 60 of these mutations result in loss of function, due to introduction of early stop codons or frameshifts [5–14]. Some mutations are more frequent in Northern Europeans, while others are more frequent in Asians populations.

It took several years to identify the first clinical relevant mutations due to the repetitive nature of the *FLG*. The endeavor toward successful identification of the first two mutations [7, 15, 16] has been described by McLean [17].

11.2.1 Sample Material

Many filaggrin genotyping reports do not state the source of genomic DNA [12, 13, 18–73]. Most studies that state information about the source use blood leukocytes [14, 15, 40, 74–119], but the use of saliva [8, 75, 76, 83, 120–122], buccal swabs [75, 92, 104, 115, 118, 120, 123–132], tissue [8], or urine [133] have also been reported. In addition, we have genotyped samples from serum, as well as formalin-fixed paraffin-embedded material. The method we developed for serum-based DNA extraction and purification begins with 0.25 mL serum and provides sufficient DNA for approximately 100 filaggrin genotyping PCRs. With formalin-fixed tissue, some kind of purification is necessary, even though simple heat treatment-based extraction can be sufficient, but in our experience it has worked with variable success.

Blood leukocytes are a convenient source for DNA, primarily due to established routines for sampling and patient identification with barcode, but any available cellular material will do. Some have advocated utilizing repeated sampling and analysis of genetic assays in order to minimize sample mix-up errors, but it has not yet been adopted for filaggrin genotyping. Keeping the sample in the original barcode-labeled sample tube and with as few analytic transfers as possible is advisable.

11.3 DNA Isolation Methods

Most reports have performed some purification, disregarding the source of DNA [14, 74, 77–80, 84–91, 109–117, 124–127, 131–134]. Only a few groups use crude DNA extracts [96, 118]. We have performed genotyping on purified DNA but only if this material already was at hand. We prefer a simple cell lysis step using potassium hydroxide, as it is simple to perform and keeps the test costs low [135]. However, crude extracts can only be used if the genotyping method is robust and does not require purified DNA. As mentioned above, isolation of DNA from serum and formalin-fixed tissue does require purification. Our method for retrieving DNA from serum samples is based on classic phenol-chloroform extraction followed by ethanol-based DNA precipitation.

11.4 Mutation Determination Methods

Virtually all methods for genotyping rely on hybridization (annealing) of a short synthetic oligonucleotide (10–40 bases) to one or the other genomic DNA strand. This is performed either at the mutation site or some distance up or downstream. In combination with an enzyme and the nucleotides, the annealed oligonucleotide is extended at the 3′ end or ligated to another oligonucleotide. This occurs using the classic Sanger's dideoxy sequencing method, all PCR, and all the other methods used for filaggrin genotyping. The challenge of these methods is to ensure correct hybridization, i.e., the non-covalent, sequence-specific binding between two complementary strands of nucleic acids. Correct hybridization depends on the melting temperature (T_m) given by the relevant DNA sequence, to break the two hydrogen bonds between A=T and three hydrogen bonds between G≡C. To ensure correct selection, high-stringency hybridization conditions are essential, and these are determined by the annealing temperature as well as the ion strength (salt concentration) and presence of additives such as DMSO.

Below is a short description of the most frequently used methods for *FLG* mutation detection. Many reports describe a mix of different methods for the different *FLG* mutations and have altered techniques over time. Some reports do not disclose information about the genotyping method applied [13, 19, 26, 33, 58]. Other methods for SNP genotyping such as high-density oligonucleotide SNP arrays (i.e., Affymetrix) and flap endonuclease (FEN) Invader assay have not been reported in filaggrin genotyping.

11.5 Unknown Mutations

The identification of previously unknown mutations requires a DNA sequencing method. As a single sequencing reaction cannot cover the whole length of the *FLG*, most methods rely on PCR-amplified smaller fragments primarily using the primers originally described [7, 15, 122]. A few reports have utilized classic cloning of fragments [97, 136], but most sequence PCR-amplified fragments. If a non-proofreading polymerase was used for the amplification, errors might be introduced in early PCR cycles, so novel found mutations need to be confirmed in separate processes. Screening for unknown mutations can be performed by some kind of melting point detection of partially denatured DNA with temperature gradient electrophoresis or LightCycler (real-time PCR). Altered melting temperature indicates a mutation, but sequencing is required for identifying the mutation.

11.6 Known Mutations

11.6.1 Sequencing

The Sanger's chain-termination sequencing developed in 1977 is still the method of choice for DNA sequencing in spite of the relatively high costs. Today most use automated fluorescence dideoxynucleotides triphosphates (ddNTP) with capillary electrophoresis size detection. Newer, cheaper methods have been developed, but have not found use in filaggrin research yet. The method is considered the gold standard and is used to validate other methods. In addition, it is required to detect previously unknown mutations and seems to be a preferred method for the detection of many Asian mutations. It has been one of the first methods for identification of *FLG* mutations and is still widely used [11, 14, 15, 23, 31, 34, 62, 66, 78, 79, 85, 86, 89, 93–95, 98, 99, 101, 102, 109–111, 113, 114, 132]. Most sequence PCR fragments, but classical cloning with subsequent sequencing has also been reported [97, 136].

11.6.2 Restriction Fragment Length Polymorphism (RFLP)

RFLP is the earliest method used to detect mutations and is simple to perform. Originally it was performed directly on genomic DNA followed by hybridization with radioactively labeled DNA fragments. More recently, RFLP analysis has been performed following PCR-based amplification of a specific region of the *FLG*. RFLP makes use of different restriction endonucleases and their high affinity to cut unique and specific DNA sequences. By performing a restriction digest on a double-stranded DNA sample and determining the resulting fragment lengths, it is possible to determine whether a specific restriction site is present. Some still use sizing of the fragment on agarose gels, but most use some kind of automated assay like capillary electrophoresis RFLP.

The advantage of the method is that mutation detection not only relies on the hybridization specificity of the PCR primer but also on the restriction enzyme as well as the correct size. Some mutations of interests do not include a restriction enzyme recognition site, but this might be artificially created by mismatches included in an oligonucleotide in a previous PCR amplification step. However, this method is potentially compromised by other genetic variations at the recognition site for the restriction enzyme that may abolish or introduce digestion, resulting in misinterpretation. The requirement for specific endonucleases, the fact that the exact mutations cannot necessarily be resolved in a single experiment, and the subsequent sizing assays make RFLP a poor choice for high-throughput analyses. In spite of this, many still use this technique, especially in detection of the 2282del4 mutation [8, 10, 12, 15, 31, 44, 47, 48, 53, 54, 57, 59, 61, 64, 66, 75, 77, 83, 85, 87, 88, 92, 93, 103, 104, 106–110, 112–115, 117, 120, 122, 124, 128, 130, 131, 133, 137–144].

11.6.3 Allele-Specific PCR

Allele-specific PCR comes in many variants. Generally, allele specificity occurs by two primers hybridizing on their 3′ end to either the wild-type or the mutant variant. With careful optimization, allele-specific amplification is possible in combination with a common primer targeting the opposite DNA strand. Many assays use a variation of this method, alone or in combination with other techniques.

11.6.4 Fluorescent PCR Fragment Length Analysis

For SNPs, the amplified PCR fragments are the same size provided both allele-specific primers have the same length, but in case of insertion or deletion variants (such as 2282del4), the fragments differ in size. Originally, sizing of products was performed by gel electrophoresis with ethidium bromide intercalating and UV light visualization, and this method is still in use. However, many reports describe the use of fluorescent nucleotides or primers to allow automated fragment sizing by capillary gel electrophoresis on instruments also used as DNA sequencers [5, 8, 10–12, 16, 18, 20, 29, 30, 46, 49–52, 54, 55, 61, 62, 64–66, 77, 80, 82, 83, 87, 91, 104, 108, 109, 112, 113, 117, 120, 125, 130, 132, 134, 143, 145, 146].

11.6.5 Competitive Allele-Specific PCR (KASP)

This method is a variant of the allelic-specific PCR combined with competitive binding of two allele-specific forward primers, each incorporating unique tail sequences corresponding to two different FRET cassettes, one labeled with FAM dye and the other with HEX dye. The assay has relative low reagent costs and does not require a specific fluorescent reader; however, it does require purified DNA, which increases test costs dramatically. The producer (LGC Genomics, Hoddesdon, UK) offers both reagents and genotyping services with assays that are either functionally or in silico validated. KASP is likely to be applied more frequently in the future due to the low test costs. It has already found its way into filaggrin genotyping [127, 128, 130, 131].

11.6.6 Suspension Array-Based Allele-Specific PCR (Luminex)

We use a suspension array of spectrally coded microbeads for our preferred genotyping method. Hybridization to the microbead-based suspension array is preceded by a multiplexed, allele-specific, asymmetric PCR using tagged primers. Detection following hybridization to the suspension array uses a dedicated flow cytometer (Luminex, Austin, TX, USA) [147]. This method is also a variant of allele-specific PCR with tagged primers, somewhat like the KASP assay but with a biotinylated common primer. The allele-specific primers are in limiting concentrations relative to the common primers to ensure the generation of biotinylated single-stranded PCR products incorporating the anti-tag DNA sequence that will hybridize to the spectrally coded microbeads carrying tag sequences as capture probes. Hybridized PCR products are made fluorescent by labeling with streptavidin-phycoerythrin, and each bead is decoded and analyzed in the dedicated flow cytometer, in our case a Bio-Plex 200 (Bio-Rad, Hercules, CA,

USA). The combination of asymmetric PCR- and allele-specific primers in limiting concentration allows for optimization of the individual PCR reactions and enables detection of multiple both wildtype and mutant variants in the same tube. As only very few beads are necessary, the test cost is relatively low and can be performed with detection of at least six mutations in 96-well plates [29, 41, 67–73, 118, 119, 135, 147–158]. The method does not require purified DNA. Hence, crude extracts from a simple alkaline lysis of blood work fine, contributing to keep test costs low.

11.6.7 5′ Nuclease (TaqMan)

This method is also a variant of allele-specific PCR. It exploits the 5′ nuclease activity of the Taq polymerase. The TaqMan assay (Life Technologies, Carlsbad, CA, USA) is performed concurrently with a PCR reaction, and the results can be read in real time [159]. The assay requires forward and reverse PCR primers that will amplify a region that includes the polymorphic site. Allele discrimination is achieved using one or two allele-specific probes that hybridize specifically to the polymorphic site. The probes will have a fluorescence resonance energy transfer (FRET) fluorophore at the 5′ end and a quencher molecule linked to their 3′ end. When the probe is intact, the quencher is in close proximity to the fluorophore, preventing the fluorophore signal. During the PCR annealing/amplification step, the allele-specific probe binds to the target DNA strand and then gets degraded by 5′ nuclease activity of the Taq polymerase as it extends the DNA from the PCR primers resulting in the separation of the fluorophore from the quencher molecule thereby generating a detectable signal. The assay can be multiplexed by up to seven SNPs in one reaction using different fluorophores. However, since the selectivity is in the different probes, multiplexing can be difficult. Also this assay is prone to errors by other sequence variations close to the mutations site. However, as amplification and detection occur simultaneously, it is easy to perform many simultaneous reactions in microtiter plates even in automated setup. The method requires relatively expensive probes and instruments, and optimization of the assay may be necessary. The method was quickly adopted by McLean and coworkers, especially for the R501X detection and many other researchers have used it since [8, 16, 20, 22, 28, 29, 35, 37, 38, 45, 46, 49–52, 54–57, 61, 62, 64–66, 74, 77, 81, 82, 85, 94–96, 100, 102, 104, 107, 116, 117, 122, 125, 129, 134, 145, 160].

11.6.8 Primer Extension (iPLEX, MALDI-TOF)

The iPLEX (Sequenom, San Diego, CA, USA) is a PCR-based method that starts by amplifying fragments of approximately 100 base pairs around the SNP of interest [161]. The generated fragments are subsequently treated with shrimp alkaline phosphatase to inactivate remaining nucleotides and then subjected to a single-base extension reaction. The adduct-forming ions are removed by adding an anion exchange resin. Matrix-assisted laser desorption/ionization time-of-flight mass spectrometry (MALDI-TOF) analysis is performed by deposition of the reaction products into silica-based chip arrays containing matrix spots (3-hydroxypicolinic acid) in a 384 format. The charged ions of various sizes are generated on the sample slide, by a potential difference between the sample slide and ground. As the potential difference is constant with respect to all ions, ions with smaller m/z value (lighter ions) and more charged ions move faster through space until they reach the detector. Consequently, the time of ion flight differs according to the mass of the generated DNA fragments. It is possible to design algorithms that allow automated design of primers for 20–30 plex reactions.

The method is useful for low- to medium-throughput analyses; however, it requires rather expensive instrumentation. The method is also influenced by primer hybridization, but this does not have effects upon the allele specificity. This method has only been used by a few groups for filaggrin genotyping [29, 38, 57, 60, 100].

Comparable alternative SNP genotyping technologies have not been reported for filaggrin genotyping, including SNPstream from Beckman Coulter (Fullerton, CA, USA) and SNPlex from ABI (Foster City, CA, USA), both of which have the capability to multiplex 12–48 SNP assays in a 384-well format.

11.6.9 GoldenGate

The GoldenGate method (Illumina, San Diego, CA, USA) is an allele-specific extension and ligation assay [162]. Allele discrimination is achieved by two primers that have a base complementary to each of the two alleles at the 3′ end. The primers will get extended from the 3′ end only if the primers match completely. Both primers have a unique sequence at the 5′ end, which binds to PCR primers a later stage. The reaction mixture contains a reverse third primer common to both alleles. This primer has two unique sequences in addition to the gene-specific 3′ end, an address tag that allows for specific binding to a solid beadchip support and a 5′ tag that will be recognized by the final PCR primer. If extension occurs, a ligase will connect the reverse primer to the extended sequence. In the final PCR reaction, a reverse PCR primer and Cy3 and Cy5 fluorescent-labeled downstream primers generate products that can hybridize to one of the 1,520 different beads on the chip, and the signal is then read by a BeadStation that has been mapped to known location and identity [38, 82].

11.7 Errors and Prevention

All assays are prone for erratic results, so the challenge is to minimize their frequency. Nearly all described methods utilize some kind of amplification step of a specific DNA fragment. This process is performed over and over, and huge quantities of the amplified DNA are generated in the laboratory. These molecules have the potential to contaminate future samples, as it was seen in the beginning of clinical DNA testing where all samples turned out positive after a few months [163, 164]. Severe preventative actions are strongly recommended, as are very stringent clean-up using a hypochlorite solution, discarding of all used reagents, UV irradiation of workspace for PCR setup in a dedicated laminar flow bench, and use of filter tips. In addition to physical separation of PCR setup facilities and PCR product analysis facilities is the use of uracil-DNA glycosylase (UNG) and substitution of dTTP with dUTP in the PCR process, an important preventive step [165]. This enables the digestion of any PCR products introduced into the sample, thus ensuring amplification of genuine DNA and eliminating false positives. Surprisingly, it seems that our laboratory is the only one reporting the use of UNG. A single group states that they do not use UNG [116], but the rest do not mention the possibility. Some groups might have used UNG without mentioning it, but many use polymerases that are known to be incapable of incorporating uracil nucleotides.

Another general problem is erratic priming. Most techniques use specific priming of a short oligonucleotide where the purine-pyrimidine base hydrogen bond pairing provides the binding strength. Several factors can influence the binding strength, such as temperature, ionic strength, and concentrations, but primarily mismatch in the pairing. Any change from the expected sequence alters the priming, and either no or false priming may occur and lead to erratic interpretation of the results. We have discovered a G->A substitution 21 base pairs upstream of the 2282del4 site at the 5′ of our allele-specific forward primer. In 2282del4 negative samples (i.e., 2282del4 wildtype) with this G->A substitution, we get a reduced wildtype signal, which does not interfere with the genotype interpretation. But with 2282del positive samples (i.e., 2282del4 heterozygotes), the reduced wildtype signal results in a reduced "wildtype signal to mutation signal ratio" that falls outside of both the wildtype and the heterozygote range. A similar finding has also been seen in other genotyping assays like the HFE gene mutations H63D, and S65C [166].

A thorough validation of each method is necessary. Copying published methods is a good start, but cannot substitute validation. We have experienced variations in assay performance depending upon, e.g., the chosen polymerase, thermocycler, and sample material. Using reagents from the same producer minimizes the problems, but we have experienced that a supplier of Taq polymerase suddenly sends an inferior product without being able to explain the change or difference. Several thermocyclers of the same brand and model sitting on the same bench may have different performance. All these variables may be a concern with only "in silico verified" methods, especially when offered commercially without any insight in methodological problems.

Magnolis et al. [38] compared an allele-specific extension/ligation beadchip assay (GoldenGate, Illumina) with a Taq 5′-exonuclease/FRET assay (TaqMan, Roche). Even though both assays gave perfect retest results, the positive predictive value for the GoldenGate assay was only 53 % assuming the TaqMan assay as the golden standard. As such they advocated for the use of the TaqMan method for *FLG* mutation detection. However, they do not provide any details for the beadchip assay, and it seems that it has been established by the commercial company without proper validation.

11.8 Other Methods

A few investigations have been performed with other genotyping methods. These include denaturing high-performance liquid chromatography [99], which uses reversed-phase HPLC to interrogate SNPs. It is easily automated, as no labeling or purification of the DNA fragments is needed, but the column temperature must be optimized for each target in order to achieve the right degree of denaturation. Others have used recording of melting temperatures of fluorescence-labeled PCR fragment [126] following allele-specific PCR [21, 83, 133]. The heterozygotes can be identified as the heteroduplexes changed the shape of the melting curve. In most cases, homozygous polymorphisms are also distinguishable.

As a curiosity, Raman spectroscopy has been used to identify filaggrin deficiency in skin extract [124, 167, 168].

Conclusion

Every laboratory testing procedure, no matter how well established, involves the possibility of generating error. This is equally true for genetic testing. For example, the PCR method for DNA amplification allows minute quantities of DNA to be multiplied in a way that facilitates testing. Yet there is a danger that the sample tested may get contaminated with extraneous genetic material, such as from previously amplified products or from the operator, thus amplifying irrelevant DNA. There are also occasional errors with the sequence fidelity of amplified products, resulting in reading errors. Careful testing according to ISO 15189 (Medical Laboratory Accreditation) or CLIA certification (Clinical Laboratory Improvement Amendments) is strongly recommended. Inclusion of samples with a known genotype in each test run is also prudent. Exchange of samples among other filaggrin genotyping laboratories is a quality-ensuring task. We encourage any laboratory that performs filaggrin genotyping to participate in the External Quality Assessment *FLG* program from the European Molecular Genetics Quality Network (EMQN). It is a free pilot program that we have initiated. At the moment, only the mutations c.1537C>T, c.2318_2321del, and c.7375C>T, commonly designated as R501X, 2282del4, and R2447X, are reported.

Conflict of Interest Neither of the authors has a conflict of interest to declare.

References

1. Shi MM, Myrand SP, Bleavins MR, de la Iglesia FA. High-throughput genotyping method for glutathione S-transferase T1 and M1 gene deletions using TaqMan probes. Res Commun Mol Pathol Pharmacol. 1999;103(1):3–15.

2. Wang L, Luhm R, Lei M. SNP and mutation analysis. Adv Exp Med Biol. 2007;593:105–16.
3. Presland RB, Haydock PV, Fleckman P, Nirunsuksiri W, Dale BA. Characterization of the human epidermal profilaggrin gene. Genomic organization and identification of an S-100-like calcium binding domain at the amino terminus. J Biol Chem. 1992;267(33):23772–81.
4. Markova NG, Marekov LN, Chipev CC, Gan SQ, Idler WW, Steinert PM. Profilaggrin is a major epidermal calcium-binding protein. Mol Cell Biol. 1993; 13(1):613–25.
5. Weidinger S, Illig T, Baurecht H, Irvine AD, Rodriguez E, Diaz-Lacava A, et al. Loss-of-function variations within the filaggrin gene predispose for atopic dermatitis with allergic sensitizations. J Allergy Clin Immunol. 2006;118(1):214–9.
6. Zhang H, Guo Y, Wang W, Shi M, Chen X, Yao Z. Mutations in the filaggrin gene in Han Chinese patients with atopic dermatitis. Allergy. 2011;66(3): 420–7.
7. Palmer CN, Irvine AD, Terron-Kwiatkowski A, Zhao Y, Liao H, Lee SP, et al. Common loss-of-function variants of the epidermal barrier protein filaggrin are a major predisposing factor for atopic dermatitis. Nat Genet. 2006;38(4):441–6.
8. Brown SJ, Relton CL, Liao H, Zhao Y, Sandilands A, Wilson IJ, et al. Filaggrin null mutations and childhood atopic eczema: a population-based case-control study. J Allergy Clin Immunol. 2008;121(4):940–6.
9. Akiyama M. FLG mutations in ichthyosis vulgaris and atopic eczema: spectrum of mutations and population genetics. Br J Dermatol. 2010;162(3):472–7.
10. Hsu CK, Akiyama M, Nemoto-Hasebe I, Nomura T, Sandilands A, Chao SC, et al. Analysis of Taiwanese ichthyosis vulgaris families further demonstrates differences in FLG mutations between European and Asian populations. Br J Dermatol. 2009;161(2):448–51.
11. Chen H, Common JE, Haines RL, Balakrishnan A, Brown SJ, Goh CS, et al. Wide spectrum of filaggrin-null mutations in atopic dermatitis highlights differences between Singaporean Chinese and European populations. Br J Dermatol. 2011;165(1):106–14.
12. Nemoto-Hasebe I, Akiyama M, Nomura T, Sandilands A, McLean WH, Shimizu H. FLG mutation p. Lys4021X in the C-terminal imperfect filaggrin repeat in Japanese patients with atopic eczema. Br J Dermatol. 2009;161(6):1387–90.
13. Osawa R, Konno S, Akiyama M, Nemoto-Hasebe I, Nomura T, Nomura Y, et al. Japanese-specific filaggrin gene mutations in Japanese patients suffering from atopic eczema and asthma. J Invest Dermatol. 2010;130(12):2834–6.
14. Li M, Liu Q, Liu J, Cheng R, Zhang H, Xue H, et al. Mutations analysis in filaggrin gene in northern China patients with atopic dermatitis. J Eur Acad Dermatol Venereol. 2013;27(2):169–74.
15. Smith FJ, Irvine AD, Terron-Kwiatkowski A, Sandilands A, Campbell LE, Zhao Y, et al. Loss-of-function mutations in the gene encoding filaggrin cause ichthyosis vulgaris. Nat Genet. 2006;38(3):337–42.
16. Sandilands A, Terron-Kwiatkowski A, Hull PR, O'Regan GM, Clayton TH, Watson RM, et al. Comprehensive analysis of the gene encoding filaggrin uncovers prevalent and rare mutations in ichthyosis vulgaris and atopic eczema. Nat Genet. 2007; 39(5):650–4.
17. McLean WH. The allergy gene: how a mutation in a skin protein revealed a link between eczema and asthma. F1000 Med Rep. 2011;3:2. doi:10.3410/M3-2.:2.
18. Filipiak-Pittroff B, Schnopp C, Berdel D, Naumann A, Sedlmeier S, Onken A, et al. Predictive value of food sensitization and filaggrin mutations in children with eczema. J Allergy Clin Immunol. 2011;128(6): 1235–41.
19. Common JE, Brown SJ, Haines RL, Goh CS, Chen H, Balakrishnan A, et al. Filaggrin null mutations are not a protective factor for acne vulgaris. J Invest Dermatol. 2011;131(6):1378–80.
20. Greisenegger E, Novak N, Maintz L, Bieber T, Zimprich F, Haubenberger D, et al. Analysis of four prevalent filaggrin mutations (R501X, 2282del4, R2447X and S3247X) in Austrian and German patients with atopic dermatitis. J Eur Acad Dermatol Venereol. 2010;24(5):607–10.
21. Gruber R, Janecke AR, Grabher D, Sandilands A, Fauth C, Schmuth M. Evidence for genetic modifiers other than filaggrin mutations in X-linked ichthyosis. J Dermatol Sci. 2010;58(1):72–5.
22. Gruber R, Wilson NJ, Smith FJ, Grabher D, Steinwender L, Fritsch PO, et al. Increased pachyonychia congenita severity in patients with concurrent keratin and filaggrin mutations. Br J Dermatol. 2009;161(6):1391–5.
23. Iga N, Tanizaki H, Endo Y, Egawa G, Fujisawa A, Tanioka M, et al. Hydrolyzed wheat protein-containing facial soap-induced wheat-dependent exercise-induced anaphylaxis in a patient without filaggrin mutations. J Dermatol. 2013;40(6):494–5.
24. Jakasa I, Koster ES, Calkoen F, McLean WH, Campbell LE, Bos JD, et al. Skin barrier function in healthy subjects and patients with atopic dermatitis in relation to filaggrin loss-of-function mutations. J Invest Dermatol. 2011;131(2):540–2.
25. Jungersted JM, Scheer H, Mempel M, Baurecht H, Cifuentes L, Hogh JK, et al. Stratum corneum lipids, skin barrier function and filaggrin mutations in patients with atopic eczema. Allergy. 2010;65(7):911–8.
26. Kaae J, Menne T, Carlsen BC, Zachariae C, Thyssen JP. The hands in health and disease of individuals with filaggrin loss-of-function mutations: clinical reflections on the hand eczema phenotype. Contact Dermatitis. 2012;67(3):119–24.
27. Kezic S, O'Regan GM, Yau N, Sandilands A, Chen H, Campbell LE, et al. Levels of filaggrin degradation products are influenced by both filaggrin genotype and atopic dermatitis severity. Allergy. 2011;66(7): 934–40.
28. O'Regan GM, Campbell LE, Cordell HJ, Irvine AD, McLean WH, Brown SJ. Chromosome 11q13.5 vari-

ant associated with childhood eczema: an effect supplementary to filaggrin mutations. J Allergy Clin Immunol. 2010;125(1):170–4.
29. Thyssen JP, Thuesen B, Huth C, Standl M, Carson CG, Heinrich J, et al. Skin barrier abnormality caused by filaggrin (FLG) mutations is associated with increased serum 25-hydroxyvitamin D concentrations. J Allergy Clin Immunol. 2012;130(5):1204–7.
30. Marenholz I, Bauerfeind A, Esparza-Gordillo J, Kerscher T, Granell R, Nickel R, et al. The eczema risk variant on chromosome 11q13 (rs7927894) in the population-based ALSPAC cohort: a novel susceptibility factor for asthma and hay fever. Hum Mol Genet. 2011;20(12):2443–9.
31. Ma L, Zhang L, Di ZH, Zhao LP, Lu YN, Xu J, et al. Association analysis of filaggrin gene mutations and atopic dermatitis in Northern China. Br J Dermatol. 2010;162(1):225–7.
32. Li M, Chen X, Chen R, Bao Y, Yao Z. Filaggrin gene mutations are associated with independent atopic asthma in Chinese patients. Allergy. 2011;66(12): 1616–7.
33. Ezzedine K, Droitcourt C, Ged C, Diallo A, Hubiche T, de Verneuil H, et al. Usefulness of a global clinical ichthyosis vulgaris scoring system for predicting common FLG null mutations in an adult caucasian population. Br J Dermatol. 2012;167(5):1165–9.
34. Iwamoto S, Ebihara N, Hori K, Funaki T, Asada Y, Yokoi N, et al. Filaggrin mutations are not associated with chronic allergic keratoconjunctivitis. Br J Ophthalmol. 2012;96(9):1272–3.
35. Margolis DJ, Apter AJ, Gupta J, Hoffstad O, Papadopoulos M, Campbell LE, et al. The persistence of atopic dermatitis and filaggrin (FLG) mutations in a US longitudinal cohort. J Allergy Clin Immunol. 2012;130(4):912–7.
36. Tan HT, Ellis JA, Koplin JJ, Matheson MC, Gurrin LC, Lowe AJ, et al. Filaggrin loss-of-function mutations do not predict food allergy over and above the risk of food sensitization among infants. J Allergy Clin Immunol. 2012;130(5):1211–3.
37. Carson CG, Rasmussen MA, Thyssen JP, Menne T, Bisgaard H. Clinical presentation of atopic dermatitis by filaggrin gene mutation status during the first 7 years of life in a prospective cohort study. PLoS One. 2012;7(11):e48678.
38. Margolis DJ, Apter AJ, Mitra N, Gupta J, Hoffstad O, Papadopoulos M, et al. Reliability and validity of genotyping filaggrin null mutations. J Dermatol Sci. 2013;70:67–8.
39. Li M, Cheng R, Shi M, Liu J, Zhang G, Liu Q, et al. Analyses of FLG mutation frequency and filaggrin expression in isolated ichthyosis vulgaris (IV) and atopic dermatitis-associated IV. Br J Dermatol. 2013;168(6):1335–8.
40. Thyssen JP, Johansen JD, Zachariae C, Menne T, Linneberg A. Xerosis is associated with atopic dermatitis, hand eczema and contact sensitization independent of filaggrin gene mutations. Acta Derm Venereol. 2013;18:10–1539.
41. Kaae J, Menne T, Thyssen JP. Severe occupational protein contact dermatitis caused by fish in 2 patients with filaggrin mutations. Contact Dermatitis. 2013;68(5):319–20.
42. Brown SJ, Relton CL, Liao H, Zhao Y, Sandilands A, McLean WH, et al. Filaggrin haploinsufficiency is highly penetrant and is associated with increased severity of eczema: further delineation of the skin phenotype in a prospective epidemiological study of 792 school children. Br J Dermatol. 2009;161(4): 884–9.
43. Stemmler S, Nothnagel M, Parwez Q, Petrasch-Parwez E, Epplen JT, Hoffjan S. Variation in genes of the epidermal differentiation complex in German atopic dermatitis patients. Int J Immunogenet. 2009;36(4):217–22.
44. Bisgaard H, Halkjaer LB, Hinge R, Giwercman C, Palmer C, Silveira L, et al. Risk analysis of early childhood eczema. J Allergy Clin Immunol. 2009; 123(6):1355–60.
45. Rodriguez S, Hall AJ, Granell R, McLean WH, Irvine AD, Palmer CN, et al. Carrier status for the common R501X and 2282del4 filaggrin mutations is not associated with hearing phenotypes in 5,377 children from the ALSPAC cohort. PLoS One. 2009;4(6):e5784.
46. Ruether A, Stoll M, Schwarz T, Schreiber S, Folster-Holst R. Filaggrin loss-of-function variant contributes to atopic dermatitis risk in the population of Northern Germany. Br J Dermatol. 2006;155(5):1093–4.
47. Stemmler S, Parwez Q, Petrasch-Parwez E, Epplen JT, Hoffjan S. Two common loss-of-function mutations within the filaggrin gene predispose for early onset of atopic dermatitis. J Invest Dermatol. 2007;127(3):722–4.
48. Palmer CN, Ismail T, Lee SP, Terron-Kwiatkowski A, Zhao Y, Liao H, et al. Filaggrin null mutations are associated with increased asthma severity in children and young adults. J Allergy Clin Immunol. 2007; 120(1):64–8.
49. Zhao Y, Terron-Kwiatkowski A, Liao H, Lee SP, Allen MH, Hull PR, et al. Filaggrin null alleles are not associated with psoriasis. J Invest Dermatol. 2007;127(8):1878–82.
50. Weichenthal M, Ruether A, Schreiber S, Nair R, Voorhees JJ, Schwarz T, et al. Filaggrin R501X and 2282del4 mutations are not associated with chronic plaque-type psoriasis in a German cohort. J Invest Dermatol. 2007;127(6):1535–7.
51. Morar N, Cookson WO, Harper JI, Moffatt MF. Filaggrin mutations in children with severe atopic dermatitis. J Invest Dermatol. 2007;127(7):1667–72.
52. Huffmeier U, Traupe H, Oji V, Lascorz J, Stander M, Lohmann J, et al. Loss-of-function variants of the filaggrin gene are not major susceptibility factors for psoriasis vulgaris or psoriatic arthritis in German patients. J Invest Dermatol. 2007;127(6):1367–70.
53. Liao H, Waters AJ, Goudie DR, Aitken DA, Graham G, Smith FJ, et al. Filaggrin mutations are genetic modifying factors exacerbating X-linked ichthyosis. J Invest Dermatol. 2007;127(12):2795–8.

54. Henderson J, Northstone K, Lee SP, Liao H, Zhao Y, Pembrey M, et al. The burden of disease associated with filaggrin mutations: a population-based, longitudinal birth cohort study. J Allergy Clin Immunol. 2008;121(4):872–7.
55. Bisgaard H, Simpson A, Palmer CN, Bonnelykke K, McLean I, Mukhopadhyay S, et al. Gene-environment interaction in the onset of eczema in infancy: filaggrin loss-of-function mutations enhanced by neonatal cat exposure. PLoS Med. 2008;5(6):e131.
56. Rice NE, Patel BD, Lang IA, Kumari M, Frayling TM, Murray A, et al. Filaggrin gene mutations are associated with asthma and eczema in later life. J Allergy Clin Immunol. 2008;122(4):834–6.
57. Weidinger S, Baurecht H, Wagenpfeil S, Henderson J, Novak N, Sandilands A, et al. Analysis of the individual and aggregate genetic contributions of previously identified serine peptidase inhibitor Kazal type 5 (SPINK5), kallikrein-related peptidase 7 (KLK7), and filaggrin (FLG) polymorphisms to eczema risk. J Allergy Clin Immunol. 2008;122(3):560–8.
58. Kang TW, Lee JS, Oh SW, Kim SC. Filaggrin mutation c.3321delA in a Korean patient with ichthyosis vulgaris and atopic dermatitis. Dermatology. 2009;218(2):186–7.
59. Kabashima-Kubo R, Nakamura M, Sakabe J, Sugita K, Hino R, Mori T, et al. A group of atopic dermatitis without IgE elevation or barrier impairment shows a high Th1 frequency: possible immunological state of the intrinsic type. J Dermatol Sci. 2012;67(1):37–43.
60. Weidinger S, Rodriguez E, Stahl C, Wagenpfeil S, Klopp N, Illig T, et al. Filaggrin mutations strongly predispose to early-onset and extrinsic atopic dermatitis. J Invest Dermatol. 2007;127(3):724–6.
61. Basu K, Palmer CN, Lipworth BJ, Irwin McLean WH, Terron-Kwiatkowski A, Zhao Y, et al. Filaggrin null mutations are associated with increased asthma exacerbations in children and young adults. Allergy. 2008;63(9):1211–7.
62. Giardina E, Paolillo N, Sinibaldi C, Novelli G. R501X and 2282del4 filaggrin mutations do not confer susceptibility to psoriasis and atopic dermatitis in Italian patients. Dermatology. 2008;216(1):83–4.
63. Novak N, Baurecht H, Schafer T, Rodriguez E, Wagenpfeil S, Klopp N, et al. Loss-of-function mutations in the filaggrin gene and allergic contact sensitization to nickel. J Invest Dermatol. 2008;128(6):1430–5.
64. Brown SJ, Sandilands A, Zhao Y, Liao H, Relton CL, Meggitt SJ, et al. Prevalent and low-frequency null mutations in the filaggrin gene are associated with early-onset and persistent atopic eczema. J Invest Dermatol. 2008;128(6):1591–4.
65. Rogers AJ, Celedon JC, Lasky-Su JA, Weiss ST, Raby BA. Filaggrin mutations confer susceptibility to atopic dermatitis but not to asthma. J Allergy Clin Immunol. 2007;120(6):1332–7.
66. Huffmeier U, Boiers U, Lascorz J, Reis A, Burkhardt H. Loss-of-function mutations in the filaggrin gene: no contribution to disease susceptibility, but to autoantibody formation against citrullinated peptides in early rheumatoid arthritis. Ann Rheum Dis. 2008;67(1):131–3.
67. Hougaard MG, Johansen JD, Linneberg A, Bandier J, Stender S, Carlsen BC, et al. Skin prick test reactivity to aeroallergens by filaggrin mutation status. J Eur Acad Dermatol Venereol. 2012;11:10–3083.
68. Thyssen JP, Linneberg A, Carlsen BC, Johansen JD, Engkilde K, Hansen T, et al. A possible association between a dysfunctional skin barrier (filaggrin null-mutation status) and diabetes: a cross-sectional study. BMJ Open. 2011;1(1):e000062.
69. Berg ND, Husemoen LL, Thuesen BH, Hersoug LG, Elberling J, Thyssen JP, et al. Interaction between filaggrin null mutations and tobacco smoking in relation to asthma. J Allergy Clin Immunol. 2012;129(2):374–80, 380.e1–2.
70. Thyssen JP, Carlsen BC, Johansen JD, Meldgaard M, Szecsi PB, Stender S, et al. Filaggrin null-mutations may be associated with a distinct subtype of atopic hand eczema. Acta Derm Venereol. 2010;90(5):528.
71. Thyssen JP, Carlsen BC, Johansen JD, Meldgaard M, Szecsi PB, Stender S, et al. Filaggrin haploinsufficiency among patients with dermatitis from a tertiary referral centre: early findings and possible phenotype. Contact Dermatitis. 2010;62(3):182–3.
72. Thyssen JP, Carlsen BC, Menne T, Linneberg A, Nielsen NH, Meldgaard M, et al. Filaggrin null mutations increase the risk and persistence of hand eczema in subjects with atopic dermatitis: results from a general population study. Br J Dermatol. 2010;163(1):115–20.
73. Thyssen JP, Johansen JD, Linneberg A, Menne T, Nielsen NH, Meldgaard M, et al. The association between null mutations in the filaggrin gene and contact sensitization to nickel and other chemicals in the general population. Br J Dermatol. 2010;162(6):1278–85.
74. Angelova-Fischer I, Mannheimer AC, Hinder A, Ruether A, Franke A, Neubert RH, et al. Distinct barrier integrity phenotypes in filaggrin-related atopic eczema following sequential tape stripping and lipid profiling. Exp Dermatol. 2011;20(4):351–6.
75. Blanchard C, Stucke EM, Burwinkel K, Caldwell JM, Collins MH, Ahrens A, et al. Coordinate interaction between IL-13 and epithelial differentiation cluster genes in eosinophilic esophagitis. J Immunol. 2010;184(7):4033–41.
76. Brown SJ, Asai Y, Cordell HJ, Campbell LE, Zhao Y, Liao H, et al. Loss-of-function variants in the filaggrin gene are a significant risk factor for peanut allergy. J Allergy Clin Immunol. 2011;127(3):661–7.
77. Poninska J, Samolinski B, Tomaszewska A, Raciborski F, Samel-Kowalik P, Walkiewicz A, et al. Filaggrin gene defects are independent risk factors for atopic asthma in a Polish population: a study in ECAP cohort. PLoS One. 2011;6(2):e16933.
78. Cascella R, Cuzzola VF, Lepre T, Galli E, Moschese V, Chini L, et al. Full sequencing of the FLG gene in Italian patients with atopic eczema: evidence of new mutations, but lack of an association. J Invest Dermatol. 2011;131(4):982–4.

79. Cheng R, Li M, Zhang H, Guo Y, Chen X, Tao J, et al. Common FLG mutation K4671X not associated with atopic dermatitis in Han Chinese in a family association study. PLoS One. 2012;7(11):e49158.
80. Cramer C, Link E, Horster M, Koletzko S, Bauer CP, Berdel D, et al. Elder siblings enhance the effect of filaggrin mutations on childhood eczema: results from the 2 birth cohort studies LISAplus and GINIplus. J Allergy Clin Immunol. 2010;125(6):1254–60.
81. Flohr C, England K, Radulovic S, McLean WH, Campbel LE, Barker J, et al. Filaggrin loss-of-function mutations are associated with early-onset eczema, eczema severity and transepidermal water loss at 3 months of age. Br J Dermatol. 2010;163(6):1333–6.
82. Gao PS, Rafaels NM, Hand T, Murray T, Boguniewicz M, Hata T, et al. Filaggrin mutations that confer risk of atopic dermatitis confer greater risk for eczema herpeticum. J Allergy Clin Immunol. 2009;124(3): 507–13, 513.e1–7.
83. Gruber R, Elias PM, Crumrine D, Lin TK, Brandner JM, Hachem JP, et al. Filaggrin genotype in ichthyosis vulgaris predicts abnormalities in epidermal structure and function. Am J Pathol. 2011;178(5):2252–63.
84. Hu Z, Xiong Z, Xu X, Li F, Lu L, Li W, et al. Loss-of-function mutations in filaggrin gene associate with psoriasis vulgaris in Chinese population. Hum Genet. 2012;131(7):1269–74.
85. Imoto Y, Enomoto H, Fujieda S, Okamoto M, Sakashita M, Susuki D, et al. S2554X mutation in the filaggrin gene is associated with allergen sensitization in the Japanese population. J Allergy Clin Immunol. 2010;125(2):498–500.
86. Kim EJ, Jeong MS, Li K, Park MK, Lee MK, Yoon Y, et al. Genetic polymorphism of FLG in Korean ichthyosis vulgaris patients. Ann Dermatol. 2011;23(2):170–6.
87. Lee DE, Park SY, Han JY, Ryu HM, Lee HC, Han YS. Association between filaggrin mutations and atopic dermatitis in Korean pregnant women. Int J Dermatol. 2013;52(6):772–3.
88. Lesiak A, Kuna P, Zakrzewski M, van Geel M, Bladergroen RS, Przybylowska K, et al. Combined occurrence of filaggrin mutations and IL-10 or IL-13 polymorphisms predisposes to atopic dermatitis. Exp Dermatol. 2011;20(6):491–5.
89. Ramesh R, Chen H, Kukula A, Wakeling EL, Rustin MH, McLean WH. Exacerbation of X-linked ichthyosis phenotype in a female by inheritance of filaggrin and steroid sulfatase mutations. J Dermatol Sci. 2011;64(3):159–62.
90. Ross-Hansen K, Linneberg A, Johansen JD, Hersoug LG, Brasch-Andersen C, Menne T, et al. The role of glutathione S-transferase and claudin-1 gene polymorphisms in contact sensitization: a cross-sectional study. Br J Dermatol. 2013;168(4):762–70.
91. Savilahti EM, Ilonen J, Kiviniemi M, Saarinen KM, Vaarala O, Savilahti E. Human leukocyte antigen (DR1)-DQB1*0501 and (DR15)-DQB1*0602 haplotypes are associated with humoral responses to early food allergens in children. Int Arch Allergy Immunol. 2010;152(2):169–77.
92. Schuttelaar ML, Kerkhof M, Jonkman MF, Koppelman GH, Brunekreef B, de Jongste JC, et al. Filaggrin mutations in the onset of eczema, sensitization, asthma, hay fever and the interaction with cat exposure. Allergy. 2009;64(12):1758–65.
93. Perusquia-Ortiz AM, Oji V, Sauerland MC, Tarinski T, Zaraeva I, Seller N, et al. Complete filaggrin deficiency in ichthyosis vulgaris is associated with only moderate changes in epidermal permeability barrier function profile. J Eur Acad Dermatol Venereol. 2013;7:10.
94. Winge MC, Suneson J, Lysell J, Nikamo P, Lieden A, Nordenskjold M, et al. Lack of association between filaggrin gene mutations and onset of psoriasis in childhood. J Eur Acad Dermatol Venereol. 2013;27(1):e124–7.
95. Winge MC, Bilcha KD, Lieden A, Shibeshi D, Sandilands A, Wahlgren CF, et al. Novel filaggrin mutation but no other loss-of-function variants found in Ethiopian patients with atopic dermatitis. Br J Dermatol. 2011;165(5):1074–80.
96. Weidinger S, O'Sullivan M, Illig T, Baurecht H, Depner M, Rodriguez E, et al. Filaggrin mutations, atopic eczema, hay fever, and asthma in children. J Allergy Clin Immunol. 2008;121(5):1203–9.
97. Xiong Z, Luo S, Xu X, Zhang L, Peng H, Li W, et al. Novel FLG mutations associated with ichthyosis vulgaris in the Chinese population. Clin Exp Dermatol. 2012;37(2):177–80.
98. Zhang H, Guo Y, Wang W, Yu X, Yao Z. Associations of FLG mutations between ichthyosis vulgaris and atopic dermatitis in Han Chinese. Allergy. 2011; 66(9):1253–4.
99. Zhang X, Liu S, Chen X, Zhou B, Liu D, Lei G, et al. Novel and recurrent mutations in the filaggrin gene in Chinese patients with ichthyosis vulgaris. Br J Dermatol. 2010;163(1):63–9.
100. Ballardini N, Kull I, Soderhall C, Lilja G, Wickman M, Wahlgren CF. Eczema severity in preadolescent children and its relation to sex, filaggrin mutations, asthma, rhinitis, aggravating factors and topical treatment: a report from the BAMSE birth cohort. Br J Dermatol. 2013;168(3):588–94.
101. Pellerin L, Henry J, Hsu CY, Balica S, Jean-Decoster C, Mechin MC, et al. Defects of filaggrin-like proteins in both lesional and nonlesional atopic skin. J Allergy Clin Immunol. 2013;131(4):1094–102.
102. Hoppe T, Winge MC, Bradley M, Nordenskjold M, Vahlquist A, Berne B, et al. X-linked recessive ichthyosis: an impaired barrier function evokes limited gene responses before and after moisturizing treatments. Br J Dermatol. 2012;167(3):514–22.
103. Nemoto-Hasebe I, Akiyama M, Nomura T, Sandilands A, McLean WH, Shimizu H. Clinical severity correlates with impaired barrier in filaggrin-related eczema. J Invest Dermatol. 2009;129(3): 682–9.
104. Enomoto H, Hirata K, Otsuka K, Kawai T, Takahashi T, Hirota T, et al. Filaggrin null mutations are associated with atopic dermatitis and elevated levels of IgE

in the Japanese population: a family and case-control study. J Hum Genet. 2008;53(7):615–21.
105. Schumann H, Has C, Kohlhase J, Bruckner-Tuderman L. Dystrophic epidermolysis bullosa pruriginosa is not associated with frequent FLG gene mutations. Br J Dermatol. 2008;159(2):464–9.
106. Sinclair C, O'Toole EA, Paige D, El BH, Robinson J, Dobson R, et al. Filaggrin mutations are associated with ichthyosis vulgaris in the Bangladeshi population. Br J Dermatol. 2009;160(5):1113–5.
107. Marenholz I, Kerscher T, Bauerfeind A, Esparza-Gordillo J, Nickel R, Keil T, et al. An interaction between filaggrin mutations and early food sensitization improves the prediction of childhood asthma. J Allergy Clin Immunol. 2009;123(4):911–6.
108. Sergeant A, Campbell LE, Hull PR, Porter M, Palmer CN, Smith FJ, et al. Heterozygous null alleles in filaggrin contribute to clinical dry skin in young adults and the elderly. J Invest Dermatol. 2009;129(4):1042–5.
109. Nomura T, Akiyama M, Sandilands A, Nemoto-Hasebe I, Sakai K, Nagasaki A, et al. Prevalent and rare mutations in the gene encoding filaggrin in Japanese patients with ichthyosis vulgaris and atopic dermatitis. J Invest Dermatol. 2009;129(5):1302–5.
110. Gruber R, Janecke AR, Fauth C, Utermann G, Fritsch PO, Schmuth M. Filaggrin mutations p. R501X and c.2282del4 in ichthyosis vulgaris. Eur J Hum Genet. 2007;15(2):179–84.
111. Hamada T, Sandilands A, Fukuda S, Sakaguchi S, Ohyama B, Yasumoto S, et al. De novo occurrence of the filaggrin mutation p.R501X with prevalent mutation c.3321delA in a Japanese family with ichthyosis vulgaris complicated by atopic dermatitis. J Invest Dermatol. 2008;128(5):1323–5.
112. Hubiche T, Ged C, Benard A, Leaute-Labreze C, McElreavey K, de Verneuil H, et al. Analysis of SPINK 5, KLK 7 and FLG genotypes in a French atopic dermatitis cohort. Acta Derm Venereol. 2007;87(6):499–505.
113. Chen H, Ho JC, Sandilands A, Chan YC, Giam YC, Evans AT, et al. Unique and recurrent mutations in the filaggrin gene in Singaporean Chinese patients with ichthyosis vulgaris. J Invest Dermatol. 2008; 128(7):1669–75.
114. Nomura T, Akiyama M, Sandilands A, Nemoto-Hasebe I, Sakai K, Nagasaki A, et al. Specific filaggrin mutations cause ichthyosis vulgaris and are significantly associated with atopic dermatitis in Japan. J Invest Dermatol. 2008;128(6):1436–41.
115. Chang YC, Wu WM, Chen CH, Hu CF, Hsu LA. Association between P478S polymorphism of the filaggrin gene and risk of psoriasis in a Chinese population in Taiwan. Arch Dermatol Res. 2008;300(3): 133–7.
116. Ekelund E, Lieden A, Link J, Lee SP, D'Amato M, Palmer CN, et al. Loss-of-function variants of the filaggrin gene are associated with atopic eczema and associated phenotypes in Swedish families. Acta Derm Venereol. 2008;88(1):15–9.
117. Oji V, Seller N, Sandilands A, Gruber R, Gerss J, Huffmeier U, et al. Ichthyosis vulgaris: novel FLG mutations in the German population and high presence of CD1a+ cells in the epidermis of the atopic subgroup. Br J Dermatol. 2009;160(4):771–81.
118. Carlsen BC, Meldgaard M, Johansen JD, Thyssen JP, Menne T, Szecsi PB, et al. Filaggrin compound heterozygous patients carry mutations in trans position. Exp Dermatol. 2013;22(9):572–5.
119. Carlsen BC, Johansen JD, Menne T, Meldgaard M, Szecsi PB, Stender S, et al. Filaggrin null mutations and association with contact allergy and allergic contact dermatitis: results from a tertiary dermatology clinic. Contact Dermatitis. 2010;63(2):89–95.
120. Epstein TG, LeMasters GK, Bernstein DI, Ericksen MB, Martin LJ, Ryan PH, et al. Genetic variation in small proline rich protein 2B as a predictor for asthma among children with eczema. Ann Allergy Asthma Immunol. 2012;108(3):145–50.
121. van den Bogaard EH, Bergboer JG, Vonk-Bergers M, van Vlijmen-Willems IM, Hato SV, van der Valk PG, et al. Coal tar induces AHR-dependent skin barrier repair in atopic dermatitis. J Clin Invest. 2013;123(2):917–27.
122. Sandilands A, O'Regan GM, Liao H, Zhao Y, Terron-Kwiatkowski A, Watson RM, et al. Prevalent and rare mutations in the gene encoding filaggrin cause ichthyosis vulgaris and predispose individuals to atopic dermatitis. J Invest Dermatol. 2006;126(8): 1770–5.
123. Janssens M, Mulder AA, van Smeden J, Pilgram GSK, Wolterbeek R, Lavrijsen APM, et al. Electron diffraction study of lipids in non-lesional stratum corneum of atopic eczema patients. Biochim et Biophys Acta (BBA) – Biomembranes. 2013;1828(8):1814–21.
124. Gonzalez FJ, Valdes-Rodriguez R, Ramirez-Elias MG, Castillo-Martinez C, Saavedra-Alanis VM, Moncada B. Noninvasive detection of filaggrin gene mutations using Raman spectroscopy. Biomed Opt Expr. 2011;2(12):3363–6.
125. Kezic S, Kammeyer A, Calkoen F, Fluhr JW, Bos JD. Natural moisturizing factor components in the stratum corneum as biomarkers of filaggrin genotype: evaluation of minimally invasive methods. Br J Dermatol. 2009;161(5):1098–104.
126. Mlitz V, Latreille J, Gardinier S, Jdid R, Drouault Y, Hufnagl P, et al. Impact of filaggrin mutations on Raman spectra and biophysical properties of the stratum corneum in mild to moderate atopic dermatitis. J Eur Acad Dermatol Venereol. 2012;26(8): 983–90.
127. Landeck L, Visser M, Kezic S, John SM. Genotype-phenotype associations in filaggrin loss-of-function mutation carriers. Contact Dermatitis. 2013;68(3): 149–55.
128. Landeck L, Visser M, Skudlik C, Brans R, Kezic S, John SM. Clinical course of occupational irritant contact dermatitis of the hands in relation to filaggrin genotype status and atopy. Br J Dermatol. 2012; 167(6):1302–9.

129. Janssens M, van Smeden J, Gooris GS, Bras W, Portale G, Caspers PJ, et al. Increase in short-chain ceramides correlates with an altered lipid organization and decreased barrier function in atopic eczema patients. J Lipid Res. 2012;53(12):2755–66.
130. Visser MJ, Landeck L, Campbell LE, McLean WH, Weidinger S, Calkoen F, et al. Impact of atopic dermatitis and loss-of-function mutations in the filaggrin gene on the development of occupational irritant contact dermatitis. Br J Dermatol. 2013;168(2): 326–32.
131. Sabolic Pipinić I, Varnai VM, Turk R, Breljak D, Kezic S, Macan J. Low frequency of filaggrin null mutations in Croatia and their relation with allergic diseases. Int J Immunogenet. 2013;40(3): 192–8.
132. Kezic S, Kemperman PM, Koster ES, de Jongh CM, Thio HB, Campbell LE, et al. Loss-of-function mutations in the filaggrin gene lead to reduced level of natural moisturizing factor in the stratum corneum. J Invest Dermatol. 2008;128(8):2117–9.
133. Gruber R, Janecke AR, Grabher D, Horak E, Schmuth M, Lercher P. Lower prevalence of common filaggrin mutations in a community sample of atopic eczema: is disease severity important? Wien Klin Wochenschr. 2010;122(19–20):551–7.
134. Barker JN, Palmer CN, Zhao Y, Liao H, Hull PR, Lee SP, et al. Null mutations in the filaggrin gene (FLG) determine major susceptibility to early-onset atopic dermatitis that persists into adulthood. J Invest Dermatol. 2007;127(3):564–7.
135. Meldgaard M, Bollen PJ, Finsen B. Non-invasive method for sampling and extraction of mouse DNA for PCR. Lab Anim. 2004;38(4):413–7.
136. Sasaki T, Kudoh J, Ebihara T, Shiohama A, Asakawa S, Shimizu A, et al. Sequence analysis of filaggrin gene by novel shotgun method in Japanese atopic dermatitis. J Dermatol Sci. 2008;51(2):113–20.
137. Clausen ML, Jungersted JM, Andersen PS, Slotved HC, Krogfelt KA, Agner T. Human Beta-defensin-2 as a marker for disease severity and skin barrier properties in Atopic Dermatitis. Br J Dermatol. 2013;6:10.
138. Aslam A, Lloyd-Lavery A, Warrell DA, Misbah S, Ogg GS. Common filaggrin null alleles are not associated with hymenoptera venom allergy in Europeans. Int Arch Allergy Immunol. 2011;154(4): 353–5.
139. Droitcourt C, Touboul D, Ged C, Ezzedine K, Cario-Andre M, de Verneuil H, et al. A prospective study of filaggrin null mutations in keratoconus patients with or without atopic disorders. Dermatology. 2011;222(4):336–41.
140. Chawes BL, Bonnelykke K, Kreiner-Moller E, Bisgaard H. Children with allergic and nonallergic rhinitis have a similar risk of asthma. J Allergy Clin Immunol. 2010;126(3):567–73.
141. Ching GK, Hon KL, Ng PC, Leung TF. Filaggrin null mutations in childhood atopic dermatitis among the Chinese. Int J Immunogenet. 2009;36(4):251–4.
142. Molin S, Vollmer S, Weiss EH, Ruzicka T, Prinz JC. Filaggrin mutations may confer susceptibility to chronic hand eczema characterized by combined allergic and irritant contact dermatitis. Br J Dermatol. 2009;161(4):801–7.
143. Nomura T, Sandilands A, Akiyama M, Liao H, Evans AT, Sakai K, et al. Unique mutations in the filaggrin gene in Japanese patients with ichthyosis vulgaris and atopic dermatitis. J Allergy Clin Immunol. 2007;119(2):434–40.
144. Betz RC, Pforr J, Flaquer A, Redler S, Hanneken S, Eigelshoven S, et al. Loss-of-function mutations in the filaggrin gene and alopecia areata: strong risk factor for a severe course of disease in patients comorbid for atopic disease. J Invest Dermatol. 2007;127(11):2539–43.
145. Marenholz I, Nickel R, Ruschendorf F, Schulz F, Esparza-Gordillo J, Kerscher T, et al. Filaggrin loss-of-function mutations predispose to phenotypes involved in the atopic march. J Allergy Clin Immunol. 2006;118(4):866–71.
146. Howell MD, Kim BE, Gao P, Grant AV, Boguniewicz M, DeBenedetto A, et al. Cytokine modulation of atopic dermatitis filaggrin skin expression. J Allergy Clin Immunol. 2009;124(3 Suppl 2):R7–12.
147. Meldgaard M, Szecsi PB, Carlsen BC, Thyssen JP, Johansen JD, Menne T, et al. A novel multiplex analysis of filaggrin polymorphisms: a universally applicable method for genotyping. Clin Chim Acta. 2012;413(19–20):1488–92.
148. Linneberg A, Fenger RV, Husemoen LL, Thuesen BH, Skaaby T, Gonzalez-Quintela A, et al. Association between loss-of-function mutations in the filaggrin gene and self-reported food allergy and alcohol sensitivity. Int Arch Allergy Immunol. 2013;161(3):234–42.
149. Bandier J, Ross-Hansen K, Carlsen BC, Menne T, Linneberg A, Stender S, et al. Carriers of filaggrin gene (FLG) mutations avoid professional exposure to irritants in adulthood. Contact Dermatitis. 2013;28:10.
150. Carlsen BC, Thyssen JP, Menne T, Meldgaard M, Linneberg A, Nielsen NH, et al. Association between filaggrin null mutations and concomitant atopic dermatitis and contact allergy. Clin Exp Dermatol. 2011;36(5):467–72.
151. Carlsen BC, Meldgaard M, Hamann D, Hamann Q, Hamann C, Thyssen JP, et al. Latex allergy and filaggrin null mutations. J Dent. 2011;39(2):128–32.
152. Ross-Hansen K, Menne T, Johansen JD, Carlsen BC, Linneberg A, Nielsen NH, et al. Nickel reactivity and filaggrin null mutations–evaluation of the filaggrin bypass theory in a general population. Contact Dermatitis. 2011;64(1):24–31.
153. Thyssen J, Johansen J, Carlsen B, Linneberg A, Meldgaard M, Szecsi P, et al. The filaggrin null genotypes R501X and 2282del4 seem not to be associated with psoriasis: results from general population study and meta-analysis. J Eur Acad Dermatol Venereol. 2012;26:782–4.

154. Thyssen J, Carlsen B, Bisgaard H, Giwercman C, Johansen J, Linneberg A, et al. Individuals who are homozygous for the 2282del4 and R501X filaggrin null mutations do not always develop dermatitis and complete long-term remission is possible. J Eur Acad Dermatol Venereol. 2012;26:386–9.
155. Thyssen JP, Linneberg A, Johansen JD, Carlsen BC, Zachariae C, Meldgaard M, et al. Atopic diseases by filaggrin mutations and birth year. Allergy. 2012; 67(5):705–8.
156. Thyssen JP, Ross-Hansen K, Johansen JD, Zachariae C, Carlsen BC, Linneberg A, et al. Filaggrin loss-of-function mutation R501X and 2282del4 carrier status is associated with fissured skin on the hands: results from a cross-sectional population study. Br J Dermatol. 2012;166(1):46–53.
157. Thyssen JP, Linneberg A, Ross-Hansen K, Carlsen BC, Meldgaard M, Szecsi PB, et al. Filaggrin mutations are strongly associated with contact sensitization in individuals with dermatitis. Contact Dermatitis. 2013;68(5): 273–6.
158. Skaaby T, Husemoen LL, Martinussen T, Thyssen JP, Melgaard M, Thuesen BH, et al. Vitamin D status, filaggrin genotype, and cardiovascular risk factors: a Mendelian randomization approach. PLoS One. 2013;8(2):e57647.
159. McGuigan FE, Ralston SH. Single nucleotide polymorphism detection: allelic discrimination using TaqMan. Psychiatr Genet. 2002;12(3):133–6.
160. Ercan H, Ispir T, Kirac D, Baris S, Ozen A, Oztezcan S, et al. Predictors of atopic dermatitis phenotypes and severity: roles of serum immunoglobulins and filaggrin gene mutation R501X. Allergol Immunopathol (Madr). 2013;41(2):86–93.
161. Ragoussis J, Elvidge GP, Kaur K, Colella S. Matrix-assisted laser desorption/ionisation, time-of-flight mass spectrometry in genomics research. PLoS Genet. 2006;2(7):e100.
162. Kim S, Misra A. SNP genotyping: technologies and biomedical applications. Annu Rev Biomed Eng. 2007;9:289–320.
163. Kwok S, Higuchi R. Avoiding false positives with PCR. Nature. 1989;339(6221):237–8.
164. Lo YM, Mehal WZ, Fleming KA. False-positive results and the polymerase chain reaction. Lancet. 1988;2(8612):679.
165. Sobek H, Schmidt M, Frey B, Kaluza K. Heat-labile uracil-DNA glycosylase: purification and characterization. FEBS Lett. 1996;388(1):1–4.
166. Jeffrey GP, Adams PC. Pitfalls in the genetic diagnosis of hereditary hemochromatosis. Genet Test. 2000;4(2):143–6.
167. Gonzalez FJ, Alda J, Moreno-Cruz B, Martinez-Escaname M, Ramirez-Elias MG, Torres-Alvarez B, et al. Use of Raman spectroscopy for the early detection of filaggrin-related atopic dermatitis. Skin Res Technol. 2011;17(1):45–50.
168. O'Regan GM, Kemperman PM, Sandilands A, Chen H, Campbell LE, Kroboth K, et al. Raman profiles of the stratum corneum define 3 filaggrin genotype-determined atopic dermatitis endophenotypes. J Allergy Clin Immunol. 2010;126(3):574–80.

Part III

Epidemiology and Environmental Exposures

Prevalence of Filaggrin Gene Mutations: An Evolutionary Perspective

12

Marten C.G. Winge and Maria Bradley

Contents

The impact of filaggrin gene (*FLG*) mutations seems to have a peculiar pattern on disease, strongly associated with certain diseases (atopic dermatitis (AD) [1], ichthyosis vulgaris (IV) [2]) and having disease-modifying effects in others (i.e., X-linked recessive ichthyosis, pachyonychia congenita) [3–5]. Nevertheless, *FLG* mutations seem to be prevalent even in the general population. The identification of disease-causing *FLG* mutations, being either nonsense or frameshift mutations in a protein-coding exon, has enabled identification of causative variants directly from sequence analysis [1, 6]. Due to the close homology between different filaggrin repeats, genotyping has been hampered until a comprehensive sequencing strategy was established [1, 2, 7]. The repetitive nature of the region is reflecting the scarce information currently available in public databases on *FLG* mutation prevalence data or haplotype-tagging SNPs in general populations. Although this is likely to change, currently *FLG* mutation prevalence data are mainly derived from individual studies, where either the entire *FLG* gene has been sequenced or, for the main part, only selected known risk variants tested [8]. Therefore, there is a risk of underestimating the prevalence rates in populations where only certain variants have been tested. Taking that into account, loss-of-function variants have been extensively studied in certain populations such as the UK, Germany, and Japan and are prevalent in the general population, as well as robust risk factors for developing disease. The prevalence of *FLG* mutations still remains largely unexplored

M.C.G. Winge, MD, PhD (✉)
Department of Dermatology,
Program in Epithelial Biology, Stanford University,
269 Campus Drive, Stanford, CA 94035, USA
e-mail: mcgwinge@stanford.edu

M. Bradley, MD, PhD
Department of Dermatology,
Institution of Medicine,
Karolinska University Hospital,
Stockholm, 17176, Sweden
e-mail: Maria.bradley@ki.se

J.P. Thyssen, H. Maibach (eds.), *Filaggrin*,
DOI 10.1007/978-3-642-54379-1_12, © Springer-Verlag Berlin Heidelberg 2014

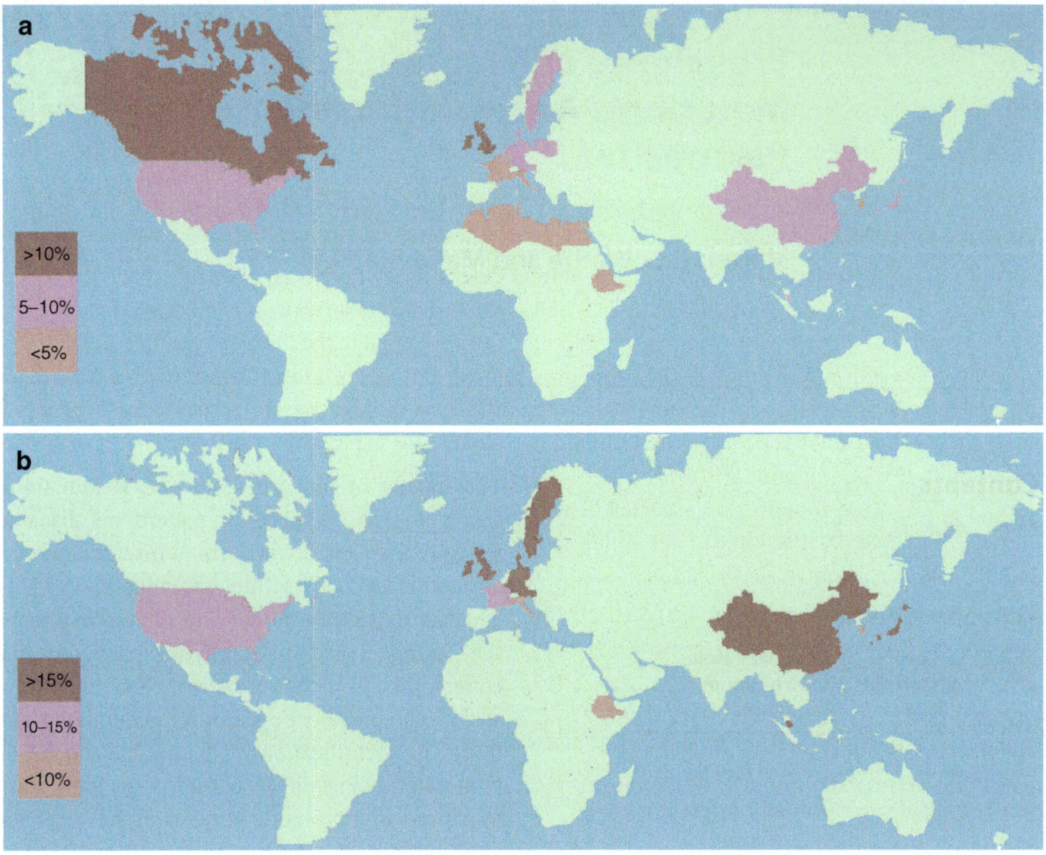

Fig. 12.1 Schematic overview over the reported distribution of *FLG* gene mutations in the general population (**a**) as well as among individuals with AD (**b**)

in many populations or entire continents, such as South America and Oceania. However, from studies conducted to date, a picture is emerging where the prevalence rate of *FLG* gene mutations clearly seems to vary between different populations, both in the general population and in individuals with IV and AD (Fig. 12.1) [9, 10].

Each ancestral population has its own unique spectrum of mutations; some are shared and some unique. Many are found on different haplotypes. Thus, individual variants can only partially mediate the contribution of *FLG* mutations to disease manifestation and thereby contribute to the heterogeneous mutation pattern conferring the genetic risk for developing AD and IV. More than 40 mutations besides the two most common ones in the European population (R501X and 2282del4) have been detected to date [10], all leading to loss of filaggrin expression. The prevalence of *FLG* loss-of-function variants has distinct differences among different ethnic groups, and each population's burden of *FLG* mutations differs, in both overall impact and the impact of each individual mutation. It is yet to be demonstrated whether, for instance, variation in intragenic gene dosage [11] follows the same pattern. The variation detected so far in intragenic gene dosage of filaggrin seems to occur through duplication of repeat eight, repeat ten, or both. For instance, an individual carrying two 12-repeat alleles (through duplication of 8 and 10) will therefore have more available filaggrin compared to an individual carrying two 10-repeat alleles. Therefore, it is possible that such variation in filaggrin amount in the skin may vary in-between populations, just as the prevalence of truncating mutations, also leading to a variation in available filaggrin. Therefore, further mapping of these largely population-specific mutations is necessary for estimating the global prevalence of *FLG* mutations in the general population as well as the combined association with AD and IV.

12.1 Europe

The initial association studies revealing *FLG* mutations as causative IV and strongly associated with AD [1, 2] were mainly including patients of Irish, Scottish, and Danish descent. Subsequently, *FLG* mutation prevalence has been extensively studied in European populations and is predicted to be present in 7–10 % of the European population [6]. Although the prevalence of *FLG*-null mutations varies across Europe, R501X and 2292del4 are the two most common ones and have consistently shown significant association with AD and IV, as well as being largely prevalent in the general population across the continent [12]. In the European population, although frequently studied, the impact of *FLG* mutations remains to be defined for several populations in Europe. Even within populations where the mutation prevalence has been reported, there is a need for additional studies, including all known variants, to fully estimate the impact of the *FLG* mutations described to date [8]. Comparing some of the prevalence data reported throughout Europe, among individuals where at least one loss-of-function allele has been reported, highlights that there is clearly variation within Europe (Table 12.1). Interestingly, the Northern countries seem to have a higher prevalence of *FLG* mutations compared to Mediterranean countries, both in the general population and among AD patients. For instance, Scandinavian countries (Sweden and Denmark) have a prevalence of 5.7–8.1 % in the general population and 12.5–19.7 % of AD patients [14, 29–31, 58–60]. Similar prevalence has been reported from the UK, 7.6–14.2 % (18.1–55.2 %) [1, 7, 26, 36, 37], and Germany, 3.8–9.6 % (15.2–22.9 %) [16–22, 25, 28, 32]. However, in the Mediterranean countries, Croatia, 2.7 % (2.6 %) [61], and Italy, 4 % (3 %), are *FLG* mutation carriers. In fact, both R501X and 2282del4 seem to be rare even among Italian AD cases (allele frequency <1 % for each), and exon and promoter sequencing in 220 AD patients only identified three additional rare mutations and no association with AD [27, 62]. The pattern of *FLG* mutations in other Mediterranean populations has not yet been examined, but the Italian and Croatian data suggest that different genetic factors may predispose to atopic dermatitis in these populations warranting further investigation [6].

12.2 Asia

FLG mutations are associated with disease in several Asian populations; however, the mutation spectrum varies [10] and is more family specific than the few presumably ancestral mutations seen in the European population. The mutation pattern in studied Asian populations is distinct and complex and is described separately in a subsequent chapter. Briefly, studies from Japan, 1.5–6.5 % of the general population (5.6–27 % with AD) [50, 51, 63–65]; Korea, 1.5 % (2.4 %) [66]; China, 0–6.5 % (15–31.4 %) [44, 67, 68]; and Taiwan, 3.8 % (14.7 %) [57], indicate that a larger number of mutations, with a more family-specific distribution pattern, give rise to a combined, strong risk of developing AD and IV. A comparison that has been made between the European population and the Singaporean Chinese well highlights these discrepancies. In the European population, two prevalent *FLG* mutations account for over 80 % of the *FLG*-null alleles, whereas in the Singaporean Chinese population there are eight different *FLG*-null mutations that account for 80 % of the spectrum of *FLG* mutations [56].

12.3 North America and Africa

In North American and African populations, less genotype information than in Europe and Asia is available. Data from a Canadian population studied in relation to peanut allergy showed that 11 %

Table 12.1 *FLG* mutation prevalence in different populations

Author	Population	GP (*n*)	*FLG* mut GP (%)	AD (*n*)	*FLG* mut AD (%)
Gruber et al. [13]	Austria	110	2.7		
Thyssen et al. [8]	Denmark	3,335	8.1	177	19.7
Thyssen et al. [14]	Denmark	2,500	7.6		
Mlitz et al. [15]	France	99	4	97	10.3
Betz et al. [16]	Germany	449	8	145	15.2
Marenholz et al. [17]	Germany	871	9.4		
Stemmler et al. [18]	Germany	667	9.6	374	15.8
Weidinger et al. [19]	Germany	2,864	7.7		
Cramer et al. [20]	Germany	2,867	6.2		
Oji et al. [21]	Germany	752	4.6		
Weichenthal et al. [22]	Germany	276	7.6		
Huffmeier et al. [23]	Germany	376	3.8		
Novak et al. [24]	Germany	1,468	7.5		
Greisenegger et al. [25]	Germany	402	7.7	462	22.9
Palmer et al. [1]	Ireland	186	8.6		55.8
Sandilands et al. [7]	Ireland	736	7.6		45.2
Zhao et al. [26]	Ireland/UK	2,117	8		
Cascella et al. [27]	Italy	201	4		3
Poninska et al. [28]	Poland	510	4.8		
Ekelund et al. [29]	Sweden			386	18.9
Ballardini et al. [30]	Sweden	1,608	6.5	286	13
Winge et al. [31]	Sweden	341	5.7		
de Jongh et al. [32]	The Netherlands	217	7.4		
Palmer et al. [1]	UK	1,008	9.3		
Barker et al. [33]	UK	1,334	8.8		42
Brown et al. [34]	UK	747	11.5		40,2
Brown et al. [35]	UK	789	14.2		18,1
Rice et al. [36]	UK	5,289	9		
Henderson et al. [37]	UK	6,971	8.8		
Van Limbergen et al. [38]	UK	944	10,9		
Gaoet al. [39]	African American	152	1.3	187	6.4
	African American	177	0.5		
Margolis et al. [40]	African American			370	5.8
Brown et al. [41]	Canadian	891	11		
Winge et al. [42]	Ethiopian	103	0	106	0,1
Margolis et al. [40]	European American	156	5.8	433	27.5
Gao et al. [39]	European American			276	27.9
Palmer et al. [1]	North African	124	0		
Li et al. [43]	China	301	4	339	26
Ma et al. [44]	China	169	6.5	160	15
Chen et al. [45]	China	160	1		
Zhang et al. [46]	China	92	0	261	31.4
Li et al. [47]	China	301	4		
Zhang et al. [48]	China	100	3		
Ching et al. [49]	China	191	0	174	2.3
Nomura et al. [50]	Japan	156	0	143	5.6
Nomura et al. [51]	Japan	133	1.5	102	11.1
Nomura et al. [52]	Japan	134	3.8	137	27

Table 12.1 (continued)

Author	Population	GP (*n*)	*FLG* mut GP (%)	AD (*n*)	*FLG* mut AD (%)
Imoto et al. [53]	Japan	1,499	6.5		
Lee et al. [54]	Korea	133	1.5	42	2.4
Chen et al. [45]	Singapore	100	0		
Common et al. [55]	Singapore	434	7.3		
Chen et al. [56]	Singapore	433	6.9	390	21.3
Wang et al. [57]	Taiwan	212	3.8	212	14.7

Table modified from Thyssen et al. [8]
[a]Prevalence data reported from selected studies on the European North American and African populations as well as an overview of studies investigating *FLG* mutations in Asian populations
Abbreviations: *GP* general population or healthy controls, *AD* atopic dermatitis, *FLG* mut filaggrin gene mutation

of their participants carried *FLG* mutations [41]. In the USA, in a cohort of subjects with AD, 16.3 % carried one or more *FLG* mutations and specifically in 27.5 % of Americans of European descent and 5.8 % of Americans of African descent [40]. Previously, the prevalence of *FLG* mutations in the general population has been estimated to be 5.8 % (27.9 % with AD) for European Americans and 0.5–1.3 % (6.4 %) for African Americans [39]. In their initial study, Palmer et al. genotyped a subset of individuals of North African descent without finding R501X or 2282del4 in any patients [1]. Until recently, *FLG-null* mutations had not been detected within African populations. A study of an AD and IV case-control material in the Ethiopian population showed that none of the four common European *FLG* mutations (R501X, 2282del4, S3247, R2447X) were prevalent. After sequencing 40 individuals and genotyping for the detected mutation in 209 individuals, only one loss-of-function mutation in one individual was identified [42].

Taken together, a strong association has been detected in Northern Europe and North America, but in Southern European (Italian and Croatian) and African populations (North African and Ethiopian), the association to disease and the overall prevalence of *FLG* mutations seem to diminish. Also, the prevalence of *FLG* mutations seems to be lower in African Americans compared to Americans of European descent [39, 40]. The highest prevalence of *FLG* mutations and association to AD seem to follow a north-south gradient, and Mediterranean and African populations seem to have a distinctively lower prevalence.

12.4 A Possible Evolutionary Role of Carrying *FLG* Mutations

One explanation could be that the *FLG* mutations prevalent today occurred after the major separation of the respective populations [6, 10, 56]. The specificity of these mutations between populations indicates that they have arisen after divergence of populations. Therefore, populations with common ancestors share common ancestral mutations, but populations without genetic admixture display a different prevalence pattern of mutations [6]. The high prevalence of *FLG* mutations in some populations, regardless of disease prevalence, makes it tempting to speculate that carrying *FLG* mutations may have or may previously have had an evolutionary advantage. The phenomenon of a "heterozygote advantage" has been discussed in, for instance, malaria-endemic areas where a type of balanced selection favoring a heterozygous state has been proposed in other conditions such as the underlying sickle-cell anemia mutations seen in malaria-endemic areas [69]. Varying susceptibility to malaria between different ethnic groups has been demonstrated by polymorphisms in several loci affecting immune response pathways [70, 71]. Such genotypic and phenotypic adaptation may also play a role in other immune-mediated pathways. For instance, it has been discussed that a more permeable barrier (such as in filaggrin-deficient skin) could confer increased immunity to infections. Repeated low-level exposure of pathogens to antigen-presenting cells in the skin might increase the immunity against infections during pandemics, such as influenza, tuberculosis, or the bubonic plague that wiped out 30–60 % of the entire European population [72]. These pandemics may not have been as widespread in populations where *FLG* mutations seem less prevalent.

Filaggrin serves several important functions in the skin, and the same is true for its degradation products. Another hypothesis of a possible evolutionary advantage of carrying *FLG* mutations involves altered levels of filaggrin-degradation metabolites. The main degradation products of filaggrin are the two organic acids: *trans-urocanic* (UCA) and pyrrolidone-5-carboxylic acids. UCA is derived from numerous histidine residues of filaggrin, which during filaggrin degradation serve as substrates for the formation of UCA [72]. UCA has been proposed to act as a UV-absorbing substance in the stratum corneum, and photoisomerization of UCA produces a molecule with an action spectrum within the UVB range [6]. Filaggrin deficiency has been associated with lower concentrations of UCA in skin cultures [73] and in vivo [74]. Also, a siRNA *FLG* knockdown model has shown that lack of filaggrin leads to an increased sensitivity to UV-induced apoptosis [73]. A result of epidermal exposure to solar UVB radiation is synthesis of vitamin D_3, which is converted in the liver to 25-OH vitamin D_3 [75]. It has been demonstrated that individuals carrying common *FLG* mutations may display up to 10 % higher serum 25-OH vitamin D_3 levels [14]. Therefore, it is possible that the mechanism mediating the significant differences in 25-OH vitamin D_3 in individuals with *FLG* mutations could involve a differentially altered UV-absorbing capacity in filaggrin-deficient skin. It remains to be clarified how filaggrin mutations affect serum 25-OH vitamin D_3 levels, if having a higher serum level could have carried an evolutionary benefit and if such mechanism is a factor underlying the high frequency of mutations in Northern countries with less UVB exposure.

Several studies, some included here, indicate that *FLG* loss-of-function variant is less common in certain populations. As *FLG* mutations clearly have a significant burden on common skin diseases, other mechanisms may be more important for the pathogenesis of IV and AD in the groups where *FLG* gene mutations do not seem to be involved in the pathomechanism. Other differential mechanisms could include, besides differential abnormal filaggrin regulation, such as cytokine-mediated downregulation [76–79], other environmental or genetic triggers of the immune response and involve other barrier-related genes. For instance, defects in tight junction genes have been linked to AD [80] and the late cornified envelope genes *LCE3B/C* to psoriasis [81]. It is also possible that IV and AD may be less common in populations with lower incidence of *FLG* mutations, which needs to be determined in further epidemiological studies. It remains to be clarified if the explanation to the high prevalence in certain populations involves an evolutionary advantage of being an *FLG* mutation carrier. Although *FLG* loss-of-function variants are a robust risk factor for developing common skin disease, the mutation prevalence displays a remarkable variation among different populations. This variation may have important clinical significance for AD and IV patients from certain populations, both considering future therapies aimed at compensating for *FLG* loss-of-function mutations and delineating the impact of other genetic and environmental risk factors on diseases associated with epidermal filaggrin deficiency.

References

1. Palmer CN, Irvine AD, Terron-Kwiatkowski A, Zhao Y, Liao H, Lee SP, et al. Common loss-of-function variants of the epidermal barrier protein filaggrin are a major predisposing factor for atopic dermatitis. Nat Genet. 2006;38(4):441–6. PubMed PMID: 16550169. Epub 2006/03/22. eng.
2. Smith FJ, Irvine AD, Terron-Kwiatkowski A, Sandilands A, Campbell LE, Zhao Y, et al. Loss-of-function mutations in the gene encoding filaggrin cause ichthyosis vulgaris. Nat Genet. 2006;38(3):337–42. PubMed PMID: 16444271. Epub 2006/01/31. eng.

3. Ramesh R, Chen H, Kukula A, Wakeling EL, Rustin MHA, McLean WHI. Exacerbation of X-linked ichthyosis phenotype in a female by inheritance of filaggrin and steroid sulfatase mutations. J Dermatol Sci. 2011;64(3):159–62.
4. Liao H, Waters AJ, Goudie DR, Aitken DA, Graham G, Smith FJ, et al. Filaggrin mutations are genetic modifying factors exacerbating X-linked ichthyosis. J Invest Dermatol. 2007;127(12):2795–8. PubMed PMID: 17657246. Epub 2007/07/28. eng.
5. Gruber R, Wilson NJ, Smith FJ, Grabher D, Steinwender L, Fritsch PO, et al. Increased pachyonychia congenita severity in patients with concurrent keratin and filaggrin mutations. Br J Dermatol. 2009;161(6):1391–5. PubMed PMID: 19785597.
6. Brown SJ, McLean WH. One remarkable molecule: filaggrin. J Invest Dermatol. 2012;132(3 Pt 2):751–62. PubMed PMID: 22158554. Pubmed Central PMCID: 3378480.
7. Sandilands A, Terron-Kwiatkowski A, Hull PR, O'Regan GM, Clayton TH, Watson RM, et al. Comprehensive analysis of the gene encoding filaggrin uncovers prevalent and rare mutations in ichthyosis vulgaris and atopic eczema. Nat Genet. 2007;39(5):650–4.
8. Thyssen JP, Godoy-Gijon E, Elias PM. Ichthyosis vulgaris – the filaggrin mutation disease. Br J Dermatol. 2013;168(6):1155–66. PubMed PMID: 23301728.
9. Irvine AD, McLean WH, Leung DY. Filaggrin mutations associated with skin and allergic diseases. N Engl J Med. 2011;365(14):1315–27. PubMed PMID: 21991953. Epub 2011/10/14. eng.
10. Akiyama M. FLG mutations in ichthyosis vulgaris and atopic eczema: spectrum of mutations and population genetics. Br J Dermatol. 2010;162(3):472–7.
11. Brown SJ, Kroboth K, Sandilands A, Campbell LE, Pohler E, Kezic S, et al. Intragenic copy number variation within filaggrin contributes to the risk of atopic dermatitis with a dose-dependent effect. J Invest Dermatol. 2012;132(1):98–104. PubMed PMID: 22071473. Pubmed Central PMCID: 3236450. Epub 2011/11/11. eng.
12. Rodriguez E, Baurecht H, Herberich E, Wagenpfeil S, Brown SJ, Cordell HJ, et al. Meta-analysis of filaggrin polymorphisms in eczema and asthma: robust risk factors in atopic disease. J Allergy Clin Immunol. 2009;123(6):1361–70.e5. PubMed PMID: 19501237.
13. Gruber R, Janecke AR, Grabher D, Horak E, Schmuth M, Lercher P. Lower prevalence of common filaggrin mutations in a community sample of atopic eczema: is disease severity important? Wiener Klinische Wochenschrift. 2010;122(19–20):551–7. PubMed PMID: 20865458.
14. Thyssen JP, Thuesen B, Huth C, Standl M, Carson CG, Heinrich J, et al. Skin barrier abnormality caused by filaggrin (FLG) mutations is associated with increased serum 25-hydroxyvitamin D concentrations. J Allergy Clin Immunol. 2012;130(5):1204–7 e2. PubMed PMID: 22921868.
15. Mlitz V, Latreille J, Gardinier S, Jdid R, Drouault Y, Hufnagl P, et al. Impact of filaggrin mutations on Raman spectra and biophysical properties of the stratum corneum in mild to moderate atopic dermatitis. J Eur Acad Dermatol Venereol. 2012;26(8):983–90. PubMed PMID: 21812836.
16. Betz RC, Pforr J, Flaquer A, Redler S, Hanneken S, Eigelshoven S, et al. Loss-of-function mutations in the filaggrin gene and alopecia areata: strong risk factor for a severe course of disease in patients comorbid for atopic disease. J Invest Dermatol. 2007;127(11):2539–43. PubMed PMID: 17581619.
17. Marenholz I, Kerscher T, Bauerfeind A, Esparza-Gordillo J, Nickel R, Keil T, et al. An interaction between filaggrin mutations and early food sensitization improves the prediction of childhood asthma. J Allergy Clin Immunol. 2009;123(4):911–6. PubMed PMID: 19348926.
18. Stemmler S, Parwez Q, Petrasch-Parwez E, Epplen JT, Hoffjan S. Two common loss-of-function mutations within the filaggrin gene predispose for early onset of atopic dermatitis. J Invest Dermatol. 2007;127(3):722–4. PubMed PMID: 17008875.
19. Weidinger S, O'Sullivan M, Illig T, Baurecht H, Depner M, Rodriguez E, et al. Filaggrin mutations, atopic eczema, hay fever, and asthma in children. J Allergy Clin Immunol. 2008;121(5):1203–9 e1. PubMed PMID: 18396323. Epub 2008/04/09. eng.
20. Cramer C, Link E, Horster M, Koletzko S, Bauer CP, Berdel D, et al. Elder siblings enhance the effect of filaggrin mutations on childhood eczema: results from the 2 birth cohort studies LISAplus and GINIplus. J Allergy Clin Immunol. 2010;125(6):1254–60 e5. PubMed PMID: 20513523. Epub 2010/06/02. eng.
21. Oji V, Seller N, Sandilands A, Gruber R, Gerss J, Huffmeier U, et al. Ichthyosis vulgaris: novel FLG mutations in the German population and high presence of CD1a+ cells in the epidermis of the atopic subgroup. Br J Dermatol. 2009;160(4):771–81. PubMed PMID: 19183181.
22. Weichenthal M, Ruether A, Schreiber S, Nair R, Voorhees JJ, Schwarz T, et al. Filaggrin R501X and 2282del4 mutations are not associated with chronic plaque-type psoriasis in a German cohort. J Invest Dermatol. 2007;127(6):1535–7. PubMed PMID: 17380114.
23. Huffmeier U, Traupe H, Oji V, Lascorz J, Stander M, Lohmann J, et al. Loss-of-function variants of the filaggrin gene are not major susceptibility factors for psoriasis vulgaris or psoriatic arthritis in German patients. J Invest Dermatol. 2007;127(6):1367–70. PubMed PMID: 17255953.
24. Novak N, Baurecht H, Schafer T, Rodriguez E, Wagenpfeil S, Klopp N, et al. Loss-of-function mutations in the filaggrin gene and allergic contact sensitization to nickel. J Invest Dermatol. 2008;128(6):1430–5. PubMed PMID: 18049447.
25. Greisenegger E, Novak N, Maintz L, Bieber T, Zimprich F, Haubenberger D, et al. Analysis of four prevalent filaggrin mutations (R501X, 2282del4,

R2447X and S3247X) in Austrian and German patients with atopic dermatitis. J Eur Acad Dermatol Venereol. 2010;24(5):607–10. PubMed PMID: 19874431.

26. Zhao Y, Terron-Kwiatkowski A, Liao H, Lee SP, Allen MH, Hull PR, et al. Filaggrin null alleles are not associated with psoriasis. J Invest Dermatol. 2007;127(8):1878–82.
27. Cascella R, Foti Cuzzola V, Lepre T, Galli E, Moschese V, Chini L, et al. Full sequencing of the FLG gene in Italian patients with atopic eczema: evidence of new mutations, but lack of an association. J Invest Dermatol. 2011;131(4):982–4. PubMed PMID: 21289640.
28. Poninska J, Samolinski B, Tomaszewska A, Raciborski F, Samel-Kowalik P, Walkiewicz A, et al. Filaggrin gene defects are independent risk factors for atopic asthma in a Polish population: a study in ECAP cohort. PLoS One. 2011;6(2):e16933. PubMed PMID: 21365004. Pubmed Central PMCID: 3041817.
29. Ekelund E, Lieden A, Link J, Lee SP, D'Amato M, Palmer CN, et al. Loss-of-function variants of the filaggrin gene are associated with atopic eczema and associated phenotypes in Swedish families. Acta Derm Venereol. 2008;88(1):15–9. PubMed PMID: 18176743.
30. Ballardini N, Kull I, Soderhall C, Lilja G, Wickman M, Wahlgren CF. Eczema severity in preadolescent children and its relation to sex, filaggrin mutations, asthma, rhinitis, aggravating factors and topical treatment: a report from the BAMSE birth cohort. Br J Dermatol. 2013;168(3):588–94. PubMed PMID: 23445315.
31. Winge MC, Suneson J, Lysell J, Nikamo P, Lieden A, Nordenskjold M, et al. Lack of association between filaggrin gene mutations and onset of psoriasis in childhood. J Eur Acad Dermatol Venereol. 2013;27(1):e124–7. PubMed PMID: 22182180.
32. de Jongh CM, Khrenova L, Verberk MM, Calkoen F, van Dijk FJ, Voss H, et al. Loss-of-function polymorphisms in the filaggrin gene are associated with an increased susceptibility to chronic irritant contact dermatitis: a case-control study. Br J Dermatol. 2008;159(3):621–7. PubMed PMID: 18637008.
33. Barker JN, Palmer CN, Zhao Y, Liao H, Hull PR, Lee SP, et al. Null mutations in the filaggrin gene (FLG) determine major susceptibility to early-onset atopic dermatitis that persists into adulthood. J Invest Dermatol. 2007;127(3):564–7. PubMed PMID: 16990802.
34. Brown SJ, Sandilands A, Zhao Y, Liao H, Relton CL, Meggitt SJ, et al. Prevalent and low-frequency null mutations in the filaggrin gene are associated with early-onset and persistent atopic eczema. J Invest Dermatol. 2008;128(6):1591–4. PubMed PMID: 18094728. Epub 2007/12/21. eng.
35. Brown SJ, Relton CL, Liao H, Zhao Y, Sandilands A, Wilson IJ, et al. Filaggrin null mutations and childhood atopic eczema: a population-based case-control study. J Allergy Clin Immunol. 2008;121(4):940–46 e3. PubMed PMID: 18313126. Epub 2008/03/04. eng.
36. Rice NE, Patel BD, Lang IA, Kumari M, Frayling TM, Murray A, et al. Filaggrin gene mutations are associated with asthma and eczema in later life. J Allergy Clin Immunol. 2008;122(4):834–6. PubMed PMID: 18760831. Pubmed Central PMCID: 2775129.
37. Henderson J, Northstone K, Lee SP, Liao H, Zhao Y, Pembrey M, et al. The burden of disease associated with filaggrin mutations: a population-based, longitudinal birth cohort study. J Allergy Clin Immunol. 2008;121(4):872–7 e9. PubMed PMID: 18325573. Epub 2008/03/08. eng.
38. Van Limbergen J, Russell RK, Nimmo ER, Zhao Y, Liao H, Drummond HE, et al. Filaggrin loss-of-function variants are associated with atopic comorbidity in pediatric inflammatory bowel disease. Inflamm Bowel Dis. 2009;15(10):1492–8. PubMed PMID: 19408338.
39. Gao P-S, Rafaels NM, Hand T, Murray T, Boguniewicz M, Hata T, et al. Filaggrin mutations that confer risk of atopic dermatitis confer greater risk for eczema herpeticum. J Allergy Clin Immunol. 2009;124(3):507–13.e7.
40. Margolis DJ, Apter AJ, Gupta J, Hoffstad O, Papadopoulos M, Campbell LE, et al. The persistence of atopic dermatitis and filaggrin (FLG) mutations in a US longitudinal cohort. J Allergy Clin Immunol. 2012;130(4):912–7. PubMed PMID: 22951058. Pubmed Central PMCID: 3462287.
41. Brown SJ, Asai Y, Cordell HJ, Campbell LE, Zhao Y, Liao H, et al. Loss-of-function variants in the filaggrin gene are a significant risk factor for peanut allergy. J Allergy Clin Immunol. 2011;127(3):661–7. PubMed PMID: 21377035. Pubmed Central PMCID: 3081065. Epub 2011/03/08. eng.
42. Winge MC, Bilcha KD, Lieden A, Shibeshi D, Sandilands A, Wahlgren CF, et al. Novel filaggrin mutation but no other loss-of-function variants found in Ethiopian patients with atopic dermatitis. Br J Dermatol. 2011;165(5):1074–80. PubMed PMID: 21692775. Epub 2011/06/23. Eng.
43. Li M, Liu Q, Liu J, Cheng R, Zhang H, Xue H, et al. Mutations analysis in filaggrin gene in northern China patients with atopic dermatitis. J Eur Acad Dermatol Venereol. 2013;27(2):169–74. PubMed PMID: 22220561.
44. Ma L, Zhang L, Di ZH, Zhao LP, Lu YN, Xu J, et al. Association analysis of filaggrin gene mutations and atopic dermatitis in Northern China. Br J Dermatol. 2010;162(1):225–7. PubMed PMID: 19863505.
45. Chen H, Ho JC, Sandilands A, Chan YC, Giam YC, Evans AT, et al. Unique and recurrent mutations in the filaggrin gene in Singaporean Chinese patients with ichthyosis vulgaris. J Invest Dermatol. 2008;128(7):1669–75. PubMed PMID: 18239616.
46. Zhang H, Guo Y, Wang W, Shi M, Chen X, Yao Z. Mutations in the filaggrin gene in Han Chinese patients with atopic dermatitis. Allergy. 2011;66(3):420–7. PubMed PMID: 21039602.

47. Li M, Chen X, Chen R, Bao Y, Yao Z. Filaggrin gene mutations are associated with independent atopic asthma in Chinese patients. Allergy. 2011;66(12):1616–7. PubMed PMID: 21923666.
48. Zhang X, Liu S, Chen X, Zhou B, Liu D, Lei G, et al. Novel and recurrent mutations in the filaggrin gene in Chinese patients with ichthyosis vulgaris. Br J Dermatol. 2010;163(1):63–9. PubMed PMID: 20222934.
49. Ching GK, Hon KL, Ng PC, Leung TF. Filaggrin null mutations in childhood atopic dermatitis among the Chinese. Int J Immunogenet. 2009;36(4):251–4. PubMed PMID: 19602001.
50. Nomura T, Sandilands A, Akiyama M, Liao H, Evans AT, Sakai K, et al. Unique mutations in the filaggrin gene in Japanese patients with ichthyosis vulgaris and atopic dermatitis. J Allergy Clin Immunol. 2007;119(2):434–40. PubMed PMID: 17291859. Epub 2007/02/13. eng.
51. Nomura T, Akiyama M, Sandilands A, Nemoto-Hasebe I, Sakai K, Nagasaki A, et al. Specific filaggrin mutations cause ichthyosis vulgaris and are significantly associated with atopic dermatitis in Japan. J Invest Dermatol. 2008;128(6):1436–41. PubMed PMID: 18200065. Epub 2008/01/18. eng.
52. Nomura T, Akiyama M, Sandilands A, Nemoto-Hasebe I, Sakai K, Nagasaki A, et al. Prevalent and rare mutations in the gene encoding filaggrin in Japanese patients with ichthyosis vulgaris and atopic dermatitis. J Invest Dermatol. 2009;129(5):1302–5. PubMed PMID: 19037238.
53. Imoto Y, Enomoto H, Fujieda S, Okamoto M, Sakashita M, Susuki D, et al. S2554X mutation in the filaggrin gene is associated with allergen sensitization in the Japanese population. J Allergy Clin Immunol. 2010;125(2):498–500 e2. PubMed PMID: 20159264.
54. Lee DE, Park SY, Han JY, Ryu HM, Lee HC, Han YS. Association between filaggrin mutations and atopic dermatitis in Korean pregnant women. Int J Dermatol. 2013;52(6):772–3. PubMed PMID: 2192369.
55. Common JE, Brown SJ, Haines RL, Goh CS, Chen H, Balakrishnan A, et al. Filaggrin null mutations are not a protective factor for acne vulgaris. J Invest Dermatol. 2011;131(6):1378–80. PubMed PMID: 21326297. Pubmed Central PMCID: 3094589.
56. Chen H, Common JE, Haines RL, Balakrishnan A, Brown SJ, Goh CS, et al. Wide spectrum of filaggrin-null mutations in atopic dermatitis highlights differences between Singaporean Chinese and European populations. Br J Dermatol. 2011;165(1):106–14. PubMed PMID: 21428977. Epub 2011/03/25. eng.
57. Wang IJ, Lin TJ, Kuo CF, Lin SL, Lee YL, Chen PC. Filaggrin polymorphism P478S, IgE level, and atopic phenotypes. Br J Dermatol. 2011;164(4):791–6. PubMed PMID: 21219289.
58. Thyssen JP, Johansen JD, Linneberg A, Menne T, Nielsen NH, Meldgaard M, et al. The association between null mutations in the filaggrin gene and contact sensitization to nickel and other chemicals in the general population. Br J Dermatol. 2010;162(6):1278–85. PubMed PMID: 20346018.
59. Thyssen JP, Ross-Hansen K, Johansen JD, Zachariae C, Carlsen BC, Linneberg A, et al. Filaggrin loss-of-function mutation R501X and 2282del4 carrier status is associated with fissured skin on the hands: results from a cross-sectional population study. Br J Dermatol. 2012;166(1):46–53. PubMed PMID: 21777221.
60. Thyssen JP, Johansen JD, Zachariae C, Menne T, Linneberg A. Xerosis is associated with atopic dermatitis, hand eczema and contact sensitization independent of filaggrin gene mutations. Acta Derm Venereol. 2013;93(4):406–10. PubMed PMID: 23420336.
61. Sabolic Pipinic I, Varnai VM, Turk R, Breljak D, Kezic S, Macan J. Low frequency of filaggrin null mutations in Croatia and their relation with allergic diseases. Int J Immunogenet. 2013;40(3):192–8. PubMed PMID: 23078034.
62. Giardina E, Paolillo N, Sinibaldi C, Novelli G. R501X and 2282del4 filaggrin mutations do not confer susceptibility to psoriasis and atopic dermatitis in Italian patients. Dermatology. 2008;216(1):83–4. PubMed PMID: 18032906.
63. Enomoto H, Hirata K, Otsuka K, Kawai T, Takahashi T, Hirota T, et al. Filaggrin null mutations are associated with atopic dermatitis and elevated levels of IgE in the Japanese population: a family and case-control study. J Hum Genet. 2008;53(7):615–21. PubMed PMID: 18521703. Epub 2008/06/04. eng.
64. Hamada T, Sandilands A, Fukuda S, Sakaguchi S, Ohyama B, Yasumoto S, et al. De novo occurrence of the filaggrin mutation p.R501X with prevalent mutation c.3321delA in a Japanese family with ichthyosis vulgaris complicated by atopic dermatitis. J Invest Dermatol. 2008;128(5):1323–5.
65. Osawa R, Akiyama M, Shimizu H. Filaggrin gene defects and the risk of developing allergic disorders. Allergol Int. 2011;60(1):1–9. PubMed PMID: 21173567. Epub 2010/12/22. Eng.
66. Kang TW, Lee JS, Oh SW, Kim SC. Filaggrin mutation c.3321delA in a Korean patient with ichthyosis vulgaris and atopic dermatitis. Dermatology. 2009;218(2):186–7.
67. Zhang H, Guo Y, Wang W, Shi M, Chen X, Yao Z. Mutations in the filaggrin gene in Han Chinese patients with atopic dermatitis. Allergy. 2011;66(3):420–7. PubMed PMID: 21039602. Epub 2010/11/03. Eng.
68. Zhang X, Liu S, Chen X, Zhou B, Liu D, Lei G, et al. Novel and recurrent mutations in the filaggrin gene in Chinese patients with ichthyosis vulgaris. Br J Dermatol. 2010;163(1):63–9.
69. Weatherall DJ. Phenotype[mdash]genotype relationships in monogenic disease: lessons from the thalassaemias. Nat Rev Genet. 2001;2(4):245–55.
70. McGuire W, Hill AV, Allsopp CE, Greenwood BM, Kwiatkowski D. Variation in the TNF-alpha promoter region associated with susceptibility to cerebral

malaria. Nature. 1994;371(6497):508–10. PubMed PMID: 7935762. Epub 1994/10/06. eng.
71. Fernandez-Reyes D, Craig AG, Kyes SA, Peshu N, Snow RW, Berendt AR, et al. A high frequency African coding polymorphism in the N-terminal domain of ICAM-1 predisposing to cerebral malaria in Kenya. Hum Mol Genet. 1997;6(8):1357–60. PubMed PMID: 9259284. Epub 1997/08/01. eng.
72. Irvine AD, McLean WHI. Breaking the (un)sound barrier: filaggrin is a major gene for atopic dermatitis. J Invest Dermatol. 2006;126(6):1200–2.
73. Mildner M, Jin J, Eckhart L, Kezic S, Gruber F, Barresi C, et al. Knockdown of filaggrin impairs diffusion barrier function and increases UV sensitivity in a human skin model. J Invest Dermatol. 2010;130(9):2286–94.
74. Kezic S, Kammeyer A, Calkoen F, Fluhr JW, Bos JD. Natural moisturizing factor components in the stratum corneum as biomarkers of filaggrin genotype: evaluation of minimally invasive methods. Br J Dermatol. 2009;161(5):1098–104. PubMed PMID: 19857209. Epub 2009/10/28. eng.
75. Holick MF. Vitamin D, deficiency. N Engl J Med. 2007;357(3):266–81. PubMed PMID: 17634462.
76. Gutowska-Owsiak D, Schaupp AL, Salimi M, Selvakumar TA, McPherson T, Taylor S, et al. IL-17 downregulates filaggrin and affects keratinocyte expression of genes associated with cellular adhesion. Exp Dermatol. 2012;21(2):104–10. PubMed PMID: 22229441.
77. Gutowska-Owsiak D, Schaupp AL, Salimi M, Taylor S, Ogg GS. Interleukin-22 downregulates filaggrin expression and affects expression of profilaggrin processing enzymes. Br J Dermatol. 2011;165(3):492–8. PubMed PMID: 21564072. Epub 2011/05/14. eng.
78. Howell MD, Kim BE, Gao P, Grant AV, Boguniewicz M, DeBenedetto A, et al. Cytokine modulation of atopic dermatitis filaggrin skin expression. J Allergy Clin Immunol. 2009;124(3 Suppl 2):R7–12. PubMed PMID: 19720210. Epub 2009/09/02. eng.
79. Hvid M, Vestergaard C, Kemp K, Christensen GB, Deleuran B, Deleuran M. IL-25 in atopic dermatitis: a possible link between inflammation and skin barrier dysfunction? J Invest Dermatol. 2011;131(1):150–7. PubMed PMID: 20861853.
80. De Benedetto A, Rafaels NM, McGirt LY, Ivanov AI, Georas SN, Cheadle C, et al. Tight junction defects in patients with atopic dermatitis. J Allergy Clin Immunol. 2011;127(3):773–86 e7.
81. de Cid R, Riveira-Munoz E, Zeeuwen PL, Robarge J, Liao W, Dannhauser EN, et al. Deletion of the late cornified envelope LCE3B and LCE3C genes as a susceptibility factor for psoriasis. Nat Genet. 2009;41(2):211–5. PubMed PMID: 19169253. Epub 2009/01/27. eng.

Filaggrin Gene Mutations in Asian Races

13

Ruhong Cheng, Ming Li, Hui Zhang, and Zhirong Yao

Contents

Currently, a total of 51 null mutations in the filaggrin gene (*FLG*) have been identified in ichthyosis vulgaris (IV), atopic dermatitis (AD), and asthma patients from seven Asian populations (mainland Chinese, Taiwanese, Singaporean Chinese, Japanese, Korean, Pakistani, and Bangladeshi). Only three *FLG* mutations, R501X, 2282del4, and E2422X, are shared by European and Asian populations. But their weight in the spectrums of the two continents populations differs significantly. The mutation 3321delA is the most prevalent mutation in five Asian populations (Japanese, Singaporean Chinese, mainland Chinese, Taiwanese, and Korean) [1–7]; however, it has not been reported in European populations. The mutation R501X, also being identified in five Asian populations (Japanese, Singaporean Chinese, mainland Chinese, Bangladeshi, and Pakistani) and common in European populations, however, was rare in Asians [4, 5, 8–10]. The mutation Q2417X is common in mainland Chinese, Taiwanese, as well as Singaporean Chinese [1–4, 11, 12]. Each of the 12 *FLG* mutations, including 7 nonsense mutations (R826X, S1302X, S1515X, E2422X, S2706X, K4671X) and 4 frameshift mutations (441delA, 2282del4, 6950del8, 7945delA), has been found in two Asian populations [1, 4, 10, 12–14].

The remaining 39 (76.5 %) *FLG* mutations are population-specific or family-specific. Five nonsense mutations (S1695X, Q1701X, S2554X, S2889X, S3296X), originally identified in Japanese, have not yet been found in other Asian populations [5, 15, 16]. A total

R. Cheng, MD • M. Li, MD • H. Zhang, MD
Z. Yao, MD (✉)
Department of Dermatology, Xin Hua Hospital Affiliated to Shanghai Jiao Tong University School of Medicine, Kongjiang Road 1665, Shanghai Shanghai 200092, China
e-mail: shirleywo6@163.com; aypyslm@163.com; c_zhanghui@sina.com; dermatology.yao@sohu.com

J.P. Thyssen, H. Maibach (eds.), *Filaggrin*,
DOI 10.1007/978-3-642-54379-1_13, © Springer-Verlag Berlin Heidelberg 2014

of 12 *FLG* mutations, including 5 frameshift mutations (1640delG, 2952delC, 4004del2, 4275del2, 8393delA), 5 nonsense mutations (G323X, Q368X, S406X, Q1745X, R4307X), and 2 insertion mutations (1249insG and 9040_9058dup), were shown to be specific to Singaporean Chinese [4, 12]. Nineteen *FLG null* variants, including 9 frameshift mutations (441-442delAG, 3222del4, 4026delT, 4271delAA, 5757del4, 6218-6219delAA, 6834del5, 7145del4, 8001del4), 9 nonsense mutations (Q1070X, R1140X, Q1256X, R1474X, Q1712X, Q1790X, Q2397X, R4306X, Q4492X), and one insertion mutation 478insA, are specific to mainland Chinese [1, 2, 17–20]. A nonsense mutation E1795X has only been found in one IV family of Taiwanese [3]. The 2767insT mutation, initially found in two IV-affected individuals from a Bangladeshi family, has not been reported in other Asian populations [10] (Fig. 13.1).

13.1 Japanese

Nomura et al. carried out the first study of *FLG* mutations in an Asian population. They demonstrated the absence of European prevalent *FLG* mutations (2282del4 and R501X) in 253 Japanese individuals and identified two novel *FLG null* mutations (3321delA and S2554X) in four Japanese families with IV. Eight (5.6 %) of 143 Japanese patients with AD were found to carry the two novel mutations (S2554X and 3321delA in six and two patients, respectively) [15]. Neither of them was found in 156 controls. In another study, the allele frequency of S2554X and 3321delA in AD patients was 3.4 and 2.7 %, respectively. Only 1 % of healthy subjects carried the two *FLG null* mutations. Mutation S2554X and combined *FLG null* variants (S2554X and 3321delA) were both associated with the development of AD significantly in the family and case-control studies [21].

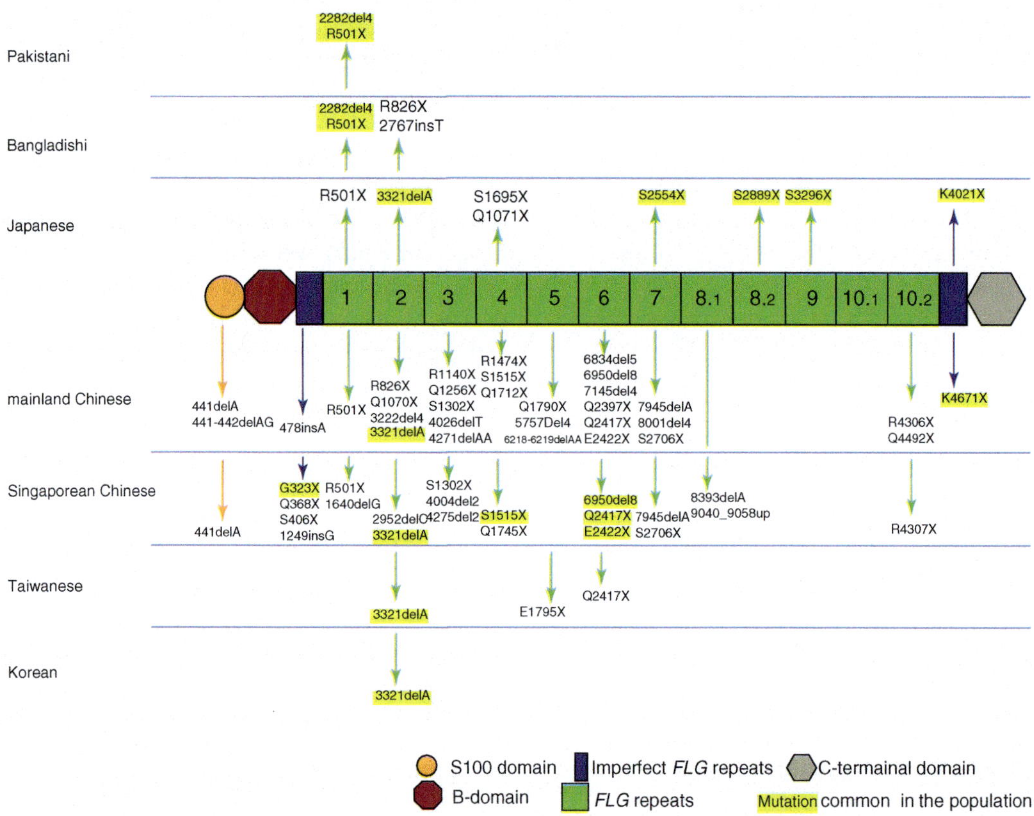

Fig. 13.1 Spectrum of filaggrin gene (*FLG*) mutations in Asia among various populations

In 2008, Nomura et al. screened the two mutations (3321delA and S2554X) in seven Japanese families with IV. Two of the families were shown to harbor neither of them. A comprehensive sequencing of *FLG* gene was subsequently performed in the two families with IV. Another two *FLG* null mutations (S2889X, S3296X) were subsequently identified. A study on the four *FLG* null mutations (3321delA, S2554X, S2889X, and S3296X) in an AD cohort showed that 10 (9.8 %), 6 (5.9 %), 4 (3.9 %), and 3 (2.9 %) of 102 unrelated Japanese patients with AD carried the mutations S2889X, S2554X, 3321delA, and S3296X, respectively. More than 20 % of Japanese AD patients were found to be carriers of the four null variants in the *FLG*, compared to 3 % in controls. A significant association between the combined *FLG* mutations (3321delA, S2554X, S2889X, and S3296X) and AD was confirmed [5].

The null variant Q1701X, found in a homozygous state in a proband with IV, was carried by 1.7 % of AD patients. The mutation S1695X, located in six amino acids upstream from Q1701X, was found to be extremely rare. About 25 % of Japanese patients with AD were carriers of one or more of the six *FLG* null mutations (R501X, 3321delA, S2554X, S2889X, S3296X, Q1701X, and S1695X), compared to 4 % of controls. A strong association between the combined *FLG* null mutations and AD was demonstrated [16].

In addition, the mutation K4021X in the C-terminal incomplete filaggrin repeat was carried by 4 (2.9 %) of 137 Japanese patients with AD [14]. A compound heterozygous carrier of both R501X and 3321delA was also found in a Japanese patient severely affected by AD and IV. Of note, the R501X mutation in the family without European ancestry was on a different haplotype from that found in Europeans. Hence, the mutation R501X was carried on one of the two 11-repeat filaggrin alleles (each carries a duplicated repeat 8) in Europeans, whereas in the Japanese family, the R501X was on one of the 12-repeat filaggrin alleles (each carries duplicated repeat 8 and duplicated repeat 10). The mutant allele in Europeans also carries a single nucleotide polymorphism p.R3564L in filaggrin repeat 10, whereas all members in the Japanese family were homozygous for a different single nucleotide polymorphism, p.R3564H, in the same codon on repeat 10. The mutation R501X is rare in Japan because it has been excluded from 4 individuals with IV, 143 individuals with AD, and 391 controls [22].

In general, eight different *FLG* mutations (R501X, 3321delA, S1695X, Q1701X, S2554X, S2889X, S3296X, and K4021X) have been found in Japanese patients with AD and IV. Mutations S2889X, S2554X, 3321delA, S3296X, and K4021X were shown to be more prevalent in AD patients in Japanese population, whereas R501X, Q1701X, and S1695X were relatively uncommon. Up to 27 % of Japanese AD patients were shown to carry one or more of the eight *FLG* null mutations compared to that 3.7 % in general controls. Hence, *FLG* null mutations are significantly associated with AD in the Japanese population [14].

In 2010, Osawa R et al. carried out a study analyzing *FLG* mutations, AD, and asthma in Japanese. They screened for the eight identified *FLG* mutations in a cohort of 172 AD patients (73 experienced complications with asthma), 134 unrelated controls, and 137 patients with asthma. It was shown that 27.4 % of patients with AD complicated with asthma carried one or more *FLG* mutations; 26.3 % of AD patients without asthma carried one or more *FLG* mutations. The *FLG* variants were also carried by 3.7 % of control individuals. In European populations, the proportion was 7.5 %. Eight percent of the asthma patients carried one or more *FLG* mutations. Of asthma patients complicated by AD, 22.2 % carried one or more *FLG* mutations; 5.9 % of asthma patients without AD carried one or more *FLG* mutations. It was confirmed that in Japan there was a strong association between *FLG* mutations and AD with or without asthma and between *FLG* mutations and asthma patients who had AD [23]. However, no association between asthma and *FLG* mutations was identified.

Imoto et al. carried out the first population-based study in Asia on the effects of *FLG* null mutations on atopic phenotypes. The study included a total of 1,575 participants. Four *FLG* mutations (3321delA, S2554X, S2889X, and S3296X) were genotyped in 1,499 participants. Of 1,093 participants who completed the asthma/rhinitis questionnaire, 98 (6.5 %) of them had at least one *FLG* null mutation, and one participant was compound heterozygote for the null mutations. Interestingly, the mutation S2554X in *FLG* was demonstrated to be statistically associated with allergen sensitization in the Japanese population [6].

13.2 Singaporean Chinese

Early in 2008, Chen et al. performed full sequencing of the *FLG* in eight unrelated Singaporean Chinese IV patients. Two of the IV patients were compound heterozygous carriers of the 441delA and 1249insG, and 1249insG and 7945delA mutations, respectively. The other four patients were heterozygote for the E2422X, Q2417X, 7945delA, and R4307X mutations, respectively. Among 100 Singaporean and 160 Chinese normal controls, only the mutation E2422X, initially reported in a Dutch IV patient, was identified in a Chinese control [12]. Subsequently in 2011, Chen et al. carried out a comprehensive *FLG* analysis of 69 individuals with IV and 23 individuals with AD and revealed 21 mutations in the cohort. Two of them were 3321delA and R501X, which were most common in Asian and European populations, respectively. In the process of screening for the 21 *FLG* mutations in 425 individuals with AD, a novel mutation 4275del2 was identified.

Ultimately, a total of 22 various *FLG null* mutations (see Fig. 13.1) have been found in Singaporean Chinese with AD and IV. The mutations G323X, 3321delA, S1515X, 6950_6957del8, Q2417X, and E2422X were shown to be more prevalent in Singaporean Chinese with AD [4]. In a study screening for all the 22 *FLG* mutations in 425 Singaporean Chinese AD patients, 17 mutations were found to be carried by 20.2 % of the participants. In contrast, 7.3 % of 440 population controls carried the 17 *FLG* null mutations. Statistical analysis showed that the *FLG* null combined genotype was significantly associated with AD [4].

13.3 Mainland Chinese

At the beginning, Ching et al. screened for five identified *FLG* mutations (R501X, 2282del4, and R2447X in Europeans; S2554X and S2889X in Japanese) in 174 Chinese patients with AD and 191 controls. Only four (2.3 %) patients with AD were found to be heterozygous carriers of the mutation R501X [8]. To figure out the spectrum of *FLG* mutations in mainland Chinese, Zhang and Li et al. carried out a comprehensive sequencing of the entire *FLG* coding region in 600 individuals in Chinese AD case series (261 and 339 patients from the south and north parts of China, respectively). A total of 26 *FLG* null mutations were found in Chinese AD patients. Eight *FLG null* mutations (441delA, R501X, R826X, 6950del8, R1140X, R1474X, S2706X, and R4306X) were specific to AD in southern China, whereas another eight *FLG null* mutations (478insA, 8001del4, Q2397X, 7145del4, Q1070X, S1515X, Q1712X, and 4026delT) were specific to AD in northern of China. The remaining 10 *FLG null* mutations (4271delAA, 3222del4, 3321delA, Q1790X, 5757del4, 6834del5, Q2417X, E2422X, 7945delA, K4671X) were identified in both the southern and northern part of China. In mainland China, 28.3 % of AD patients were *FLG null* mutation carriers. The combined *FLG null* mutations were significantly associated with AD in both southern and northern Chinese [1, 2].

Two mutations, 3321delA and K4671X, were most common in a cohort of Chinese AD patients. 11.8 % of 600 Chinese Han with AD were carriers of the mutation 3321delA, whereas 10.3 % of AD patients were shown to harbor the mutation K4671X. The frequencies of 3321delA and K4671X in 301 healthy controls were 1.00 and 2.66 %, respectively. The association of AD with either 3321delA or K4671X was statistically significant in case-control studies [1, 2]. A family-based association study on a cohort of 100 family trios with an AD proband confirmed the association between 3321delA and AD. Mutation K4671X, named K4021X in Japanese AD patients with a frequency of 2.9 %, was not identified in Taiwanese and Singaporean Chinese individuals with AD or IV. Moreover, the family-based association study indicated that the mutation K4671X was not overtransmitted to AD probands and that the P value did not reach statistical significance in a transmission disequilibrium test. To further clarify the paradoxical results from family studies and case-control studies, filaggrin expression at both mRNA and protein levels was measured in the skin of three subjects with K4671X mutations and a normal control, suggesting that K4671X was not a pathogenic mutation. The association between 3321delA and AD was confirmed by the family-based study [24].

A series of studies on *FLG* mutations in Chinese IV patients were carried out. In total, 12 *FLG* mutations have been found in isolated mainland Chinese IV cases. The mutations 3321delA and K4671X were the most common mutations, which is consistent with the observation in AD patients. Zhang et al. showed that 8 of 11 Chinese families with IV harbored the mutation 3321delA, whereas the mutation E2422X was present in only two of them. The mutation Q1256X was merely found in one of the families [19]. Xiong et al. sequenced the entire coding region of the *FLG* in probands of three pedigrees with IV. They found 3321delA in all the families and identified two novel mutations (478insA and 6218-6219delAA) [18]. In 2011, Zhang et al. carried out a comprehensive sequencing of *FLG* gene in 21 unrelated individuals with IV. Thirteen (61.9 %) of the 21 IV patients were *FLG null* mutation carriers. A total of four *FLG null* mutations, 3321delA, K4671X, Q2417X, and Q1790X, were found in the patients, with a frequency of 33.3, 19.1, 4.8, and 4.8 %, respectively [11]. Subsequently in 2013, Li et al. conducted another comprehensive sequencing of the *FLG* gene in 21 IV pedigrees and 33 patients with sporadic isolated IV. In this study, a total of seven *FLG null* mutations were identified, including 3321delA, Q1790X, 5757del4, S2706X, and K4671X and two novel mutations 441-442delAG and Q4492X. Twenty-six (48.14 %) of 54 IV patients were carriers of the mutation 3321delA, whereas 9 (16.66 %) of 54 IV patients carried the mutation K4671X. Only 1.85 % of IV patients were shown to carry one of the other 5 *FLG null* mutations [17]. Interestingly, the frequency of *FLG null* mutations in individuals with AD and IV was significantly lower than that among patients with IV alone in both studies [11, 17], which illustrates that other factors than *FLG* mutations in AD can result in an ichthyosiform phenotype.

Li et al. also investigated the association between *FLG null* mutations and atopic asthma (AA) without coexisting AD. A comprehensive sequencing of the *FLG* in 121 AA patients without AD was conducted. Four *FLG null* mutations (3321delA, S1302X, Q2397X, and K4671X) were identified in 19 AA patients without AD (15.7 %). The mutations 3321delA and K4671X were carried by 7 (5.79 %) and 10 (8.26 %) of 121 patients. Only one (0.83 %) of them carried the mutation S1302X and Q2397X, respectively. The association between the combined *FLG* mutations and AA without AD was statistically significant.

13.4 Taiwanese

A total of 3 *FLG null* mutations (3321delA, Q2417X, and E1795X) have been revealed in Taiwanese. The corresponding study was a comprehensive sequencing of the entire *FLG* coding region in 12 individuals from 4 unrelated Taiwanese IV families. The mutation 3321delA was found in three IV families, whereas Q2417X was detected in one of the four families. The mutation E1795X, being unique in the Taiwanese population, was found in a family where the proband was a compound heterozygous carrier of E1795X and 3321delA [3]. It can be deduced that mutation 3321delA, prevalent in mainland Chinese, Singaporean Chinese, and Japanese, is also the common mutation among Taiwanese IV patients. The mutation Q2417X is relatively prevalent in Chinese population since it has been found in Singaporean Chinese and mainland Chinese.

13.5 Korean

To date, 3321delA was the only *FLG* mutation found in Koreans. It was first found in a 35-year-old Korean patient with IV and AD in a heterozygous state [25]. Later, Lee, DE et al. revealed that the frequency of the mutation 3321delA was 2.4 and 1.5 % in AD patients and healthy controls, respectively. The mutation S2554X, relatively common in Japan, was not found in 42 unrelated AD patients and 133 healthy controls [7]. An extensive case-control study may be required to identify novel *FLG* mutations in the Korean population and further illustrates the role of *FLG* mutations play in the pathogenesis of IV and AD in Koreans.

13.6 Bangladeshi and Pakistani

In 2009, Sinclair et al. screened a cohort of 80 Bangladeshi families (an Indian subcontinent population residing in East London) with siblings affected by AD for the *FLG* mutations R501X and 2282del4. Six of the families were shown to carry the two *FLG* mutations, although only a weak association between the two *FLG* mutations and AD was observed. Furthermore, sequencing of the *FLG* in two individuals with the IV phenotype from two unrelated families was performed and revealed another two mutations, R826X and 2767insT. These two mutations were not detected within other Bangladeshi families [10]. In 2010, Samdani et al. succeeded in demonstrating the presence of a 1.5 kb *FLG* gene fragment in 5 out of 7 IV families, including 16 affected and 19 unaffected members. In 2011, Naz et al. successfully targeted and amplified an 811 bp *FLG* fragment in 7 unrelated Pakistani IV families with 16 affected and 19 unaffected members. Both studies suggest the possibility of the R501X and 2282del4 mutations as being the major causes of IV in Pakistanis [9, 13].

The spectrum of *FLG* mutations in Bangladeshi and Pakistanis is different from that observed in Eastern and Southeastern Asian populations. Two *FLG* mutations (R501X and 2282del4) prevalent in the European population were found to be the most common in AD patients from Bangladesh and IV patients from Pakistan, which can be explained by the fact that Bangladeshi and Pakistanis from South Asia share common lineage with South Europeans.

References

1. Zhang H, Guo Y, Wang W, Shi M, Chen X, Yao Z. Mutations in the filaggrin gene in Han Chinese patients with atopic dermatitis. Allergy. 2011;66:420–7.
2. Li M, Liu Q, Liu J, Cheng R, Zhang H, Xue H, et al. Mutations analysis in filaggrin gene in northern China patients with atopic dermatitis. J Eur Acad Dermatol Venereol. 2013;27:169–74.
3. Hsu CK, Akiyama M, Nemoto-Hasebe I, Nomura T, Sandilands A, Chao SC, et al. Analysis of Taiwanese ichthyosis vulgaris families further demonstrates differences in FLG mutations between European and Asian populations. Br J Dermatol. 2009;161:448–51.
4. Chen H, Common JE, Haines RL, Balakrishnan A, Brown SJ, Goh CS, et al. Wide spectrum of filaggrin-null mutations in atopic dermatitis highlights differences between Singaporean Chinese and European populations. Br J Dermatol. 2011;165:106–14.
5. Nomura T, Akiyama M, Sandilands A, Nemoto-Hasebe I, Sakai K, Nagasaki A, et al. Specific filaggrin mutations cause ichthyosis vulgaris and are significantly associated with atopic dermatitis in Japan. J Invest Dermatol. 2008;128:1436–41.
6. Imoto Y, Enomoto H, Fujieda S, Okamoto M, Sakashita M, Susuki D, et al. S2554X mutation in the filaggrin gene is associated with allergen sensitization in the Japanese population. J Allergy Clin Immunol. 2010;125:498–500 e2.
7. Lee DE, Park SY, Han JY, Ryu HM, Lee HC, Han YS. Association between filaggrin mutations and atopic dermatitis in Korean pregnant women. Int J Dermatol. 2013;52(6):772–3.
8. Ching GK, Hon KL, Ng PC, Leung TF. Filaggrin null mutations in childhood atopic dermatitis among the Chinese. Int J Immunogenet. 2009;36:251–4.
9. Samdani AJ, Naz N, Ahmed N. Molecular studies of ichthyosis vulgaris in Pakistani families. J Coll Physicians Surg Pak. 2010;20:644–8.
10. Sinclair C, O'Toole EA, Paige D, El Bashir H, Robinson J, Dobson R, et al. Filaggrin mutations are associated with ichthyosis vulgaris in the Bangladeshi population. Br J Dermatol. 2009;160:1113–5.
11. Zhang H, Guo Y, Wang W, Yu X, Yao Z. Associations of FLG mutations between ichthyosis vulgaris and atopic dermatitis in Han Chinese. Allergy. 2011;66:1253–4.
12. Chen H, Ho JC, Sandilands A, Chan YC, Giam YC, Evans AT, et al. Unique and recurrent mutations in the filaggrin gene in Singaporean Chinese patients with ichthyosis vulgaris. J Invest Dermatol. 2008;128:1669–75.
13. Naz N, Samdani AJ. Detection of filaggrin gene mutation (2282del4) in Pakistani Ichthyosis vulgaris families. J Coll Physicians Surg Pak. 2011;21:382–3.
14. Nemoto-Hasebe I, Akiyama M, Nomura T, Sandilands A, McLean WH, Shimizu H. FLG mutation p.Lys4021X in the C-terminal imperfect filaggrin repeat in Japanese patients with atopic eczema. Br J Dermatol. 2009;161:1387–90.

15. Nomura T, Sandilands A, Akiyama M, Liao H, Evans AT, Sakai K, et al. Unique mutations in the filaggrin gene in Japanese patients with ichthyosis vulgaris and atopic dermatitis. J Allergy Clin Immunol. 2007;119:434–40.
16. Nomura T, Akiyama M, Sandilands A, Nemoto-Hasebe I, Sakai K, Nagasaki A, et al. Prevalent and rare mutations in the gene encoding filaggrin in Japanese patients with ichthyosis vulgaris and atopic dermatitis. J Invest Dermatol. 2009;129:1302–5.
17. Li M, Cheng R, Shi M, Liu J, Zhang G, Liu Q, et al. Analyses of FLG mutation frequency and filaggrin expression in isolated ichthyosis vulgaris and atopic dermatitis associated ichthyosis vulgaris. Br J Dermatol. 2013;168(6):1335–8.
18. Xiong Z, Luo S, Xu X, Zhang L, Peng H, Li W, et al. Novel FLG mutations associated with ichthyosis vulgaris in the Chinese population. Clin Exp Dermatol. 2012;37:177–80.
19. Zhang X, Liu S, Chen X, Zhou B, Liu D, Lei G, et al. Novel and recurrent mutations in the filaggrin gene in Chinese patients with ichthyosis vulgaris. Br J Dermatol. 2010;163:63–9.
20. Li M, Chen X, Chen R, Bao Y, Yao Z. Filaggrin gene mutations are associated with independent atopic asthma in Chinese patients. Allergy. 2011;66:1616–7.
21. Enomoto H, Hirata K, Otsuka K, Kawai T, Takahashi T, Hirota T, et al. Filaggrin null mutations are associated with atopic dermatitis and elevated levels of IgE in the Japanese population: a family and case-control study. J Hum Genet. 2008;53:615–21.
22. Hamada T, Sandilands A, Fukuda S, Sakaguchi S, Ohyama B, Yasumoto S, et al. De novo occurrence of the filaggrin mutation p.R501X with prevalent mutation c.3321delA in a Japanese family with ichthyosis vulgaris complicated by atopic dermatitis. J Invest Dermatol. 2008;128:1323–5.
23. Osawa R, Konno S, Akiyama M, Nemoto-Hasebe I, Nomura T, Nomura Y, et al. Japanese-specific filaggrin gene mutations in Japanese patients suffering from atopic eczema and asthma. J Invest Dermatol. 2010;130:2834–6.
24. Cheng R, Li M, Zhang H, Guo Y, Chen X, Tao J, et al. Common FLG mutation K4671X not associated with atopic dermatitis in Han Chinese in a family association study. PLoS One. 2012;7:e49158.
25. Kang TW, Lee JS, Oh SW, Kim SC. Filaggrin mutation c.3321delA in a Korean patient with ichthyosis vulgaris and atopic dermatitis. Dermatology. 2009;218:186–7.

14 Filaggrin Gene Loss-of-Function Mutations and Exposure to Endocrine Disrupting Environmental Chemicals

Ulla N. Joensen and Jacob P. Thyssen

Contents

A cornerstone in the treatment of atopic and xerotic skin is improvement and maintenance of skin hydration through daily use of personal care products. These include moisturizing creams (emollients) and lotions, which almost inevitably contain several classes of chemicals not directly needed for the treatment of xerotic skin, such as preservatives, perfumes, or other additive compounds [1]. It is well known that at least some of these can penetrate the epidermis and be absorbed systemically, where they potentially can affect body functions [2]. It is assumed that the skin of filaggrin gene (*FLG*) mutation carriers is more permeable to allergens than the skin of wild-type carriers [3, 4], but the evidence for increased permeability to many types of chemicals is currently very limited. It can therefore be hypothesized that skin moisturizing therapy in *FLG* mutation carriers may protect from symptoms related to the skin barrier abnormality through an improvement of skin hydration but also increase dermal exposure to several classes of chemicals and allergens, causing secondary disease. Thus, *FLG* loss-of-function mutation carriers could be expected to have higher internal exposure to several classes of additive chemicals for two reasons: increased use of personal care products to treat xerotic and inflamed skin and/or increased absorption of chemicals through a compromised skin barrier.

The topic is largely unexplored, but the first results are now emerging to suggest that *FLG* mutation carriers may indeed constitute a group

U.N. Joensen, MD, PhD (✉)
Department of Growth and Reproduction, Copenhagen University Hospital (Rigshospitalet), Blegdamsvej 9, Copenhagen DK-2100, Denmark
e-mail: ulla.nordstroem.joensen@regionh.dk

J.P. Thyssen, MD, PhD
National Allergy Research Centre, Department of Dermatology and Allergology, Gentofte Hospital, University of Copenhagen, Niels Andersensvej 65, Hellerup 2900, Denmark
e-mail: jacob.p.thyssen@regionh.dk

J.P. Thyssen, H. Maibach (eds.), *Filaggrin*,
DOI 10.1007/978-3-642-54379-1_14, © Springer-Verlag Berlin Heidelberg 2014

that is more exposed to some chemicals due to more frequent application of moisturizers to treat xerotic skin [5]. Previous studies have supported the notion that atopic skin, which is strongly associated with *FLG* mutations, has facilitated penetration of chemicals when compared to non-atopic skin [6].

Recently, we evaluated a group of healthy young Danish men presenting for a military service health check and performed genotyping for the R501X, 2282del4, and R2447X *FLG* mutations. The men provided urine samples, which were evaluated for several phthalate metabolites as an assessment of the levels of internal exposure to these chemicals. It was found that the group of men with loss-of-function *FLG* mutations had significantly higher urinary concentrations of most measured phthalate metabolites, being up to 33 % higher than in the group with the normal genotype [7]. This was especially shown for the low-molecular-weight phthalates, which are more water-soluble and are more often used in personal care products. Notably, atopic dermatitis was almost nonexistent in the Danish cohort, suggesting that the difference in urinary phthalate concentrations was rather explained by the genotype. Phthalates, the diesters of 1,2-benzenedicarboxylic acid, are man-made chemicals used in a wide range of consumer products, including moisturizers and other cosmetics. They are rapidly metabolized and excreted in the urine after absorption, but humans are continuously exposed by skin contact with personal care products, solvents, and plastics [8], as well as through the diet and from inhalation. Phthalates are known reproductive toxicants in animals [9], causing developmental abnormalities of the reproductive tract as well as inhibition of testicular testosterone production in pre- or perinatally exposed animals [10, 11]. Results from epidemiological studies have been divergent but point to some effect of phthalates on reproductive function in humans [12]. Biomonitoring studies show a large regional and individual variation in internal exposure to phthalates, and *FLG* loss-of-function mutation carriers constitute the first genetic group in which a higher internal exposure has been shown.

In the 1990s, several studies showed a decline in human semen quality, with decreasing sperm concentration and total sperm counts over the second half of the twentieth century [13, 14]. The results were controversial and the subject of much debate because of the retrospective design applied in these studies. However, the results were supported by strong evidence of a simultaneous worldwide increase in the incidence of testicular germ cell cancer; a disease known to be linked to decreased semen quality [15, 16]. The rapid decrease in semen quality and the increase of testicular cancer, as well as regional differences in ethnically similar countries with good-quality registry data, suggested that some unknown environmental factors were likely to be primarily responsible for the rise in the disorders. Interestingly, atopic diseases have also become increasingly prevalent in the same time period, pointing to a change in environmental factors as well.

Little is known about the possible association between atopic disorders and testicular function. However, patients with testicular cancer were found in one study to have a higher prevalence of atopy at the time of cancer diagnosis [17], and patients with moderate to severe atopy appear to have lower testosterone and estradiol levels and higher LH levels than healthy controls [18]. In men attending an infertility clinic, one study showed that atopy and cryptorchidism were statistically associated [19]. Although these findings partly support the hypothesis that increased chemical exposure in atopic skin can affect reproductive functions, it remains unknown whether atopy and poor testicular function are truly associated. Since *FLG* loss-of-function mutations are associated with an early onset of atopic dermatitis, a severe course, and persistence into adulthood, it is indeed possible that atopic disorders are associated with decreased testicular function due to increased chemical exposure. Pertinently, dermatitis causes secondary downregulation of filaggrin proteins, likely in a dose-dependent manner, which will further weaken the barrier [20].

In our study on healthy young men on *FLG* genotype and phthalate exposure, testicular function was also assessed by measuring semen quality and serum levels of reproductive hormones

[7]. No associations between *FLG* mutation status and reproductive hormones or semen quality were found, but the study may not have had sufficient statistical power to detect such an association with only 65 *FLG* loss-of-function mutation carriers included and semen quality assessed from a single semen sample only.

Regulation of chemicals such as phthalates in personal care products is based on expected exposure of healthy individuals. Some human studies have started to incorporate measures of combined exposure to multiple compounds [21] in relation to various outcomes, and there is concern that the cumulative exposure to many different phthalates may exceed the maximum recommended levels for some individuals. It is indeed possible that *FLG* loss-of-function mutation carriers can fall above the accepted limits of the combined exposure to several chemicals.

Phthalates constitute only one group of chemicals where dermal uptake is an important determinant of exposure. It is unknown, but likely, that individuals with an impaired skin barrier may also be more internally exposed to other chemicals used in topical products used for the treatment of inflamed and xerotic skin. In individuals with filaggrin-related skin conditions, it is important to ensure that harmful chemicals are not overlooked as a possible side effect of the treatment.

Another important topic that is completely unexplored regards topically applied pharmaceuticals. It is unknown whether individuals with *FLG* loss-of-function mutations may receive a higher internal exposure of the same dermally applied dose when compared to individuals with a normal skin barrier function. Pharmacokinetic studies are usually carried out on healthy individuals with normal skin, but in the dawning age of personalized medicine, genetic variations or skin barrier dysfunctions should likely be taken into account to ensure correct dosage of medication on an individual basis.

In conclusion, special attention to transepidermal absorption of chemicals and medication may be warranted in *FLG* mutation carriers. Future research in the area of chemical exposure and skin barrier function is important to clarify whether there is cause for extra care for those individuals with various diseases of the skin, whether *FLG* loss-of-function mutation carriers should be particularly informed about a chemical hazard, and whether there is a need for dosage adjustment for some topically applied pharmaceuticals in genetically susceptible groups.

References

1. Dodson RE, Nishioka M, Standley LJ, Perovich LJ, Brody JG, Rudel RA. Endocrine disruptors and asthma-associated chemicals in consumer products. Environ Health Perspect. 2012;120(7):935–43.
2. Janjua NR, Frederiksen H, Skakkebæk NE, Wulf HC, Andersson AM. Urinary excretion of phthalates and paraben after repeated whole-body topical application in humans. Int J Androl. 2008;31(2):118–30.
3. Novak N, Baurecht H, Schafer T, Rodriguez E, Wagenpfeil S, Klopp N, et al. Loss-of-function mutations in the filaggrin gene and allergic contact sensitization to nickel. J Invest Dermatol. 2008;128(6): 1430–5.
4. Thyssen JP, Linneberg A, Engkilde K, Menne T, Johansen JD. Contact sensitization to common haptens is associated with atopic dermatitis: new insight. Br J Dermatol. 2012;166(6):1255–61.
5. Sergeant A, Campbell LE, Hull PR, Porter M, Palmer CN, Smith FJ, et al. Heterozygous null alleles in filaggrin contribute to clinical dry skin in young adults and the elderly. J Invest Dermatol. 2009;129(4):1042–5.
6. Jakasa I, de Jongh CM, Verberk MM, Bos JD, Kezic S. Percutaneous penetration of sodium lauryl sulphate is increased in uninvolved skin of patients with atopic dermatitis compared with control subjects. Br J Dermatol. 2006;155(1):104–9.
7. Joensen UN, Jørgensen N, Meldgaard M, Frederiksen H, Andersson AM, Menné T, et al. Associations of filaggrin gene loss-of-function variants with phthalate metabolites and testicular function in young Danish Men. Environ Health Perspect. 2014 Jan 3. [Epub ahead of print].
8. Wittassek M, Koch HM, Angerer J, Bruning T. Assessing exposure to phthalates – the human biomonitoring approach. Mol Nutr Food Res. 2011;55(1):7–31.
9. Fisher JS, Macpherson S, Marchetti N, Sharpe RM. Human 'testicular dysgenesis syndrome': a possible model using in-utero exposure of the rat to dibutyl phthalate. Hum Reprod. 2003;18(7):1383–94.
10. Boberg J, Christiansen S, Axelstad M, Kledal TS, Vinggaard AM, Dalgaard M, et al. Reproductive and behavioral effects of diisononyl phthalate (DINP) in perinatally exposed rats. Reprod Toxicol. 2011;31(2): 200–9.
11. Gray Jr LE, Ostby J, Furr J, Price M, Veeramachaneni DN, Parks L. Perinatal exposure to the phthalates DEHP, BBP, and DINP, but not DEP, DMP, or DOTP,

alters sexual differentiation of the male rat. Toxicol Sci. 2000;58(2):350–65.
12. Joensen UN, Frederiksen H, Jensen MB, Lauritsen MP, Olesen IA, Lassen TH, et al. Phthalate excretion pattern and testicular function: a study of 881 healthy Danish men. Environ Health Perspect. 2012;120(10):1397–403.
13. Carlsen E, Giwercman A, Keiding N, Skakkebæk NE. Evidence for decreasing quality of semen during past 50 years. BMJ. 1992;305(6854):609–13.
14. Swan SH, Elkin EP, Fenster L. The question of declining sperm density revisited: an analysis of 101 studies published 1934–1996. Environ Health Perspect. 2000;108(10):961–6.
15. Schmiedel S, Schuz J, Skakkebæk NE, Johansen C. Testicular germ cell cancer incidence in an immigration perspective, Denmark, 1978 to 2003. J Urol. 2010;183(4):1378–82.
16. Chia VM, Quraishi SM, Devesa SS, Purdue MP, Cook MB, McGlynn KA. International trends in the incidence of testicular cancer, 1973–2002. Cancer Epidemiol Biomarkers Prev. 2010;19(5):1151–9.
17. Swerdlow AJ, Huttly SR, Smith PG. Testicular cancer and antecedent diseases. Br J Cancer. 1987;55(1):97–103.
18. Ebata T, Itamura R, Aizawa H, Niimura M. Serum sex hormone levels in adult patients with atopic dermatitis. J Dermatol. 1996;23(9):603–5.
19. Weidinger S, Schafer T, Malek B, von Schmiedeberg S, Schill WB, Ring J, et al. Association between atopy and cryptorchidism. J Allergy Clin Immunol. 2004;114(1):192–3.
20. Vestergaard C, Hvid M, Johansen C, Kemp K, Deleuran B, Deleuran M. Inflammation-induced alterations in the skin barrier function: implications in atopic dermatitis. Chem Immunol Allergy. 2012;96:77–80.
21. Wolff MS, Teitelbaum SL, Pinney SM, Windham G, Liao L, Biro F, et al. Investigation of relationships between urinary biomarkers of phytoestrogens, phthalates, and phenols and pubertal stages in girls. Environ Health Perspect. 2010;118(7):1039–46.

Part IV

Filaggrin and Atopic Disease

Filaggrin Mutations and the Atopic March

15

Jennifer R. Heimall and Jonathan M. Spergel

Contents

J.R. Heimall, MD
Division of Allergy and Immunology, Department of Pediatrics, The Children's Hospital of Philadelphia, 3550 Market St; 3rd Floor, Philadelphia, PA 19104, USA
e-mail: heimallj@email.chop.edu

J.M. Spergel, MD, PhD (✉)
Allergy Section, Division of Allergy and Immunology, The Children's Hospital of Philadelphia, 3550 Market Street, Philadephia, PA 19104, USA
e-mail: spergel@email.chop.edu

15.1 Epidemiology of Atopic Disease

Atopic disease is increasing in prevalence throughout the world, affecting up to 20 % of individuals, particularly those in industrialized nations [1]. The first manifestation of atopic disease is often atopic dermatitis (AD), which is the most common chronic cutaneous inflammatory disease of early childhood. AD affects approximately 15–20 % of children worldwide, with significant impairment of quality of life attributable to the associated symptoms and secondary infections [2]. In the United States, it is estimated that 17 % of children aged 5–9 years suffer from AD. This disorder is also considered the first step of the atopic march, which is the hypothesis that atopic disease occurs as a progression of symptoms with age. In the atopic march, AD is associated with later development of allergic rhinitis, food allergy, and allergic asthma. While food allergy also commonly presents early in life, allergic rhinitis and allergic asthma with aeroallergen sensitization are more commonly seen in later childhood. By the end of the first 6 years of life, 75 % of children with AD go on to develop allergic rhinitis, while 50 % go on to develop asthma [3].

15.2 Filaggrin

Filaggrin is a S100 calcium-binding protein with key roles in establishing the skin's epidermal barrier function. It is expressed in the skin and in the

J.P. Thyssen, H. Maibach (eds.), *Filaggrin*,
DOI 10.1007/978-3-642-54379-1_15, © Springer-Verlag Berlin Heidelberg 2014

nose up to the transitional epithelium as well as the keratinized epithelium of the conjunctiva and oral cavity. Filaggrin is not found in the deeper mucosal surfaces of the nose and lung and only minimally in most parts of the gut. Functional filaggrin binds to keratin filaments and contributes to compaction and barrier integrity at the outer layer of the skin known as the cornified envelope. As the cornified envelope forms, filaggrin ensures appropriate barrier function by facilitating collapse of the keratinocyte cytoskeleton to create a dense collection of proteins and lipids which are cross-linked by transglutaminases. Filaggrin metabolites are a significant component of the "natural moisturizing factor" (NMF), which is a key contributor to maintenance of hydration of the outermost layer of the skin known as the stratum corneum. In addition, these acids are significant contributors to maintaining a slightly acidic normal cutaneous pH at about 6 or less. NMF is also thought to play a crucial role in maintaining barrier integrity and skin antimicrobial defense [3, 4].

15.3 Filaggrin in Atopic Dermatitis

Skin barrier dysfunction has been shown to have a substantive role in the predisposition of AD. In addition, a family history of AD or atopy is highly predictive of increased risk of developing AD [5]. Taken together, these features lead to a quest to identify a genetic marker that might predispose certain individuals to the development of AD. Filaggrin gene (*FLG*) mutations were initially described in patients with ichthyosis vulgaris (IV), a skin condition of disordered skin keratinization and scaling. IV has been linked to loss-of-function *FLG* mutations inherited in a semidominant fashion [6]. The first identified *FLG* mutations were R501X and 2282del4, but there are now 23 described mutations in patients of European descent and 30 described mutations in patients of Asian descent [7, 8]. Although there was no predominant mutation identified in a recent large cohort of Asian patients, only the c3321delA mutation was seen across patient populations from China, Singapore, Japan, Korea, and Taiwan. In contrast, 80 % of European patients with *FLG* mutations carry the hotspot mutations R501X or 2282del4. In patients of African descent, the R501X and 2282del4 mutations are not common, but have been described. There is a carrier frequency of just below 10 % for *FLG* mutations in European populations [9]. Thus, *FLG* mutations have been identified in many populations worldwide and with a fairly high carrier rate.

Once the association of IV to *FLG* mutations was solidified, this led to the consideration that filaggrin deficiency might play a role in AD as well. There is significant overlap of these diseases, with 35–40 % of individuals afflicted by IV also suffering from atopic disease. The drive to investigate a link between *FLG* mutations and atopic disease, and AD, in particular, was further encouraged by previous studies demonstrating linkage of AD to the region of the epidermal differentiation complex. The epidermal differentiation complex is the region of chromosome 1 encoding a cluster of genes crucial to formation of the stratum corneum of the epidermis [10, 11].

The two most common *FLG* null mutations (R501X and 2282del14) have been the most reproducible genetic factor in many large European cohorts for increased risk of AD. Overall, between 15 and 20 % of AD patients of European descent are estimated to carry a mutation in the *FLG*. Further indicative of the clinical significance of the function of this protein, *FLG* null mutations have been identified in as many as 50 % of patients with moderate to severe AD in European populations and 20 % in Asian populations [7, 12, 13]. The identified mutations are all nonsense mutations that either prevent or significantly decrease filaggrin expression in the skin. The development of AD in *FLG* mutation carriers is not dependent upon allergic sensitization. However, the odds ratio for AD as compared to non-AD is somewhat higher [14, 15]. In several studies describing patients with AD and *FLG* mutations, the dermatitis in patients with *FLG* mutation is typically more severe, based on SCORAD assessment, presents early in life and often persists to adulthood. In addition, it manifests with hyperlinearity of the palms and is more strongly associated with asthma than in patients with AD who do not have these genetic abnormalities [15–17]. The NMF is of significantly different composition in patients with AD who carry *FLG* mutations versus those who do not carry the mutations [12]. Filaggrin expression is decreased in keratinocytes cultured in the presence of IL-4 and IL-13 and in active lesions of skin from patients with AD even without mutations in the *FLG* [18].

Patients with *FLG* null mutations have been shown to have increased transepidermal water loss (TEWL) even prior to the onset of clinical cutaneous disease, supporting the concept that the barrier dysfunction is induced by poor filaggrin function rather than inflammation causing the skin barrier defects [19]. In addition, the degree of TEWL correlated with the severity of the patients' AD as determined by SCORAD [20, 21]. This is not surprising, since dry skin is inherently more pruritic. As patients experience more pruritus, they will reflexively scratch more, perpetuating the itch-scratch cycle that drives poorly controlled AD.

15.4 Skin Barrier Dysfunction in Allergic Sensitization

Defects in skin barrier function are thought to allow transcutaneous sensitization to both food and environmental allergens. Barrier dysfunction is also associated with systemic responses such as increased total IgE and airway hyperreactivity, possibly due to these sensitizations. In mouse models of *FLG* null mutations, increased transcutaneous sensitization [22–24] and allergen-specific airway inflammation upon non-cutaneous reexposure have been demonstrated. This suggests development of functional allergen-specific antibodies in the mouse model. There is an increased risk (odds ratio of 1.6) of allergic sensitization in *FLG* carriers with a history of AD. AD severity as determined by SCORAD has been found to have a correlation with the specific IgE for house dust, mite allergen, and cat dander in patients with *FLG* null mutations. This correlation has not been seen in AD patients with wild-type *FLG* alleles [20]. The R501X mutation is associated with higher serum IgE levels overall when compared with the other common mutation 2282del4 due to a higher rate of penetrance [5].

Allergic sensitization was associated with *FLG* mutations only in those patients who also had AD [9, 14]. It therefore follows that infants with severe AD with defects in epidermal barrier function, such as those induced by *FLG* mutations, have an increased risk of allergic sensitization. Maintaining and repairing defects in the epidermal barrier are likely to be crucial in atopy prone individuals to prevent progression along the atopic march.

The common immunologic features of atopic disease, including elevations in total IgE, and specific IgE to allergens, Th2 cytokine skewing, eosinophilia, and epithelial dysfunction, led to the theory that these diseases represent a progression disease that has its root in a shared initiating insult. The atopic march is the purported stepwise progression of patients' atopic disease development over the course of a lifetime. According to the atopic march theory, atopic disease first presents with the presentation of AD in infancy. If left unchecked, the disease process then permits further atopic development due to atopic sensitization with aeroallergens and clinical presentations of allergic rhinitis and allergic asthma. This is supported by observations that the incidence of AD peaks in the first years of life and then declines, while the incidence of asthma and allergic rhinitis is low in infancy but increases with age. In addition, the risk for developing asthma correlates with the severity of AD. While 30 % of patients with mild AD go on to develop asthma, in patients with severe AD, 70 % go on to develop asthma [25].

In one cohort of over 1,000 children followed from birth until age 7 years, 21 % experienced AD in the first 2 years of life. In these children with early onset of AD and allergic sensitization to wheat, cat, dust mite, soy, or birch, there was an association with early wheezing, which was categorized as any episodes of wheezing under age 3 years. Children with this combination of symptoms experienced more significant overall impairment in lung function [26]. In another cohort of 94 children with AD followed prospectively for the first 7 years of life, 43 % went on to develop asthma and 45 % went on to develop allergic rhinitis. Severe AD in infancy was associated with increased rates of food-specific IgE and early-onset AD was associated with an increased risk of inhalant sensitization [27]. There are also animal models supportive of the association between skin barrier dysfunction, AD, and later development of other atopic disease.

Tape stripping followed by allergen application to the skin of mice induced IgE, IgG1, and IL-4 expression as well as a predominantly eosinophilic infiltrate. Conversely, application of allergen to intact skin led to Th1 skewing in draining lymph nodes. This supports the concept that disruption in the barrier function of the skin can predispose to a development of a typical atopic milieu of IgE and Th2-type cytokine production [28]. In tape-stripped mice exposed to ovalbumin epicutaneously, ovalbumin-specific IgE was detectable, indicating that sensitization had occurred. This was not seen in mice exposed to ovalbumin via intraperitoneal injection. When the epicutaneously exposed mice were later challenged with inhaled ovalbumin, there was a significant increase in eosinophils in BAL obtained after the challenge [29].

The Notch pathway is a key signaling pathway in the skin keratinocyte structural organization. In mice with a keratinocyte-specific knockdown of the Notch signaling pathway, there was a phenotypic expression of chronic skin barrier defects with increased TEWL and development of a skin disorder similar to AD [30]. The genetically engineered mice had increased IgE levels and systemic Th2 cell expansion. These mice were also found to have increased expression of thymic stromal lymphoprotein (TSLP) in their keratinocytes and had increased susceptibility to allergic asthma characterized by severe bronchospasm response to OVA challenge, which was not seen in wild-type controls. However, when the TSLP receptor was deleted, the development of asthma as defined by OVA-induced bronchospasm was blocked. Further, in outbred transgenic mice that continued to have high TSLP expression but relative absence of a skin barrier defect, a severe asthma phenotype developed [30]. Taken together, these data suggest a critical role for TSLP in the progression of the atopic march from AD to allergic asthma.

TSLP is an IL-7-like cytokine, produced in epithelial cells of the skin, lung, and gut, that has been associated with AD and allergic asthma. When antigen-presenting cells, particularly dendritic cells, are cocultured with T cells in a TSLP-rich media, the ensuing T-cell response is predominantly of the Th2 polarization, with an increased expression of IL-4, IL-5, IL-13, and TNF-alpha. TSLP can also independently directly promote naïve T-cell differentiation toward Th2 phenotype in culture [31]. Human epithelial cells from lesional skin of AD patients have been found to express higher levels of TSLP than healthy skin and non-lesional skin of AD patients [32]. However, it is important to note that it has not yet been demonstrated that TSLP serum concentrations are elevated in atopic individuals compared to healthy controls. It is theorized that the increased TSLP expression of keratinocytes leads to an overall Th2 skewing in patients with AD, which then feeds into the cycle of the atopic march to promote the development of allergic rhinitis and allergic asthma. This theory has been supported by two studies [33, 34].

In a large study of *FLG* mutations in atopic disease, which was inclusive of over 3,000 European children of German descent, the presence of one of five known *FLG* null alleles increased the risk of eczema, particularly AD (OR 4.5) [9]. In this study, associations were observed between *FLG* mutations and all three of the disease phenotypes (asthma, allergic rhinitis, and AD) that we associate with the atopic march. The children studied were aged 9–11 years of age. Skin prick testing was performed to six aeroallergens (*Dermatophagoides pteronyssinus*, *Dermatophagoides farina*, *Alternaria tenuis*, cat, mixed grass, and tree pollen). Skin prick testing was determined to be positive if the wheal was 3 mm greater than the negative control. AD was defined as a physician diagnosis of AD and positive skin prick testing to at least one aeroallergen.

Asthma was defined as physician-diagnosed asthma or at least two episodes of spastic bronchitis on a patient questionnaire and categorized as allergic if there was presence of at least one positive skin prick test to an aeroallergen. Allergic rhinitis was defined as physician-diagnosed hay fever in the setting of at least one positive skin prick test. In addition, *FLG* null alleles were associated with increased risk of sensitization (positive skin test), and this was strongest in individuals with eczema [9]. In a study of a cohort of over 900 Dutch children, three common *FLG* mutations (R501X, 2282del4, and R2447X) were found to be associated with early eczema that persisted until later childhood, asthma, and aeroallergen sensitization [35].

15.5 Hygiene Hypothesis and the Atopic March

An alternative theory is that microbes have a role in allergic sensitization and for the atopic march. The hygiene hypothesis is the concept that microbial exposures early in life are protective against development of atopic disease by inducing the development of Th1- rather than Th2-type CD4 cells. Th2 cells are commonly associated with atopy. Toll-like receptors are pattern recognition molecules found on many cell types including keratinocytes which lead to activation of the innate immune systems via induction of cytokine and other cell signaling molecule production after exposure to bacterial and viral proteins, DNA and RNA. Dermal exposure to TLR2 and TLR 4 agonists led to a decrease in airway hyperresponsiveness and mucous production [36], suggesting a protective role for TLR activation to stop the atopic march.. However, in keratinocytes with decreased filaggrin expression, increased IL-6 and TSLP were detected following poly I:C stimulation of TLR3 [37]. As noted previously, TSLP derived from the skin has been implicated in the development of atopic asthma. These data suggest that decreased filaggrin expression in the skin can detract from the potential protective effect of microbial exposure and can further drive the atopic march.

15.6 Filaggrin in Airway Disease: Allergic Rhinitis and Asthma

Since approximately 40 % of infants with AD go on to later develop asthma and/or allergic rhinitis, then the next logical quest was the search to determine if there is an association between *FLG* mutations and airway atopic disease. Filaggrin is not expressed in the mucosal surfaces of the nose, upper airway, or lungs; therefore, it would seem unlikely to demonstrate an association between decreased expression of this protein and the development of allergic rhinitis and allergic asthma. Somewhat surprisingly, then, one of the initial studies to assess the association of the R501X and 2282del4 mutations with susceptibility to atopic disease found a significant association between the presence of these mutations with the phenotype of allergic rhinitis and AD and allergic sensitization plus AD [14]. However, in the same large study of German children discussed previously, *FLG* mutations were associated with a twofold increased risk of allergic rhinitis irrespective of prior AD history [9]. This association was dependent upon AD history in another large cohort [16].

Filaggrin has also been proposed as a major genetic risk factor for the development of allergic asthma. Patients with *FLG* null alleles have been demonstrated to have an increased need for asthma medication and greater obstructive lung function impairment [38]. In addition, mild persistent asthma patients who are carriers of a *FLG* null allele have a nearly twofold higher risk of exacerbation with a need for oral steroid therapy compared to asthma patients with wild-type filaggrin alleles [39]. This association is strongest when the patients have both asthma and AD, which further supports the role *FLG* mutations can play in permitting the atopic march. When asthma was used as the inclusion criteria irrespective of AD history, the strength of the association with *FLG* mutations dropped notably [15]. However, this may reflect the heterogeneity of patients diagnosed with asthma.

The above associations are thought to be primarily due to transcutaneous sensitizations to allergens through the impaired skin barrier observed in AD. The presence of allergic sensitization leads to an overall Th2-type skewing, which then triggers asthma and rhinitis symptoms triggered by these allergens. But this cannot be the only link between allergic rhinitis, asthma, AD, and *FLG* mutations, as some patients developed symptoms in the absence of demonstrable allergic sensitization [9].

15.7 Filaggrin in Food Allergy and Gastrointestinal Disease

In studies of the eosinophilic esophagitis transcriptome, the highest concentration of genetic abnormalities has been described in the area of the epidermal differentiation complex, where filaggrin is located. In addition, patients with eosinophilic esophagitis commonly also suffer from AD. Thus far, no significant differences in filaggrin expression have been noted in patients with atopic eosinophilic esophagitis as compared to those eosinophilic esophagitis patients without atopic sensitization [40].

Patients with *FLG* null mutations have been found to have an increased rate of peanut sensitization, with an odds ratio of 3.8 when controlled for the presence of AD [41]. Transcutaneous sensitization early in life in the absence of enteral exposure has been postulated to be one potential etiology of the increased rate of peanut sensitization and resultant allergy [42].

15.8 Treatments

Filaggrin is a crucial component of the barrier function in the layers of the skin and epithelium where it is found. Mutations in the *FLG* causes functional deficits, leading to the rather obvious conclusion that therapy aimed to address this defect must address barrier dysfunction first and foremost.

AD is traditionally perceived as an immunologic disease, to be treated with immunosuppressants. However, the association of decreased barrier function with increased risk of allergic sensitization and bacterial and viral infection associated with *FLG* mutations supports the growing body of evidence that effective therapy must enhance skin barrier functions. In the skin, this will mean aggressive proactive, rather than reactive, use of topical compounds to enhance barrier function, which will restore normal skin pH and minimize the risk of allergic sensitization and infection [41]. Infectious agents such as herpesvirus and *Staphylococcus aureus* are common triggers for flares and must also be actively controlled. In addition, *FLG* mutations are associated with a decreased resistance to these organisms [44, 45, 47].

Moisturization is an important component of the barrier function as evidenced by the increased TEWL seen in patients with *FLG* mutations. Basic cleansing is important, and use of a pH-balanced (ph=6) cleansing product can be helpful to restore the skin to its normal physiologic acid–base balance, which is crucial in the host defense role of the skin barrier. In an effort to control *Staphylococcal* infection and superantigen stimulation causing severely flared AD, dilute bleach baths are also often recommended, as is the use of topical antimicrobial agents such as mupirocin [46]. After the bath unscented and dye-free emollients should be applied immediately, while the skin is still damp. Avoidance of products containing perfumes and alcohols is also crucial as these components can contribute to further drying and irritation of the skin. Lipid-containing emollients help to supplement the barrier dysfunction of the skin [45]. Emollients are recommended to be used at least twice a day with increasing frequency as needed to keep the skin moist and limit pruritus. A typical child will require 150–200 g of emollient on a weekly basis, and adults may require up to 500 g. Emollients containing potential allergens such as peanut, oat, and shea nut should be avoided since the skin can serve as entry for allergic sensitization [46]. A recent pilot study of 15 AD patients using a hydrogel compound under occlusion demonstrated similar efficacy to use of triamcinolone 0.1 %, supporting the concept that restoration of barrier function is an imperative treatment objective [49].

Since the inherent problem of the *FLG* mutations gives chronic skin barrier dysfunctions that are not resolved with resolution of an acute flare, a proactive approach to minimize future exacerbations has been proposed [43, 48]. Acute flares of AD are typically treated in the traditional fashion, with higher-potency topical steroids and removal of identifiable triggers. Once acute flares are resolved, prophylactic twice-weekly application of either topical steroids or calcineurin inhibitors in areas of previous flares can be beneficial to achieve optimal control of chronic AD [50]. Prophylactic use of fluticasone 0.05 % four times weekly for 4 weeks followed by twice weekly for 16 weeks was associated with a significant decrease in AD relapse compared to placebo without evidence of skin atrophy [51]. Tacrolimus 0.1 % ointment has also been used for twice-weekly prophylaxis with significant decrease in frequency of exacerbations and significant improvement in quality of life. These regiments have been used in patients with varying severity of disease with similar benefits [41]. These prophylactic regimens may decrease the overall burden of exposure to high-potency steroids by allowing for a stable steady state of control of AD as a chronic disease.

Treatment of AD with pimecrolimus compared to betamethasone led to similar improvements in TEWL, filaggrin expression, and decreased dye penetration, which was used as a proxy for functionality of the outside-inside barrier and risk of transcutaneous sensitization [52]. However, betamethasone was associated with a more significant decrease in epidermal proliferation, which was also associated with epidermal thinning. The effects of betamethasone were primarily attributed to its reduction in inflammation and blood flow. While this is beneficial in the short term, the long-term risk of skin atrophy is substantial. Pimecrolimus was associated with regular lipid bilayers, which are a key component to normal barrier function of the skin. This, when added to the other shared benefits with high-potency topical steroids, may make pimecrolimus and the calcineurin inhibitors as a whole better suited to the long-term therapy needed to address barrier dysfunction in most patients with AD [50]. Topical pimecrolimus has been shown to be both safe and effective in infants as young as 3 months of age [53]. However, the use of pimecrolimus 1 % cream in this study did not lead to a decreased rate of later atopic disease development such as allergic rhinitis or allergic asthma. The more rapid reduction in symptoms and measurable parameters associated with topical steroids, however, supports their continued stepwise use for acute flares [54]. Effective treatment of inflammation can also help to decrease *Staphylococcal* carriage.

Conclusion

Over the last 5 years, *FLG* loss-of-function mutations have emerged as a significant genetic marker for AD, which can be expected to be more severe, more highly associated with allergic sensitization, and persist later in life than in patients with other forms of eczema. It is noteworthy that filaggrin expression is also lower in the lesional skin of AD patients without loss-of-function mutations. This is one indicator of a uniform disease pathway of impaired skin barrier function in all eczema patients, which underscores the growing understanding of the important role of the intact skin epithelium in prevention of allergic sensitization early in life. The theory of impaired barrier function and elevated TSLP expression is a hallmark of the atopic march hypothesis. AD has long been implicated as the first step of the atopic march, and further studies have linked *FLG* mutations to increased risk for asthma in patients who suffer from concomitant AD, as well as allergic rhinitis and peanut allergy. These other atopic diseases are thought to be due to transcutaneous sensitization to both food and aeroallergens, which then allows development of lifelong atopic disease precipitated by predominant skewing of the naïve T cells toward a Th2 phenotype. As we are better able to define the subgroups of asthmatic and allergic rhinitis patients into those who have truly progressed along the route of the atopic march, the associations between *FLG* loss-of-function mutations and other atopic disease may become clearer. By addressing the barrier dysfunction of AD proactively with the use of effective and preventative emollients and targeted use of immunomodulators, it is possible that we may optimize our patients' chances to stop the atopic march, lower their overall burden of atopy, and enhance their quality of life.

References

1. Spergel J. From atopic dermatitis to asthma: the atopic march. Ann Allergy Asthma Immunol. 2010;105: 99–106.
2. Laughter D, Istvan JA, Tofte SJ. The prevalence of atopic dermatitis in Oregon school children. J Am Acad Dermatol. 2000;43:649–55.
3. O'Regan G, Sandilands A, McLean I, Irvine A. Filaggrin in atopic dermatitis. J Allergy Clin Immunol. 2008;122:689–93.
4. Rawlings AV, Harding CR. Moisturization and skin barrier function. Dermatol Ther. 2004;17(suppl I):43–8.
5. Weidinger S, Illig T, Baurecht H, Irvine AD, Rodriguez E, Diaz-Lacava A, et al. Loss of function variations within the filaggrin gene predispose for atopic dermatitis with allergic sensitization. J Allergy Clin Immunol. 2006;118:214–9.
6. Smith F, Irvine A, Terron-Kwaitkowski A, Sandilands A, Campbell LE, Zhao Y, et al. Loss-of-function mutations in the gene encoding filaggrin cause ichthyosis vulgaris. Nat Genet. 2006;38:337–42.
7. Chen H, Common JEA, Haines RL, Balakrishnan A, Brown SJ, Goh CS, et al. Wide spectrum of filaggrin null mutations in atopic dermatitis highlights differences between Singaporean Chinese and European populations. Brit J Dematol. 2011;165:106–14.
8. Irvine AD, McLean I, Leung D. Filaggrin mutations associated with skin and allergic diseases. N Engl J Med. 2011;365:1315–27.
9. Weidinger S, O'Sullivan M, Illig T, Baurecht H, Depner M, Rodriguez E, et al. Filaggrin mutations, atopic eczema, hay fever and asthma in children. J Allergy Clin Immunol. 2008;121:1203–9.
10. Cookson W. The immunogenetics of asthma and eczema- a new focus on the epithelium. Nat Rev Immunol. 2004;4:978–88.
11. Miscke D, Korge BP, Marenholz I, Volz A, Ziegler A. Genes encoding structural proteins of epidermal cornification and S100 calcium-binding proteins form a gene complex ("epidermal differentiation complex") on human chromosome 1q21. J Invest Dermatol. 1996;106:989–92.
12. O'Regan G, Kemperman P, Sandilands A, Chen H, Campbell LE, Kroboth K, et al. Raman profiles of the stratum corneum define 3 filaggrin genotype-determined atopic dermatitis endophenotypes. J Allergy Clin Immunol. 2010;126:574–80.
13. Palmer C, Irvine A, Terron-Kwaitkowski A, Zhao Y, Liao H, Lee SP, et al. Common loss-of-function variants of the epidermal barrier protein filaggrin are a major predisposing factor for atopic dermatitis. Nat Genet. 2006;38:441–6.
14. Marenholz I, Nickel R, Ruschendorf F, Schulz F, Esparza-Gordillo J, Kerscher T, et al. Filaggrin loss-of-function mutations predispose to phenotypes involved in the atopic march. J Allergy Clin Immunol. 2006;118:866–71.
15. Rodriguez E, Baurecht H, Herberich E, Wagenpfeil S, Brown SJ, Cordell HJ, et al. Meta-analysis of filaggrin polymorphisms in eczema and asthma: robust risk factors in atopic disease. J Allergy Clin Immunol. 2009;123:1361–70.
16. Henderson J, Northstone K, Lee SP, Liao H, Zhao Y, Pembrey M, et al. The burden of disease associated with filaggrin mutations; a population based longitudinal birth cohort study. J Allergy Clin Immunol. 2008;121:872–7.
17. Irvine A. Fleshing out filaggrin phenotypes. J Invest Dermatol. 2007;127:504–7.
18. Howell MD, Kim BE, Gao P, Grant AV, Boguniewicz M, DeBenedetto A, Schneider L, et al. Cytokine modulation of atopic dermatitis filaggrin skin expression. J Allergy Clin Immunol. 2007;120:150–5.
19. Jakasa I, Koster E, Calkoen F, McLean WH, Campbell LE, Bos JD, et al. Skin barrier function in healthy subjects and patients with atopic dermatitis in relation to filaggrin loss-of-function mutations. J Invest Dermatol. 2011;131:540–2.
20. Nemoto-Hasebe I, Akiyama A, Nomura T, Sandilands A, McLean W, Shimizu H. Clinical severity correlates with impaired barrier in filaggrin-related eczema. J Invest Dermatol. 2009;129:682–9.
21. Flohr C, England K, Radulvic S, McLean WH, Campbel LE, Barker J, et al. Filaggrin loss-of-function mutations are associated with early-onset eczema, eczema severity and transepidermal water loss at 3 months of age. Br J Dermatol. 2010;163:1333–6.
22. Fallon PG, Sasaki T, Sandilands A, Campbell LE, Saunders SP, Mangan NE, et al. A homozygous frameshift mutation in the mouse Flg gene facilitates enhanced percutaneous allergen priming. Nat Genet. 2009;124:494–525.
23. Oyoshi M, Geha R. Filaggrin deficient mice exhibit Th17-dominated skin inflammation and increased susceptibility to inoculation with Vaccinia Virus. J Allergy Clin Immunol. 2010;125:AB123.
24. Kawasaki H, Nagao K, Kubo A, Hata T, Shimizu A, Mizuno H, Yamata T, et al. Altered stratum corneum barrier and enhanced percutaneous responses in filaggrin-null mice. J Allergy Clin Immunol. 2012; 129:1538–46.
25. Spergel JM, Paller A. Atopic dermatitis and the atopic march. J Allergy Clin Immunol. 2003;112:S118–27.
26. Illi S, von Mutils E, Lau S, Nickel R, Gruber C, Niggemann B. The natural course of atopic dermatitis from birth to age seven years and the association with asthma. J Allergy Clin Immunol. 2004;113(5):925–31.
27. Gustafsson D, Sjoberg D, Foucard T. Development of allergies and asthma in infants and young children

with atopic dermatitis- a prospective follow up to seven years of age. Allergy. 2000;55(3):240–5.
28. Kondo H, Ichikawa Y, Imokawa G. Percutaneous sensitization with allergens through barrier disrupted skin elicits a Th2 dominant cytokine response. Eur J Immunol. 1998;28:769–79.
29. Spergel JM, Mizoguchi E, Brewer JP, Martin TR, Bhan AK, Geha RS. Epicutaneous sensitization to protein antigen induces localized allergic dermatitis hyperresponsiveness to methacholine after single exposure to aerosolized antigen in mice. J Clin Invest. 1999;101:1614–22.
30. Demehri S, Morimoto M, Holtzman M, Kopan R. Skin-derived TSLP triggers progression from epidermal barrier defects to asthma. PLoS Biol. 2009;7(5):e1000067. doi:10.1371/journal.pbio./1000067.
31. Ziegler S, Artis D. Sensing the outside world: TSLP regulates barrier immunity. Nat Immunol. 2010; 11(4):289–93.
32. Soumelis V, Recche P, Kanzler H, Yuan W, Edward G, Homey B, et al. Human epithelial cells trigger dendritic cell mediated allergic inflammation by producing TSLP. Nat Immunol. 2002;3:673–80.
33. Levya-Castillo JM, Hener P, Jiang H, Li M. TSLP produced by keratinocytes promotes allergen sensitization through skin and thereby triggers atopic march in mice. J Invest Dermatol. 2013;133(1):154–63.
34. Jiang H, Hener P, Li J, Li M. Skin thymic stromal lymphopoietin promotes airway sensitization to inhalant house dust mites leading to allergic asthma in mice. Allergy. 2012;67(8):1078–82.
35. Schuttelaar MLA, Kerkhof M, Jonkman MF, Koppelman GH, Brunekreef B, de Jongste JC, et al. Filaggrin mutations in the onset of eczema, sensitization, asthma, hay fever and the interaction with cat exposure. Allergy. 2009;64:1758–65.
36. Haapakoski R, Karisola P, Fyhrquist N, Yuan W, Edward G, Homey B, et al. Toll-like receptor activation during cutaneous sensitization blocks development of asthma through IFN-gamma-dependent mechanism. J Invest Dermatol. 2013;133(4):964–72.
37. Lee HK, Cho KA, Kim JY, Baek JH, Woo S, Kim JW. Filaggrin knockdown and Toll-like receptor 3 (TLR3) stimulation enhanced the production of thymic stromal lymphopoietin (TSLP) from epidermal layers. Exp Dermatol. 2011;20(2):149–51.
38. Palmer C, Ismail T, Lee SP, Terron-Kwiatkowski A, Zhao Y, Liao H, et al. Filaggrin null mutations are associated with increased asthma severity in children and young adults. J Allergy Clin Immunol. 2007;120:64–8.
39. Basu K, Palmer C, Lipworth B, Irwin McLean WH, Terron-Kwiatkowski A, Zhao Y, et al. Filaggrin null mutations are associated with increased asthma exacerbations in children and young adults. Allergy. 2008;63:1211–7.
40. Sherrill J, Rothenberg M. Genetic dissection of eosinophilic esophagitis provides insight into disease pathogenesis and treatment strategies. J Allergy Clin Immunol. 2011;128:23–32.
41. Brown S, Asai Y, Cordell H, Campbell LE, Zhao Y, Liao H, et al. Loss of function variants in the filaggrin gene are a significant risk factor for peanut allergy. J Allergy Clin Immunol. 2011;127:661–7.
42. Lack G, Fox D, Northstone K, Golding J. Factors associated with the development of peanut allergy in childhood. N Engl J Med. 2003;348(11): 977–85.
43. Wollenberg A, Bieber T. Proactive therapy of atopic dermatitis- an emerging concept. Allergy. 2009;64: 276–8.
44. Miajlovic H, Fallon P, Irvine A, Foster T. Effect of filaggrin breakdown products on growth of and protein expression by Staphylococcus aureus. J Allergy Clin Immunol. 2010;126:1184–90.
45. Gao P, Rafaels NM, Hand T, Murray T, Boguniewicz M, Hata T, et al. Filaggrin mutations that confer risk of atopic dermatitis confer greater risk for eczema herpeticum. J Allergy Clin Immunol. 2009;124: 507–13.
46. Huang JT, Abrams M, Tlougan B, Rademaker A, Paller AS. Treatment of Staphylococcus aureus colonization in atopic dermatitis decreases disease severity. Pediatrics. 2009;123(5):e808–14.
47. Jung T, Stingl G. Atopic dermatitis: therapeutic concepts evolving from new pathophysiologic insights. J Allergy Clin Immunol. 2008;122:1074–81.
48. Darsow U, Wollenberg A, Simon D, Taïeb A, Werfel T, Oranje A, et al. ETFAD/EADV eczema task force position paper on diagnosis and treatment of atopic dermatitis. J Eur Acad Dermatol Venereol. 2010;24:317–28.
49. Park K, Kamangar F, Heller M, Lee E, Bhutani T, Busse K, et al. Instant barrier repair: a pilot study investigating occlusion with a new hydrogel patch for the treatment of atopic dermatitis. J Dermatolog Treat. 2013;24(2):119–21.
50. Boguniewicz M, Eichenfield L, Hultsch T. Current management of atopic dermatitis and interruption of the atopic march. J Allergy Clin Immunol. 2003;112:S140–50.
51. Jensen J, Pfeiffer S, Witt M, Bräutigam M, Neumann C, Weichenthal M, et al. Different effects of pimecrolimus and betamethasone on the skin barrier in patients with atopic dermatitis. J Allergy Clin Immunol. 2009;123:1124–33.
52. Hanifin, JM, Boguniewicz, M, Eichenfield, LF, Schneider LC, Paller AS, Preston JA, et al. long-term study of safety and allergic comorbidity development in a randomized trial of pimecrolimus cream in infants with atopic dermatitis. 2010 Annual meeting of the Society for Investigative Dermatology. Atlanta, 5–8 May 2010 (Abstract and Poster).
53. Boguniewicz M, Leung D. Recent insights into atopic dermatitis and implications for management of infectious complications. J Allergy Clin Immunol. 2010;125:4–13.
54. Patrizi A, Pileri A, Bellini F, Raone B, Neri I, Ricci G. Atopic dermatitis and the atopic march: what is new? J Allergy (Cairo). 2011;2011:279425.

Atopic Dermatitis

16

Teresa Tsakok and Carsten Flohr

Contents

T. Tsakok, MA, BM, BCh (Oxon)
Core Medical Trainee/NIHR Academic Clinical Fellow, St John's Institute of Dermatology, Guy's and St. Thomas' Hospitals NHS Foundation Trust, Westminster Bridge Road, London SE1 7EH, UK
e-mail: teresa.tsakok@kcl.ac.uk

C. Flohr, MD, MA, MPhil, MSc, PhD (✉)
Department of Paediatric Dermatology and Children's Allergies, St. John's Institute of Dermatology, Guy's and St. Thomas' Hospitals NHS Foundation Trust and King's College London, Westminster Bridge Road, London SE1 7EH, UK
e-mail: carsten.flohr@kcl.ac.uk

16.1 Introduction

Atopic dermatitis (AD; synonyms atopic eczema and eczema) is the commonest pediatric inflammatory skin disease, with considerable impact on children's quality of life and costs to families and society at large [1, 2]. It is a chronically relapsing dermatitis, and dry, itchy skin is a cardinal feature in both lesional areas and elsewhere. The xerosis is primarily caused by a loss of integrity in the stratum corneum and an associated increase in transepidermal water loss (TEWL) [3]. The key gene involved in skin barrier function is the filaggrin gene (*FLG*), found in the epidermal differentiation complex on chromosome 1q21 [4]. Filaggrin is an important component of the granular cell layer of the epidermis. It aggregates keratin filaments, leading to keratinocyte compaction and formation of the stratum corneum [3]. Up to 50 % of patients with moderate to severe AD carry at least one *FLG* loss-of-function mutation [5].

We comprehensibly review the association between the inheritance of *FLG* mutations and AD from a clinical perspective – in particular, disease onset, chronicity, severity, and the association with other allergic diseases. We performed a systematic online search in PubMed from inception until May 2013 to identify papers within the remit of this chapter, using the terms atopic dermatitis, atopic eczema, eczema, filaggrin, and *FLG* (Table 16.1).

J.P. Thyssen, H. Maibach (eds.), *Filaggrin*,
DOI 10.1007/978-3-642-54379-1_16, © Springer-Verlag Berlin Heidelberg 2014

Table 16.1 Summary of main papers referenced (in alphabetical order)

Authors, year, country [Ref.]	Study type	Number of participants	Age	Outcomes/exposures	Key results and conclusions
Ballardini et al., 2013, Sweden [27]	Birth cohort (BAMSE)	3,301 (1,854 genotyped)	0–13 years	AD prevalence and severity; *FLG* mutation status (R501X, R2447X, and 2282del4)	*FLG* mutations did not influence AD severity. Prevalence of rhinitis and asthma was associated with AD severity
Barker et al., 2007, UK [10]	Case–control	163 cases (adults with persistent childhood-onset AD); 1,463 controls	16–82 years	Allele frequencies of two common *FLG* mutations (R501X and 2282del4)	Combined allele frequency was 0.270 versus population frequency 0.046, representing an OR of 7.7 (95 % CI 5.3–10.9)
Bisgaard et al., 2008, Denmark [48]	Birth cohort (COPSAC and MAAS)	411 (COPSAC); 940 (MAAS)	0–5 years	Age of onset/pet ownership; *FLG* mutation status (R501x and 2282del4)	Data was obtained in early life in a Danish high-risk birth cohort (COPSAC) and findings replicated in an unselected UK birth cohort (MAAS). FLG mutation increased AD risk during the first year of life (COPSAC: HR 2.26, 95 % CI 1.27–4.00; MAAS: HR 1.95, 95 % CI 1.13–3.36), with a further increase in risk related to cat exposure at birth (COPSAC: HR 11.11, 95 % CI 3.79–32.60; MAAS: HR 3.82, 95 % CI 1.35–10.81). Dog exposure was moderately protective (COPSAC: HR 0.49, 95 % CI 0.24–1.01; MAAS: HR 0.59, 95 % CI 0.16– 2.20)
Broccardo et al., 2011, USA [39]	Comparative proteomic profiling study	65 (29 with AD without previous eczema herpeticum, 21 with AD with a history of eczema herpeticum, 15 nonatopic controls)	1–80 years	Expression levels of proteins related to skin barrier and generation of natural moisturizing factor (using skin taping and mass spectrometry)	Expression levels of proteins including filaggrin-2 were significantly lower in lesional versus non-lesional sites of patients with AD with and without a history of eczema herpeticum
Brown et al., 2008, UK [28]	Case–control	190 cases; 621 controls	7–9 years	Allele frequencies of 5 common *FLG* mutations (R501X, 2282del4, R2447X, S3247X, and 3702delG) and AD risk	The combined null genotype (carriage of ≥1 *FLG* mutation) was significantly associated with AD ($p = 1.2 \times 10^{-4}$). The OR for individuals carrying 2 null mutations was 26.9 (95 % CI, 3.3–216.1), but heterozygote carriers showed no significant increase in risk (OR, 1.2; 95 % CI, 0.7–1.9). Asthma in the context of AD showed significant association with FLG null mutations ($p = 7.1 \times 10^{-4}$)

Brown et al., 2008, UK [13]	Case–control	186 cases (adults with persistent childhood-onset AD); 1,035 controls	16–82 years	Allele frequencies of 5 common *FLG* mutations (R501X, 2282del4, R2447X, and S3247X and 3702delG), age of onset and persistence of AD	One novel, previously unreported single base-pair deletion (3673delC) was identified in one of the AD cases but none of the controls. Three of the common *FLG* variants were strongly and independently associated with early-onset persistent AD: R501X (OR 5.6, 95 % CI 3.7–8.3), 2282del4 (OR 4.1, 95 % CI 2.7–6.4), and R2447X (OR 7.3, 95 % CI 2.8–18.7). The combined null genotype of all six mutations showed a highly significant association (OR 5.6, 95 % CI 4.1–7.8)
Brown et al., 2009, UK [23]	Cohort	792	7–9 years	Dermatologist-assessed features of ichthyosis vulgaris, AD, and xerosis; AD severity; allele frequencies of 6 common *FLG* mutations (R501X, 2282del4, R2447X, S3247X, 3702delG, 3673delC)	Summating skin features associated with ichthyosis, keratosis pilaris, palmar hyperlinearity, and flexural AD showed a penetrance of 100 % in children with two *FLG* mutations, 87.8 % in children with one *FLG* mutation, and 46.5 % in wild-type individuals ($p<0.0001$). *FLG* null mutations were associated with more severe AD ($p=0.0042$)
Brown et al., 2011, UK [31]	Case–control	71 cases (food challenge-proven peanut allergy); 1,000 controls. Replication tested in 390 cases Canadian patients with peanut allergy; 891 population controls	0–18 years	Association between *FLG* null alleles (R501X, 2282del4, R2447X, and S3247X), AD, and peanut allergy	FLG loss-of-function mutations showed a strong and significant association with peanut allergy in the food challenge-positive patients (OR 5.3, 95 % CI 2.8–10.2), and this association was replicated in the Canadian study (OR 1.9, 95 % CI 1.4–2.6). This association remained significant ($p=0.0008$) after controlling for coexistent AD. There was also a strong association of AD with peanut allergy in the food challenge-positive patients, with an OR of 7.4 (95 % CI, 4.1–13.7)
Brown et al., 2012, UK [11]	Case–control	876 cases (Irish pediatric AD patients); 928 controls	3 years (mean)	Intragenic copy number variation; *FLG* mutation status (R501X, 2282del4, S3247X, and R2447X) and allele frequencies	Having excluded *FLG* mutation carriers, the control group had a significantly higher number of repeats than cases ($p=0.043$), and the odds of having AD was reduced by 22 % for each additional intragenic copy number (OR 0.88; 95 % CI 0.78–0.98)

(continued)

Table 16.1 (continued)

Authors, year, country [Ref.]	Study type	Number of participants	Age	Outcomes/exposures	Key results and conclusions
Carson et al., 2012, Denmark [18]	Birth cohort	397	0–7 years	AD diagnosis, clinical distribution, severity, and age of onset; *FLG* mutation status (R501X and 2282del4)	43 % children developed AD. The R501X and/or 2282del4 *FLG* null mutations were present in 15 % of children with AD. *FLG* mutations were associated with a higher number of unscheduled visits (3.6 vs. 2.7; $p=0.04$) and AD that was more severe (moderate-severe SCORAD 44 % vs. 31 %; $p=0.14$) and widespread (10 % vs. 6 % of the body area, $p=0.001$), with an earlier age at onset (246 vs. 473 days, $p=0.0001$) compared to wild type
Cramer et al., 2010, Germany [49]	Birth cohort (LISAplus and GINIplus)	1,039 (LISAplus) 1.828 (GINIplus)	0–6 years	*FLG* mutation status (R501X, 2282del4); questionnaire data on AD symptoms and elder siblings	Children with FLG mutations had a significantly higher risk of AD if they had elder siblings: OR 3.27 (95 % CI, 1.14–9.36) in LISAplus and OR 2.41 (95 % CI, 1.06–5.48) in GINIplus. Attending day care centers lessened this effect
Ekelund et al., 2008, Sweden [22]	Family-based cohort	1,514	29 years (mean)	*FLG* mutation status (R501X and 2282del4); dermatologist-diagnosed AD and severity score; physician-diagnosed allergic asthma and rhinoconjunctivitis; IgE levels	There was an association between *FLG* gene variants and AD ($p=9.5\times10^{-8}$). The highest OR for the combined allele (4.73, 95 % CI 1.98–11.29) was found for the subgroup with severe AD. *FLG* mutations were also associated with raised allergen-specific IgE, allergic asthma, and allergic rhinoconjunctivitis occurring in the context of AD
Flohr et al., 2010, UK [15]	Birth cohort	88	3 months	AD diagnosis and severity; TEWL; *FLG* mutation status (R501X, 2282del4, R2447X and S3247X)	17 % children carried at least one FLG mutation. FLG mutation carriers were significantly more likely to have clinically dry skin, even in the absence of AD (OR 8.5, 95 % CI 1.09–66.58). *FLG* mutation carriers were more likely to have AD by 3 months of age (OR 4.26, 95 % CI 1.34–13.57). *FLG* mutations were significantly associated with higher median TEWL (all children, *FLG* "yes" 21.59 vs. *FLG* "no" 11.24, $p<0.001$), even without clinical AD (*FLG* "yes" 15.99 vs. *FLG* "no" 10.82, $p=0.01$)

Flohr et al., 2013, UK [16]	Birth cohort	619 (exclusively breastfed 3-month-old infants)	3 months	AD diagnosis, distribution, and severity; TEWL; *FLG* mutation status; skin-prick testing to 6 foods (cow's milk, egg, sesame, peanut, cod fish, and wheat); FLG mutation status (R501X, 2282del4, R2447X, S3247X, 3673delC, and 3702delG)	Skin barrier impairment (raised TEWL) was positively associated with food sensitization risk, even after adjustment for FLG status and AD (adjusted OR 2.23; 95 % CI 1.04–4.76). Children with AD were even more likely to be food-sensitized (OR 6.18, 95 % CI 2.94–12.98), and this association was disease severity dependent. There was no significant association between *FLG* carriage, AD phenotype (flexural vs. non-flexural), and food sensitization
Gao et al., 2009, USA [33]	Case–control	278 cases (AD patients, 112 of these had eczema herpeticum); 157 controls (nonatopic)	17–41 years (mean)	Association between *FLG* null alleles (R501X and 2282del4) and risk of eczema herpeticum	Significant associations were observed for both the R501X and 2282del4 mutations and AD among European American subjects ($p = 1.46 \times 10^{-5}$, 3.87×10^{-5}, respectively), but the frequency of the R501X mutation was 3 times higher (25 % vs. 9 %) for AD with eczema herpeticum than for AD without eczema herpeticum (OR 3.4, 95 % CI 1.7–6.8). Associations with AD with eczema herpeticum were stronger for the combined null mutations (OR 10.1, 95 % CI 4.7–22.1)
Henderson et al., 2008, UK [20]	Birth cohort (ALSPAC)	6,971	0–11 years	*FLG* mutation status (R501X and 2282del4); questionnaire data on eczema, asthma, and allergic rhinitis; skin-prick testing to common allergens including mixed grasses, house dust mite, and cat	*FLG* mutations were significantly associated with AD (OR 2.73, 95 % CI 1.87–3.99); *FLG*-associated AD presents in early life and is more persistent (HR for AD resolution for *FLG* mutations vs. *FLG* wild type = 0.67, 95 % CI 0.58–0.77). *FLG* mutations conferred an asthma risk of 1.80 (95 % CI, 1.34–2.41); asthma risk was especially high in the context of AD (OR 3.16, 95 % CI 2.25–4.43). Strong associations were also identified with sensitization to grass, house dust mite, and cat dander; the strongest was for sensitization to all three allergens (OR 2.12, 95 % CI 1.03–4.37)

(continued)

Table 16.1 (continued)

Authors, year, country [Ref.]	Study type	Number of participants	Age	Outcomes/exposures	Key results and conclusions
Marenholz et al., 2006, Germany [7]	Cohort (GENUFAD family-based cohort and MAS birth cohort)	1,092 cases (children with AD); 321 controls	0–10 years	*FLG* mutation status (R501X and 2282del4); AD diagnosis and severity; IgE levels; history of asthma or allergic rhinitis	There was a highly significant association of the *FLG* null mutations with AD and concomitant asthma; this was replicated in the second cohort. *FLG* mutations predisposed to asthma (OR 6.21, 95 % CI 2.6–14.8), allergic rhinitis (OR 4.79, 95 % CI 2.0–11.6), and allergic sensitization (OR 3.84, 95 % CI 1.9–7.7) only in the presence of AD. The presence of two *FLG* null alleles was an independent risk factor for asthma in children with AD, and the two investigated mutations accounted for about 11 % of AD cases in the German population
Palmer et al., 2006, UK [6]	Case–control	52 cases (Irish children with dermatologist-diagnosed eczema); 189 controls. Replicated in Scottish population cohort, BREATHE cohort, and COPSAC cohort	1–16 years	*FLG* mutation status (R510X and 2282del4); diagnosis of asthma	Two independent loss-of-function genetic variants in the *FLG* gene were found to be strong predisposing factors for AD ($p=3\times10^{-17}$ for combined genotype). These variants also showed a highly significant association with asthma occurring in the context of AD ($p=6\times10^{-12}$ for combined genotype)
Rodriguez et al., 2009, UK [26]	Meta-analysis	5,791 AD cases; 26,454 controls; 1,951 families. 3,138 asthma cases; 17,164 controls; 1,511 offspring	–	Association between *FLG* mutations and AD/asthma	*FLG* null mutations increased AD risk (OR 3.12, 95 % CI 2.57–3.79) and were associated with more severe and dermatologist-diagnosed disease. *FLG* mutations were also significantly associated with asthma (OR 1.48, 95 % CI 1.32–1.66). However, although strong effects were observed for the compound phenotype asthma plus AD (OR 3.29, 95 % CI 2.84–3.82), there appeared to be no association with asthma in the absence of AD

Schuttelaar et al., 2009, Netherlands [17]	Birth cohort (PIAMA)	934	1–8 years	*FLG* mutation status (R501X, 2282del4 and R2447X); association with AD, sensitization, asthma, allergic rhinitis, and cat exposure	Combined *FLG* mutations were significantly associated with AD at all ages when occurring in the first year of life (OR 2.0, 95 % CI 1.4–2.8). *FLG* mutations combined with AD in the first year of life were associated with later development of asthma (OR 3.7, 95 % CI 1.8–7.5) and allergic rhinitis (OR 4.0, 95 % CI 1.2–13.6). The *FLG* 2282del4 mutation alone was significantly associated with the development of AD during the first year, having AD up to 8 years and sensitization at the age of 8 years, which was enhanced by early-life cat exposure (OR 8.2, 95 % CI 2.6–25.9; OR 6.0, 95 % CI 3.2–11.3; and OR 5.4, 95 % CI 1.2–23.6, respectively)
Stemmler et al., 2007, Germany [14]	Case–control	378 cases (dermatologist-diagnosed AD); 700 controls	0.5–72 years	*FLG* mutation status (R501X, 2282del4); age of AD onset	Combined allele frequency of two *FLG* mutations was 5.1 % in healthy controls, 8.7 % in all AD patients ($p=0.001$), 10.3 % in patients with age of onset in childhood ($p=3.6\times10^{-5}$), and 12.1 % in AD patients with age of onset before 2 years ($p=7.6\times10^{-7}$)
Thyssen et al., 2011, Germany [21]	1. Cross-sectional study using a population-based sample	Random sample of 3,335 subjects from Copenhagen	18–69 years	*FLG* mutation status (2282del4 and R501X)	*FLG* homozygous/compound heterozygous individuals accounted for 0.3 % of adults, 3 % of AD patients, and 0.7 % of children. 1/9 adults and 1/3 children had never experienced AD. Yearlong complete remission was observed in half of homozygous patients
	2. Cross-sectional study using hospital outpatients	499 patients seen in AD clinic	–		
	3. Birth cohort (COPSAC)	411 children born to mothers with history of asthma	0–11 years		

(continued)

Table 16.1 (continued)

Authors, year, country [Ref.]	Study type	Number of participants	Age	Outcomes/exposures	Key results and conclusions
van den Oord and Sheikh, 2009, UK [12]	Meta-analysis	24 studies	–	*FLG* mutation status; AD; food allergy; asthma; allergic rhinitis; anaphylaxis; relevant immunological variables relating to risk of allergic sensitization and disease	*FLG* mutations increased risk of developing allergic sensitization, AD, and allergic rhinitis. FLG mutations were associated with more severe and persistent AD. FLG mutations also increased the risk of asthma in people with preceding AD
Weidinger et al., 2006, Germany [9]	Family-based cohort	476 parents; 196 offspring	54.71 years (mean for parents); 22.12 years (mean for offspring)	*FLG* mutation status (R501X and 2282del4); dermatologist-diagnosed AD; IgE; questionnaire data on allergies	There were prominent associations between two FLG mutations and AD ($p=5.1\times10^{-8}$ for combined genotype). FLG mutations were also associated with the extrinsic subtype of AD, characterized by high IgE levels ($p=9.8\times10^{-8}$ for combined genotype) and allergic sensitizations ($p=2.3\times10^{-7}$ for combined genotype). FLG mutations were significantly associated with palmar hyperlinearity ($p=5.9\times10^{-6}$ for combined genotype)
Weidinger et al., 2008, Germany [29]	Cross-sectional (ISAAC II)	3,099	9.6 years (mean)	*FLG* mutation status (R501X, 2282del4, R2447X, S3247X, and 3702delG); AD; allergic rhinitis; asthma	*FLG* variants increased the risk for AD more than threefold (OR 3.12, 95 % CI 2.33–4.173; population-attributable risk 13.5 %). Independent of AD, *FLG* mutations conferred a risk for allergic rhinitis (OR 2.64, 95 % CI 1.76–4.00; population-attributable risk 10.8 %). In contrast, the association with asthma (OR 1.79, 95 % CI 1.19–2.68) was restricted to asthma occurring in the context of AD, and there was a significant association with the complex phenotype AD plus asthma (OR 3.49, 95 % CI 2.00–6.08)

16.2 General Associations

In 2006, Palmer et al. published the seminal finding that two loss-of-function mutations in the *FLG* gene were strongly associated with AD [6]. These have since been widely replicated, not only in white European but also in East Asian populations [5, 7–10]. Furthermore, intragenic copy number variation within *FLG* (typically 10–12 repeats) affects AD risk independently of *FLG* null mutations, with approximately 12 % reduction in AD risk for each additional copy number (OR = 0.88, 95 % CI 0.78–0.98) [11]. *FLG* loss-of-function mutations represent the strongest known genetic risk factor for AD, with odds ratios ranging from 3.12 to 4.78 [1, 2, 9, 12] and a population-attributable risk of 11 % [3, 7].

16.3 Disease Onset

Given an impaired skin barrier from birth in those with *FLG* null alleles, it would follow that the associated AD phenotype is of early onset. This theory has been extensively investigated. For instance, Barker et al. studied 163 adults with persistent childhood-onset AD, reporting that 42 % carried one or both of the two commonest *FLG* null alleles (R501X and 2282del4) compared to 8.8 % of the general population ($p = 1.7 \times 10^{-53}$) [4, 10]. This work was extended by Brown et al. in a population-based setting, reporting a significant and independent association with early-onset AD persisting into adulthood, with the combined null genotype of the six commonest *FLG* mutations showing a highly significant association (OR 5.6, 95 % CI 4.1–7.8) [3, 13]. Similarly, Stemmler et al. found the strongest association between *FLG* mutations and AD in the subgroup with AD onset <2 years of age, with a carrier frequency of 21.3 % versus 9.6 % in controls ($p = 8.45 \times 10^{-6}$) [5, 14]. More recently, Flohr et al. showed in a cohort of over 600 3-month-old infants that *FLG* mutation carriers were more likely to have AD than wild-type subjects (OR 3.55, 95 % CI 2.16–5.84) [6, 15, 16]. Furthermore, where children have been followed up longitudinally, such as in the Dutch PIAMA cohort ($n = 934$), *FLG* mutation carriage (R501X, 2282del4 and R2447X) has been linked with AD onset during the first year of life (OR 2.6, 95 % CI 1.6–4.2, OR 2282del4 alone 8.2, 95 % CI 2.6–25.9), but not AD commencing after the first birthday [1, 2, 17]. More specifically, Carson et al. reported in a cohort of 411 children born to asthmatic mothers that *FLG* mutations were associated with a mean onset at 246 days, compared to wild-type individuals at 473 days ($p < 0.0001$) [3, 18].

16.4 Disease Course

Several studies, some of which have been outlined above, have also suggested that *FLG* mutation inheritance predisposes to more persistent AD [4, 10, 13, 19]. One paper looked at the influence of *FLG* mutations upon disease trajectory in a large population-based birth cohort study, using survival analysis of children with AD established at 42 months [3, 20]. The mean time that AD persisted was 76.7 months in children with *FLG* mutations compared to 65.6 months in wild-type individuals, reflected in a hazard ratio of 0.67 (95 % CI 0.58–0.77). Since then, Thyssen et al. have explored the natural history of AD in subjects homozygous for *FLG* loss-of-function mutations in a large sample from three cohorts, where *FLG* homozygotes or compound heterozygotes accounted for 0.7 % of children, 0.3 % of adults, and 3 % of AD patients [5, 21]. Despite a complete absence of epidermal filaggrin in these individuals, one in three children and one in nine adults reported never suffering from AD. Furthermore, year-long complete remission was observed in half of the *FLG* homozygotes – even among patients from a hospital dermatology clinic – suggesting that the natural course of AD in patients homozygous for the *FLG* null genotype can be surprisingly favorable.

16.5 Disease Severity

Another important question is whether *FLG* null mutation inheritance predisposes to more severe AD. Ekelund et al. conducted the first analysis of *FLG* mutation variants in a Swedish population, reporting that the highest odds ratio for the combined R501X and 2282del4 *FLG* genotype was for the subgroup with severe disease (OR = 4.73, 95 % CI 1.98–11.29). This was defined as a score of ≥ 4 using an index comprising age at onset, need for hospitalization, number of body sites affected, and total serum IgE levels [6, 22]. Subsequently, Brown et al. conducted a prospective study of 792 school children with a predominance of mild-moderate AD, reporting that *FLG* mutations were associated with a more severe phenotype ($p = 0.004$) [5, 7–10, 23]. However, although the results were statistically significant, the mean difference was only 2 points on the Three-Item Severity (TIS) score index [11, 24]. This disconnect between a statistically significant association and clinically meaningful significance had already been highlighted by Morar et al. [9, 12, 25] and was more recently confirmed in a UK cohort of 3-month-old infants, where a small but significant difference in AD severity between *FLG* mutation and wild-type carriers with AD was observed (median SCORAD *FLG* mutation "yes" vs. "no" 12.5 vs. 7.5, $p < 0.001$) [7, 15]. Furthermore, the same group showed that moderate-severe disease was associated with a higher TEWL compared to mild AD ($r = 0.55$, $p < 0.001$) [10, 16]. There was also a significantly higher TEWL in children with *FLG* loss-of-function mutations versus wild-type children, even if they did not have AD, suggesting that skin barrier impairment precedes the AD phenotype [13, 15]. Similarly, the Danish COPSAC study showed significant but clinically marginal effects using different AD severity parameters (number of unscheduled visits *FLG* mutation "yes" vs. "no" 3.6 vs. 2.7, $p = 0.04$; moderate-severe SCORAD *FLG* mutation "yes" vs. "no" 44 % vs. 31 %, $p = 0.14$; body surface area involvement *FLG* mutation "yes" vs. "no" 10 % vs. 6 %, $p < 0.001$) [14, 18]. Finally, two meta-analyses confirmed a significant association between *FLG* mutation inheritance and AD severity [12, 14, 16, 26]. However, a more recent study from Sweden (BAMSE cohort, $n = 3{,}301$) did not show an association between AD severity and *FLG* phenotype in preadolescent children [27].

16.6 Other Disease Associations

In addition to the influence of *FLG* mutations on AD onset, severity, and chronicity, a number of specific disease associations have been explored, partly to delineate AD subphenotypes.

16.6.1 Clinical Presentation

Palmar hyperlinearity shows a strong association with *FLG* null mutations. This was first reported by Weidinger et al. ($p = 5.4 \times 10^{-6}$) in a collection of 476 German families with an AD history [9]. Brown et al. then performed a population-based case–control study in which palmar hyperlinearity was found to be significantly associated with the combined *FLG* null genotype, as well as with the four most prevalent *FLG* mutations when analyzed individually; this resulted in a positive predictive value of 71 % and a negative predictive value of 90 % for palmar hyperlinearity in association with *FLG* mutation inheritance [28]. The same group subsequently made the observation of three distinct patterns of palmar hyperlinearity after detailed physical examination of 484 children, with a cross-hatched pattern predominating overall and in AD, compared to a linear pattern more commonly found in ichthyosis vulgaris [23]. Another important clinical feature of *FLG* inheritance is keratosis pilaris; like palmar hyperlinearity, this is found in both ichthyosis vulgaris and AD.

Little work has been done on whether AD in *FLG* mutation carriers follows a different body distribution compared to wild-type children with AD. Flohr et al. found no link between FLG mutation inheritance and a flexural versus non-flexural pattern [16], but Carson et al. recently suggested that *FLG* null mutations predispose to AD in air-exposed sites, especially the cheeks and the back of the hands [18]. This may be important, as allergic sensitization – for instance, to aeroallergens – and other allergic diseases also show significant association with *FLG* mutation inheritance.

16.6.2 Allergic Sensitization: Food and Respiratory Allergies

Palmer et al. were the first to make the observation that *FLG* mutation carriers have an increased risk of developing asthma – but only in the context of current or previous AD [6]. Large population-based studies have consistently replicated this observation, with overall risk ratios ranging from 1.48 to 1.79 [12, 20, 29]. Allergic sensitization and hay fever are also independently associated with *FLG* mutations [12, 29]. One meta-analysis synthesized data from 24 studies of *FLG* defects and allergic disease, reporting odds ratios for allergic sensitization (1.91, 95 % CI 1.44–2.54), allergic rhinitis (2.84, 95 % CI 2.08–3.88), and asthma (2.79, 95 % CI 1.77–4.41) [12]. Given that filaggrin is not expressed in respiratory epithelia, the most likely explanation is that transcutaneous sensitization to aeroallergens occurs via an impaired epidermal barrier.

A similar mechanism has been suggested for food sensitization, based on the observation that children exposed to arachis oil creams (containing peanut protein) in early life have a significantly higher risk of developing challenge-proven peanut allergy [30] compared to unexposed atopic and healthy controls. Furthermore, a case–control study comparing 71 challenge-proven patients with peanut allergy with 1,000 non-sensitized controls found an almost fourfold increase in peanut allergy risk in children with at least one *FLG* mutation compared to wild-type children, even after adjustment for AD (OR = 3.8, 95 % CI 1.7–8.3) [31]. The risk of peanut allergy almost doubled where children had AD in early life (OR = 7.4, 95 % CI 4.1–13.7). However, the study relied on parent-reported AD, only looked at one food, and did not include an assessment of skin barrier function. These issues were subsequently addressed in an investigation by Flohr et al., who showed independent links between skin barrier impairment (raised TEWL), AD phenotype, and food sensitization as early as 3 months of age [16]. AD increased the risk of food sensitization more than sixfold overall (OR = 6.18, 95 % CI 2.94–12.98), and the association was particularly strong for sensitization to cow's milk (OR = 9.11, 95 % CI 2.27–36.59), egg (OR = 9.48, 95 % CI 3.77–23.83), and peanut (OR = 4.09, 95 % CI 1.00–16.76). There was also a stepwise increase in food sensitization risk with more severe AD ($OR_{SCORAD<20} = 3.91$, 95 % CI 1.79–9.00 vs. $OR_{SCORAD \geq 20} = 25.60$, 95 % CI 9.03–72.57, $p_{trend} < 0.001$), even after adjustment for *FLG* mutation inheritance and TEWL. This further implies that a defective skin barrier enhances transcutaneous antigen exposure and allergic priming, particularly where eczematous skin inflammation is present – an effect that is likely to be mediated by antigen-presenting cells in the epidermis. This in turn may lead to systemic sensitization and distal allergic inflammation in the respiratory tract or gut.

The wealth of clinical evidence in humans is also supported by animal work, with Fallon et al. performing topical application of ovalbumin in flaky tail mice homozygous for a mutation analogous to human *FLG* mutations. This resulted in cutaneous inflammatory infiltrates and enhanced cutaneous allergen priming, associated with an upregulation in allergen-specific IgE and Th2 cytokines [32].

16.6.3 Skin Infections

FLG mutation inheritance also impacts upon susceptibility to infectious complications of AD, as shown by Gao et al., who reported that AD patients who developed eczema herpeticum had a three times higher frequency of the R501X mutation versus wild-type children (25 % vs. 9 %, $p=0.0002$) [33]. Indeed, it is well established that the antimicrobial barrier is compromised in AD, with 90 % of AD patients colonized with *Staphylococcus aureus* (SA) compared to 5–30 % of healthy individuals [34]. There is also a reduction in antimicrobial peptide expression in the skin of AD sufferers, further enhancing susceptibility to skin infections [35]. The pH-lowering effect of filaggrin breakdown products in the skin may likewise play a role, as evidenced in vitro by Miajlovic et al., who showed that SA growth rates and protein expression were reduced in the presence of urocanic acid and pyrrolidone carboxylic acid [36]. Several studies have shown that filaggrin deficiency is associated with elevated skin-surface pH [37, 38], so it appears likely that this disrupted "acid mantle" of the stratum corneum facilitates the growth of SA and other microbial agents.

A final study used skin taping and mass spectrometry to analyze protein expression in AD patients and healthy controls, then classified subjects according to history of eczema herpeticum and *Staphylococcus aureus* colonization. Interestingly, this proteomic profiling revealed that proteins related to the skin barrier (including filaggrin-2) and generation of natural moisturizing factor (of which filaggrin is a part) were in fact expressed at lower levels in lesional versus non-lesional sites of AD subjects both with and without a history of eczema herpeticum. The authors suggested that lower expression of such proteins could further exacerbate barrier defects and perpetuate water loss from the skin. No significant differences were found between patients with AD with and without a history of eczema herpeticum [39].

16.7 Gene-Gene and Gene-Environment Interactions

Gene-gene interactions have been a major focus of interest, with epistatic effects postulated for a number of candidates including the serine protease inhibitor SPINK5 and the kallikrein 7 gene KLK7, which also contributes to the regulation of the desquamation process in the epidermis [40–42]. However, a substantial study involving more than 2,500 cases and 10,000 controls failed to confirm this [43].

FLG is also thought to exhibit gene-environment interactions. This is a seemingly intuitive concept given the rise in AD prevalence in recent decades, which cannot be explained by genetic mutations alone, and the *FLG*-associated population-attributable risk of only around 10 %. As argued above, a fundamental aspect of the *FLG* story is that the atopic march is initiated by a weakened epidermal barrier and enhanced penetration of environmental food and aeroallergens, leading to allergic sensitization and potentially flares in existing AD, contributing to more severe and chronic disease. Any environmental factor that has a detrimental effect on skin barrier integrity should consequently enhance AD risk (Fig. 16.1). For instance, frequent use of detergents as well as higher chlorine and calcium carbonate levels in tap water has been associated with AD in population-based settings. This is potentially explained by a reduction in natural moisturizing factor, increase in skin pH, and a subsequent upregulation in protease activity [44–47]. Cat ownership in early life only increases AD risk in individuals carrying at least one *FLG* loss-of-function mutation [17, 48]. Sibship size is another potential environmental modulator, with two large German birth cohort studies reporting that children with *FLG* null mutations have a greater risk of AD if they have an older sibling. However, there is currently no explanation as to why this might be the case, and attendance at a day care center lessened this effect [49], potentially mediated by microbial exposure. Furthermore, there are no studies at present that have assessed the effect of the constituents of the skin microbiome on skin barrier function in the context of *FLG* genotype and AD risk. Finally, environmental pollution – such as NO_2 levels, maternal smoking, and passive smoking by the child – has been also been associated with an increase in AD in the offspring [50–55]. However, the latter does not appear linked to the *FLG* genotype [20], and no study

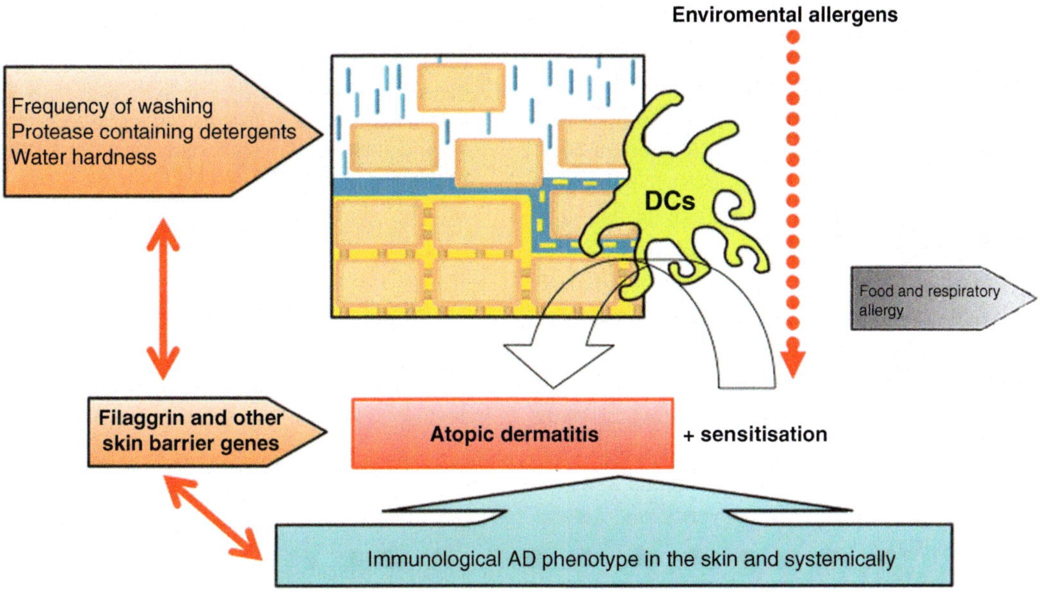

Fig. 16.1 The interplay between skin barrier-related environmental and genetic as well as immunological factors in the development of AD

has so far assessed gene-environment interactions between fine particle air pollution, *FLG* mutations, skin barrier function, and AD risk.

16.8 Therapeutic Implications

The evolving *FLG* story has led to a paradigm shift in the proposed pathoetiology of AD, from a primarily immuno-centric view to one in which skin barrier function takes center stage. The potential therapeutic implications are significant. For instance, much research into novel therapeutics is currently aimed at improving skin barrier function [56]. Immunomodulation is a promising approach, given that inflammation appears to reduce filaggrin expression via upregulation in Th2 cytokines IL-4, IL-13 [57], and IL-22 [58]. Indeed, topical anti-inflammatories have already been shown to reverse the reduced filaggrin expression seen in lesional AD [59]. Another strategy is high-throughput screening to identify novel compounds able to upregulate filaggrin expression. Gentamicin and PTC124 (a potent nonsense mutation inhibitor) have both been put forward as candidates [60]. However, efforts here have been hampered by the fact that filaggrin is not normally expressed in monolayer keratinocyte cultures, leading to difficulties in developing a high-throughput screening assay. Finally, it is possible to target filaggrin protein translation, thereby enabling "read through" of stop codons and restoring full protein expression. This is an approach already under investigation for genetic disorders such as cystic fibrosis and Duchenne's muscular dystrophy, although it is tempting to postulate that skin conditions may represent more successful targets because of the potential for direct topical drug application.

Conclusion

The closer we look, the more complex AD becomes, and we might well be dealing with several distinct entities that clinically manifest in a similar way, rather than one disease. Here our improved understanding of skin barrier function, as well as genetic and immunological biomarkers, will help to delineate AD subphenotypes, with the discovery of the *FLG* gene making a major contribution. Despite this, half of AD patients, even with moderate-severe disease, do not carry a *FLG* mutation, so it is important that the search for further genetic loci continues [61]. It is also a puzzle as to why the clinical presentations of ichthyosis vulgaris and AD are so different, given the joint association with *FLG* loss-of-function mutations. With a population-attributable risk of only 10 % for *FLG* mutations in AD, it is vital that we continue to explore early-life *FLG*-environment interactions in population-based study designs. These include determination of phenotypic skin barrier function and immunological markers, partly to explain discrepancies in previous epidemiological study results, but also to explore how these factors influence the effect of environmental risk factors on AD development, severity, and natural history [62].

References

1. McKenna SP, Doward LC, Meads DM, Tennant A, Lawton G, Grueger J. Quality of life in infants and children with atopic dermatitis: addressing issues of differential item functioning across countries in multinational clinical trials. Health Qual Life Outcomes. 2007;5:45.
2. Williams HC. Clinical practice. Atopic dermatitis. N Engl J Med. 2005;352(22):2314–24.
3. McGrath JA, Uitto J. The filaggrin story: novel insights into skin-barrier function and disease. Trends Mol Med. 2008;14(1):20–7.
4. Sandilands A, Terron-Kwiatkowski A, Hull PR, O'Regan GM, Clayton TH, Watson RM, et al. Comprehensive analysis of the gene encoding filaggrin uncovers prevalent and rare mutations in ichthyosis vulgaris and atopic eczema. Nat Genet. 2007;39(5): 650–4.
5. Baurecht H, Irvine AD, Novak N, Illig T, Buhler B, Ring J, et al. Toward a major risk factor for atopic eczema: meta-analysis of filaggrin polymorphism data. J Allergy Clin Immunol. 2007;120(6): 1406–12.
6. Palmer CN, Irvine AD, Terron-Kwiatkowski A, Zhao Y, Liao H, Lee SP, et al. Common loss-of-function variants of the epidermal barrier protein filaggrin are a major predisposing factor for atopic dermatitis. Nat Genet. 2006;38(4):441–6.
7. Marenholz I, Nickel R, Rüschendorf F, Schulz F, Esparza-Gordillo J, Kerscher T, et al. Filaggrin loss-of-function mutations predispose to phenotypes involved in the atopic march. J Allergy Clin Immunol. 2006;118(4):866–71.
8. Ruether A, Stoll M, Schwarz T, Schreiber S, Folster-Holst R. Filaggrin loss-of-function variant contributes to atopic dermatitis risk in the population of Northern Germany. Br J Dermatol. 2006;155(5): 1093–4.
9. Weidinger S, Illig T, Baurecht H, Irvine AD, Rodriguez E, Diaz-Lacava A, et al. Loss-of-function variations within the filaggrin gene predispose for atopic dermatitis with allergic sensitizations. J Allergy Clin Immunol. 2006;118(1):214–9.
10. Barker JN, Palmer CN, Zhao Y, Liao H, Hull PR, Lee SP, et al. Null mutations in the filaggrin gene (FLG) determine major susceptibility to early-onset atopic dermatitis that persists into adulthood. J Invest Dermatol. 2007;127(3):564–7.
11. Brown SJ, Kroboth K, Sandilands A, Campbell LE, Pohler E, Kezic S, et al. Intragenic copy number variation within filaggrin contributes to the risk of atopic dermatitis with a dose-dependent effect. J Invest Dermatol. 2012;132(1):98–104.
12. van den Oord RA, Sheikh A. Filaggrin gene defects and risk of developing allergic sensitization and allergic disorders: systematic review and meta-analysis. BMJ. 2009;339:b2433.
13. Brown SJ, Sandilands A, Zhao Y, Liao H, Relton CL, Meggitt SJ, et al. Prevalent and low-frequency null mutations in the filaggrin gene are associated with early-onset and persistent atopic eczema. J Invest Dermatol. 2008;128(6):1591–4.
14. Stemmler S, Parwez Q, Petrasch-Parwez E, Epplen JT, Hoffjan S. Two common loss-of-function mutations within the filaggrin gene predispose for early onset of atopic dermatitis. J Invest Dermatol. 2007; 127(3):722–4.
15. Flohr C, England K, Radulovic S, McLean WH, Campbell LE, Barker J, et al. Filaggrin loss-of-function mutations are associated with early-onset eczema, eczema severity and transepidermal water loss at 3 months of age. Br J Dermatol. 2010;163(6): 1333–6.
16. Flohr C, Perkin M, Logan K, Marrs T, Radulovic S, Campbell LE, et al. Atopic dermatitis and disease severity are the main risk factors for food. J Invest Dermatol. 2014;134(2):345–50.
17. Schuttelaar ML, Kerkhof M, Jonkman MF, Koppelman GH, Brunekreef B, de Jongste JC, et al. Filaggrin mutations in the onset of eczema, sensitization, asthma, hay fever and the interaction with cat exposure. Allergy. 2009;64(12):1758–65.
18. Carson CG, Rasmussen MA, Thyssen JP, Menne T, Bisgaard H. Clinical presentation of atopic dermatitis by filaggrin gene mutation status during the first 7 years of life in a prospective cohort study. PLoS One. 2012;7(11):e48678.
19. Weidinger S, Rodriguez E, Stahl C, Wagenpfeil S, Klopp N, Illig T, et al. Filaggrin mutations strongly predispose to early-onset and extrinsic atopic dermatitis. J Invest Dermatol. 2007;127(3):724–6.
20. Henderson J, Northstone K, Lee SP, Liao H, Zhao Y, Pembrey M, et al. The burden of disease associated with filaggrin mutations: a population-based, longitudinal birth cohort study. J Allergy Clin Immunol. 2008;121(4):872–9.

21. Thyssen JP, Carlsen BC, Bisgaard H, Giwercman C, Johansen JD, Linneberg A, et al. Individuals who are homozygous for the 2282del4 and R501X filaggrin null mutations do not always develop dermatitis and complete long-term remission is possible. J Eur Acad Dermatol Venereol. 2011;26(3):386–9.
22. Ekelund E, Lieden A, Link J, Lee SP, D'Amato M, Palmer CN, et al. Loss-of-function variants of the filaggrin gene are associated with atopic eczema and associated phenotypes in Swedish families. Acta Derm Venereol. 2008;88(1):15–9.
23. Brown SJ, Relton CL, Liao H, Zhao Y, Sandilands A, McLean WH, et al. Filaggrin haploinsufficiency is highly penetrant and is associated with increased severity of eczema: further delineation of the skin phenotype in a prospective epidemiological study of 792 school children. Br J Dermatol. 2009;161(4):884–9.
24. Oranje AP, Glazenburg EJ, Wolkerstorfer A, de Waard-van der Spek FB. Practical issues on interpretation of scoring atopic dermatitis: the SCORAD index, objective SCORAD and the three-item severity score. Br J Dermatol. 2007;157(4):645–8.
25. Morar N, Cookson WO, Harper JI, Moffatt MF. Filaggrin mutations in children with severe atopic dermatitis. J Invest Dermatol. 2007;127(7):1667–72.
26. Rodriguez E, Baurecht H, Herberich E, Wagenpfeil S, Brown SJ, Cordell HJ, et al. Meta-analysis of filaggrin polymorphisms in eczema and asthma: robust risk factors in atopic disease. J Allergy Clin Immunol. 2009;123(6):1361–7.
27. Ballardini N, Kull I, Soderhall C, Lilja G, Wickman M, Wahlgren CF. Eczema severity in preadolescent children and its relation to sex, filaggrin mutations, asthma, rhinitis, aggravating factors and topical treatment: a report from the BAMSE birth cohort. Br J Dermatol. 2013;168(3):588–94.
28. Brown SJ, Relton CL, Liao H, Zhao Y, Sandilands A, Wilson IJ, et al. Filaggrin null mutations and childhood atopic eczema: a population-based case–control study. J Allergy Clin Immunol. 2008;121(4):940–3.
29. Weidinger S, O'Sullivan M, Illig T, Baurecht H, Depner M, Rodriguez E, et al. Filaggrin mutations, atopic eczema, hay fever, and asthma in children. J Allergy Clin Immunol. 2008;121(5):1203–1209e1.
30. Lack G, Fox D, Northstone K, Golding J, Avon Longitudinal Study of Parents and Children Study Team. Factors associated with the development of peanut allergy in childhood. N Engl J Med. 2003;348(11):977–85.
31. Brown SJ, Asai Y, Cordell HJ, Campbell LE, Zhao Y, Liao H, et al. Loss-of-function variants in the filaggrin gene are a significant risk factor for peanut allergy. J Allergy Clin Immunol. 2011;127(3):661–7.
32. Fallon PG, Sasaki T, Sandilands A, Campbell LE, Saunders SP, Mangan NE, et al. A homozygous frameshift mutation in the mouse Flg gene facilitates enhanced percutaneous allergen priming. Nat Genet. 2009;41(5):602–8.
33. Gao PS, Rafaels NM, Hand T, Murray T, Boguniewicz M, Hata T, et al. Filaggrin mutations that confer risk of atopic dermatitis confer greater risk for eczema herpeticum. J Allergy Clin Immunol. 2009;124(3):507–13, 513e1–7.
34. Wolf R, Wolf D. Abnormal epidermal barrier in the pathogenesis of atopic dermatitis. Clin Dermatol. 2012;30(3):329–34.
35. Bieber T. Atopic dermatitis. N Engl J Med. 2008; 358(14):1483–94.
36. Miajlovic H, Fallon PG, Irvine AD, Foster TJ. Effect of filaggrin breakdown products on growth of and protein expression by Staphylococcus aureus. J Allergy Clin Immunol. 2010;126(6):1184–90.e3.
37. Kezic S, O'Regan GM, Yau N, Sandilands A, Chen H, Campbell LE, et al. Levels of filaggrin degradation products are influenced by both filaggrin genotype and atopic dermatitis severity. Allergy. 2011;66(7):934–40.
38. Jungersted JM, Scheer H, Mempel M, Baurecht H, Cifuentes L, Hogh JK, et al. Stratum corneum lipids, skin barrier function and filaggrin mutations in patients with atopic eczema. Allergy. 2010;65(7):911–8.
39. Broccardo CJ, Mahaffey S, Schwarz J, Wruck L, David G, Schlievert PM, et al. Comparative proteomic profiling of patients with atopic dermatitis based on history of eczema herpeticum infection and Staphylococcus aureus colonization. J Allergy Clin Immunol. 2011;127(1):186–93, 193.e1–11.
40. Walley AJ, Chavanas S, Moffatt MF, Esnouf RM, Ubhi B, Lawrence R, et al. Gene polymorphism in Netherton and common atopic disease. Nat Genet. 2001;29(2):175–8.
41. Vasilopoulos Y, Cork MJ, Murphy R, Williams HC, Robinson DA, Duff GW, et al. Genetic association between an AACC insertion in the 3'UTR of the stratum corneum chymotryptic enzyme gene and atopic dermatitis. J Invest Dermatol. 2004;123(1):62–6.
42. Hubiche T, Ged C, Benard A, Léauté-Labrèze C, McElreavey K, de Verneuil H, et al. Analysis of SPINK 5, KLK 7 and FLG genotypes in a French atopic dermatitis cohort. Acta Derm Venereol. 2007; 87(6):499–505.
43. Weidinger S, Baurecht H, Wagenpfeil S, Henderson J, Novak N, Sandilands A, et al. Analysis of the individual and aggregate genetic contributions of previously identified serine peptidase inhibitor Kazal type 5 (SPINK5), kallikrein-related peptidase 7 (KLK7), and filaggrin (FLG) polymorphisms to eczema risk. J Allergy Clin Immunol. 2008;122(3):560–4.
44. Sherriff A, Golding J, Alspac Study Team. Hygiene levels in a contemporary population cohort are associated with wheezing and atopic eczema in preschool infants. Arch Dis Child. 2002;87(1):26–9.
45. McNally NJ, Williams HC, Phillips DR, Smallman-Raynor M, Lewis S, Venn A, et al. Atopic eczema and domestic water hardness. Lancet. 1998;352(9127): 527–31.
46. Cork MJ, Robinson DA, Vasilopoulos Y, Ferguson A, Moustafa M, MacGowan A, et al. New perspectives on epidermal barrier dysfunction in atopic dermatitis: gene-environment interactions. J Allergy Clin Immunol. 2006;118(1):3–21; quiz22–3.

47. Chaumont A, Voisin C, Sardella A, Bernard A. Interactions between domestic water hardness, infant swimming and atopy in the development of childhood eczema. Environ Res. 2012;116:52–7.
48. Bisgaard H, Simpson A, Palmer CN, Bonnelykke K, Mclean I, Mukhopadhyay S, et al. Gene-environment interaction in the onset of eczema in infancy: filaggrin loss-of-function mutations enhanced by neonatal cat exposure. PLoS Med. 2008;5(6):e131.
49. Cramer C, Link E, Horster M, Koletzko S, Bauer CP, Berdel D, et al. Elder siblings enhance the effect of filaggrin mutations on childhood eczema: results from the 2 birth cohort studies LISAplus and GINIplus. J Allergy Clin Immunol. 2010;125(6):1254–5.
50. Morgenstern V, Zutavern A, Cyrys J, Brockow I, Koletzko S, Krämer U, et al. Atopic diseases, allergic sensitization, and exposure to traffic-related air pollution in children. Am J Respir Crit Care Med. 2008;177(12):1331–7.
51. Annesi-Maesano I, Moreau D, Caillaud D, Lavaud F, Le Moullec Y, Taytard A, et al. Residential proximity fine particles related to allergic sensitization and asthma in primary school children. Respir Med. 2007;101(8):1721–9.
52. Krämer U, Sugiri D, Ranft U, Krutmann J, von Berg A, Berdel D, et al. Eczema, respiratory allergies, and traffic-related air pollution in birth cohorts from small-town areas. J Dermatol Sci. 2009;56(2): 99–105.
53. Tanaka K, Miyake Y, Sasaki S, Ohya Y, Hirota Y, Osaka Maternal and Child Health Study Group. Maternal smoking and environmental tobacco smoke exposure and the risk of allergic diseases in Japanese infants: the Osaka Maternal and Child Health Study. J Asthma. 2008;45(9):833–8.
54. Lee Y-L, Li C-W, Sung F-C, Yu H-S, Sheu H-M, Guo YL. Environmental factors, parental atopy and atopic eczema in primary-school children: a cross-sectional study in Taiwan. Br J Dermatol. 2007;157(6):1217–24.
55. Jedrychowski W, Perera F, Maugeri U, Mrozek-Budzyn D, Miller RL, Flak E, et al. Effects of prenatal and perinatal exposure to fine air pollutants and maternal fish consumption on the occurrence of infantile eczema. Int Arch Allergy Immunol. 2011;155(3): 275–81.
56. Elias PM. Therapeutic implications of a barrier-based pathogenesis of atopic dermatitis. Ann Dermatol. 2010;22(3):245–54.
57. Howell MD, Kim BE, Gao P, Grant AV, Boguniewicz M, DeBenedetto A, et al. Cytokine modulation of atopic dermatitis filaggrin skin expression. J Allergy Clin Immunol. 2009;124(3 Suppl 2):R7–12.
58. Gutowska-Owsiak D, Schaupp AL, Salimi M, Taylor S, Ogg GS. Interleukin-22 downregulates filaggrin expression and affects expression of profilaggrin processing enzymes. Br J Dermatol. 2011;165(3):492–8.
59. Jensen J-M, Pfeiffer S, Witt M, Bräutigam M, Neumann C, Weichenthal M, et al. Different effects of pimecrolimus and betamethasone on the skin barrier in patients with atopic dermatitis. J Allergy Clin Immunol. 2009;123(5):1124–33.
60. Tan L, Narayan SB, Chen J, Meyers GD, Bennett MJ. PTC124 improves readthrough and increases enzymatic activity of the CPT1A R160X nonsense mutation. J Inherit Metab Dis. 2011;34(2):443–7.
61. Ellinghaus D, Baurecht H, Esparza-Gordillo J, Rodríguez E, Matanovic A, Marenholz I, et al. High-density genotyping study identifies four new susceptibility loci for atopic dermatitis. Nat Genet. 2013; 45(7):808–12.
62. Bieber T, Cork M, Reitamo S. Atopic dermatitis: a candidate for disease-modifying strategy. Allergy. 2012;67(8):969–75.

Asthma

17

Simon F. Thomsen

Contents

S.F. Thomsen, MD, PhD
Department of Dermatology, Bispebjerg Hospital,
Bispebjerg Bakke 23, Copenhagen
DK-2400, Denmark
e-mail: simonfrancisthomsen@gmail.com

17.1 Introduction

Asthma is a chronic inflammatory respiratory disease affecting up to one-fourth of the population in industrialized countries [1]. Moreover, asthma remains undiagnosed in a considerable proportion of the population [2]. During the second half of the last century, the prevalence of asthma has risen to epidemic proportions, mainly in developed countries, but, more recently, also in developing countries. After a period, primarily between the 1960s and the 1990s, with a high number of new asthma cases, the incidence now seems to have reached a plateau; in some countries with a formerly high incidence, the incidence seems to have declined [1].

The reasons for the rapid increase in asthma prevalence are imperfectly understood, but since it has occurred more rapidly than changes to the genome sequence would allow, environmental factors have been evoked. According to *the hygiene hypothesis* formulated by British epidemiologist David Strachan in 1989, lifestyle changes associated with a decreased microbial diversity in industrialized societies during the second half of the last century such as higher standards of personal cleanliness, better household amenities, and increased use of antibiotics are responsible [3]. These factors are thought to have affected asthma susceptibility in the population through a complex developmental interplay with genetic factors. Of particular interest, a recent population study of 3,335 Danish adults, 18–69 years of age, suggested that carriers of

J.P. Thyssen, H. Maibach (eds.), *Filaggrin*,
DOI 10.1007/978-3-642-54379-1_17, © Springer-Verlag Berlin Heidelberg 2014

loss-of-function mutations (the variants *R501X* and *2282del4*) in the filaggrin gene (*FLG*) have been more susceptible to such environmental changes, exemplified by the observation that the prevalence of atopic diseases among successive generations in this population increased more among carriers than among noncarriers [4].

17.2 Clinical Characteristics of Asthma

Wheezing, shortness of breath, chest tightness, and cough are typical asthma symptoms. However, these symptoms occur with variable frequency between patients, and some patients might only have one or a few different symptoms—for example, patients with *cough-variant asthma* [5]. Symptoms may occur during daytime as well as nighttime, which indicate uncontrolled, severe asthma.

Asthma can be divided into intermittent and persistent disease, the latter which, according to the Global Initiative for Asthma (GINA) guidelines, can be divided further into mild, moderate, and severe [6]. This definition is based on symptomatic frequency and degree of lung function impairment, but possibly constitutes a pathophysiological continuum of disease severity, with the more severe cases having a higher degree of airway inflammation dominated by eosinophils and mast cells resulting in chronic airflow obstruction.

There is no single diagnostic test that definitely can decide whether a person has asthma. Consequently, the diagnosis is made on the basis of a history of characteristic recurrent airway symptoms concomitantly with an objective verification of airflow limitation, such as decline in lung function (e.g., forced expiratory volume in the first second, FEV_1), reversible airflow obstruction, airway hyperresponsiveness, or peak expiratory flow rate variability.

It is recognized that asthma is not a single disease but probably constitutes several subtypes of disease that probably have different causes [7]. Already in 1947, Rackeman divided asthma into *extrinsic/atopic* and *intrinsic/nonatopic* subtypes, characterized, respectively, by a concomitant positive and negative skin prick test to aeroallergens [8]. Unlike classical atopic asthma, which usually has an early onset, nonatopic asthma seems to be associated with later onset, female preponderance, obesity, greater symptomatology, decreased sensitivity to inhaled corticosteroids, and a relatively low degree of eosinophilic inflammation [9].

There is an increase in the number of T helper (Th) 2 cells in the airways of asthmatic patients, whereas in normal airways Th1 cells predominate. By secreting the cytokines IL-4 and IL-13, which drive IgE production by B cells, IL-5, which is responsible for eosinophil differentiation in the bone marrow, and IL-9, which attracts and drives the differentiation of mast cells, Th2 cells play a central role in the asthmatic inflammation [10].

17.3 Asthma Is a Multifactorial Disorder

Asthma runs in families. However, the disease is not caused by a single mutation in one gene, and therefore, the transmission of the asthma does not follow simple Mendelian inheritance. Rather, asthma is a *multifactorial disorder*, which means that many factors—both genetic and environmental—contribute to its development. This makes prediction of asthma for a given genotype, or combination of genotypes, difficult.

The individual risk of asthma depends on the degree of genetic relatedness between that individual and the relative with the disease (Fig. 17.1). For example, the recurrence risk of asthma in children with one affected parent is around 25 %, whereas the risk if both parents are affected is around 50 %. Twin studies also support that asthma has a strong genetic component, as the recurrence risk of asthma in monozygotic twins is much higher than in dizygotic twins [11]. However, the fact that monozygotic twins are not always concordant for asthma implies that environmental risk factors also play an important role.

Over a hundred different asthma susceptibility genes have been identified; however, each contributes only marginally to disease risk. These relate chiefly to (1) mucosal biology and

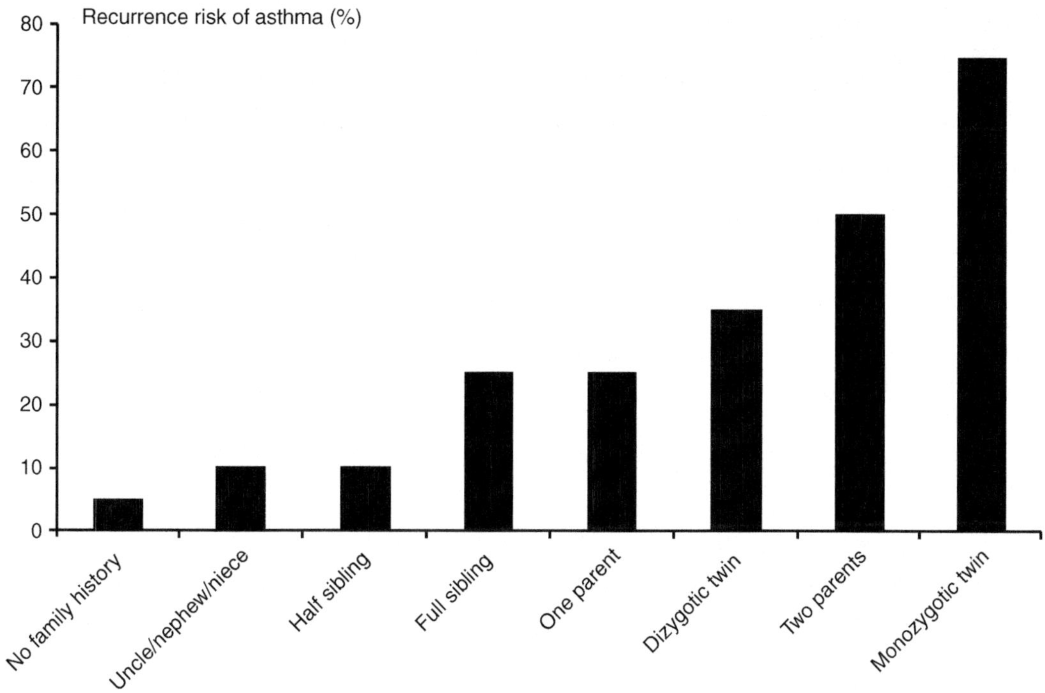

Fig. 17.1 Recurrence risk of asthma

function, (2) functioning of the immune system, and (3) lung function and disease expression [12]. Although molecular genetic studies of asthma have been conducted for several decades, they have translated only sparsely into new treatments or modifiable options for asthma. Notably, most known asthma susceptibility genes probably increase the risk of disease by a factor of about only 1.2 or even less [13].

17.4 Natural History of Asthma

The atopic march concerns the development of atopic dermatitis (AD) and concomitant sensitization to food and aeroallergens in early childhood, progressing to asthma and allergic rhinitis in later childhood [14]. Typically, the child develops AD in the first months of life accompanied by sensitization to milk and egg. Sensitization to indoor allergens such as house dust mite, cockroach, and furred pets follows, and within the first 2 years of life, the child develops recurrent episodes of wheezing, mostly in conjunction with viral respiratory tract infections. After this age, wheezing episodes become more frequent and occur in between respiratory infections. Later in childhood, allergy to outdoor allergens develops, and allergic rhinoconjunctivitis occurs in relation to exposure to grass and tree pollen. Despite the teenage years being a time when asthma symptoms may disappear or become less pronounced, after some symptom-free years, skin and respiratory symptoms return in a subset.

Although the skin sensitization occurring in AD appears to be the trigger for the subsequent development of the other allergic conditions, the progression is not uniform in all atopic children. Allergic manifestations can develop at any point in life. Further, many will experience only one or perhaps two atopic manifestations, and the development of these can be interspaced by several years. In some children the sequence of events is reversed, so that asthma precedes the development of eczema, and sometimes symptoms occur simultaneously, rendering the age at onset of the different disorders indistinguishable. Accordingly, the severity of the atopic syndrome varies highly between affected individuals, and the course of the disease depends on a dynamic interplay between many innate and triggering factors.

17.5 Why *FLG* Mutations Are Thought to Increase the Risk of Asthma

Individuals with AD have a marked increased risk of later asthma. Interestingly, this increased risk is probably dose dependent so that individuals with early onset AD or more severe eczema have an even higher risk of later development of asthma [15]. The advent of *FLG* mutations has led to the hypothesis that the skin acts as the site of primary sensitization through defects in the epidermal barrier with secondary reactivity in the airways [14]. Filaggrin is critical for skin barrier function. Thus, deficient levels of filaggrin lead to epidermal defects, to increased transepidermal water loss, and possibly to increased penetration of antigens into the skin, allowing skin-resident antigen-presenting cells such as Langerhans or dendritic cells to capture environmental antigens [16]. In addition, barrier-disrupted keratinocytes release immune adjuvants that activate and mature these innate immune cells, as well as affect their ability to direct naive T-cell polarization and thereby affect the character of the T-cell response.

Filaggrin is not expressed in the lower airway epithelium [16]. Therefore, the assumption is that AD is a causal risk factor for asthma and systemic allergen sensitization in the context of *FLG* mutations. Consequently, *FLG* mutations are unlikely to directly affect barrier function and allergen sensitization in the lungs. Instead, filaggrin deficiency-driven primary percutaneous allergic sensitization is speculated to lead secondarily to hyperactive airways and asthma [17].

17.6 Association Between *FLG* Mutations and Asthma

The association between *FLG* null variants and asthma has been studied in independent populations from several countries such as Germany, the United Kingdom, the United States, Austria, Croatia, Poland, Sweden, Denmark, the Netherlands, Japan, and China. The main part of these studies has included children, whereas some have been performed in adult populations. Some of these studies were limited to study the overall risk of asthma attributable to *FLG* mutations, irrespective of eczema, and did not explicitly stratify on eczema status. Moreover, while some have used a questionnaire definition of disease, based, for example, on parent report, most of the studies have employed a physician's diagnosis of disease (e.g., based on hospital records). Notably, case-control and family studies have more often used clinical definitions of disease, whereas population studies, which are more inclined to include larger populations of individuals, have tended to use self-reported measures of disease. In fact, the earliest studies of *FLG* mutations and asthma preferably used case-control or family designs, whereas later studies were more often population-based. These latter studies, which recruited individuals from the background population, can be assumed to report the least biased risk estimates of asthma in *FLG* mutation carriers since disease status, and accordingly, *FLG* carrier status, is not affected by the selection method of the study participants.

The collective evidence from studies of *FLG* and asthma points to an overall increased risk of asthma in *FLG* mutation carriers of about 1.5 relative to noncarriers, irrespective of AD status [18]. In contrast, the risk of asthma in individuals *with* AD carrying *FLG* null mutations appears to be increased about three times relative to noncarriers, whereas their risk of eczema-free asthma seems not to be increased [19, 20]. These results indicate that AD is a compulsory precursor for development of asthma in *FLG* mutation carriers.

17.7 *FLG* Mutations Predispose to Asthma *Selectively* in Patients with AD

The study by Palmer and colleagues from 2006 of families from Ireland, Scotland, and the United States with ichthyosis vulgaris and of selected clinical cohorts of children from Ireland,

Scotland, and Denmark with AD, asthma, and maternal asthmatic predisposition, respectively, was first to report an association between *FLG* loss-of-function mutations and AD [21]. Interestingly, asthma was also highly associated with *FLG* null carrier status (the variants *R501X* and *2282del4*) in these individuals, but *exclusively* in those with coexisting AD. In contrast, the risk of asthma in the absence of AD was not increased. This finding spawned the theory of the deficient epidermal barrier as an important component in the pathogenesis of atopic asthma.

17.7.1 Case-Control and Family Studies

The risk of eczema-associated asthma in *FLG* mutation carriers, estimated in case-control and family studies, tends to be higher than that estimated in general population studies: all published case-control and family studies find a significantly and markedly higher risk of eczema-associated asthma in *FLG* mutation carriers relative to noncarriers. For example, in the original discovery cohort described by Palmer and colleagues, the risk of asthma in eczema patients associated with *FLG* null status was increased almost 13 times: OR = 12.94 (5.05–35.93) [21].

Moreover, a study of European, mainly German, children from 2006 by Marenholz et al. found that the risk of asthma was increased more than six times in individuals with AD and *FLG* null mutations: reported OR = 6.21 (2.6–14.8). The risk of asthma with eczema was even more increased in patients with two mutant alleles, OR = 11.76 (1.2–116.3). The study estimated that the proportion of cases of eczema-associated asthma in the (German) population attributable to the mutant *FLG* alleles was ~21 % [22].

Another study from 2006 by Weidinger and colleagues of 476 German families with AD found that the risk of eczema-associated asthma was increased by a factor of 3.4 [20, 23]. Moreover, a later study by Weidinger et al. of 274 adults with AD found an OR of eczema-associated asthma of 6.26 (3.12–12.57) [20, 24]. For a complete overview of case-control and family studies, please see Table 17.1.

17.7.2 Population Studies

The markedly increased asthma risk reported in case-control and family studies is somewhat in contrast to the generally lower estimates obtained from several random population studies (see Table 17.2 for a complete overview of population-based studies). Notably, Brown and co-workers studied 811 English children aged 7–9 years and found that the overall population risk of asthma in individuals with *FLG* null status was not increased in heterozygotes, OR = 1.0 (0.6–1.6), whereas it was in homozygotes, OR = 3.6 (1.0–13.7) [36]. Of children with eczema *and* asthma, 23.2 % carried one or more of *FLG* null mutations compared with 11.8 % of noncarriers. The risk of atopic eczema-associated asthma in *FLG* mutation carriers was increased 1.5 times (0.8–2.8), whereas the OR for carriers of two null mutations was as high as 11.9 (3.1–45.6).

Henderson et al. studied a large population-based birth cohort comprising 6,971 children also from the United Kingdom and found an overall population risk of asthma in *FLG* mutation carriers of 1.80 (1.34–2.41), whereas *FLG* mutations conferred an increased asthma risk in the context of eczema of 3.16 (2.25–4.43) [37].

In a population-based study of 5,289 older adults again from the United Kingdom, Rice et al. found that the risk of currently symptomatic asthma associated with *FLG* mutations was increased a little under 1.5 times with a reported OR of 1.45 (1.04–2.03) [38]. Moreover, there was a borderline association with wheezing in the previous 12 months, OR = 1.26 (1.00–1.59). It was not possible to estimate the risk of asthma in individuals with and without AD, respectively, in that study, due to only few individuals reporting both conditions simultaneously ($n = 41$). Interestingly, the *R501X* mutation was independently associated with symptomatic asthma, OR = 1.59 (1.05–2.40), whereas *2282del4* was not.

Weidinger et al. examined 3,099 German children recruited as part of the International Study of Asthma and Allergies in Childhood II (ISAAC II) and found a significant overall association with asthma, OR = 1.79 (1.19–2.68). However, the association with asthma was restricted to asthma

Table 17.1 Case-control and family studies reporting an association between filaggrin gene (*FLG*) mutations and asthma

Study	Year	Country	Population	Main finding in relation to asthma
Palmer et al. [21]	2006	Several[a]	Families, children[b]	*FLG* mutations increase the risk of eczema-associated asthma
Weidinger et al. [23]	2006	Germany	476 families with AD[c]	*FLG* mutations increase the risk of eczema-associated asthma
Marenholz et al. [22]	2006	Germany[d]	1,092 children with AD	*FLG* mutations increase the risk of eczema-associated asthma
Morar et al. [25]	2007	United Kingdom	426 families with AD	*FLG* mutations increase the risk of eczema-associated asthma
Rogers et al. [26]	2007	United States	646 children with asthma	*FLG* mutations increase the risk of eczema-associated asthma
Weidinger et al. [24]	2007	Germany	274 adults with AD	*FLG* mutations increase the risk of eczema-associated asthma
Hubiche et al. [27]	2007	France	99 patients with AD	*FLG, SPINK5,* and *KLK7* mutations increase the risk of eczema-associated asthma additively
Ekelund et al. [28]	2008	Sweden	406 families with AD	*FLG* mutations increase the risk of eczema-associated asthma
Müller et al. [29]	2009	Several[e]	496 children/ families with AD	*FLG* mutations increase the risk of eczema-associated asthma
Greisenegger et al. [30]	2010	Germany, Austria	462 adults with AD	*FLG* mutations do not increase the risk of eczema-associated asthma
Lesiak et al. [31]	2011	Poland	163 patients with AD	*FLG* mutations increase the risk of eczema-associated asthma
Li et al. [32]	2011	China	121 children with atopic asthma	*FLG* mutations increase the risk of eczema-free asthma
Osawa et al. [33]	2011	Japan	172 patients with AD[f]	*FLG* mutations increase the risk of eczema-associated asthma
Wang et al. [34]	2011	China	116 children with AD	*FLG* mutation increase the risk of eczema-associated asthma
Zhang et al. [35]	2011	China	261 patients with AD	*FLG* mutation do not increase the risk of eczema-associated asthma

[a]Ireland, Scotland, the United States, and Denmark
[b]Extended pedigrees with ichthyosis vulgaris and clinical cohorts with atopic diseases
[c]AD is atopic dermatitis
[d]Children mainly from Germany but also from Italy, Sweden, the Netherlands, and Poland
[e]Twelve European countries and Canada
[f]Also included patients with asthma

occurring in the context of eczema with a strong association with eczema plus asthma: reported OR = 3.49 (2.00–6.08) [39].

17.7.3 Other Risk Alleles

Most of the studies of populations of European ancestry have analyzed the risk of asthma according to carrier status of the two commonest *FLG* variants *R501X* and *2282del4*. The main part of these studies reports risk estimates of asthma according to the carrier status of both the individual and the combined genotypes, and although there is some heterogeneity between studies for these estimates, it seems that the risk of eczema-associated asthma according to the individual genotypes *R501X* and *2282del4* can be assumed to be equivalent. This uniform risk according to the different mutations is expected as studies have demonstrated that each of these null mutations produces truncated forms of profilaggrin, which results in a marked reduction or absence of processed filaggrin when present in the homozygote or compound heterozygote state [45, 46].

Table 17.2 Population studies reporting an association between filaggrin gene (*FLG*) mutations and asthma

Study	Year	Country	Population	Main finding in relation to asthma
Brown et al. [36]	2008	United Kingdom	811 children	*FLG* mutations increase the risk of eczema-associated asthma
Henderson et al. [37]	2008	United Kingdom	6,971 children	*FLG* mutations increase the risk of eczema-associated asthma
Rice et al. [38]	2008	United Kingdom	5,289 older adults	*FLG* mutations increase the risk of asthma
Weidinger et al. [39]	2008	Germany	3,099 children	*FLG* mutations do not increase the risk of eczema-associated asthma
Schuttelaar et al. [40]	2009	Netherlands	934 children	*FLG* mutations increase the risk of eczema-associated asthma
Poninska et al. [41]	2011	Poland	3,802 children and adults	*FLG* mutations increase the risk of eczema-free asthma
Marenholz et al. [42]	2011	United Kingdom	9,395 children	No synergistic effect of *FLG* mutations and rs7927894 (11q13) on the risk of asthma
Berg et al. [43]	2012	Denmark	3,471 adults	*FLG* mutations modify the effect of smoking on the risk of (nonatopic) asthma
Sabolic Pipinic et al. [44]	2013	Croatia	440 students	*FLG* mutations do not increase the risk of asthma (not estimated because of too few cases)

As more *FLG* mutations have been discovered during the past few years, these have been included in several of the more recent studies. For example, in the general population study of 3,099 German children by Weidinger and colleagues, the five mutations *R501X*, *2282del4*, *R2447X*, *S3247X*, and *3702delG* (there were no carriers of *3702delG*) were examined [39]. However, individual genotype risks were not computed in that study.

Greisenegger et al. studied the four *FLG* variants *R501X*, *2282del4*, *R2447X*, and *S3247X* in 462 Austrian and German adult AD patients and 402 control individuals [30]. Mutation carrier status of the combined genotype was not significantly different in patients with asthma compared to patients without asthma (26 % vs. 22 %), OR = 1.27 (0.79–2.05). Risks pertaining to the individual genotypes were not computed for asthma in that study.

In the population-based study of 811 children from the United Kingdom, Brown et al. also studied the five alleles *R501X*, *2282del4*, *R2447X*, *S3247X*, and *3702delG* [36]. The mutations were considered together in the analyses and were associated with a small excess risk of eczema-associated asthma of 1.5 (0.8–2.8).

Schuttelaar et al. studied the genotypes *R501X*, *2282del4*, and *R2447X* in 934 children from the Netherlands as part of the Prevention and Incidence of Asthma and Mite Allergy (PIAMA) birth cohort [40]. The combined genotype was associated with asthma at age 0–8 years in the context of eczema in the first year of life, OR = 3.2 (1.2–8.5), but not in children without eczema in the first year, OR = 1.8 (0.5–6.1). Specifically, the *2282del4* mutation was reported to be significantly associated with asthma at the age of 0–8 years, OR = 5.2 (2.4–11.5).

Sabolic Pipinic and colleagues studied the four alleles *R501X*, *2282del4*, *R2447X*, and *S3247X* among 440 Croatian students [44]. However, the relationship between these mutations and asthma was not determined due to a low carrier frequency (2.6 %) among the Croatian population for these genotypes. In fact, no carriers of *R2447X* and *S3247X* mutations were identified, consistent with a low frequency of *FLG* mutations in the Slavic population.

Table 17.3 Studies of the association between filaggrin gene (*FLG*) mutations and asthma severity and prognosis

Study	Year	Country	Population	Main finding in relation to asthma
Palmer et al. [49]	2007	Scotland	874 children with asthma	*FLG* mutations increase the severity of asthma
Basu et al. [50]	2008	Scotland	1,135 children with asthma	*FLG* mutations increase the risk of asthma exacerbations
Marenholz et al. [48]	2009	Germany	871 children	*FLG* mutations predict future asthma in children with eczema and food sensitizations
Bønnelykke et al. [51]	2010	Denmark	411 children	*FLG* mutations increase the risk of asthma exacerbations

17.7.4 Populations of Non-European Descent

A few studies of Asian populations have studied *FLG* mutations in relation to asthma. Notably, Zhang et al. studied in total 18 *FLG* mutations of which 10 were novel (*R826X*, *3222del4*, *R1140X*, *4271delAA*, *Q1790X*, *5757del4*, *6834del5*, *6950del8*, *S2706X*, and *K4671X*) and 8 were previously reported (*441delA*, *R501X*, *3321delA*, *R1474X*, *Q2417X*, *E2422X*, *7945delA*, and *R4306X*). These mutations were present in 31.4 % of 261 patients with AD but were not associated with asthma [35].

Also Li et al. [32] and Wang et al. [34] studied *FLG* mutations in the Chinese population but, in contrast to the study by Zhang et al., found a positive association with asthma. Particularly the study by Wang et al. found that the *P478S GG* genotype significantly increased the risk of developing asthma in patients with AD, OR=4.68 (1.37–16.03). Moreover, Osawa and co-workers studied *FLG* mutations in patients with AD and asthma, respectively, and found an increased risk of asthma in Japanese-specific *FLG* mutation carriers [33].

17.7.5 Interaction with Environmental Exposures

A study of 3,471 Danish adults found that carriers of *FLG* null mutations had a significantly higher prevalence of asthma [43]. Interestingly, *FLG* null mutations modified the effect of smoking, rendering smokers more susceptible to asthma, and reduced pulmonary function with a dose-dependent effect of tobacco consumption. The effect was observed irrespective of AD but seemed to be confined to nonatopic asthma. The biological mechanism underlying this association is not fully clear, but the finding still implicates that *FLG* mutation carriers may be more vulnerable to certain environmental exposures, particularly smoking.

An interaction between *FLG* null mutations and exposure to cat on the risk of atopic diseases has been investigated in two other studies of Danish [47] and Dutch [40] children, respectively. These studies found an increased risk of AD and allergic sensitization to cat, but *not* asthma, among *FLG* mutation carriers in the context of cat exposure.

In a study by Marenholz et al. of 871 children from Germany, the presence of the *FLG* mutations *R501X*, *2282del4*, and/or *R2447X* was shown to predict future asthma in children with eczema and food sensitizations, indicating that early subgroup-specific interventions could prevent the progression from eczema to asthma [48].

17.7.6 *FLG* Mutations and Asthma Severity

Apart from influencing asthma susceptibility, mutations in *FLG* also influence several qualities of asthma, such as the risk of exacerbations and the symptomatic severity (Table 17.3). Notably, in a study of 411 Danish children with maternal asthmatic predisposition followed from birth until school age, carriers of *R501X* and *2282del4* had an almost two times increased risk of asthma exacerbations, which was expressed within the first 1.5 years of life and, furthermore, had a marked and persistent increase in acute severe asthma exacerbations from 1 year of age [51].

Moreover, in a large study of children and young adults ($n=874$) with physician-diagnosed asthma attending primary and secondary clinics in 18 primary care practices and a secondary care asthma clinic in Scotland, *R501X* and *2282del4* carriers had a significantly increased disease burden, both in terms of lung function, with a greater airway obstruction in the *FLG* null carriers, and in the intensity of medication required for disease control [49]. Intriguingly, the *FLG* mutations were associated with asthma disease severity even in the absence of a history of eczema [49]. The individual contribution to the overall effect of the *2284del4* allele was lower than that observed for *R501X*, suggesting that patients carrying the *R501X* mutation have more severe asthma. Studying the same population (extended to include 1,135 individuals) showed that *FLG* mutations were associated with a two-fold greater risk of exacerbations in children with asthma [50]. Similarly, exacerbations were significant for the *R501X*, but not the *2282del4* mutation, and the combined genotype compared to the wild type, respectively. Individuals with *FLG* null alleles were more likely to require oral steroids (31 % vs. 20 %) for their exacerbations. There was also an increased risk of school absence owing to asthma exacerbations in asthmatic individuals with *FLG* null mutations relative to wild-type carriers (43 % vs. 30 %) [50].

17.7.7 *FLG* Mutations and Intermediate Asthma Phenotypes

Several studies have examined the association between *FLG* mutations and *intermediate asthma phenotypes*, such as lung function and bronchial hyperresponsiveness [26, 37, 40, 43, 48, 49]. Notably, Rogers and colleagues studied 646 children with asthma from the United States and found that, respectively, individuals with and without *FLG* mutations did *not* differ in terms of bronchial reactivity to methacholine, postalbuterol FEV_1, or FEV_1/FVC ratio. This lack of association was observed even in the subset with AD and in the subset with European ancestry [26].

Contrary to this, in a large random population of children from the United Kingdom studied by Henderson et al., subjects heterozygous for *R501X* had increased bronchial responsiveness (0.44 % decrease in FEV_1 per micromole of methacholine) compared with that seen in wild-type homozygotes [37]. Further, subjects who were heterozygous for one of the deficiency alleles (*R501X* or *2282del4*) had increased bronchial responsiveness (0.29 % FEV_1 per micromole of methacholine), whereas the combination of homozygotes and compound heterozygotes did not support a linear effect on bronchial responsiveness per deficiency allele carried.

Berg and colleagues studied 3,471 Danish adults from the general population and found that *FLG* null status (*R501X* or *2282del4*) was significantly associated with a decreased FEV_1/FVC ratio but not with a low FEV_1 (<80 % of expected), or with the level of exhaled nitric oxide (a marker of airway inflammation) [43].

17.7.7.1 Is There an Association Between *FLG* Mutations and Asthma Independent of AD?

A few studies have found a significant association between *FLG* null variants and asthma in the *absence* of AD. First, a Polish study of 3,802 individuals, 6–44 years of age, from the general population found an increased risk of asthma, notably atopic asthma, independent of the presence of AD with a reported OR of 2.02 (1.07–3.81) [41]. The association with asthma independent of eczema was present also in the subgroup of children aged 6–7 years (whose parents presumably would remember their children having had the disease) and in the subgroup of patients who denied a history of broadly defined inflammatory skin disease, OR=2.30 (1.07–4.93). However, the association with asthma disappeared when the analysis was restricted to individuals with questionnaire-defined asthma, OR=1.15 (0.58–2.32). Moreover, the association was only valid for the *2282del4* allele and the combined genotype (*2282del4* and/or *R501X*), not for the *R501X*

Table 17.4 Studies of the association between filaggrin gene (*FLG*) mutations and asthma in other patient populations

Study	Year	Country	Population	Main finding in relation to asthma
Betz et al. [52]	2007	Belgium, Germany	449 patients with alopecia areata	*FLG* mutations increase the risk of eczema-associated asthma in patients with alopecia areata
Van Limbergen et al. [53]	2009	Scotland	403 children with IBD[a]	*FLG* mutations increase the risk of eczema-associated asthma in patients with IBD
Landeck et al. [54]	2013	Germany	459 patients with contact eczema	*FLG* mutations do not increase the risk of asthma in patients with contact eczema of the hands

[a]IBD is inflammatory bowel disease

genotype alone, and, furthermore, was only present among those who were unaware of having asthma (i.e., who were diagnosed for the first time with asthma by a physician as part of the study). The carrier frequency of *FLG* variants in the Polish population was low, consistent with a low carrier frequency in populations from Eastern Europe: only 3.76 and 0.8 % had the *2282del4* allele and the *R501X* allele, respectively. This made the number of asthma patients carrying a mutation very low, which could have influenced the precision of the results. Also, the participation rate in the study was only 25.5 %, which could have resulted in selection bias.

A study from China found that the mutation frequency in children with atopic asthma, of at least one of several *FLG* variants common in individuals of Asian descent (*3321delA*, *3222del4*, *K4671X*, *S1302X*, and *Q2397X*), was 15.7 % compared with only 4.0 % among healthy controls [32]. This corresponded to an increased risk of eczema-free asthma of almost four for the combined genotype, OR = 3.9 (2.0–7.9). Moreover, significant associations were also found for the individual alleles: *3321delA*, OR = 5.8 (1.5–22.1), and *K4671X*, OR = 3.1 (1.3–7.7). However, the study included only 121 cases and 301 controls, which provided low statistical power to detect an effect on the individual genotypic level. Moreover, cases and controls were inadequately age matched: the median age was 5.0 years in cases versus a mean age in controls of 16.5 years, which could have led to spurious associations.

Several methodological factors alluding to study design and patient characteristics might have influenced the outcome of these studies; therefore, the reliability and implications of a positive association between *FLG* variants and eczema-free asthma are currently not resolved.

17.7.8 Association Between *FLG* Mutations and Asthma in Other Patient Populations

A study of 449 Belgian and German patients with the atopic disease-associated hair disorder alopecia areata showed that *FLG* mutations *R501X* and *2282del4* conferred an excess risk of eczema-associated asthma of 4.63 (2.30–9.35) [52] (Table 17.4).

Moreover, a study of 403 Scottish children with inflammatory bowel disease (i.e., Crohn's disease or ulcerative colitis) showed that the risk of asthma in the context of AD was increased almost four times, OR = 3.86 (1.65–9.08), in carriers of *R501X* or *2282del4* [53].

Lastly, in a cohort of 459 German patients with occupational irritant contact eczema of the hands, 68 patients (14.8 %) carried one or more of four *FLG* null alleles tested (*R501X*, *2282del4*, *R2447X*, and/or *S3247X*). Among these patients, classical features of AD/ichthyosis vulgaris, such as flexural eczema, dry skin, pityriasis alba, dirty neck, pulpitis sicca, hyperlinear palms, keratosis pilaris, family history of eczema, and serum total IgE were significantly associated with *FLG* mutations, whereas asthma was not significantly so, OR = 1.47 (0.81–2.66) [54].

17.8 Other Genotypes Interacting with *FLG* Mutations May Increase the Risk of Asthma

A few studies have examined the combined action of *FLG* and other genotypes in the susceptibility to asthma. Interestingly, in a hospital-based sample of 99 children and adults with AD from France, the risk of asthma increased linearly according to the number of mutated alleles in *SPINK5* (associated with the keratinization disorder Netherton syndrome [55]), *KLK7*, and *FLG*: having no variants corresponded to a risk of asthma of 20 %, whereas having one, two, or three variants, respectively, corresponded to a prevalence of asthma of 41, 50, and 75 % [27]. Although the number of examined patients was small, the study still suggests an additive effect of compound alleles other than *FLG* on the risk of asthma in AD patients.

A study by Lesiak and colleagues of 163 patients with AD and 204 healthy control individuals from Poland examined the combined action of *FLG* mutations and *IL-4*, *IL-10*, and *IL-13* polymorphisms [31]. Significant interactions were demonstrated between *2282del4* and the CT genotype for *IL-13* or GG genotype for *IL-10* and a higher risk of developing AD, but not asthma.

Marenholz et al. studied 9,395 children from the United Kingdom as part of the Avon Longitudinal Study of Parents and Children (ALSPAC) birth cohort and found no synergistic effect on the risk of asthma of the *FLG* mutations *R501X* and *2282del4* and a polymorphism on 11q13 (*rs7927894*) that has previously been associated with a higher risk of eczema [42].

Conclusion

There is strong supporting evidence that *FLG* mutations may be the fundamental predisposing factor not only for the development of eczema but also for initial sensitization and progression to asthma. Of particular note is the association between *FLG* defects and asthma selectively in patients with AD, which supports the hypothesis that asthma is secondary to allergic sensitization occurring after epidermal skin barrier disruption. However, this course may be relevant only for certain types of asthma, particularly classical atopic asthma with early onset, whereas adult-onset asthma and nonatopic asthma may result from different pathways. Further work is needed within larger well-characterized populations, comprising children, adults, and diverse ethnic groups, and with measured lung inflammatory markers, environmental exposures, and genetic variants other than *FLG* mutations in order to further delineate the importance of *FLG* mutations in the clinical expression and prognosis of asthma.

References

1. Asher MI, Montefort S, Björkstén B, Lai CK, Strachan DP, Weiland SK, Williams H. ISAAC Phase Three Study Group. Worldwide time trends in the prevalence of symptoms of asthma, allergic rhinoconjunctivitis, and eczema in childhood: ISAAC Phases One and Three repeat multicountry cross-sectional surveys. Lancet. 2006;368:733–43.
2. Nolte H, Nepper-Christensen S, Backer V. Unawareness and undertreatment of asthma and allergic rhinitis in a general population. Respir Med. 2006;100:354–62.
3. Strachan DP. Hay fever, hygiene, and household size. BMJ. 1989;299:1259–60.
4. Thyssen JP, Linneberg A, Johansen JD, Carlsen BC, Zachariae C, Meldgaard M, et al. Atopic diseases by filaggrin mutations and birth year. Allergy. 2012;67: 705–8.
5. Lougheed MD, Turcotte SE, Fisher T. Cough variant asthma: lessons learned from deep inspirations. Lung. 2012;190:17–22.
6. Global Initiative for Asthma. www.ginasthma.org.
7. Agache I, Akdis C, Jutel M, Virchow JC. Untangling asthma phenotypes and endotypes. Allergy. 2012;67: 835–46.
8. Rackeman FM. A working classification of asthma. Am J Med. 1947;3:601–6.
9. Lugogo NL, Kraft M, Dixon AE. Does obesity produce a distinct asthma phenotype? J Appl Physiol. 2010;108:729–34.
10. Barnes PJ. Immunology of asthma and chronic obstructive pulmonary disease. Nat Rev Immunol. 2008;8:183–92.
11. Thomsen SF, van der Sluis S, Kyvik KO, Skytthe A, Backer V. Estimates of asthma heritability in a large twin sample. Clin Exp Allergy. 2010;40:1054–61.

12. Vercelli D. Discovering susceptibility genes for asthma and allergy. Nat Rev Immunol. 2008;8:169–82.
13. Gibson G. Decanalization and the origin of complex disease. Nat Rev Genet. 2009;10:134–40.
14. Spergel JM. From AD, to asthma: the atopic march. Ann Allergy Asthma Immunol. 2010;105:99–106.
15. Lowe AJ, Carlin JB, Bennett CM, Hosking CS, Abramson MJ, Hill DJ, Dharmage SC. Do boys do the atopic march while girls dawdle? J Allergy Clin Immunol. 2008;121:1190–5.
16. De Benedetto A, Kubo A, Beck LA. Skin barrier disruption: a requirement for allergen sensitization? J Invest Dermatol. 2012;132:949–63.
17. Kubo A, Nagao K, Amagai M. Epidermal barrier dysfunction and cutaneous sensitization in atopic diseases. J Clin Invest. 2012;122:440–7.
18. Irvine AD, McLean WH, Leung DY. Filaggrin mutations associated with skin and allergic diseases. N Engl J Med. 2011;365:1315–27.
19. van den Oord RA, Sheikh A. Filaggrin gene defects and risk of developing allergic sensitisation and allergic disorders: systematic review and meta-analysis. BMJ. 2009;339:b2433.
20. Rodriguez E, Baurecht H, Herberich E, Wagenpfeil S, Brown SJ, Cordell HJ, et al. Meta-analysis of filaggrin polymorphisms in eczema and asthma: robust risk factors in atopic disease. J Allergy Clin Immunol. 2009;123:1361–70.
21. Palmer CN, Irvine AD, Terron-Kwiatkowski A, Zhao Y, Liao H, Lee SP, et al. Common loss-of-function variants of the epidermal barrier protein filaggrin are a major predisposing factor for AD. Nat Genet. 2006;38:441–6.
22. Marenholz I, Nickel R, Rüschendorf F, Schulz F, Esparza-Gordillo J, Kerscher T, et al. Filaggrin loss-of-function mutations predispose to phenotypes involved in the atopic march. J Allergy Clin Immunol. 2006;118:866–71.
23. Weidinger S, Illig T, Baurecht H, Irvine AD, Rodriguez E, Diaz-Lacava A, et al. Loss-of-function variations within the filaggrin gene predispose for AD with allergic sensitizations. J Allergy Clin Immunol. 2006;118:214–9.
24. Weidinger S, Rodríguez E, Stahl C, Wagenpfeil S, Klopp N, Illig T, et al. Filaggrin mutations strongly predispose to early onset and extrinsic AD. J Invest Dermatol. 2007;127:724–6.
25. Morar N, Cookson WO, Harper JI, Moffatt MF. Filaggrin mutations in children with severe AD. J Invest Dermatol. 2007;127:1667–72.
26. Rogers AJ, Celedón JC, Lasky-Su JA, Weiss ST, Raby BA. Filaggrin mutations confer susceptibility to AD but not to asthma. J Allergy Clin Immunol. 2007;120: 1332–7.
27. Hubiche T, Ged C, Benard A, Léauté-Labrèze C, McElreavey K, de Verneuil H, et al. Analysis of SPINK 5, KLK 7 and FLG genotypes in a French AD cohort. Acta Derm Venereol. 2007;87:499–505.
28. Ekelund E, Liedén A, Link J, Lee SP, D'Amato M, Palmer CN, et al. Loss-of-function variants of the filaggrin gene are associated with atopic eczema and associated phenotypes in Swedish families. Acta Derm Venereol. 2008;88:15–9.
29. Müller S, Marenholz I, Lee YA, Sengler C, Zitnik SE, Griffioen RW, et al. Association of Filaggrin loss-of-function-mutations with AD and asthma in the Early Treatment of the Atopic Child (ETAC) population. Pediatr Allergy Immunol. 2009;20:358–61.
30. Greisenegger E, Novak N, Maintz L, Bieber T, Zimprich F, Haubenberger D, et al. Analysis of four prevalent filaggrin mutations (R501X, 2282del4, R2447X and S3247X) in Austrian and German patients with AD. J Eur Acad Dermatol Venereol. 2010;24:607–10.
31. Lesiak A, Kuna P, Zakrzewski M, van Geel M, Bladergroen RS, Przybylowska K, et al. Combined occurrence of filaggrin mutations and IL-10 or IL-13 polymorphisms predisposes to AD. Exp Dermatol. 2011;20:491–5.
32. Li M, Chen X, Chen R, Bao Y, Yao Z. Filaggrin gene mutations are associated with independent atopic asthma in Chinese patients. Allergy. 2011;66: 1616–7.
33. Osawa R, Konno S, Akiyama M, Nemoto-Hasebe I, Nomura T, Nomura Y, et al. Japanese-specific filaggrin gene mutations in Japanese patients suffering from atopic eczema and asthma. J Invest Dermatol. 2010;130:2834–6.
34. Wang IJ, Lin TJ, Kuo CF, Lin SL, Lee YL, Chen PC. Filaggrin polymorphism P478S, IgE level, and atopic phenotypes. Br J Dermatol. 2011; 164:791–6.
35. Zhang H, Guo Y, Wang W, Shi M, Chen X, Yao Z. Mutations in the filaggrin gene in Han Chinese patients with AD. Allergy. 2011;66:420–7.
36. Brown SJ, Relton CL, Liao H, Zhao Y, Sandilands A, Wilson IJ, et al. Filaggrin null mutations and childhood atopic eczema: a population-based case-control study. J Allergy Clin Immunol. 2008;121:940–6.
37. Henderson J, Northstone K, Lee SP, Liao H, Zhao Y, Pembrey M, et al. The burden of disease associated with filaggrin mutations: a population-based, longitudinal birth cohort study. J Allergy Clin Immunol. 2008;121:872–7.
38. Rice NE, Patel BD, Lang IA, Kumari M, Frayling TM, Murray A, et al. Filaggrin gene mutations are associated with asthma and eczema in later life. J Allergy Clin Immunol. 2008;122:834–6.
39. Weidinger S, O'Sullivan M, Illig T, Baurecht H, Depner M, Rodriguez E, et al. Filaggrin mutations, atopic eczema, hay fever, and asthma in children. J Allergy Clin Immunol. 2008;121:1203–9.
40. Schuttelaar ML, Kerkhof M, Jonkman MF, Koppelman GH, Brunekreef B, de Jongste JC, et al. Filaggrin mutations in the onset of eczema, sensitization, asthma, hay fever and the interaction with cat exposure. Allergy. 2009;64:1758–65.
41. Poninska J, Samolinski B, Tomaszewska A, Raciborski F, Samel-Kowalik P, Walkiewicz A, et al. Filaggrin gene defects are independent risk factors for atopic asthma in a Polish population: a study in ECAP cohort. PLoS One. 2011;6:e16933.

42. Marenholz I, Bauerfeind A, Esparza-Gordillo J, Kerscher T, Granell R, Nickel R, et al. The eczema risk variant on chromosome 11q13 (rs7927894) in the population-based ALSPAC cohort: a novel susceptibility factor for asthma and hay fever. Hum Mol Genet. 2011;20:2443–9.
43. Berg ND, Husemoen LL, Thuesen BH, Hersoug LG, Elberling J, Thyssen JP, et al. Interaction between filaggrin null mutations and tobacco smoking in relation to asthma. J Allergy Clin Immunol. 2012;129:374–80.
44. Sabolic Pipinic I, Varnai VM, Turk R, Breljak D, Kezic S, Macan J. Low frequency of filaggrin null mutations in Croatia and their relation with allergic diseases. Int J Immunogenet. 2013;40:192–8.
45. Smith FJ, Irvine AD, Terron-Kwiatkowski A, Sandilands A, Campbell LE, Zhao Y, et al. Loss-of-function mutations in the gene encoding filaggrin cause ichthyosis vulgaris. Nat Genet. 2006;38:337–42.
46. Sandilands A, Terron-Kwiatkowski A, Hull PR, O'Regan GM, Clayton TH, Watson RM, Carrick T, et al. Comprehensive analysis of the gene encoding filaggrin uncovers prevalent and rare mutations in ichthyosis vulgaris and atopic eczema. Nat Genet. 2007;39:650–4.
47. Bisgaard H, Simpson A, Palmer CN, Bønnelykke K, McLean I, Mukhopadhyay S, et al. Gene-environment interaction in the onset of eczema in infancy: filaggrin loss-of-function mutations enhanced by neonatal cat exposure. PLoS Med. 2008;5:e131.
48. Marenholz I, Kerscher T, Bauerfeind A, Esparza-Gordillo J, Nickel R, Keil T, et al. An interaction between filaggrin mutations and early food sensitization improves the prediction of childhood asthma. J Allergy Clin Immunol. 2009;123:911–6.
49. Palmer CN, Ismail T, Lee SP, Terron-Kwiatkowski A, Zhao Y, Liao H, et al. Filaggrin null mutations are associated with increased asthma severity in children and young adults. J Allergy Clin Immunol. 2007;120:64–8.
50. Basu K, Palmer CN, Lipworth BJ, Irwin McLean WH, Terron-Kwiatkowski A, Zhao Y, et al. Filaggrin null mutations are associated with increased asthma exacerbations in children and young adults. Allergy. 2008;63:1211–7.
51. Bønnelykke K, Pipper CB, Tavendale R, Palmer CN, Bisgaard H. Filaggrin gene variants and atopic diseases in early childhood assessed longitudinally from birth. Pediatr Allergy Immunol. 2010;21:954–61.
52. Betz RC, Pforr J, Flaquer A, Redler S, Hanneken S, Eigelshoven S, et al. Loss-of-function mutations in the filaggrin gene and alopecia areata: strong risk factor for a severe course of disease in patients comorbid for atopic disease. J Invest Dermatol. 2007;127:2539–43.
53. Van Limbergen J, Russell RK, Nimmo ER, Zhao Y, Liao H, Drummond HE, et al. Filaggrin loss-of-function variants are associated with atopic comorbidity in pediatric inflammatory bowel disease. Inflamm Bowel Dis. 2009;15:1492–8.
54. Landeck L, Visser M, Kezic S, John SM. Genotype-phenotype associations in filaggrin loss-of-function mutation carriers. Contact Dermatitis. 2013;68:149–55.
55. Chavanas S, Bodemer C, Rochat A, Hamel-Teillac D, Ali M, Irvine AD, et al. Mutations in SPINK5, encoding a serine protease inhibitor, cause Netherton syndrome. Nat Genet. 2000;25:141–2.

Filaggrin Mutations, Skin Barrier Dysfunction, and Sensitization in Allergic Rhinitis

18

Tatiana A. Vogel and Marie-Louise A. Schuttelaar

Contents

T.A. Vogel, MD • M.-L.A. Schuttelaar, MD, PhD (✉)
Department of Dermatology, University Medical Centre Groningen, Hanzeplein 1, Groningen 9700 RB, The Netherlands
e-mail: m.l.a.schuttelaar@umcg.nl

18.1 Introduction

Allergic rhinitis (AR) or hay fever is a common chronic inflammatory disease of the nasal mucosa with an estimated prevalence in adults from Western Europe that ranges from 3 to 19 % [1]. Research has shown a relatively low rate of physician-based diagnosis of about 13 % [2], and it was presumed that around 45 % of all AR patients remain undiagnosed. This same phenomenon was found in multiple different studies over time. Therefore, the expected prevalence of AR in adults from Western Europe is around 30 %. Children are slightly more often affected than adults [2, 3].

This very common disease was first described in 1819 by the physician John Bostock in his case report "Periodical affection of the eyes and the chest" [4]. In this first report about AR, the classical symptoms were described of itchy, runny nose and eyes accompanied by a slumping feeling occurring in spring- and summertime. At that time, therapeutic advice included obtaining fresh air and the avoidance of a moist and closed atmosphere. Unfortunately, the reported patient experienced no improvement of his complaints. However, during an unusually warm summer season, he stayed indoors almost all summer and experienced a relief of his complaints. A few decades later, Charles Harrison Blackley discovered that the mechanism behind AR was caused by pollen, and therefore, he named it "hay fever" [5].

J.P. Thyssen, H. Maibach (eds.), *Filaggrin*,
DOI 10.1007/978-3-642-54379-1_18, © Springer-Verlag Berlin Heidelberg 2014

18.2 Definition and Clinical Features

Nowadays, AR is more precisely characterized as irritation and inflammation of the nasal and ocular mucosal epithelium caused by environmental triggers (OMIM 607154). These include airborne allergens such as pollen and house dust mite and can cause typical symptoms of sneezing, coughing, and itchy and runny nose and eyes. However, AR can also be accompanied with very disabling headaches, fatigue, and cognitive impairment. A classification of the severity and duration of rhinitis symptoms was described in the Allergic Rhinitis and Its Impact on Asthma (ARIA) criteria in 2001 [6, 7]. AR was subdivided into mild and moderate to severe, based on the impact on social life, absence from school or work, and sleep deprivation [3]. The duration of symptoms was appointed as persistent if complaints were present ≥4 days a week and ≥4 weeks consecutively [2]. The European Academy of Allergy and Clinical Immunology (EAACI) published the revised nomenclature for allergy in which rhinitis was divided into allergic and non-allergic rhinitis. They reported that AR is an immunologically mediated hypersensitivity reaction in the nose, while non-AR is not [6]. In 2008, the ARIA workgroup updated their criteria [8]. They defined rhinitis as an inflammation of the lining of the nose, characterized by symptoms including rhinorrhea, itchy, stuffy nose, and sneezing, occurring more than 1 h a day, during two or more consecutive days. AR was classified as a form of rhinitis characterized by an IgE-mediated response against aeroallergens often associated with ocular symptoms. AR can be subdivided into intermittent AR (IAR) and persistent AR (PAR). Allergens like pollen from grass or trees during spring and summer are the main cause of IAR, while PAR is caused by allergens present in the home, such as house dust mite and cat dander.

18.3 Diagnosis

In daily practice, the diagnosis AR is based on the clinical history and physical examination. To confirm the allergic origin of rhinitis symptoms or to decide whether immunotherapy is indicated, allergy tests are performed [9]. For population-based studies and clinical trials, different questionnaires have been used. The questions "Have you ever had allergic rhinitis?" or "Has a doctor ever told you that you suffer from allergic rhinitis?" seemed the most efficient and unambiguous ones [10]. Additional diagnostic tests, such as a skin prick test or measuring total or specific IgE, did not appear to contribute to diagnosing AR.

18.4 Genetics

AR often occurs in the context of the atopic syndrome, together with atopic dermatitis, AD (atopic eczema, eczema), and asthma. The atopic syndrome is characterized by abnormal elevated levels of total IgE and elevated levels of specific IgE against different allergens [11]. It is known that 10–40 % of all patients with AR also have asthma and that almost all asthma patients have concomitant AR [12]. In AD it is presumed that based on genetic defects, the immune system is disturbed and the epithelial barrier function is diminished, which causes sensitization and leads to allergic diseases such as AR and asthma.

AR is strongly associated with asthma and chronic rhinosinusitis. It is therefore assumed that these diseases share pathophysiological mechanisms and are probably caused by the same genetic defects [11, 13]. Genetic defects specific for AR are not described, but mutations leading to asthma and concomitant AR are extensively studied, as asthma and AR are both heritable diseases with the same complex pathology of environmental and genetic factors. The most extensively studied genes in the context of AR and asthma are interleukin-13 (IL-13) and the interleukin-4 receptor α-chain (IL-4R). As reported by Bottema et al. in 2010, IL-13 showed a significant association with AR ($P=0.04$) and IL-4R mutations showed a positive trend ($P=0.06$) [14].

18.5 Filaggrin and the Atopic Syndrome

Another gene that might be associated with AR is the filaggrin gene (*FLG*). Filaggrin, coded on chromosome 1q21, is a *fil*ament-*agg*regated prote*in* that is able to bind to keratin intermediate filaments in human epithelial cells in order to create a proper skin barrier. The *FLG* null mutations, R501X and 2282del4, are a substantial molecular cause of moderate to severe ichthyosis vulgaris (IV) [15]. These mutations in the *FLG* are characterized by a premature mRNA to protein translation stop, which causes absence of the filaggrin protein in the skin in homozygous or compound heterozygous patients, and are thus called null mutations.

As IV and AD often appear together and AD is also linked to chromosome 1q21, it was suggested in 2006 that *FLG* mutations are probably associated with AD as well [16]. Two *FLG* null mutations in particular, namely, R501X and 2282del4, have been extensively studied in patients with AD in the Western population and are very strong predisposing factors for this disease [17, 18]. Other mutations seem to be of importance as well (R2447X, S3247X, and 3702delG). These *FLG* mutations are associated with early onset, severe, and persistent AD. Whether these *FLG* mutations are related to atopic diseases other than AD, such as AR, is less obvious [17, 19]. However, in the last 7 years, interest in the association between *FLG* and these other atopic diseases has been rising.

18.6 Research on *FLG* Mutations and Allergic Rhinitis in Perspective

18.6.1 Association Between *FLG* Variants and Allergic Rhinitis

The first report of an assumed association between *FLG* mutations and AR was written by Marenholtz et al. in 2006 [20]. Since that time, a few more scientific papers have been published. These include three case-control studies [20, 21], one family study [19], one cross-sectional study [22], and two cohort studies [23, 24]. Contents of these articles are summarized in Table 18.1. Moreover, three review articles have been written wherein associations between *FLG* mutations and AR were described [25–27]. In three studies, the expression of filaggrin in mucosal tissue or vestibulum of the nose has been investigated [22, 28, 29].

Marenholtz et al. first described that a possible correlation between *FLG* null mutations and AR can be identified [20]. In this case-control study within the Multicentre Allergy Study (MAS) cohort, 871 children with AD (SCORAD >15 or >20 % affected body surface) were genotyped for the mutations R501X and 2282del4 within the *FLG*. In the presence of AD, AR was significantly associated with *FLG* null mutations R501X and 2282del4 (OR, 4.79; 95 % CI, 2.0–11.6; $P=1.5\times10^{-5}$). Furthermore, in the absence of AD this link was not significant anymore ($P=0.99$).

A comparable odds ratio was found for the relation in adults between AR and *FLG* null mutations R501X and 2282del4 by Weidinger et al. in 2007 (OR, 4.04; 95 % CI, 2.11–7.72; $P=2.4\times10^{-5}$) [21]. In this case-control study, 274 adults with AD (35.9 ± 10.8 years) were compared with 252 healthy adults (39.4 ± 16.1 years). They also demonstrated that the *FLG* mutations R501X and 2282del4 are a strong risk factor for AD and concomitant atopic diseases.

The link between *FLG* mutations and AR was also investigated in a cross-sectional study by Weidinger et al. (2008) including 3,099 German school children aged 9–11 years [22]. In this study, they did not limit the investigation to the mutations R501X and 2282del4 but also investigated the more recently discovered mutations R2447X, S3247X, and 3702delG. In this study, associations between *FLG* mutations and atopic phenotypes were analyzed. They reported a significant correlation between *FLG* mutations and AR (OR, 26.4; 95 % CI, 1.76–3.96; $P=2.5\times10^{-6}$). The most interesting finding in this study was that this correlation was persistent in the absence of AD (OR, 1.93; 95 % CI, 1.21–3.05; $P=5.3\times10^{-3}$).

In contrast to the findings of Weidinger et al. in 2008, Brown et al. (2008) found no association at all between *FLG* mutations and AR (OR, 1.2; 95 % CI, 0.7–2.0; $P=0.66$) [17]. In this British case-control study, the same mutations were investigated (R501X, 2282del4, R2447X, S3247X, and 3702delG) in 784 schoolchildren aged 7–9 years.

Ekelund et al. studied the relation between *FLG* mutations R501X and 2282del4 and AR in a family

Table 18.1 Studies on the association between *FLG* mutations and allergic rhinitis

Marenholtz et al., case-control study, 2006, Germany, *n* = 871, 0–10 years		**MAS cohort**	
Eczema		Pediatrician's diagnosis/parents report/observed eczema	
Atopic dermatitis		Eczema with specific IgE to ≥ 1 allergen (≥0.70 kU/l)	
AR, hay fever		Pediatrician's diagnosis	
Phenotype (MAS)	*P value*	*OR (95 % CI)*	*Null alleles %*
Eczema (*n* = 186)	3.5×10^{-5}	3.7 (2.0–7.0)	16.7
Atopic dermatitis (*n* = 117)	6×10^{-5}	3.8 (1.9–7.7)	17.1
AR and eczema (*n* = 44)	1.5×10^{-5}	4.79 (2.0–11.6)	20.5
AR, no eczema (*n* = 76)	0.99	1.1 (0.3–3.2)	5.3
Weidinger et al., case-control study, 2007, Germany, cases 274/controls 252, 35.9/39.4 years		**Cases from outpatient clinics University of Bonn and Technical University of Munich**	
Eczema		Skin examination	
Atopic dermatitis		Eczema and sensitization and/or IgE levels ≥100 kU/l	
AR, hay fever		Physician's diagnosis	
Phenotype	*P value*	*OR (95 % CI)*	*Null alleles %*
Eczema (*n* = 274)	4.9×10^{-5}	3.53 (1.92–6.48)	21.1
Atopic dermatitis (*n* = 196)	4.6×10^{-5}	3.66 (1.96–6.83)	n.a.
AR and eczema (*n* = 172)	2.4×10^{-5}	4.04 (2.11–7.72)	n.a.
Weidinger et al., cross-sectional study, 2008, Germany, *n* = 3,099, 9–11 years		**Part of ISAAC II, Munich and Dresden**	
Eczema		Physician's diagnosis	
Atopic dermatitis		Physician's diagnosis and ≥ 1 positive SPT to allergen	
AR, hay fever		Physician's diagnosis and ≥ 1 positive SPT to allergen	
Current allergic rhinitis symptoms		Itchy, runny nose and sneezing in the last 12 months in the absence of a cold	
Phenotype	*P value*	*OR (95 % CI)*	*Null alleles %*
Eczema (*n* = 540)	2.5×10^{-14}	3.12 (2.33–4.17)	16.3
Atopic dermatitis (*n* = 193)	3.2×10^{-14}	4.56 (3.08–6.74)	n.a.
AR (*n* = 214)	2.5×10^{-6}	2.64 (1.76–3.96)	15.9
Current allergic rhinitis symptoms (*n* = 441)	2.6×10^{-3}	1.69 (1.20–2.37)	n.a.
AR adjusted for eczema (*n* = n.a.)	1×10^{-4}	2.25 (1.48–3.41)	n.a.
AR adjusted for atopic dermatitis (*n* = n.a.)	5.3×10^{-3}	1.93 (1.21–3.05)	n.a.
Current allergic rhinitis symptoms adjusted for eczema (*n* = n.a.)	0.061	1.40 (0.99–2.00)	n.a.
Brown et al., case-control study, 2008, Northwest England, *n* = 784, 7–9 years		**Population cohort 1996–2003, hospital-based, West Cumberland Hospital, Whitehaven**	
Eczema		UK diagnostic criteria over a 12-month period and skin examination	
AR, hay fever		Questionnaire	
Phenotype	*P value*	*OR (95 % CI)*	*Null alleles %*
Eczema (*n* = 195)	1.8×10^{-3}	26.9 (3.3–217.1)	18.4
AR, no eczema (*n* = 152)	0.66	1.2 (0.7–2.0)	n.a.

Table 18.1 (continued)

Ekelund et al., family study, 2008, Sweden, n = 1,514, 29 years (mean), R501X/2282del4		**Families from outpatient clinics Karolinska University Hospital and Danderyd Hospital, Stockholm**	
Eczema		Physician's diagnosis and skin examination	
Atopic dermatitis		Eczema and elevated specific IgE level (>0.35 kU/l)	
AR, hay fever		Physician's diagnosis	
Phenotype	*P value*	*OR (95 % CI)*	*Null alleles %*
Eczema (n=921)	1.3×10^{-6}	1.81 (1.31–2.50)	n.a.
Atopic dermatitis (n=588)	9.5×10^{-8}	2.21 (1.50–3.25)	n.a.
AR and eczema (n=613)	7.0×10^{-7}	2.03 (1.39–2.97)	n.a.
Henderson et al., longitudinal population-based cohort, 2008, United Kingdom, n=6,971, 0–11 years		**ALSPAC cohort**	
Eczema		Reported rash at 6, 18, 30, 42, 57, and 81 months, observed flexural dermatitis at 7–11 years	
Atopic dermatitis		Eczema and ≥ 1 positive SPT for grass/HDM/cat	
AR, hay fever		Question: allergic to pollen at 54 months, detailed questionnaire at 81 months	
Phenotype	*P value*	*OR (95 % CI)*	*Null alleles %*
Eczema (n=1,445)	1.19×10^{-22}	2.46 (2.02–2.99)	14.8
Atopic dermatitis (n=193)	3.96×10^{-20}	2.73 (1.87–3.99)	20.7
AR at 54 months (n=5,833)	0.010	1.66 (1.01–2.71)	14.1
AR at 81 months (n=5,358)	0.010	1.31 (0.96–1.78)	21.8
AR at 11 years (n=4,628)	0.032	1.22 (0.94–1.58)	10.1
Schuttelaar et al., birth cohort, 2009, Netherlands, n=934, 0–8 years		**PIAMA cohort**	
Eczema		UK diagnostic criteria at 4 years	
Atopic dermatitis		Eczema and sensitization at 4 years (=specific IgE against food/inhalant allergens)	
AR, hay fever		Questionnaire at 3–8 years	
Phenotype	*P value*	*OR (95 % CI)*	*Null alleles %*
Eczema at 4 years (n=110)	4×10^{-6}	3.3 (1.9–5.5)	20.9
Atopic dermatitis at 4 years (n=31)	4×10^{-3}	3.5 (1.4–8.7)	22.6
AR at 3–5 years (n=36)	0.39	1.4 (0.6–3.3)	12.3
AR at 6–8 years (n=30)	0.1	2.3 (0.9–6.0)	16.7

Abbreviation: *AR* allergic rhinitis

study, which included 406 families of Swedish origin [19]. Besides the important finding that *FLG* null mutations were a risk factor for severe AD, they found that *FLG* mutations may be susceptibility factor for AR and other disorders within the atopic syndrome. An odds ratio of 2.03 for the combined genotype was reported, which confirms a link between *FLG* mutations and AR in the presence of AD (OR, 2.03; 95 % CI, 1.39–2.97; $P=7\times10^{-7}$). After correcting for AD, this link did not persist.

Between 1991 and 1992, the group of Henderson et al., from the United Kingdom, performed a large birth cohort study on 14,541 pregnant women residing in Avon [24]. The children (n=6,971) of these mothers were genotyped for mutations R501X and 2282del4 and were analyzed for symptoms of hay fever at 54 and 81 months and at the age of 11 years. Positive associations between these *FLG* mutations and AR at all three time points were found, although a significant association was only found at 54 months (resp. OR, 1.66; 95 % CI, 1.01–2.71; $P=0.010$; OR, 1.31; 95 % CI, 0.96–1.78; $P=0.010$; OR, 1.22; 95 % CI, 0.94–1.58; $P=0.032$).

The most recent study that investigated an association between *FLG* mutations (R501X, 2282del4, and R2447X) and hay fever was by Schuttelaar et al. in 2009 [23]. From the Dutch Prevention and Incidence of Asthma and Mite Allergy (PIAMA) birth cohort, a selection of 934 children up to 8 years old, who were included based on the atopic history of their mothers, were investigated in this study. The combined genotype was not significantly associated with hay fever. The 2282del4 variant was associated with the first occurrence (annual incidence) of hay fever at the age of 6–8 years only (OR, 3.9; 95 % CI, 1.5–10.5; $P=0.006$). The link between *FLG* mutation 2282del4 with hay fever at 6–8 years was only significant in children with eczema in the first year of life (OR, 4.0; 95 % CI, 1.2–13.6; $P=0.03$), but not in children without eczema in the first year of life (OR, 1.7; 95 % CI, 0.2–13.3; $P=0.60$).

The three review articles, not included in the table, are mostly based on previous research articles [25–27]. They all confirm the association between *FLG* mutations and AR with coexisting AD. *FLG* mutations not only are a serious risk factor of AD but also cause sensitization of patients and progression of allergic disease and thus the so-called atopic march.

18.6.2 The Lack of Expression of Filaggrin in Mucosal Tissue

Apart from the investigations on the association between *FLG* mutations and AR, a few studies on the expression of filaggrin in mucosal tissue were performed.

In 2006, Ying et al. performed the first study on the expression of filaggrin protein in mucosal tissue [29]. They hypothesized that filaggrin would be expressed in human bronchial epithelium, because this kind of tissue expresses many proteins present in human keratinocytes. Their major finding was the total absence of filaggrin in bronchial epithelium of both asthmatic and non-asthmatic individuals.

Weidinger et al. analyzed the expression of filaggrin protein in the mouth, nasal vestibulum, transitional mucosa, and respiratory mucosa using immunohistochemical staining of biopsies [22]. In this study, the expression of filaggrin could not be detected in the respiratory tract, although the epithelium of the mouth and the nasal vestibulum up to the transitional epithelium was positive for filaggrin.

To investigate whether filaggrin had a direct effect on the barrier function of this epithelium, De Benedetto et al. studied the expression of filaggrin in the nasal and esophageal epithelium [28]. In both nasal and esophageal epithelium, filaggrin was not present. Therefore, it is unlikely that *FLG* mutations have a direct effect on the epithelial barrier of the nasal mucosa in the context of AR.

18.7 Discussion

From the aforementioned studies, one can conclude that *FLG* mutations are strongly associated with the development of AD at a young age. Unfortunately, the associations between *FLG* mutations and AR have not been extensively investigated. Nearly all scientific papers published on this topic show a correlation. However, in most studies, this was only significant in the presence of AD. In the absence of AD, an association between FLG mutations and AR was not found except for the study of Weidinger et al. in 2008 [22]. They reported an association between *FLG* mutations and a parent's report of a doctor's or physician's diagnosis of AR in combination with a positive skin prick test in the absence of AD in schoolchildren, aged 9–11 years (OR, 2.32; 95 % CI, 1.59–3.36; $P=1.0\times10^{-5}$). However, they could not demonstrate an association between *FLG* mutations and current AR symptoms (OR, 1.40; 95 % CI, 0.99–2.00, $P=0.061$). These so-called current symptoms of AR were defined as the presence of an itchy, runny nose and sneezing in the absence of a cold in the last 12 months. One could presume that children who were diagnosed with AR in the past should also have had current symptoms in the last 12 months. It is unlikely that symptoms of AR have completely disappeared in these 9–11-year-old children [30].

Therefore, it is atypical that an association with current symptoms was not found. Moreover, none of the affected AR patients within this study were homozygous or compound heterozygous for the *FLG* mutations, which weakens this association even more.

The increased risk of AR in patients with *FLG* mutations cannot be explained by a supposed increased permeability of the nasal mucosa to allergens, due to the absence of filaggrin expression in the epithelium of the mucosa of the airways and the nose. The fact that *FLG* mutations and AR are not associated in the absence of eczema indicates that the higher risk of AR in patients with *FLG* mutations can only be explained by the higher risk of eczema. The described association could be based on an impaired skin barrier in eczema, which probably facilitates sensitization. This sensitization is subsequent in progression of AD to asthma and AR, otherwise called the atopic march [31, 32].

18.7.1 The Outside-In Hypothesis

According to the "outside-in" hypothesis by Elias et al. [33, 34], skin barrier dysfunction (e.g., by *FLG* mutations) precedes sensitization via the skin and subsequent atopic diseases. O'Regan et al. stated that sensitization to aeroallergens in AR and asthma, in the context of AD, occurs through the skin [35]. This cutaneous sensitization has led to a systemic T_H2 response, caused by local processing of allergens by Langerhans cells (LHC) and migration to draining lymph nodes. This phenomenon has been described in mice, and it is assumed that in humans the mechanism is similar.

Another important aspect of the outside-in hypothesis is the increased irritability of filaggrin-deficient skin. Cutaneous irritation and concomitant inflammation account for higher concentrations of T_H2-associated cytokines, IL-4 and IL-13. These cytokines, in turn, lead to a decreased expression of filaggrin in the skin. This decreased expression leads to increased permeability and decreased moisturization, as filaggrin is a barrier protein as well as a moisturizer [36, 37].

18.7.2 Atopic Dermatitis and the Atopic Syndrome Versus Ichthyosis Vulgaris

Although *FLG* mutations certainly have an important and probably initiating role in the development of AD and the progression into the atopic march, these mutations are not exclusively a problem in the atopic syndrome. *FLG* mutations were first found in patients with another cutaneous disease, ichthyosis vulgaris (IV). All patients with IV have an epithelial barrier defect, but in contrast to AD patients, they only have a slightly higher specific IgE compared to the normal population [15]. Furthermore, the epidermis in the mouse model of IV is also characterized by inflammation [38, 39]. Moreover, animal research on epithelial barrier dysfunction has shown that a mechanical disruption of the epithelium results in cytokine production and inflammation [37].

Nowadays one of the main questions in filaggrin research is in which way can identical mutations in the *FLG* lead to such different phenotypes as IV and the atopic syndrome. This question still remains unanswered, but the increased penetration of allergens as well as irritants through the epidermal barrier and reduced inflammatory thresholds caused by *FLG* mutations can lead to percutaneous priming as well as an increased irritation [40]. In response to this ingress of allergens or irritants, a T_H2 reaction is provoked. This event, which for unknown reasons might not occur in IV, may cause sensitization through the skin in atopic patients [41]. This is supported by the findings of Thyssen et al. in 2008. They found an increased prevalence of *FLG* mutations in patients with AD and suggested that these mutations made an individual more susceptible for AD caused by environmental influences [42].

18.7.3 From Cutaneous Sensitization to Systemic Allergy

Cutaneous sensitization allows allergens to be captured and processed by LHC in the epidermis. These LHC migrate to the draining lymph nodes, in which they can interact with naive T cells to initiate T_H2 immunity against these topical applied allergens. In this manner, it is suspected that systemic sensitization to both aeroallergens and food allergens can occur in atopic patients [32, 43–45].

Although mechanical disruption of the epithelial barrier of the skin in a filaggrin-deficient mouse model results in inflammation and increased cytokine production, it fails to induce a permanent atopic state. As a result of this critical finding, O'Regan et al. supposed that there has to be a link between *FLG* mutations and the T_H2 polarity of the immune system causing the atopic state [35]. The immune system has to be programmed to initiate a T_H2-mediated immune response rather than no immune response in reaction to allergens in atopic individuals. This mechanism does not occur in IV [36].

18.7.4 Atopy Patch Testing Strengthens Percutaneous Priming Hypothesis

Sensitization, initiated by percutaneous priming of individuals to specific allergens, is supported by the concept of the atopy patch test (APT). Since 1989, this test has been used in patients with AD to demonstrate cutaneous delayed-type hypersensitivity reactions to inhalation allergens when topically applied [46, 47]. During APT, allergens, known to induce a strong IgE-mediated allergic reaction, are applied to the skin in the same manner as epicutaneous patch tests for allergic contact dermatitis. Frequently used allergens are house dust mite, cat dander, and grass and birch pollen but also food allergens. These topically applied allergens can elicit eczematous skin reactions in about 50 % of all atopic individuals, confirming their atopic state [48, 49].

Although atopy patch tests have never been used solely as a diagnostic tool and are considered controversial in the United States, they can prove the clinical relevance of a sensitization. A positive reaction to an allergen in the APT could make the diagnosis of AR more plausible, if the results are combined with a positive specific IgE against this specific allergen [49, 50]. Fuiano et al. demonstrated that individuals with respiratory symptoms and current or past AD had significantly more positive APT compared to individuals with respiratory symptoms and no current or past AD [51]. However, the test conditions in APT are far from physiological; concentrations of the causative allergen are up to 10,000 times higher, and it is unlikely that airborne allergens penetrate the stratum corneum [52].

18.7.5 Filaggrin Null Mutations and Peanut Allergy

Another phenomenon that enhances plausibility of sensitization through percutaneous priming is the association between *FLG* mutations and peanut allergy. Peanut allergy is a strongly heritable allergy that can cause dangerous anaphylaxis. The prevalence of this allergy in children is between 1.2 and 1.6 % [53]. In 72–81 % of all presentations of peanut allergy, it looked as if it occurred at the first exposure to peanut allergens. This extremely high percentage suggests unperceived previous priming to peanut allergens. A probable cause is sensitization through cutaneous exposure, as is described in rodents [44]. This exposure can be caused by the use of peanut oil (*Arachis* oil) in emollients, as frequently prescribed in AD, but also by cutaneous contact with peanut-eating family members in households where large quantities of peanuts are consumed [54, 55].

The association between peanut allergy and *FLG* mutations has recently been shown by Brown et al. [56]. In their article, a highly significant association was described in four different populations, Dutch, English, Irish, and Canadian (OR, 5.3; 95 % CI, 2.8–10.2; $P = 3.0 \times 10^{-6}$). After correcting for AD, this link still remained significant (OR, 3.8; 95 % CI, 1.7–8.3; $P = 1.0 \times 10^{-4}$).

18.7.6 Oral Tolerance

A low dose of peanut allergens exposed to the skin could cause allergic sensitization in patients with an impaired epithelial barrier, due to *FLG* mutations, and early onset AD. On the other hand, early oral exposure to peanut-containing products prevents sensitization to this specific allergen. A similar phenomenon called oral tolerance has been investigated in nickel allergy as well, although this tolerance induction is considered controversial [57–59]. Oral tolerance seems to prevent an individual from developing allergic sensitization.

18.7.7 The Perfect Prevention for the Onset of Allergic Rhinitis

One can conclude that the diminished epithelial barrier in AD, caused by *FLG* mutations, predisposes an individual to sensitization through the skin. This cutaneous sensitization can lead to systemic allergies and will initiate the atopic march. However, this progression to AR is only present in patients who had eczema. Therefore, one can state that asthma and AR are both secondary events caused by an epithelial barrier defect of the skin in eczema patients combined with a T_H2 polarity of the immune system.

The main allergens causing symptoms in AR are pollen, house dust mite, and cat dander. These allergens provide a positive APT reaction in atopic individuals, indicating an individual's sensitization. Oral tolerance has proven to be an effective tool to prevent systemic peanut and possibly nickel allergies by administration of a low dose of allergens to unsensitized persons who are at risk of developing systemic allergies. Presumably the same mechanism is applicable to the prevention of allergies to aeroallergens causing AR. Oral exposure to a low dose of these allergens at a young age could prevent or at least reduce the development of systemic sensitization and the progression of an atopic march into AR. This probably effective prevention of sensitization has to be investigated in order to make it applicable in a clinical setting.

Although oral tolerance seems a promising tool to prevent the progression into an atopic march, the diminished skin barrier, caused by a lack of filaggrin, should be restored using emollients, to preclude allergic sensitization and other atopic diseases like AR.

References

1. Skoner DP. Allergic rhinitis: definition, epidemiology, pathophysiology, detection, and diagnosis. J Allergy Clin Immunol. 2001;108(1):S2–8.
2. Bauchau V, Durham S. Prevalence and rate of diagnosis of allergic rhinitis in Europe. Eu Respir J. 2004;24(5):758–64.
3. Izquierdo-Domínguez A, Valero AL, Mullol J. Comparative analysis of allergic rhinitis in children and adults. Curr Allergy Asthma Rep. 2013;13:142–51.
4. Bostock J. Case of a periodical affection of the eyes and chest. Med Chir Trans. 1819;10(Pt 1):161.
5. Taylor G, Walker J. Charles Harrison Blackley, 1820–1900. Clin Exp Allergy. 1973;3(2):103–8.
6. Johansson SGO, Hourihane JOB, Bousquet J, Bruijnzeel-Koomen C, Dreborg S, Haahtela T, et al. A revised nomenclature for allergy: an EAACI position statement from the EAACI nomenclature task force. Allergy. 2001;56(9):813–24.
7. Johansson SGO, Bieber T, Dahl R, Friedmann PS, Lanier BQ, Lockey RF, et al. Revised nomenclature for allergy for global use: Report of the Nomenclature Review Committee of the World Allergy Organization, October 2003. J Allergy Clin Immunol. 2004;113(5):832–6.
8. Bousquet J, Schünemann HJ, Samolinski B, Demoly P, Baena-Cagnani CE, Bachert C, et al. Allergic Rhinitis and Its Impact on Asthma (ARIA): achievements in 10 years and future needs. J Allergy Clin Immunol. 2012;130:1049–62.
9. Tschopp J, Sistek D, Schindler C, Leuenberger P, Perruchoud A, Wüthrich B, et al. Current allergic asthma and rhinitis: diagnostic efficiency of three commonly used atopic markers (IgE, skin prick tests, and Phadiatop). Allergy. 1998;53(6):608–13.
10. Bousquet J, Khaltaev N, Cruz AA, Denburg J, Fokkens WJ, Togias A, et al. Allergic rhinitis and its impact on asthma (ARIA) 2008*. Allergy. 2008;63(s86):8–160.
11. Kariyawasam HH, Rotiroti G. Allergic rhinitis, chronic rhinosinusitis and asthma: unravelling a complex relationship. Curr Opin Otolaryngol Head Neck Surg. 2013;21(1):79–86.
12. Caimmi D, Marseglia A, Pieri G, Benzo S, Bosa L, Caimmi S, et al. Nose and lungs: one way, one disease. Ital J Pediatr. 2012;38(1):1–5.
13. Dold S, Wjst M, von Mutius E, Reitmeir P, Stiepel E. Genetic risk for asthma, allergic rhinitis, and atopic dermatitis. Arch Dis Child. 1992;67(8):1018–22.
14. Bottema RW, Nolte IM, Howard TD, Koppelman GH, Dubois AE, de Meer G, et al. Interleukin 13 and interleukin 4 receptor-alpha polymorphisms in rhinitis and asthma. Int Arch Allergy Immunol. 2010;153(3):259–67.
15. Thyssen J, Godoy-Gijon E, Elias P. Ichthyosis vulgaris-the filaggrin mutation disease. Br J Dermatol. 2013;168:1155–66.
16. Palmer CNA, Irvine AD, Terron-Kwiatkowski A, Zhao Y, Liao H, Lee SP, et al. Common loss-of-function

variants of the epidermal barrier protein filaggrin are a major predisposing factor for atopic dermatitis. Nat Genet. 2006;38(4):441–6.
17. Brown SJ, Relton CL, Liao H, Zhao Y, Sandilands A, Wilson IJ, et al. Filaggrin null mutations and childhood atopic eczema: a population-based case-control study. J Allergy Clin Immunol. 2008;121(4):940–6.
18. Rodríguez E, Baurecht H, Herberich E, Wagenpfeil S, Brown SJ, Cordell HJ, et al. Meta-analysis of filaggrin polymorphisms in eczema and asthma: robust risk factors in atopic disease. J Allergy Clin Immunol. 2009;123(6):1361–70.
19. Ekelund E, Liedén A, Link J, Lee SP, D'Amato M, Palmer CNA, et al. Loss-of-function variants of the filaggrin gene are associated with atopic eczema and associated phenotypes in Swedish families. Acta Derm Venereol. 2008;88(1):15–9.
20. Marenholz I, Nickel R, Rüschendorf F, Schulz F, Esparza-Gordillo J, Kerscher T, et al. Filaggrin loss-of-function mutations predispose to phenotypes involved in the atopic march. J Allergy Clin Immunol. 2006;118(4):866.
21. Weidinger S, Rodríguez E, Stahl C, Wagenpfeil S, Klopp N, Illig T, et al. Filaggrin mutations strongly predispose to early-onset and extrinsic atopic dermatitis. J Invest Dermatol. 2006;127(3):724–6.
22. Weidinger S, O'Sullivan M, Illig T, Baurecht H, Depner M, Rodriguez E, et al. Filaggrin mutations, atopic eczema, hay fever, and asthma in children. J Allergy Clin Immunol. 2008;121(5):1203–9.
23. Schuttelaar M, Kerkhof M, Jonkman M, Koppelman G, Brunekreef B, De Jongste J, et al. Filaggrin mutations in the onset of eczema, sensitization, asthma, hay fever and the interaction with cat exposure. Allergy. 2009;64(12):1758–65.
24. Henderson J, Northstone K, Lee SP, Liao H, Zhao Y, Pembrey M, et al. The burden of disease associated with filaggrin mutations: a population-based, longitudinal birth cohort study. J Allergy Clin Immunol. 2008;121(4):872–7.
25. Brown SJ, McLean WHI. One remarkable molecule: filaggrin. J Invest Dermatol. 2011;132:751–62.
26. Spergel JM. From atopic dermatitis to asthma: the atopic march. Ann Allergy Asthma Immunol. 2010;105(2):99–106.
27. Van Den Oord RAHM, Sheikh A. Filaggrin gene defects and risk of developing allergic sensitisation and allergic disorders: systematic review and meta-analysis. Br Med J. 2009;339:b2433.
28. De Benedetto A, Qualia CM, Baroody FM, Beck LA. Filaggrin expression in oral, nasal, and esophageal mucosa. J Invest Dermatol. 2008;128(6):1594–7.
29. Ying S, Meng Q, Corrigan CJ, Lee TH. Lack of filaggrin expression in the human bronchial mucosa. J Allergy Clin Immunol. 2006;118(6):1386.
30. de Bot C, Röder E, Pols D, Bindels P, van Wijk RG, van der Wouden JC, et al. Sensitisation patterns and association with age, gender, and clinical symptoms in children with allergic rhinitis in primary care: a cross-sectional study. Prim Care Respir J. 2013;22:155–60.
31. Spergel JM. Atopic march: link to upper airways. Curr Opin Allergy Clin Immunol. 2005;5(1):17–21.
32. Zheng T, Yu J, Oh MH, Zhu Z. The atopic march: progression from atopic dermatitis to allergic rhinitis and asthma. Allergy Asthma Immunol Res. 2011;3(2):67–73.
33. Elias PM, Hatano Y, Williams ML. Basis for the barrier abnormality in atopic dermatitis: outside-inside-outside pathogenic mechanisms. J Allergy Clin Immunol. 2008;121(6):1337.
34. Elias PM, Steinhoff M. "Outside-to-inside" (and now back to "outside") pathogenic mechanisms in atopic dermatitis. J Invest Dermatol. 2008;128(5):1067–70.
35. O'Regan G, Irvine A. The role of filaggrin in the atopic diathesis. Clin Exp Allergy. 2010;40(7):965–72.
36. McPherson T, Sherman V, Aslam A, Crack L, Chan H, Lloyd-Lavery A, et al. Filaggrin null mutations associate with increased frequencies of allergen-specific CD4+ T-helper 2 cells in patients with atopic eczema. Br J Dermatol. 2010;163(3):544–9.
37. Howell MD, Kim BE, Gao P, Grant AV, Boguniewicz M, DeBenedetto A, et al. Cytokine modulation of atopic dermatitis filaggrin skin expression. J Allergy Clin Immunol. 2007;120(1):150–5.
38. Fallon PG, Sasaki T, Sandilands A, Campbell LE, Saunders SP, Mangan NE, et al. A homozygous frameshift mutation in the mouse Flg gene facilitates enhanced percutaneous allergen priming. Nat Genet. 2009;41(5):602–8.
39. Scharschmidt TC, Man MQ, Hatano Y, Crumrine D, Gunathilake R, Sundberg JP, et al. Filaggrin deficiency confers a paracellular barrier abnormality that reduces inflammatory thresholds to irritants and haptens. J Allergy Clin Immunol. 2009;124(3):496–506.
40. Hougaard MG, Johansen JD, Linneberg A, Bandier J, Stender S, Carlsen BC, et al. Skin prick test reactivity to aeroallergens by filaggrin mutation status. J Eur Acad Dermatol Venereol. 2012. [Epub ahead of print].
41. Weidinger S, Illig T, Baurecht H, Irvine AD, Rodriguez E, Diaz-Lacava A, et al. Loss-of-function variations within the filaggrin gene predispose for atopic dermatitis with allergic sensitizations. J Allergy Clin Immunol. 2006;118(1):214–9.
42. Thyssen JP, Linneberg A, Johansen JD, Carlsen BC, Zachariae C, Meldgaard M, et al. Atopic diseases by filaggrin mutations and birth year. Allergy. 2012;67(5):705–8.
43. De Benedetto A, Kubo A, Beck LA. Skin barrier disruption: a requirement for allergen sensitization? J Invest Dermatol. 2012;132:949–63.
44. Strid J, Hourihane J, Kimber I, Callard R, Strobel S. Disruption of the stratum corneum allows potent epicutaneous immunization with protein antigens resulting in a dominant systemic Th2 response. Eur J Immunol. 2004;34(8):2100–9.
45. Kubo A, Nagao K, Amagai M. Epidermal barrier dysfunction and cutaneous sensitization in atopic diseases. J Clin Invest. 2012;122(2):440–7.
46. Langeveld-Wildschut EG, Bruijnzeel PL, Mudde GC, Versluisa C, Van Ieperen-Van Dijka AG, Biharia IC,

et al. Clinical and immunologic variables in skin of patients with atopic eczema and either positive or negative atopy patch test reactions. J Allergy Clin Immunol. 2000;105(5):1008–16.
47. Guler N, Kirerleri E, Tamay Z, Ones U. Atopy patch testing in children with asthma and rhinitis symptoms allergic to house dust mite. Pediatr Allergy Immunol. 2006;17(5):346–50.
48. Lipozenčić J, Wolf R. The diagnostic value of atopy patch testing and prick testing in atopic dermatitis: facts and controversies. Clin Dermatol. 2010;28(1):38–44.
49. Fuiano N, Incorvaia C. The atopy patch test: is it time to redefine its significance? Ann Allergy Asthma Immunol. 2011;106(4):278–82.
50. Darsow U, Laifaoui J, Kerschenlohr K, Wollenberg A, Przybilla B, Wüthrich B, et al. The prevalence of positive reactions in the atopy patch test with aeroallergens and food allergens in subjects with atopic eczema: a European multicenter study. Allergy. 2004;59(12):1318–25.
51. Fuiano N, Fusilli S, Incorvaia C. House dust mite-related allergic diseases: role of skin prick test, atopy patch test, and RAST in the diagnosis of different manifestations of allergy. Eur J Pediatr. 2010;169(7):819–24.
52. Tupker R, Monchy J, Coenraads P. House-dust mite hypersensitivity, eczema, and other nonpulmonary manifestations of allergy. Allergy. 1998;53(s48):92–6.
53. Thompson RL, Miles LM, Lunn J, Devereux G, Dearman RJ, Strid J, et al. Peanut sensitisation and allergy: influence of early life exposure to peanuts. Br J Nutr. 2010;103(9):1278.
54. Lack G, Fox D, Northstone K, Golding J. Factors associated with the development of peanut allergy in childhood. N Engl J Med. 2003;348(11):977–85.
55. Fox AT, Sasieni P, du Toit G, Syed H, Lack G. Household peanut consumption as a risk factor for the development of peanut allergy. J Allergy Clin Immunol. 2009;123(2):417–23.
56. Brown SJ, Asai Y, Cordell HJ, Campbell LE, Zhao Y, Liao H, et al. Loss-of-function variants in the filaggrin gene are a significant risk factor for peanut allergy. J Allergy Clin Immunol. 2011;127(3):661–7.
57. Van Hoogstraten IMW, Andersen KE, Blomberg BME, Boden D, Bruynzeel DP, Burrows D, et al. Reduced frequency of nickel allergy upon oral nickel contact at an early age. Clin Exp Immunol. 1991;85(3):441–5.
58. Van Hoogstraten IM, Boden D, von Blomberg BME, Kraal G, Scheper RJ. Persistent immune tolerance to nickel and chromium by oral administration prior to cutaneous sensitization. J Invest Dermatol. 1992;99(5):608–16.
59. Thyssen JP, Carlsen BC, Menne T. Nickel sensitization, hand eczema, and loss-of-function mutations in the filaggrin gene. Dermatitis. 2008;19(6):303–7.

Filaggrin and Non-cutaneous Disease: Food Allergy

19

Hern Tze Tina Tan and Katrina J. Allen

Contents

H.T.T. Tan, PhD
School of Medical Sciences, Health Campus, Universiti Sains Malaysia, Kubang Kerian, Kota Bharu, Kelantan 16150, Malaysia
e-mail: tina.tan08@yahoo.com

K.J. Allen, MD, PhD (✉)
Population Health, Murdoch Childrens Research Institute, Melbourne, VIC, Australia

Department of Allergy and Immunology, Royal Children's Hospital, Melbourne, VIC, Australia

Department of Paediatrics, The University of Melbourne, Flemington Road, Parkville, Melbourne, VIC 30536, Australia
e-mail: katrina.allen@rch.org.au

19.1 Introduction

With the description of the filaggrin gene (*FLG*) and its role in skin barrier function, interest has recently turned to whether filaggrin may play a role in the integrity of other epithelial barriers including the gastrointestinal tract. The role of eczema in the development of food sensitization, the first step on the atopic march pathway, could, in theory, be genetically linked by a common pathway such as skin barrier function in which filaggrin is a key player. Furthermore, since eczema and food allergy commonly co-associate in early infancy, there has been recent interest in

J.P. Thyssen, H. Maibach (eds.), *Filaggrin*,
DOI 10.1007/978-3-642-54379-1_19, © Springer-Verlag Berlin Heidelberg 2014

whether *FLG* mutations are associated with an increased risk of food allergy over and above that of eczema. This chapter aims to provide emerging evidence for the role of filaggrin in the development of food sensitization and potentially food allergy and the atopic march in early infancy.

19.2 What Is Food Allergy and Is It on the Rise?

Food allergy is an adverse reaction to food due to intolerance mediated by an immunological mechanism, while food intolerance is the non-immunological form [1, 2]. The most dangerous type of adverse food reaction is the IgE-mediated food allergy, which is initiated by an impairment of normal oral tolerance to food in allergy-predisposed (atopic) individuals [1].

The prevalence of infantile food allergy may be higher than previously thought. A recent randomized, cross-sectional survey in the United States reported food allergy prevalence in children of 8.0 %, with nearly a third of the food-allergic children having multiple food allergies [3]. Australia has recently reported the highest rates with more than 10 % of 1-year-old infants having challenge-proven IgE-mediated food allergy in the Melbourne population, the largest, most southern city [4].

As the rise in allergic disease and anaphylaxis has occurred more rapidly than changes to the genome can occur and because allergic disease appears significantly more prevalent in "Westernized countries," it has been hypothesized that factors associated with progressive lifestyle changes within affected regions are responsible for the changes in prevalence. The current most popular candidate lifestyle factors associated with the allergy epidemic include declining microbial exposure [5], increased pro-inflammatory modern diet [6], and inhaled pollutants associated with motor vehicles and the rise in traffic. Although clear and progressive changes in these factors have been long implicated in the rise of asthma and inhalant sensitization, they have not yet been proven to be causative possibly because of the multifactorial aspect of lifestyle risk factors. Furthermore, food allergy continues to rise, as asthma reaches a plateau and even enters a decline in highly westernized regions [7], suggesting that different risk factors are at play. Additional factors implicated specifically in the rise of food allergy in addition to the "hygiene hypothesis" [8] and "old friend hypothesis" include changes to infant-feeding practices [9] as well as vitamin D insufficiency [10]. The potential genetic, epigenetic, and environmental factors contributing to an increase in IgE-mediated food allergy, in prenatal, perinatal, and postnatal phases is depicted in Fig. 19.1 [11].

Because food allergy is more common in developed than in developing countries and migrants seem to acquire the incident risk of allergy of their adopted country, it seems likely that the cause of the rise is linked to the "modern lifestyle." Although environmental factors, including those associated with the hygiene hypothesis and dietary factors, have been found to be associated with the development of eczema and atopy, there has been little investigation into whether they may also play a role in the development of food allergy. In addition to factors previously linked to allergic disease in general, there may well be some factors that are specific to food allergy. These might include changing methods of food manufacturing and preparation, advent of widespread use of antacids and proton pump inhibitors, use of medicinal creams containing food allergens, or later introduction of allergenic foods into the diet of infants [12].

Although it is considered that environmental rather than genetic factors play a major role in the apparent rise in food allergy [13, 14], it is important to understand underlying genetic risk factors in order to both optimize targeting of high-risk groups for prevention and also to further understand gene–environment interactions that might be important contributory risk factors for food allergy development.

19.3 The Role of Genetics in Food Allergy Development

Although the rise in food allergy prevalence cannot be explained by genetic factors alone, there is strong evidence that susceptibility to food allergy is at least partly determined by genetics. Twin

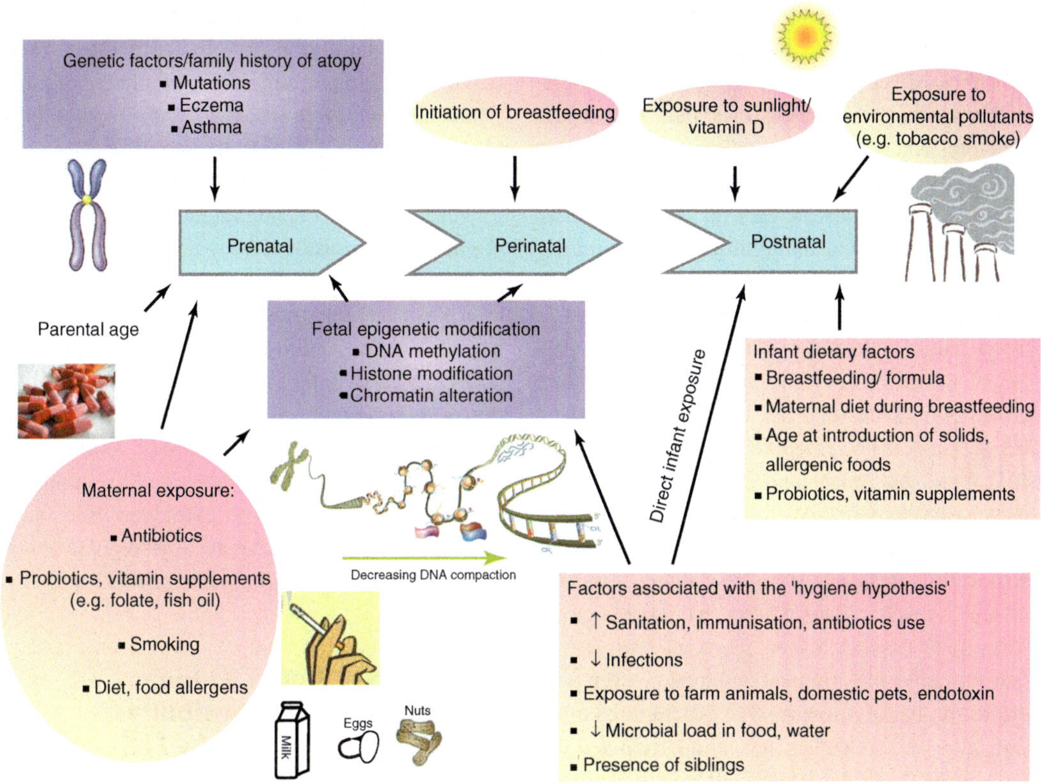

Fig. 19.1 The potential genetic, epigenetic, and environmental factors contributing to an increase in IgE-mediated food allergy in prenatal, perinatal, and postnatal phases. Genetic factors or family history of atopy, parental age, and maternal exposure to antibiotics, probiotics, supplementary food, food allergens, and smoking may act at the prenatal stage. Epigenetic modification may play a role during prenatal and perinatal periods, while breastfeeding starting at the perinatal stage and infant's dietary factors and exposure to sunlight, environmental pollutants, and factors related to the hygiene hypothesis may have an effect on food allergy development at the postnatal stage (Reprinted from Tan et al. [11], with permission from John Wiley and Sons)

studies have shown that the concordance rate for peanut allergy was much higher among monozygotic (64.3 %) than dizygotic (6.8 %; $P<0.0001$) twin pairs [15]. However, a recent study of familial aggregation observed the heritability of common food allergies (sesame, peanut, wheat, milk, egg white, soy, walnut, shrimp, and codfish) to be between 0.15 and 0.35, suggesting that genetic factors are not as important as other factors, such as environmental exposures, in the development of food allergies. Eczema and food allergy seem to be closely linked, with eczematous infants having a greatly increased risk of food allergy [16]. There is increasing interest in the role of loss-of-function mutations in the *FLG*, which lead to defects in skin barrier permeability, in the pathogenesis of eczema. Null mutations (R501X and 2282del4) in the *FLG* are associated with an increased susceptibility to eczema [17]. Individuals with two null alleles in the *FLG* have been shown to be four to seven times more likely to have eczema than those without [18]. One recent case-control study investigated whether *FLG* mutations were also associated with peanut allergy, reporting that those with peanut allergy were around five times more likely to have *FLG* loss-of-function mutations [19]. This result, along with previous studies suggesting that peanut sensitization may occur through damaged skin [20–22], indicates that epithelial barrier dysfunction may play a role in the pathogenesis of peanut allergy.

There is evidence that Asian populations may be more susceptible to allergic disease when living in "Westernized" environments [23, 24]. Earlier studies of respiratory disease observed that both allergic symptoms and sensitization were more common in Asian Australians than in non-Asian Australians [23]. Rates were also higher in Australian-born Asians than in Asian immigrants, with the prevalence increasing with length of stay in Australia [23]. Recently, the HealthNuts study (2011) found a higher rate of food-sensitized eczema among children of Asian descent [25]. Interestingly, on examination of risk factors for eczema development (a losely associated infantile allergic disease), the study found that Asian children were not only more likely to have eczema than their non-Asian counterparts but that their parents were less likely to have allergic disease than non-Asian parents. This was particularly so for Asian parents, who had migrated to Australia less than 5 years previously, suggesting a strong gene–environment interaction even over and above that of a migrational generational effect [26]. More recent studies have similarly noted that nonwhite races are more susceptible to food allergy. In the United States, the 2007 National Health Interview Survey found that non-Hispanic children had higher rates of reported food allergy compared with Hispanic children [27]. Asian populations in particular are also highly susceptible to food allergy [24], suggesting a strong genetic propensity that is amplified by a Western environment. This is consistent with earlier work indicating evolutionary differences in genetic polymorphisms affecting candidate genes [28].

19.4 Why Look at the *FLG* in the Role of Non-cutaneous Diseases Including Food Allergy?

Eczema is thought to be the starting point for the "atopic march" [29], a term used to describe the progression from eczema and food allergies to asthma and allergic rhinitis. The mechanism by which allergic diseases progress, however, is not well understood, although it is thought to involve both genetic and environmental factors, with potentially a differential effect of risk factors on those with a family history of allergic disease versus those without family history of allergic disease. Since eczema often coexists with food allergy [16, 26, 30] and *FLG* null mutations are the most common and widely replicated genetic risk factors for eczema [31], *FLG* null mutations also provide a possible mechanistic pathway through which other allergic disorders may progress. The filaggrin protein plays an important role in maintaining skin barrier function, and it has been hypothesized that an *FLG* null mutation-induced defective skin barrier may be a route via which an individual can become sensitized [32]. In addition, skin inflammation that is usually characterized by upregulation of pro-inflammatory cytokines like IL-4, IL-13, IL-17, and IL-22 may downregulate filaggrin expression and degrade filaggrin [33–36].

19.5 What Would Be the Biological Plausibility for the Role of the *FLG* in Food Allergy or Food Sensitization Development?

It has been hypothesized that sensitization to environmental allergens may occur at the skin, while tolerance occurs via the gastrointestinal tract [32]. Since filaggrin defects can cause skin barrier impairment and impaired skin barrier may increase the risk of an individual being sensitized via cutaneous exposure, filaggrin defects may be a risk for food sensitization. Some studies have shown that *FLG* mutations increase the risk of total IgE levels or allergic sensitization (defined as the presence of specific IgE antibodies against common aeroallergens or food allergens) in the presence of eczema [18, 37–39]. The aeroallergens studied include grass, birch pollen, cat dander, dog, house dust mite, and mold. Marenholz et al. [18] included food allergens of egg, milk, wheat, and soy, while Morar et al. [38] also included egg white and cow's milk. *FLG* mutations were later found to confer an increased risk for allergic sensitization independently of eczema, as measured by skin prick tests to six aeroallergens (*Dermatophagoides pteronyssinus* (European house dust mite), *Dermatophagoides farinae* (American house dust mite), *Alternaria tenuis* (fungus), cat dander, mixed grass, and tree pollen), in a population study of schoolchildren in Germany with an odds ratio of 1.46 (95 % CI, 1.07–1.99) after adjusting for eczema [40]. Meta-analysis showed that the odds of developing allergic sensitization were 1.91 (95 % CI, 1.44–2.54) in the family studies and 1.57 (95 % CI, 1.20–2.07) in the case-control studies [41].

Besides the skin, filaggrin immune reactivity has been observed in orthokeratinized and parakeratinized areas of human oral epithelium (hard palate, buccal mucosa) and the conjunctiva [42–44]. However, immunohistochemical staining showed that it is absent in the upper and lower airway or esophageal epithelium [45, 46]. It remains to be elucidated at what point in the transition from oral to nasal cavity filaggrin immune reactivity is lost and the extent to which filaggrin is expressed in these areas [46]. More recently, filaggrin mRNA expression assessed by real-time PCR showed that it was downregulated 16-fold in esophageal biopsies of eosinophilic esophagitis patients compared with normal individuals [47], suggesting there is filaggrin expression in esophagus.

This suggests that disturbance of barrier function in the oral cavity may be the route of sensitization to food allergens, and it is possible that *FLG* mutations may indirectly play a role in the development of atopic disorders in distant organs [33]. It has also been proposed that increased allergen and pathogen penetration through the stratum corneum, followed by stimulation of keratinocyte-derived thymic stromal lymphopoietin (TSLP), an interleukin-7-like cytokine, in inflamed epidermis may lead to distal effects in the lung [48].

19.6 What Is the Evidence That the *FLG* Might Be Involved in Food Allergy Development?

The association between *FLG* mutations and food allergy has not been as well studied as for other allergic conditions. A study by Van Limbergen et al. found a 4.5-fold increased risk of food allergy in carriers of *FLG* mutations (R501X and 2282del4) in a cohort of children with pediatric inflammatory bowel disease (IBD) susceptibility and coexistent atopy (median age 11.3 years) [49]. The food allergy status was parent reported, and food sensitization status was unknown. Skin prick tests, food challenges, and specific IgE assays were not performed as food allergy was not part of the original study protocol [49].

Recently, a multisite case-control study of peanut allergy found that *FLG* null mutations (R501X and 2282del4) were associated with a significantly increased risk of peanut allergy with an overall odds ratio of 5.3 and a residual odds ratio of 3.8 when corrected for eczema [19]. Seventy-one English, Dutch, and Irish oral food challenged-positive patients with peanut allergy were compared to 1,000 non-peanut-sensitized English population controls. Additionally, 100 Dutch population control samples were obtained from adult blood donors, and 100 Irish adults population controls were obtained from a population-based biobank. The age of the participants in the study by Brown et al. was variable. English cases and controls were recruited from a birth cohort that was followed up for at least 7 years, but the time point of food challenge was not reported, while the age range for the Dutch and Irish patients was from 3 to 14 years (mean 7.5 years) and from 1 to 18 years (mean 10.5 years), respectively.

Since food sensitization must occur prior to the development of food allergy and is present concurrently with food allergy, the association between *FLG* mutations and peanut allergy in the Brown et al. noted above could be due to peanut sensitization. However, Brown et al. did not report the effect of sensitization on this relationship. In addition, eczema, food sensitization, and food allergy status were not known for all population controls, which undermined the quality of the comparison with a control group. Furthermore, only peanut status was assessed in their study, while egg allergy and cow's milk allergy are more prevalent in young infants [4, 50].

Another study reported that *FLG* null mutations (R501X or 2282del4) were significantly associated with self-reported food allergy and alcohol sensitivity, but not with oral allergy syndrome (OAS), supporting the role of skin barrier functions in the development of food allergy [51].

19.7 What Is the Evidence That It Has a Role in Food Sensitization But Not Food Allergy?

Using data from a population-based food allergy cohort of 1-year-old infants, the HealthNuts study assessed the relationship between *FLG* mutations, food sensitization, and food allergy with a view to more carefully dissecting the role of filaggrin in this step of the atopic march. With the availability of age-matched negative controls, food-sensitized tolerant, and food-sensitized allergic participants, all of whom underwent skin prick testing and food challenges to several major food allergens and were examined for eczema, the study was able to specifically investigate whether *FLG* mutations increased the risk of food allergy over and above the risk of food sensitization.

HealthNuts is a single-center, population-based, cross-sectional study of food allergy in 1-year-old infants using food challenges to confirm food allergy in Melbourne, Australia. Overall, 5,276 infants aged 11–15 months inclusive were recruited from council-run immunization sessions across Melbourne, Australia, between June 2008 and August 2011. Parents or guardians provided written informed consent and completed a detailed questionnaire. Skin prick testing (SPT) to four foods including egg white, peanut, sesame, and either shrimp or cow's milk was performed by a nurse. All infants with any detectable wheals to the foods greater than negative control were invited to attend the study clinic for repeat SPT to the same four foods as well as to an extended panel of food allergens: egg white, peanut, sesame, shrimp, cow's milk, cashew, almond, hazelnut, soy, and wheat. A formal oral food challenge (OFC) was offered to all clinic participants irrespective of wheal size or history of previous ingestion reaction unless the latter was recent (reacted in less than 1 month for egg; reacted in less than 2 months for peanut and/or sesame) and clearly objective (immediate urticaria, angioedema, vomiting, or anaphylaxis as per predetermined stopping criteria) [52]. Further random samples of community-recruited participants with negative SPT to all four foods were also invited for repeat SPT and OFC as negative controls. Blood samples were collected from participants who attended the clinic after OFC with the parent or guardian's written consent.

19.7.1 Association of *FLG* Mutations and Food Allergy Among Those Food Sensitized

Because those with IgE-mediated food allergy must have a positive IgE antibody test result, the same study analyzed whether having an *FLG* mutation increased the risk of food allergy over and above that of being food sensitized among those who were sensitized. Among food-sensitized infants, there was no significant difference in *FLG* mutations when comparing food-tolerant infants with those with food allergy (OR, 0.7; 95 % CI, 0.3–1.7; $p=0.478$; Table 19.1) [53].

In order to assess whether infants recruited without a screening SPT to cow's milk might have influenced the association, a sensitivity analysis on a subset of participants recruited after introducing cow's milk SPT screening in the community was undertaken. This sensitivity analysis showed stronger associations with food sensitization and food allergy, but *FLG* mutations still did not further increase the risk of food allergy over food sensitization, which is in concordance with the above findings. Specific analyses by individual foods are described below.

19.7.2 Egg Sensitization and Egg Allergy

When analyzing food-specific sensitization, *FLG* mutations were significantly associated with having any egg sensitization (aOR, 3.1, 95 % CI, 1.0–9.0; $p=0.042$) and any egg allergy (aOR, 3.0; 95 % CI, 1.0–9.1; $p=0.046$). However, when the association with egg allergy was further analyzed among egg-sensitized infants, *FLG*

Table 19.1 Number of carriers of *FLG* mutations in negative controls, food sensitized, food sensitized but tolerant, and food sensitized and allergic infants

Group[a]		Negative control (*N*=126)	Any food sensitization (*N*=428)	Food sensitized but tolerant[b] (*N*=45)	Food sensitized and allergic[b] (*N*=321)
No *FLG* mutation	*n* (%)	121 (96.0)	374 (87.4)	38 (84.4)	283 (88.2)
FLG mutation	*n* (%)	5 (4.0)	54 (12.6)	7 (15.6)	38 (11.8)
Comparing negative control to other groups	OR (95 % CI)	1.0 (reference)	3.5 (1.4–8.9)	4.5 (1.3–14.9)	3.2 (1.2–8.5)
	p-value	–	0.009	0.015	0.016
	Adjusted[c] OR (95 % CI)	1.0 (reference)	3.0 (1.0–8.7)	5.0 (1.4–18.6)	2.9 (1.0–8.6)
	Adjusted[c] *p*-value	–	0.043	0.015	0.055
Comparing food sensitized but tolerant to food allergic	OR (95 % CI)			1.0 (reference)	0.7 (0.3–1.7)
	p-value			–	0.478

Reprinted from Tan et al. [53], with permission from Elsevier

[a]Infants with detectable wheal size during recruitment but had SPT < 2 mm in study clinic and infants invited to participate as negative control (no wheal size during recruitment) but had detectable wheal size in study clinic were excluded from the analysis (*n* = 146)

[b]These groups are subgroups of "any food sensitization" group. All "food sensitized but tolerant" infants were challenge-confirmed. Infants with SPT ≥ 2 mm to any of the ten foods tested who did not have OFC outcome (inconclusive or OFC not offered for the seven other foods and SPT <8 mm) were excluded from "food sensitized but tolerant" group (*n* = 62)

[c]Adjusted for presence of eczema

mutations were not associated with an increased risk of egg allergy among egg-sensitized infants (OR, 0.8; 95 % CI, 0.3–2.0; $p=0.645$). Again, *FLG* mutations showed stronger association with the egg-sensitized but tolerant group (aOR, 4.4; 95 % CI, 1.1–17.2; $p=0.034$).

19.7.3 Peanut Sensitization and Peanut Allergy

There was a trend towards association of *FLG* mutations with having any peanut sensitization after adjusting for eczema (aOR, 3.0 95 % CI, 0.9–9.6; $p=0.060$) but not with any peanut allergy (aOR, 1.8; 95 % CI, 0.5–7.0; $p=0.383$). The association with peanut sensitized but tolerant was significant (aOR, 4.9; 95 % CI, 1.4–17.0; $p=0.011$), where peanut-allergic participants were excluded from this group. Among peanut-sensitized infants, there was no significant difference in *FLG* mutations when comparing peanut-tolerant infants to peanut-allergic infants (OR, 0.5 95 % CI, 0.2–1.3; $p=0.140$).

19.7.4 Sesame Sensitization and Sesame Allergy

FLG mutations were not associated with having any sesame sensitization (aOR, 2.1, 95 % CI, 0.4–10.4; $p=0.385$), but there was a trend for association with sesame-sensitized but tolerant group, although the association was not statistically significant (aOR, 4.5; 95 % CI, 0.8–24.2; $p=0.082$). *FLG* mutations were not associated with having any sesame allergy (OR, 1.1; 95 % CI, 0.1–9.9; $p=0.932$) (this analysis could not be adjusted for eczema because all infants with sesame allergy had eczema). Among sesame-sensitized infants, there was no significant difference in *FLG* mutations when comparing sesame-tolerant infants to sesame-allergic infants (OR, 0.3 95 % CI, 0.0–2.9; $p=0.278$).

19.7.5 Cow's Milk Allergy

Cow's milk sensitization and allergy are of great interest to many as cow's milk is one of the major food allergens in infants. The data on cow's milk sensitization and allergy were included in the food-sensitized and food-allergic groups. Ethics approval was not granted for cow's milk OFC as part of the original study design, and screening SPT for cow's milk during recruitment was only performed on the participants recruited mid-study after approval was obtained. Those with cow's milk SPT$\geq$8 mm measured in clinic were categorized as allergic ($n=3$). Since the numbers were too small to be sufficiently powered for separate analysis, they were analyzed with other foods collectively.

Among those who came into the clinic, the number of cow's milk-sensitized infants that had egg sensitization was 67/76 (88.2 %), showing a high rate of overlap between these two conditions and therefore a high pickup rate of cow's milk sensitization from participants recruited into the clinic with egg SPT. However, some cases would have been missed from being recruited since cow's milk SPT was not available during recruitment screening in the first half of the study. Therefore, a sensitivity analysis on a subset (second half of the study from which population data for both egg and cow's milk sensitization are available) of participants recruited after introducing cow's milk SPT screening in the community was undertaken to assess whether infants recruited without a screening SPT to cow's milk might have influenced the association. This sensitivity analysis showed statistically significant association with food allergy; however, further analyses on the data showed that among those who were already food-sensitized, there was no association between *FLG* mutations and food allergy when compared to those who were food-sensitized but tolerant, consistent with the unrestricted analysis.

Taken together, these results suggest that *FLG* mutations may predispose to the development of food sensitization but additional factors are required to convert from clinically asymptomatic food sensitization to food allergy, and *FLG* mutations do not play a role in this conversion. Food-specific analysis for egg, peanut, and sesame showed similar trends of association as all foods tested combined; where after adjusting for eczema, there were significant associations between having a *FLG* mutation and egg-sensitized but tolerant and peanut-sensitized but tolerant groups, while there was a similar trend, although not statistically significant, for the sesame-sensitized but tolerant group, probably due to lack of statistical power. In all the food-specific analysis, *FLG* mutations did not further increase the risk of the specific food allergy among those sensitized.

The HealthNuts study was unique in that it was able to examine challenge-proven food allergy within a group of food sensitized (to the four major allergens) individuals and was able to specifically investigate whether *FLG* mutations increase the risk of food allergy over and above the risk of food sensitization. Using a case-control study design, Brown et al. previously identified that *FLG* mutations confer an increased risk of peanut allergy [19]. However, they were unable to interrogate the effect of sensitization on this relationship, as there was no comparison group of sensitized tolerant participants. In addition, in the study by Brown et al., not all eczema, food sensitization, and food allergy status data were available for population controls, which undermined the quality of the comparison with a control group. The age of participants in the study by Brown et al. was variable: the English patients and controls were recruited from a birth cohort that was followed up for at least 7 years but the time point of food challenge was not reported, while the age range for the Dutch and Irish patients were from 3 to 14 years (mean 7.5 years) and from 1 to 18 years (mean 10.5 years), respectively. Furthermore, only peanut status was assessed in their study, and it is not known whether the patients have multiple food sensitization or allergy such as egg allergy, which occurs more commonly in infants than peanut allergy.

One of the limitations of the HealthNuts study is potential participation bias among negative controls as a higher than expected number of participants that volunteered as negative controls had an immediate family history of asthma, eczema, hay fever, or food allergy. However, the selection of negative controls at high risk of allergic disease may in fact strengthen the findings as those with a family history of allergic disease could be expected to have a higher prevalence of *FLG* mutations than the general population (i.e., such a bias would be more likely to create a false-negative than a false-positive association). Despite this, *FLG* mutations were still more common in the food-sensitized and food-allergic groups compared with negative controls.

Another limitation is the young age of the children in the HealthNuts study. It is possible that some children might develop food allergy as they get older and this study, by looking at only 1-year-olds, might be missing a stronger association that would appear with time. However, most childhood food allergy is early onset at the time of first exposure. Since all children at risk of food allergy received a formal oral food challenge, it is highly likely that the HealthNuts data uncovered the majority of clinically relevant food allergy phenotypes.

A very recent study on Japanese infants ($n=116$) supported the role of *FLG* mutations in food sensitization, where common *FLG* variants in the Japanese population (R501X, 3321delA, S1695X, Q1701X, S2554X, S2889X, S3296X, and K4022X) were associated with having any food sensitization to egg white, cow's milk, wheat, soybean, and/or peanut ($p=0.039$) [54]. In addition, this study also found that a minor allele of rs1933064, a single nucleotide polymorphism near *FLG*, is protective for food sensitization independently of atopic eczema, although how this works remains unknown [54].

19.8 What Is the Potential Biological Role of the *FLG* Playing in Food Allergy Development?

Skin barrier impairment measured based on transepidermal water loss (TEWL) can precede clinical manifestation of eczema in infants, as elevated TEWL was previously reported in *FLG* mutation carriers unaffected by eczema in 3-month-old infants [55]. Therefore, *FLG* mutations might increase the risk of food sensitization through impaired skin barrier function caused by filaggrin deficiency regardless of eczema status. This also supports the hypothesis that sensitization to food allergen can occur through the skin, but not necessarily through inflamed skin [32] (Fig. 19.2). In addition, sensitization has been found to precede the development of eczema in some infants, while eczema precedes the development of sensitization in others [56].

The HealthNuts work above supports Lack's hypothesis that there could be a different pathogenesis for food sensitization and food allergy. That is, the epidermal barrier dysfunction due to filaggrin deficiency might cause food sensitization regardless of clinical eczema status, but filaggrin-induced skin barrier dysfunction might not play a further role in the progression to food allergy from sensitization. These results therefore suggest that although filaggrin deficiency might provide a mechanism for the development of sensitization, a second factor (or factors), either environmental or genetic (or both), is important for converting food-sensitized infants to food-allergic status. There are many potential interactions between genes and environmental factors, including the role of epigenetics, which studies how environmental factors affect gene expression that may shed light on the development of food allergy. Further research into understanding the development of food allergy or food tolerance could also focus on other factors such as those that regulate immune cells or gut permeability.

19.9 What Is the Evidence That Epigenetic Changes in the *FLG* Might Be Important in the Development of Food Allergy?

Environmental factors with a long-lasting impact on phenotype expression are likely to occur through epigenetic changes to the genome [57]. Epigenetics refers to the "structural adaptations of chromosomal regions so as to register, signal, or perpetuate altered activity states" [58]. Epigenetic mechanisms regulate gene expression changes accompanying cell differentiation and represent a major mechanism underpinning the majority of environmentally mediated changes to gene expression. The epigenome can change and adapt to environmental stimuli over a relatively short timescale and is also

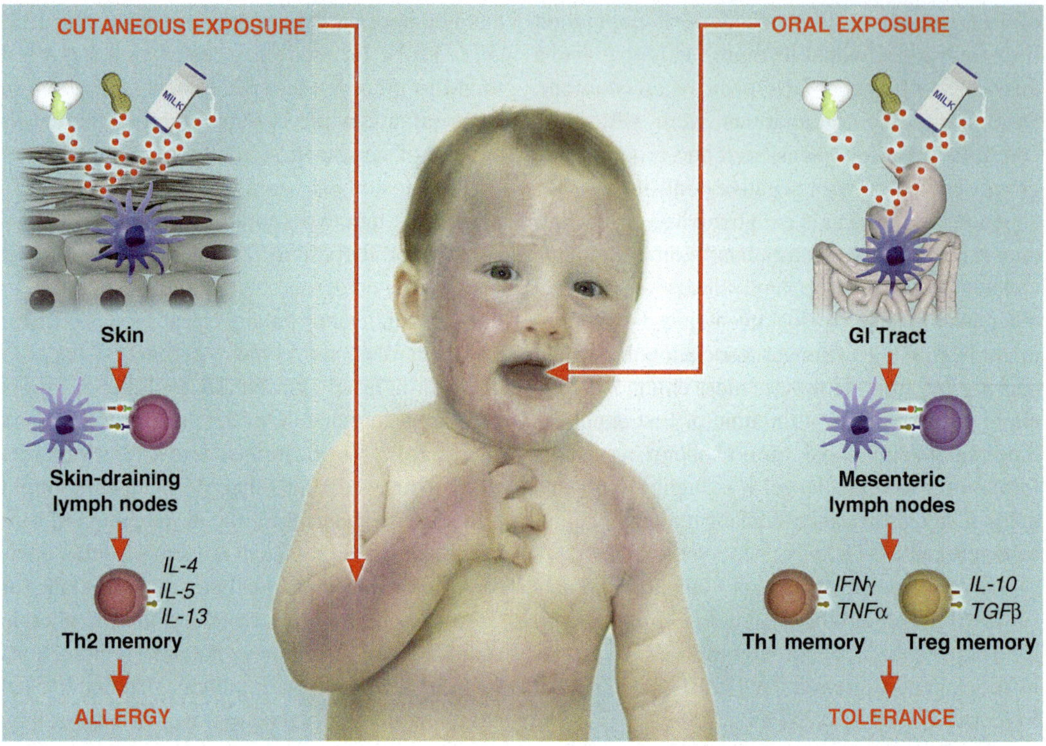

Fig. 19.2 Dual-allergen-exposure hypothesis for pathogenesis of food allergy. Tolerance occurs as a result of oral exposure to food, and allergic sensitization results from cutaneous exposure. *GI* gastrointestinal (Reprinted from Lack [32], with permission from Elsevier)

subject to epigenetic "drift" over the life course in response to both environmental and stochastic factors [59]. Epigenetic alterations can influence gene expression differentially throughout the lifespan as they are believed to occur not only prenatally or shortly after birth but also during later developmental periods. This may help to explain later disease development associated with early exposure [60].

Mounting evidence suggests that early development (in utero and early postnatal) represents an especially sensitive time for epigenetic disruption [58]. One of the most studied maternal environments with regard to epigenetics is diet, which may be an important factor in the development of food allergy. There is now evidence that nutrition is an important environmental modifier of methylation. Low methionine intake leads to changes in methylation in mice and humans, while increased consumption of folic acid, vitamin B12, choline, and betaine during pregnancy has been shown to increase methylation in mice offspring, affecting the expression of a coat color gene [61]. Besides methyl donors, other dietary factors like polyphenols, isothiocyanates, genistein, selenium, and butyrate, which mediate effects independent of folate metabolism, may also influence epigenetic patterns [62, 63]. Famine during the prenatal period leads to methylation changes in the insulin-like growth factor 2 (IGF2) gene, which persist into adulthood [64]. Methylation is also associated with smoking [65, 66], another important modifiable lifestyle factor. In children exposed to maternal smoking in utero, global methylation levels are decreased [67], with differences in global methylation being associated with the glutathione S-transferase mu 1 (GSTM1) genotype.

Unfortunately, there are no studies to date that have formally investigated the role of epigenetics in food allergy and in particular no investigations into the role of epigenetic changes in the *FLG* and its impact on allergic disease development. Many believe that the disruption of appropriate allergic pathways such as Th cell differentiation (and the corresponding cytokine expression profile) in the early postnatal period represents an important potential pathway to the development of allergy. Several studies have demonstrated that DNA methylation and the resulting changes in chromatin structure are the primary regulators of naïve Th cell differentiation. At birth, there is a skew towards a pro-allergic Th2 profile with its associated cytokine profile (including IL-4) [68], but this normally does not persist, with increasing differentiation of naive Th0 cells down the Th1 lineage (with a concomitant increasing expression of IFN-gamma). This is associated with an increasing level of methylation of the IL-4 gene and loss of methylation at IFN-gamma regulatory regions in Th1 cells. Conversely, methylation of the IFN-gamma gene attenuates transcription and is associated with Th2 polarization [69].

19.10 The Future

In summary, *FLG* mutations increase the risk of food sensitization but do not further increase the risk of food allergy over and above that of food sensitization among 1-year-old infants. However, with the high co-manifestation between eczema and food sensitization, the independence of the association of *FLG* mutations and food sensitization from the association with eczema could not be established. These results support the biologically plausible hypothesis that decreased skin barrier function increases the risk of food sensitization in early life, but other as yet undetermined factors are important in the conversion from food sensitization to food allergy. Other genomic changes that can affect filaggrin production might also be potential candidates involved in the development of food allergy, for example, changes to DNA methylation that may impact gene expression.

References

1. Ortolani C, Pastorello EA. Food allergies and food intolerances. Best Pract Res Clin Gastroenterol. 2006;20(3):467–83.
2. Wuthrich B. Food allergy, food intolerance or functional disorder? Praxis (Bern 1994). 2009;98(7):375–87.
3. Gupta RS, Springston EE, Warrier MR, Smith B, Kumar R, Pongracic J, et al. The prevalence, severity, and distribution of childhood food allergy in the United States. Pediatrics. 2011;128(1):e9–17.
4. Osborne NJ, Koplin JJ, Martin PE, Gurrin LC, Lowe AJ, Matheson MC, et al. Prevalence of challenge-proven IgE-mediated food allergy using population-based sampling and predetermined challenge criteria in infants. J Allergy Clin Immunol. 2011;127(3):668–76.e1–2.
5. Strachan DP. Family size, infection and atopy: the first decade of the "hygiene hypothesis". Thorax. 2000;55 Suppl 1:S2–10.
6. Black PN, Sharpe S. Dietary fat and asthma: is there a connection? Eur Respir J. 1997;10(1):6–12.
7. Robertson CF, Roberts MF, Kappers JH. Asthma prevalence in Melbourne schoolchildren: have we reached the peak? Med J Aust. 2004;180(6):273–6.
8. Koplin JJ, Dharmage SC, Ponsonby AL, Tang ML, Lowe AJ, Gurrin LC, et al. Environmental and demographic risk factors for egg allergy in a population-based study of infants. Allergy. 2012;67(11):1415–22.
9. Koplin JJ, Allen KJ. Optimal timing for solids introduction ? Why are the guidelines always changing? Clin Exp Allergy. 2013;43:826–34.
10. Allen KJ, Koplin JJ, Ponsonby AL, Gurrin LC, Wake M, Vuillermin P, et al. Vitamin D insufficiency is associated with challenge-proven food allergy in infants. J Allergy Clin Immunol. 2013;131(4):1109–16.e6.
11. Tan TH, Ellis JA, Saffery R, Allen KJ. The role of genetics and environment in the rise of childhood food allergy. Clin Exp Allergy. 2012;42(1):20–9.
12. Allen KJ, Koplin JJ. The epidemiology of IgE-mediated food allergy and anaphylaxis. Immunol Allergy Clin North Am. 2012;32(1):35–50.
13. Hadley C. Food allergies on the rise? Determining the prevalence of food allergies, and how quickly it is increasing, is the first step in tackling the problem. EMBO Rep. 2006;7(11):1080–3.
14. Branum AM, Lukacs SL. Food allergy among U.S. children: trends in prevalence and hospitalizations. NCHS Data Brief. 2008;(10):1–8.
15. Sicherer SH, Furlong TJ, Maes HH, Desnick RJ, Sampson HA, Gelb BD. Genetics of peanut allergy: a twin study. J Allergy Clin Immunol. 2000;106(1 Pt 1):53–6.
16. Hill DJ, Hosking CS, de Benedictis FM, Oranje AP, Diepgen TL, Bauchau V, et al. Confirmation of the association between high levels of immunoglobulin E food sensitization and eczema in infancy: an international study. Clin Exp Allergy. 2008;38(1):161–8.
17. Irvine AD. Fleshing out filaggrin phenotypes. J Invest Dermatol. 2007;127(3):504–7.
18. Marenholz I, Nickel R, Rüschendorf F, Schulz F, Esparza-Gordillo J, Kerscher T, et al. Filaggrin loss-of-function mutations predispose to phenotypes involved in the atopic march. J Allergy Clin Immunol. 2006;118(4):866–71.
19. Brown SJ, Asai Y, Cordell HJ, Campbell LE, Zhao Y, Liao H, et al. Loss-of-function variants in the filaggrin gene are a significant risk factor for peanut allergy. J Allergy Clin Immunol. 2011;127(3):661–7.
20. Lack G, Fox D, Northstone K, Golding J. Avon Longitudinal Study of Parents and Children Study

Team. Factors associated with the development of peanut allergy in childhood. N Engl J Med. 2003;348(11):977–85.
21. Fox AT, Sasieni P, du Toit G, Syed H, Lack G. Household peanut consumption as a risk factor for the development of peanut allergy. J Allergy Clin Immunol. 2009;123(2):417–23.
22. Strid J, Hourihane J, Kimber I, Callard R, Strobel S. Epicutaneous exposure to peanut protein prevents oral tolerance and enhances allergic sensitization. Clin Exp Allergy. 2005;35(6):757–66.
23. Leung RC, Carlin JB, Burdon JG, Czarny D. Asthma, allergy and atopy in Asian immigrants in Melbourne. Med J Aust. 1994;161(7):418–25.
24. Sicherer SH, Wood RA, Stablein D, Lindblad R, Burks AW, Liu AH, et al. Maternal consumption of peanut during pregnancy is associated with peanut sensitization in atopic infants. J Allergy Clin Immunol. 2010;126(6):1191–7.
25. Dharmage SC, Martin PE, Osborne NJ, Koplin JJ, Gurrin LC, Ponsonby A, et al. The epidemiology of food sensitization-associated eczema in infancy in HealthNuts, a population-based study. J Allergy Clin Immunol. 2011;127(2):AB35.
26. Martin PE, Koplin JJ, Eckert JK, Lowe AJ, Ponsonby A, Osborne NJ, et al. The prevalence and socio-demographic risk factors of clinical eczema in infancy: a population-based observational study. Clin Exp Allergy. 2013;43:642–51.
27. Branum AM, Lukacs SL. Food allergy among children in the United States. Pediatrics. 2009;124(6):1549–55.
28. Le Souef PN, Goldblatt J, Lynch NR. Evolutionary adaptation of inflammatory immune responses in human beings. Lancet. 2000;356(9225):242–4.
29. Spergel JM. From atopic dermatitis to asthma: the atopic march. Ann Allergy Asthma Immunol. 2010;105(2):99–106; quiz 107–9, 117.
30. Hill DJ, Heine RG, Hosking CS, Brown J, Thiele L, Allen KJ, et al. IgE food sensitization in infants with eczema attending a dermatology department. J Pediatr. 2007;151(4):359–63.
31. Irvine AD, McLean WH, Leung DY. Filaggrin mutations associated with skin and allergic diseases. N Engl J Med. 2011;365(14):1315–27.
32. Lack G. Epidemiologic risks for food allergy. J Allergy Clin Immunol. 2008;121(6):1331–6.
33. Howell MD, Kim BE, Gao P, Grant AV, Boguniewicz M, Debenedetto A, et al. Cytokine modulation of atopic dermatitis filaggrin skin expression. J Allergy Clin Immunol. 2007;120(1):150–5.
34. Gutowska-Owsiak D, Schaupp AL, Salimi M, Taylor S, Ogg GS. Interleukin-22 downregulates filaggrin expression and affects expression of profilaggrin processing enzymes. Br J Dermatol. 2011;165(3):492–8.
35. Gutowska-Owsiak D, Schaupp AL, Salimi M, Selvakumar TA, McPherson T, Taylor S, et al. IL-17 downregulates filaggrin and affects keratinocyte expression of genes associated with cellular adhesion. Exp Dermatol. 2012;21(2):104–10.
36. Pellerin L, Henry J, Hsu CY, Balica S, Jean-Decoster C, Mechin MC, et al. Defects of filaggrin-like proteins in both lesional and nonlesional atopic skin. J Allergy Clin Immunol. 2013;131(4):1094–102.
37. Weidinger S, Illig T, Baurecht H, Irvine AD, Rodriguez E, Diaz-Lacava A, et al. Loss-of-function variations within the filaggrin gene predispose for atopic dermatitis with allergic sensitizations. J Allergy Clin Immunol. 2006;118(1):214–9.
38. Morar N, Cookson WO, Harper JI, Moffatt MF. Filaggrin mutations in children with severe atopic dermatitis. J Invest Dermatol. 2007;127(7):1667–72.
39. Rogers AJ, Celedon JC, Lasky-Su JA, Weiss ST, Raby BA. Filaggrin mutations confer susceptibility to atopic dermatitis but not to asthma. J Allergy Clin Immunol. 2007;120(6):1332–7.
40. Weidinger S, O'Sullivan M, Illig T, Baurecht H, Depner M, Rodriguez E, et al. Filaggrin mutations, atopic eczema, hay fever, and asthma in children. J Allergy Clin Immunol. 2008;121(5):1203–9.e1.
41. van den Oord RA, Sheikh A. Filaggrin gene defects and risk of developing allergic sensitisation and allergic disorders: systematic review and meta-analysis. BMJ. 2009;339:b2433.
42. Smith SA, Dale BA. Immunologic localization of filaggrin in human oral epithelia and correlation with keratinization. J Invest Dermatol. 1986;86(2):168–72.
43. Reibel J, Clausen H, Dale BA, Thacher SM. Immunohistochemical analysis of stratum corneum components in oral squamous epithelia. Differentiation. 1989;41(3):237–44.
44. Krenzer KL, Freddo TF. Cytokeratin expression in normal human bulbar conjunctiva obtained by impression cytology. Invest Ophthalmol Vis Sci. 1997;38(1):142–52.
45. Ying S, Meng Q, Corrigan CJ, Lee TH. Lack of filaggrin expression in the human bronchial mucosa. J Allergy Clin Immunol. 2006;118(6):1386–8.
46. De Benedetto A, Qualia CM, Baroody FM, Beck LA. Filaggrin expression in oral, nasal, and esophageal mucosa. J Invest Dermatol. 2008;128(6):1594–7.
47. Blanchard C, Stucke EM, Burwinkel K, Caldwell JM, Collins MH, Ahrens A, et al. Coordinate interaction between IL-13 and epithelial differentiation cluster genes in eosinophilic esophagitis. J Immunol. 2010;184(7):4033–41.
48. Ziegler SF, Artis D. Sensing the outside world: TSLP regulates barrier immunity. Nat Immunol. 2010;11(4):289–93.
49. Van Limbergen J, Russell RK, Nimmo ER, Zhao Y, Liao H, Drummond HE, et al. Filaggrin loss-of-function variants are associated with atopic comorbidity in pediatric inflammatory bowel disease. Inflamm Bowel Dis. 2009;15(10):1492–8.
50. Sicherer SH, Sampson HA. Food allergy. J Allergy Clin Immunol. 2010;125(2 Suppl 2):S116–25.
51. Linneberg A, Fenger RV, Husemoen LL, Thuesen BH, Skaaby T, Gonzalez-Quintela A, et al. Association between loss-of-function mutations in the filaggrin gene

and self-reported food allergy and alcohol sensitivity. Int Arch Allergy Immunol. 2013;161(3):234–42.

52. Koplin JJ, Tang ML, Martin PE, Osborne NJ, Lowe AJ, Ponsonby AL, et al. Predetermined challenge eligibility and cessation criteria for oral food challenges in the HealthNuts population-based study of infants. J Allergy Clin Immunol. 2012;129(4):1145–7. Epub 2011 Nov 12.
53. Tan HT, Ellis JA, Koplin JJ, Matheson MC, Gurrin LC, Lowe AJ, et al. Filaggrin loss-of-function mutations do not predict food allergy over and above the risk of food sensitization among infants. J Allergy Clin Immunol. 2012;130(5):1211–3.e3.
54. Nomura T, Tsuge I, Inuo C, Nakajima Y, Kondo Y, Sugiura S, et al. Food sensitization in Japanese infants is associated with a common Filaggrin variant. Ann Allergy Asthma Immunol. 2013;110(5):388–90.e1.
55. Flohr C, England K, Radulovic S, McLean WH, Campbel LE, Barker J, et al. Filaggrin loss-of-function mutations are associated with early-onset eczema, eczema severity and transepidermal water loss at 3 months of age. Br J Dermatol. 2010;163(6):1333–6.
56. Lowe AJ, Abramson MJ, Hosking CS, Carlin JB, Bennett CM, Dharmage SC, et al. The temporal sequence of allergic sensitization and onset of infantile eczema. Clin Exp Allergy. 2007;37(4):536–42.
57. Vercelli D. Genetics, epigenetics, and the environment: switching, buffering, releasing. J Allergy Clin Immunol. 2004;113(3):381–6; quiz 387.
58. Bird A. Perceptions of epigenetics. Nature. 2007; 447(7143):396–8.
59. Jirtle RL, Skinner MK. Environmental epigenomics and disease susceptibility. Nat Rev Genet. 2007;8(4):253–62.
60. Miller RL, Ho SM. Environmental epigenetics and asthma: current concepts and call for studies. Am J Respir Crit Care Med. 2008;177(6):567–73.
61. Waterland RA, Jirtle RL. Transposable elements: targets for early nutritional effects on epigenetic gene regulation. Mol Cell Biol. 2003;23(15): 5293–300.
62. Johnson IT, Belshaw NJ. Environment, diet and CpG island methylation: epigenetic signals in gastrointestinal neoplasia. Food Chem Toxicol. 2008;46(4): 1346–59.
63. de Haan JB, Gevers W, Parker MI. Effects of sodium butyrate on the synthesis and methylation of DNA in normal cells and their transformed counterparts. Cancer Res. 1986;46(2):713–6.
64. Heijmans BT, Tobi EW, Stein AD, Putter H, Blauw GJ, Susser ES, et al. Persistent epigenetic differences associated with prenatal exposure to famine in humans. Proc Natl Acad Sci U S A. 2008;105(44):17046–9.
65. Belinsky SA. Gene-promoter hypermethylation as a biomarker in lung cancer. Nat Rev Cancer. 2004;4(9):707–17.
66. Vaissiere T, Hung RJ, Zaridze D, Moukeria A, Cuenin C, Fasolo V, et al. Quantitative analysis of DNA methylation profiles in lung cancer identifies aberrant DNA methylation of specific genes and its association with gender and cancer risk factors. Cancer Res. 2009;69(1):243–52.
67. Breton CV, Byun HM, Wenten M, Pan F, Yang A, Gilliland FD. Prenatal tobacco smoke exposure affects global and gene-specific DNA methylation. Am J Respir Crit Care Med. 2009;180(5):462–7.
68. Agarwal S, Rao A. Modulation of chromatin structure regulates cytokine gene expression during T cell differentiation. Immunity. 1998;9(6):765–75.
69. Jones B, Chen J. Inhibition of IFN-gamma transcription by site-specific methylation during T helper cell development. EMBO J. 2006;25(11): 2443–52.

Filaggrin in Canine Skin

20

Gunnaporn Suriyaphol, Sirin Theerawatanasirikul, and Prapat Suriyaphol

Contents

G. Suriyaphol, DVM, MSc, Dr. rer. nat. (✉)
Biochemistry Unit, Department of Veterinary Physiology, Faculty of Veterinary Science, Chulalongkorn University, 39 Henri Dunant Rd., Pathumwan, Bangkok 10330, Thailand
e-mail: gunnaporn.v@chula.ac.th

S. Theerawatanasirikul, DVM, PhD
Department of Anatomy, Faculty of Veterinary Medicine, Kasetsart University, 50 Ngamwongwan Rd., Ladyaow, Chatuchak, Bangkok 10900, Thailand
e-mail: fvetsrth@ku.ac.th

P. Suriyaphol, BSc, PhD
Bioinformatics and Data Management for Research Unit, Office for Research and Development, Faculty of Medicine, Siriraj Hospital, Mahidol University, 2 Prannok Rd., Bangkoknoi, Bangkok 10700, Thailand
e-mail: prapat.sur@mahidol.ac.th

20.1 Introduction

The study of filaggrin in dogs (canine filaggrin) began by immunohistochemistry (IHC) staining in the stratum granulosum (S. granulosum) of circumanal glands, identified by a monoclonal anti-human filaggrin antibody (Ab) in 1998 [1]. Since then, there have been no publications on canine filaggrin until the "boom" of filaggrin gene (*FLG*) studies in humans following the discovery of *FLG* mutations association with human atopic dermatitis (HAD) [2]. HAD shares several features with the second most common skin disease in dog, canine atopic dermatitis (CAD). CAD is a "genetically predisposed inflammatory and pruritic allergic skin disease with characteristic clinical features associated with IgE Abs most commonly directed against environmental allergens" [3]; approximately 10 % of the canine population is affected [4]. CAD and HAD have many analogous aspects such as young age of onset (1–3 years in dogs and <1–5 years in humans) and affected skin areas on the face and in the skin folds, as well as acute and chronic skin lesions with secondary infection [5, 6]. In addition, the emergence of the predicted *cFLG* sequence in canine genome database (canFam2 May 2005) makes the study of *cFLG* for the gene mutation and gene expression at the mRNA and protein levels of great interest.

In this chapter, we will discuss the predicted canine *FLG* (*cFLG*) and protein structures in dogs and compare them with those reported in

J.P. Thyssen, H. Maibach (eds.), *Filaggrin*,
DOI 10.1007/978-3-642-54379-1_20, © Springer-Verlag Berlin Heidelberg 2014

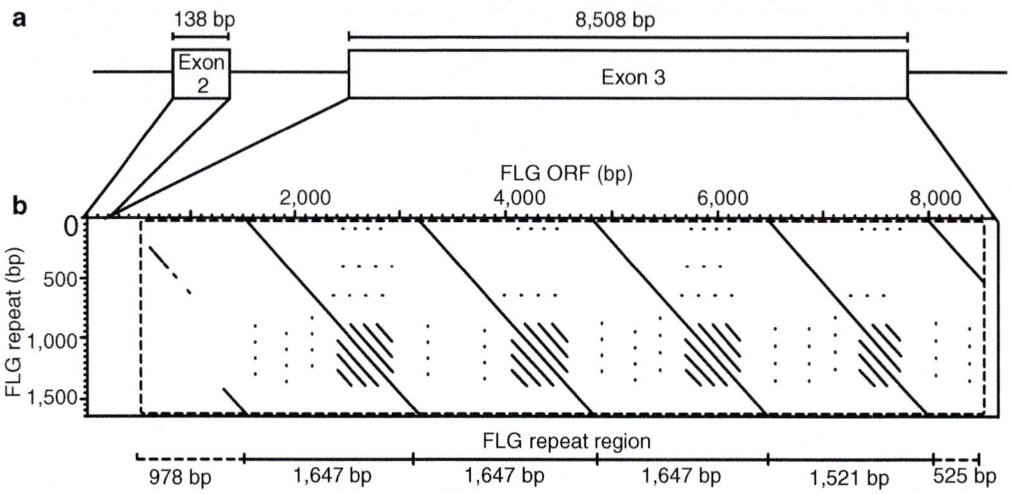

Fig. 20.1 Features of the canine filaggrin gene (*cFLG*). (**a**) Deduced structure of *cFLG*. (**b**) The dot-matrix analysis between the canine pro-filaggrin coding sequence and the *FLG* repeat 1 sequence (Adapted with permission from Kanda et al. [9]. © 2013 The Authors. Veterinary Dermatology © 2013)

humans (*hFLG*) and mice *FLG* (*mFLG*). Filaggrin expression at the mRNA and protein levels in normal dogs, dogs with CAD, and dogs with *FLG* mutations and CAD will be reviewed. In addition, the challenge of studying filaggrin in dogs, such as breed effects and the uncertain annotation of *FLG*, will be included.

20.2 Canine Filaggrin Gene and Protein Structures

From the open access draft Ensembl canine genome database (CGD), Canis_familiaris LATESTGP genome database [7], which is based on the whole genome shotgun (WGS) assembly canFam2, May 2005 [8], the canine *FLG* sequence (ENSCAFG00000023034) was predicted to be on chromosome 17 at locations 64,270,430–64,305,299 (34,870 bp). The latest work of *FLG* structure was performed by Kanda et al. [9]. Based on the same WGS assembly canFam2, the *FLG* structure was annotated to consist of three exons. Exon 1 is a noncoding exon similar to those of humans and mice. Exons 2 and 3 consist of 138 and 8,508 bp, respectively. Four units of the *FLG* repeat regions, existing on exon 3, are flanked by two incomplete repeats (Fig. 20.1) [9]. We manually realigned a part of the sequence, and we observed a 126 bp gap that did not appear in this CGD. The gap started at location 64,297,250 followed by locations 64,297,125–64,297,249 on chromosome 17 (Fig. 20.2) [10]. The gap was finally found in the latest canFam3 with 100 % homology at the locations 61,219,675–61,219,800, which is in the *FLG2* (ENSCAFT00000049887).

For the protein sequence, canine pro-filaggrin was predicted to consist of 2,882 amino acids with three complete filaggrin monomers (549 amino acids or 59 kDa) and one incomplete unit (507 amino acids or 54 kDa), flanked by truncated monomers (Fig. 20.3).

Using CLUSTAL W 1.81 and SCRATCH programs [11, 12], we aligned the pro-filaggrin amino acid sequences among different species, and we found that approximately 86 amino acids at the N-terminal were shown to be highly conserved among dogs, humans, mice, and rats with similar helical structure regions (Fig. 20.4). Sequence analysis revealed a calcium-binding domain at the N-terminal with high homology to the S100 family of calcium-binding proteins [13, 14]. This domain is probably involved in calcium signal transduction in keratinocyte differentiation or in providing calcium to calcium-dependent

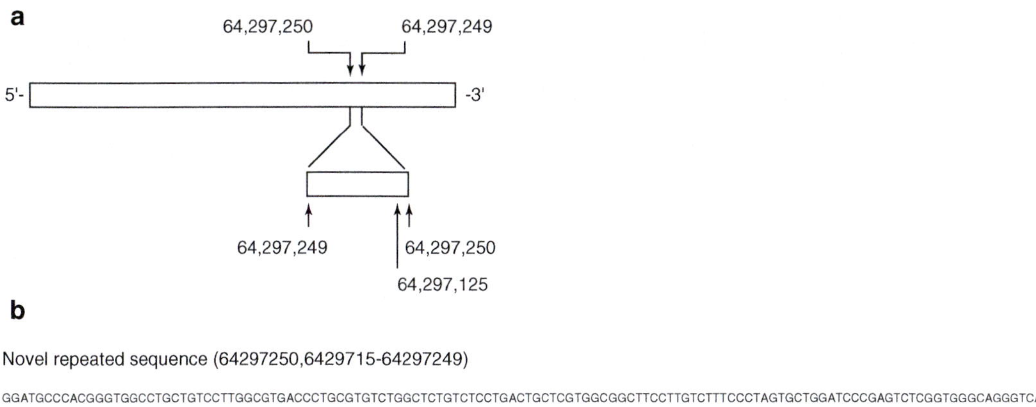

Fig. 20.2 Novel repeated sequence of canine *FLG* (*cFLG*). (**a**) Minus strand of *cFLG* (canFam2). (**b**) Nucleotide sequence of a new repeat (Reprinted with permission from Suriyaphol et al. [10]. Copyright © 2013 TJVM)

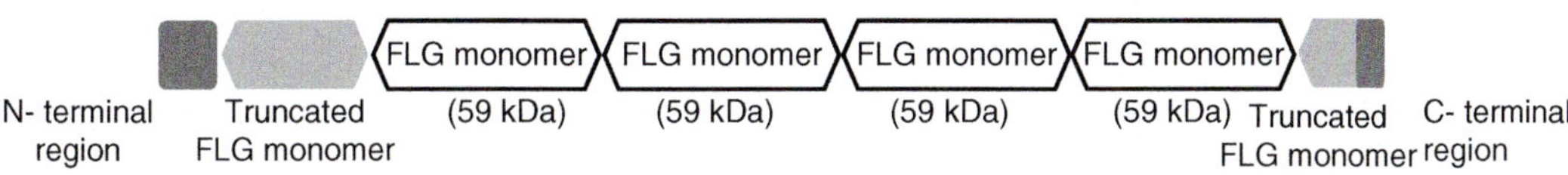

Fig. 20.3 Structure of the canine pro-filaggrin protein (Adapted with permission from Kanda et al. [9]. © 2013 The Authors. Veterinary Dermatology © 2013)

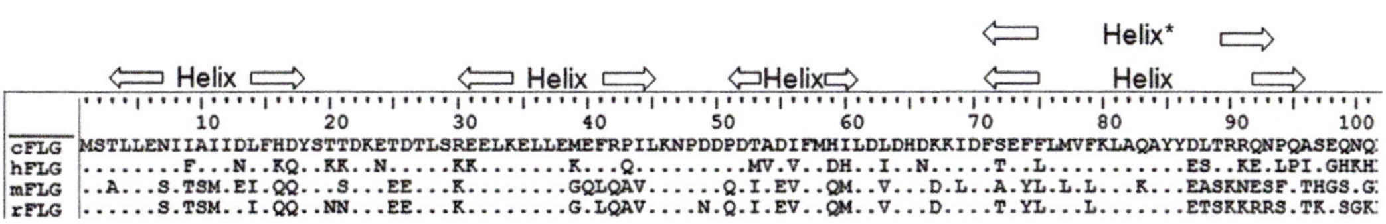

Fig. 20.4 Alignment of amino acid sequences of canine filaggrin (*cFLG*), human filaggrin (*hFLG*), mouse filaggrin (*mFLG*), and rat filaggrin (*rFLG*) at the N-terminal, using CLUSTAL W 1.81 Multiple Sequence Alignment program. The secondary structures were predicted using SCRATCH program. * shows different regions between *cFLG* (no *) and a group of *hFLG*, *mFLG*, and *rFLG* (with *). (.) indicates similar amino acids

enzymes, e.g., peptidylarginine deiminase and transglutaminase [13, 14]. The pro-filaggrin peptide structures of dogs, humans, and mice are shown in Table 20.1.

20.3 Filaggrin Expression in Canine Disease

20.3.1 At the Protein Level

A major concern about canine filaggrin immunological studies is the Ab, including the specificity of Ab and synthesized peptide antigens that are supposed to be the epitopes on a filaggrin sequence. In fact, Ab against canine filaggrin has not yet been well established. Therefore, we will mention the type of Ab used in each study and suggest that the reader will please use his or her own consideration for the validity of the data. At the time of writing, two publications on anti-canine filaggrin polyclonal Abs have been published. The most conspicuous work of filaggrin immunostaining in normal dogs was revealed by Kanda et al. [9]. This group used a rabbit anti-canine filaggrin Ab, raised against synthetic peptide SRHSRTGHGSGNSKHR, which is located in the filaggrin monomer [9]. Two bands were

Table 20.1 Pro-filaggrin peptide structures of dogs, humans, and mice

Species	N-terminal region (aa)	FLG repeat unit (aa)	Number of FLG repeat units	C-terminal region (aa)	ProFLG (aa)
Human	292	324–325	10–12	157	4,062[a]
Dog	188	507–549	4	39	2,882
Mouse[b]	283	246–255	17[b]	26	3,658

Reprinted with permission from Kanda et al. [9]. © 2013 The Authors. Veterinary Dermatology © 2013 ESVD and ACVD
Abbreviations: *aa* amino acid, *FLG* filaggrin
[a]Allele with 10 units of FLG repeats
[b]C57BL/6 J mouse

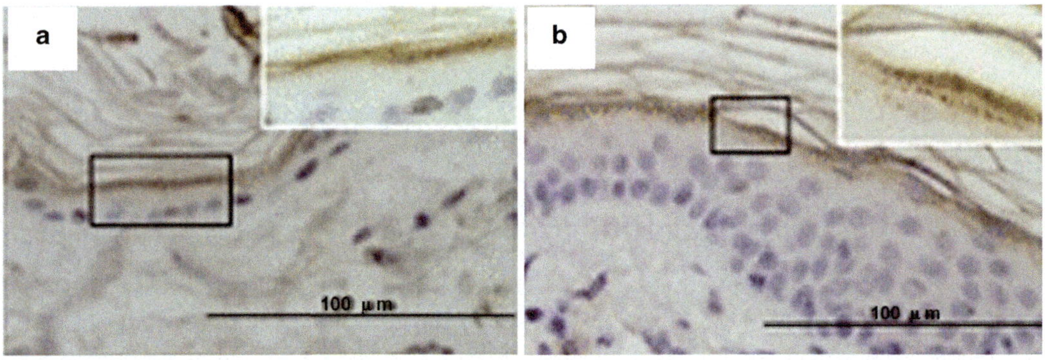

Fig. 20.5 Localization of canine filaggrin in the dorsal neck (**a**) and axilla (**b**) skin of a healthy dog, using an anti-canine filaggrin antibody for immunohistochemistry. Granular and cytoplasmic staining patterns in the stratum granulosum and stratum corneum are shown (*insets*). (Adapted with permission from Kanda et al. [9]. © 2013 The Authors. Veterinary Dermatology © 2013)

detected by Western blotting at 54 and 59 kDa, corresponding to the sizes of the predicted filaggrin monomer. Inactive pro-filaggrin is a cytoplasmic protein, localized within the keratohyalin granules in the S. granulosum. It is proteolyzed to active filaggrin proteins at the S. corneum [15]. From the IHC result, the antibody could bind both pro-filaggrin at the S. granulosum and the active filaggrin at the S. corneum of the dorsal neck and axilla (Fig. 20.5). The result corresponds to that of human filaggrin (Fig. 20.6) [16]. Active human filaggrin will form cornified cell envelopes with other proteins such as keratin, involucrin, etc., and enforce a protective skin barrier [15].

We aligned the synthetic peptide and we found that it corresponded to amino acid residues 890–905, 1439–1454, 1967–1982, and 2432–2447 in canine pro-filaggrin sequence (ENSCAFG00000023034).

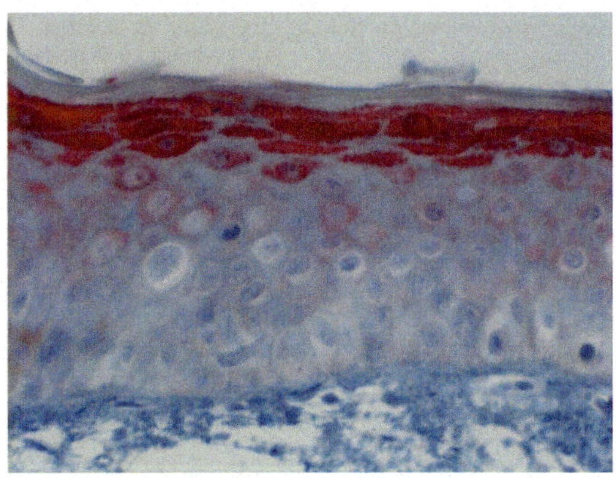

Fig. 20.6 Localization of filaggrin in the epidermis of healthy human, using a monoclonal anti-mouse filaggrin antibody for immunohistochemistry (Adapted by permission from Macmillan Publishers Ltd: Gruber et al. [16], copyright 2007)

The other group that proposed anti-canine filaggrin Ab was Chervet et al. [17]. This group compared commercially available rabbit anti-mouse filaggrin polyclonal Ab (Abcam, Cambridge, UK) (N-terminal Ab) and rabbit anti-canine filaggrin polyclonal Ab (C-terminal Ab) in immunofluorescence of non-lesional

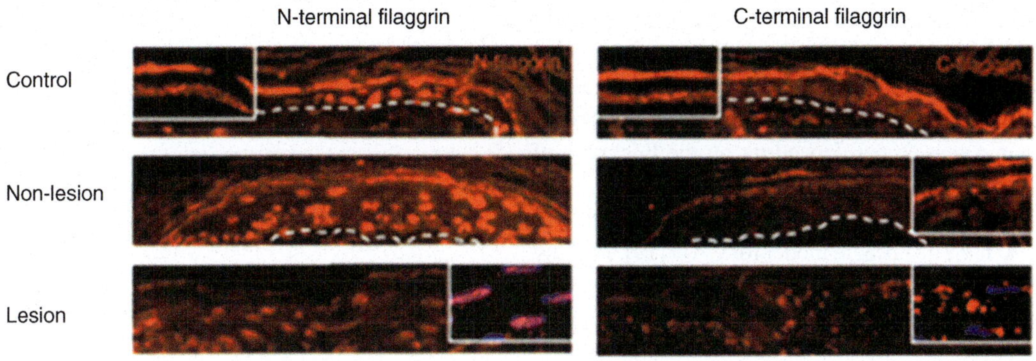

Fig. 20.7 Localization of canine filaggrin in the epidermis of healthy dog, non-lesional and lesional skin of CAD, using an anti-N-terminal (*left panel*) and C-terminal filaggrin antibodies (*right panel*) for immunofluorescence. *Insets* represent a three times higher magnification with adapted exposure time of a representative area to show the continuous and granular filaggrin lines, respectively, in control, cytoplasmic vesicles in non-lesional skin, and their localization outside of nuclei (*right panel*) in contrast to the nuclear localization (*left panel*) in lesional skin (Adapted with permission from Chervet et al. [17]. © 2010 John Wiley & Sons A/S)

and lesional CAD skin at the dorsal thorax. The N-terminal Ab was raised against synthetic peptide DSQVHSGVQVEGRRGH of the N-terminal conserved region of mouse filaggrin, which cross-reacted with human filaggrin and probably recognized the conformational epitope DSSRHSGSH of predicted canine filaggrin sequence [17]. The C-terminal Ab recognized the DSVFVQSQNGSRSHD peptide sequence, which was addressed by the authors to be located at the C-terminal of the canine pro-filaggrin sequence. Both antibodies could bind the pro-filaggrin in keratohyalin granules at the S. granulosum and the active filaggrin form at the S. corneum. However, the C-terminal Ab showed more specificity without staining the nuclei (Fig. 20.7). Data from some dogs with lesional AD skin (erythema) showed obvious decreased filaggrin expression when using both Abs. However, in some dogs, only C-terminal Ab stained less in lesional skin as mentioned by the authors (data not shown); hence, truncated filaggrin protein probably appeared. However, the cause of this truncated filaggrin has not yet been identified. It is probably due to mutation of the gene, transcription, translation, or posttranslational events. The work on mRNA and genomic levels should help, at least in some part, answer this question.

We aligned the epitopes that were mentioned in the study with the canine filaggrin peptide sequence (ENSCAFG00000023034); we found that the DSSRHSGSH epitope was on the locations 461–469, 989–997, 1538–1546, 2066–2074, and 2519–2527 of the pro-canine filaggrin sequence. Hence, the Ab is probably not specific to the N-terminal. For the DSVFVQSQNGSRSHD sequence, it did not match well with the canine filaggrin sequence. Whether or not it worked as a conformational epitope needs to be further investigated.

Since the anti-mouse filaggrin polyclonal Ab that Chervet et al. [17] used is commercially available, it has been widely used to study canine filaggrin protein expression. Theerawatanasirikul et al. [18] did IHC from dog skin of different breeds and coat types, including short coats (pug), medium-length coats (Labrador retriever), long coats with no undercoat (poodle and Shih Tzu), and long coats with a dense undercoat (golden retriever) at various sites of the body, including the ear pinnae (concave and convex), periocular, muzzle, axilla, ventral neck, ventral thorax, lateral thorax, ventral abdomen, dorsal back, lateral flank, inguinal area, perineum, forelimb, hind limb, and digital areas. The Ab was found to bind throughout the epidermal layers. Significant decreases in filaggrin expression were demonstrated in the ventral (axilla) and the weight-bearing (digital web) areas, the susceptible sites for CAD. However, when the epidermal thickness was compared among these sites, the digital web showed the thickest, whereas the axilla showed the thinnest epidermis. It can be concluded that the skin at axilla and digital web is probably a "weak" point and easy to develop barrier defect in skin diseases, not depending on the thickness of the epidermis (Fig. 20.8).

The anti-mouse filaggrin polyclonal Ab was also used in a beagle model for CAD [19]. Healthy beagles and beagles with AD were sensitized with house dust mites (HDM) for 3 h per day, for 3 consecutive days. IHC of the skin from the dogs, pre- and post-allergen exposure, was demonstrated. Similar to that observed by Theerawatanasirikul et al. [18], the positive staining was observed not only in the S. granulosum but also in lower

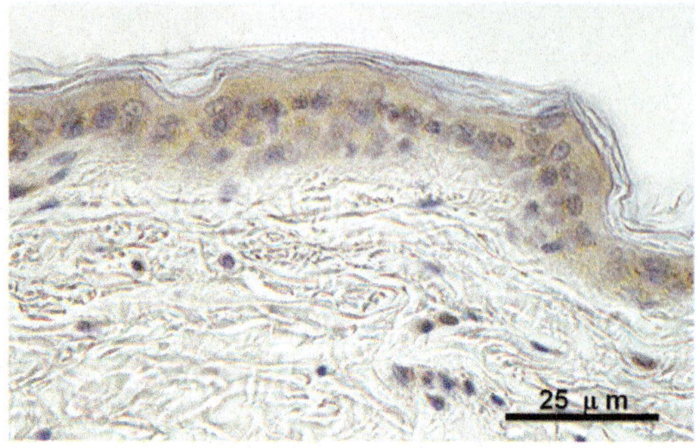

Fig. 20.8 Localization of canine filaggrin in the ventral abdomen epidermis of healthy dog, using a polyclonal anti-mouse filaggrin antibody for immunohistochemistry (Adapted with permission from Theerawatanasirikul et al. [18])

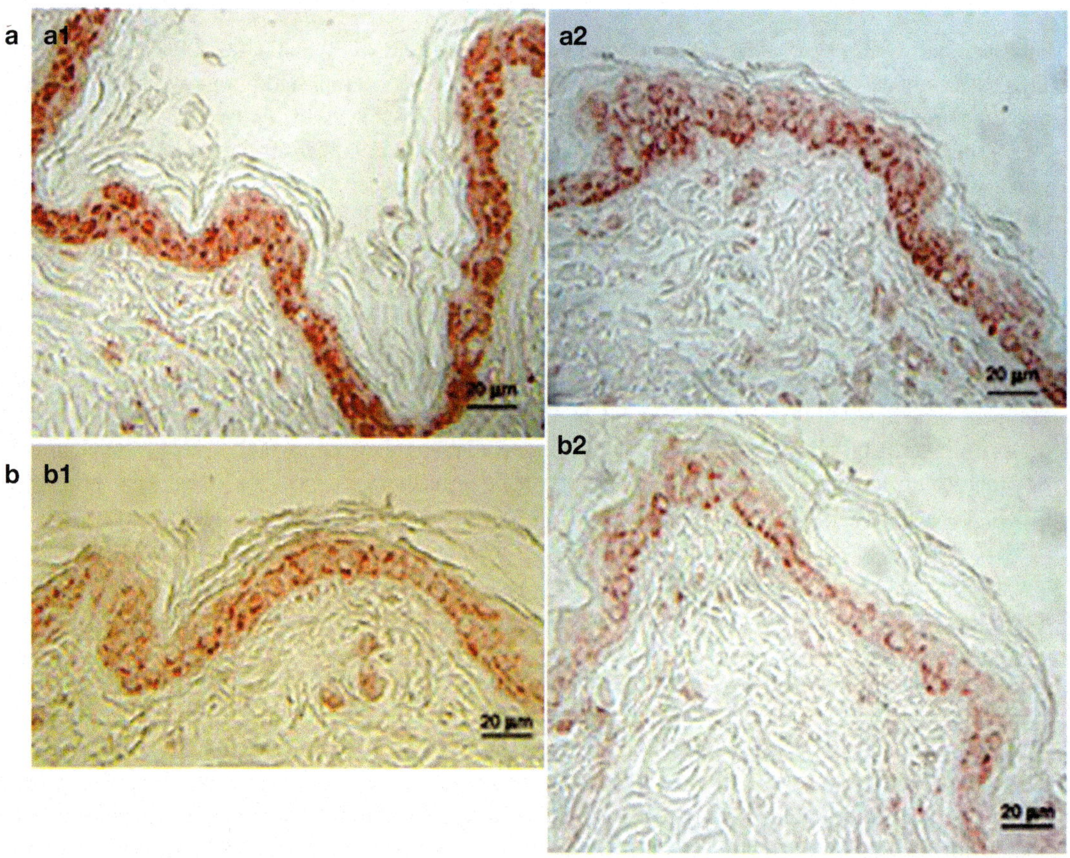

Fig. 20.9 Localization of canine filaggrin in the epidermis of healthy dog (**a**), pre-challenge (**a1**) and post-challenge (**a2**) with house dust mite, and of canine atopic dermatitis (**b**), pre-challenge (**b1**) and post-challenge (**b2**) with house dust mite, using a polyclonal anti-mouse filaggrin antibody for immunohistochemistry (Reprinted with permission from Marsella et al. [19]. © 2009 The Authors. Journal compilation © 2009 ESVD and ACVD)

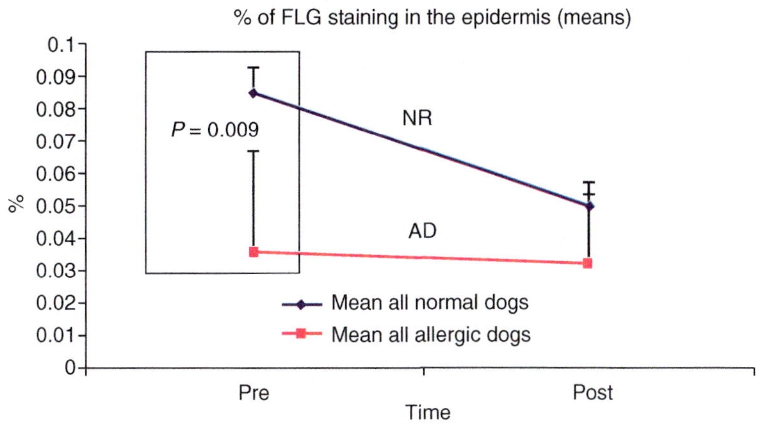

Fig. 20.10 Percentage of immunopositive pixels in the epidermis of normal (*NR*) and atopic dog (*AD*) pre- and post-challenge with house dust mite (Adapted with permission from Marsella et al. [19]. © 2009 The Authors. Journal compilation © 2009 ESVD and ACVD)

epidermal layers of all groups (Fig. 20.9). This is probably due to cross-reactivity with other proteins (e.g., keratins). However, overall filaggrin staining percentage in epidermis of all groups was quite low, with a maximum value of less than 0.09 % in normal dogs (preexposure to HDM) (Fig. 20.10). There was no correlation with the filaggrin staining percentage in any group at any timepoint with the Canine Atopic Dermatitis Extent and Severity Index 03 (CADESI-03) scores, which were used to evaluate clinical severity of CAD [19, 20]. Hence, filaggrin expression might not be associated with the severity of CAD, or perhaps other disease grading systems, such as the Canine Atopic Dermatitis Lesion Index (CADLI), should be used [21]. It is noted that in dogs with AD (pre- and postexposure to HDM) and in normal dogs (postexposure to HDM), epidermal hyperplasia, an indicator of chronic disease that is normally found in natural CAD [22], was not observed. Lately, a similar experiment of HDM-treated beagles was performed by the same group, but using a specific anti-canine filaggrin Ab. The correlation between staining score and a modified CADESI-03 was observed only on day 1 of HDM challenge [23]. The results may be due to the variety of dog breeds, and the beagle is not a susceptible breed for CAD. Hence, the clinical association of filaggrin and CAD needs to be further investigated.

The study of naturally occurring CAD was reported, using the same anti-mouse filaggrin Ab [24]. A distinct pattern of filaggrin staining in lesional CAD skin was compared with that of non-lesional and normal skin. The epidermal hyperplasia, often found in chronic CAD [22], was obviously observed in lesional skin. The Ab discontinuously bound to the S. corneum and S. granulosum, but the staining was not observed in young cells at the S. spinosum and S. basale. However, all skin layers were stained in normal and non-lesional skin, which are significantly thinner than that of lesional skin (Fig. 20.11). The binding of filaggrin Ab at the S. granulosum and S. corneum of CAD lesion with epidermal hyperplasia was similar to that observed in normal and affected HAD [16] (see Fig. 20.6).

Filaggrin expression was studied by IHC in canine circumanal glands. The Ab used was a mouse monoclonal Ab against filaggrin (Biogenesis, UK). Among the four layers (basal layer, polyhedral or spinous layer, granular layer, and horny layer) of circumanal glands, filaggrin was stained in the granular layer, the spinous layer, and in the hair follicle neck of the external root sheath. The inability to detect filaggrin at the S. corneum layer may be due to the thinness of S. corneum of that region [1] (Fig. 20.12).

Other protein expressions that related to filaggrin were reported in CAD, including involucrin and lymphoepithelial Kazal-type inhibitor (LEKTI). Involucrin is another marker protein of terminal epidermal differentiation that cross-links with filaggrin. Using an anti-mouse monoclonal Ab against involucrin (SY5), which cross-reacts with canine involucrin (Abcam,

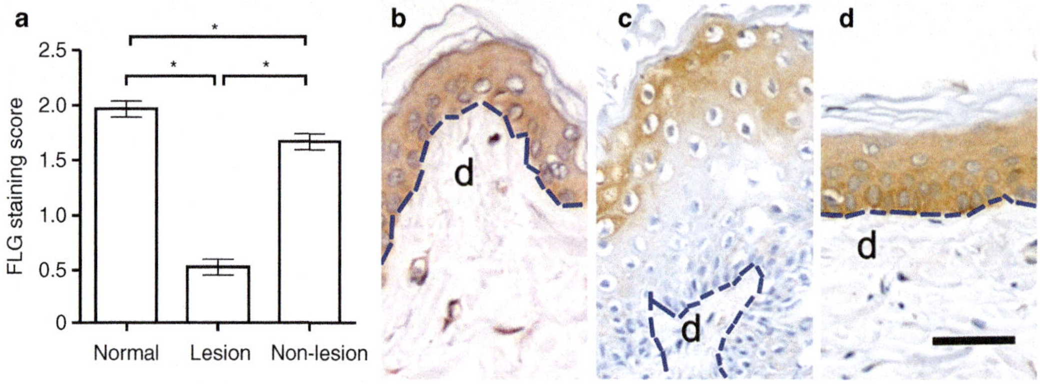

Fig. 20.11 Immunohistochemical staining for canine filaggrin. (**a**) Filaggrin staining scores in normal, lesional, and non-lesional skin. Bars indicate mean ± SD data with significant p-values $*p<0.05$. Protein expression of filaggrin in normal (scale bars = 30 μm), lesional (scale bars = 50 μm), and non-lesional skin (scale bars = 50 μm) (**b**–**d**, respectively). *d* dermis (Reprinted with permission Theerawatanasirikul et al. [24]. Copyright © 2013 TJVM)

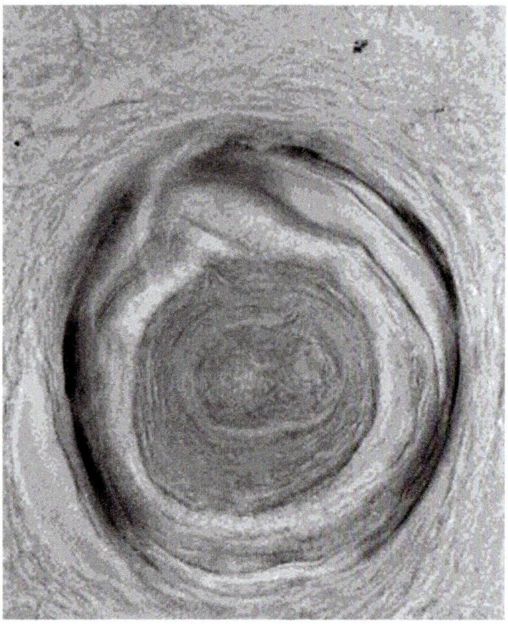

Fig. 20.12 Immunohistochemical staining for canine filaggrin in the circumanal gland epidermis of healthy dog, using a monoclonal anti-mouse filaggrin antibody (Adapted with permission from Atoji et al. [1]. Copyright © 1998 Wiley-Liss, Inc.)

Cambridge, UK), the involucrin intensity was significantly decreased in lesional skin similar to that of filaggrin [24]. Incidentally, pro-filaggrin converts to active filaggrin by several proteases in balance with the protease inhibitors. The lymphoepithelial Kazal-type inhibitor (LEKTI) is a protease inhibitor that inhibits cleavage of profilaggrin into filaggrin [25]. It is encoded by the serine protease inhibitor Kazal-type 5 gene (SPINK5) [26]. It is involved in regulation of proteolysis in epithelia formation and keratinocyte terminal differentiation [27]. The expression of LEKTI was studied in CAD by IHC with rabbit polyclonal antihuman LEKTI Ab (Santa Cruz Biotechnology, Santa Cruz, CA). In normal skin, it was obviously expressed in S. corneum. In CAD, the expression was expanded from the lower part of S. corneum to S. spinosum with great intensity in spongiotic and parakeratotic areas. The expression of LEKTI was increased in lesional and non-lesional skin compared to the normal controls corresponding to the decreased filaggrin expression [24]. This study supported the role of LEKTI on filaggrin expression in CAD.

20.3.2 At the mRNA Level

Since several questions about the role of filaggrin in CAD can be answered only by filaggrin protein expression studies due to the limitations of available filaggrin Ab, the emergence of draft Ensembl and NCBI dog genome databases [7, 28] has made the study of filaggrin at the mRNA level an intriguing subject. Interestingly, although filaggrin protein expression is decreased in CAD in several publications, the expression at the mRNA level does not seem to be in accordance. In the naturally occurring CAD, Theerawatanasirikul et al. [29] demonstrated upregulation of filaggrin mRNA expression in lesional CAD skin of small dog breeds (poodle, Shih Tzu, and pug) with weak correlation with involucrin (IVL), keratin 5 (KRT5), KRT14, and KRT17 [29]. On the other hand, in non-lesional AD skin of West Highland white terriers, decreased filaggrin expression was shown [30]. However, no significant change was observed in the lesional skin and in other breeds studied (beagle and Cavalier King Charles spaniel). For the CAD beagle model, Santoro et al. [31] showed, in the NAVDF 2010, decreased filaggrin mRNA expression in AD dog model. However, after the dogs were challenged with HDM for 3 consecutive days, filaggrin mRNA expression was increased on days 3 and 10, whereas protein expression was not different between the groups at any timepoint. The increased mRNA expression may be due to a compensatory mechanism in response to the decreased protein expression or the number of isoforms of the gene [32, 33]. In HAD, the mutation of the *FLG* can lead to reduced protein expression without any effect on the transcription levels [2, 25]. Actually, there is yet another filaggrin protein expressed in human skin and functioned in epidermal barrier similar to human filaggrin-2 [34]. In dog, predicted *FLG2* sequences were also shown in the GenBank database (Accession number: XM_540329), Archive!Ensembl database (ENSCAFG00000024442), and e!Ensembl database (ENSCAFG00000030938). However, the role of canine filaggrin-2 needs to be investigated further.

20.4 Filaggrin Gene Mutations in CAD

Several groups have tried to find the *FLG* mutation in CAD. However, the definitive mutation location in risk breeds has not yet been confirmed. The most conspicuous work of *FLG* single nucleotide polymorphism (SNP) searching was performed in 242 AD dogs of a large variety of breeds (Boxer, German shepherd, Labrador retriever, golden retriever, Shiba Inu, Shih Tzu, pit bull, and West Highland white terrier) and 417 normal controls from three locations (the UK, the USA, and Japan). One SNP within the *FLG* was associated with CAD at the location 64,297,022 (rs22588227) on canine chromosome 17 with corrected p-value = 0.009 and odds ratio = 5.6 but only in Labrador retriever from the UK [35]. In small dog breeds (poodle, Shih Tzu, and pug), an association between *FLG* mutations and CAD has been suggested at the locations 64,297,000 ($p = 0.041$, odds ratio = 3.920) and 64,297,126 ($p = 0.043$, odds ratio = 3.706) [10]. However, due to the small number of samples (12 affected dogs and 39 normal controls), the data need to be confirmed in a larger population. No associations between *FLG* mutations and AD in West Highland white terriers were found in Australia (49 CAD and 30 normal controls) and the US (30 CAD, 50 normal controls, and 28 of undetermined status) [30, 36]. The lack of *FLG* mutations in dogs is probably due to the lack of a genetic survival mechanism from such mutations. A heterozygous advantage is likely in humans, but this may not have been generally found in dogs. One major problem of *c* filaggrin expression study is the variety of dog breeds. Several dog breeds suffered from CAD, but the *cFLG* sequences, even the predicted ones, are only from some breeds. In addition, tools for canine genetic study are not as well developed as the tools for human study. Hence, *cFLG* annotation was changed from time to time in canFam2 and finally is excluded from the latest draft dog genome database (canFam3).

20.5 Where Is the Filaggrin Gene in the Latest Draft Dog Genome Database (CanFam3, September 2011)?

A new draft dog genome database version (canFam3) was launched in September 2011 without *cFLG* gene in it. In the old canFam2 (May 2005), the inconsistent *cFLG* annotation made it difficult to design primers to get the expected PCR product sizes. Suriyaphol et al. [10] showed a new fragment of the repeated *FLG* sequence that did not appear in canFam2. However, the *FLG* was not included in the final gene set for the canFam3 because this new annotation uses a different approach from the previous one (canFam2). *cFLG* was classified as "bad similarity" with low score together with RNASeq models with coverage between 50 and 75 %. This score is calculated by taking into account the individual score for each exon and intron and some parameters related to short exon and short intron penalties. Moreover, the change in the UniProt set (since 2005) could have affected the result. In the latest canFam, the *FLG* has been merged with the *FLG2* gene since they were classified as "good similarity" when they were merged. They were then prioritized (together with RNASeq models with coverage between 75 and 100 %) over genes that were not classified as good similarity and then included in the final annotation of the genome.

20.6 Conclusion and Suggestion for Future Work

Decreased filaggrin protein expression was observed in normal dog skin at the susceptible sites for CAD, compared to other sites of the body, and also in dogs with AD. Filaggrin Ab, which binds specifically to canine filaggrin epitopes, would provide more information about the potential role of canine filaggrin in any skin disease. At the mRNA level, the results from past studies have varied, which probably implies that inconstant mechanisms underlined the protein expression. The studies on the CAD model still have a long way to go. Attempts to identify *FLG* mutations have not yet been successful, since the disease occurs in several dog breeds with different genomes and filaggrin itself shows different expression patterns among the CAD groups. Pertinently, the *FLG* does not appear in the latest draft dog genome (canFam3). However, in the future, when a new annotation pipeline gets improved and fine-tuned in dogs, with more data on each dog breed and more functional data of the protein, we believe that the *FLG* will return to a dog genome database.

Acknowledgments We would like to express our thanks to Prof. Dr. Roongroje Thanawongnuwech, Assoc. Prof. Dr. Achariya Sailasuta, Assoc. Prof. Dr. Anudep Rungsripipat, and Assist. Prof. Dr. Sayamon Sukjumlong, Faculty of Veterinary Science, Chulalongkorn University. Special thanks to Ms. Nongluk Plongthongkum for providing data.

References

1. Atoji Y, Yamamoto Y, Komatsu T, Suzuki Y, Tsubota T. Circumanal glands of the dog: a new classification and cell degeneration. Anat Rec. 1998;250:251–67.
2. Palmer CN, Irvine AD, Terron-Kwiatkowski A, Zhao Y, Liao H, Lee SP, et al. Common loss-of function variants of the epidermal barrier protein filaggrin are a major predisposing factor for atopic dermatitis. Nat Genet. 2006;38:441–6.
3. Halliwell R. Revised nomenclature for veterinary allergy. Vet Immunol Immunopathol. 2006;114: 2007–8.
4. Scott DW. Skin immune system and allergic skin diseases. In: Muller and Kirk's small animal dermatology. Scott DW, Miller WH, Griffin CE, editors. 6th ed. Philadelphia: WB Saunders; 2001. p. 543–666.
5. Leung DYM. Atopic dermatitis: the skin as a window into the pathogenesis of chronic allergic diseases. J Allergy Clin Immunol. 1995;96:302–19.
6. Marsella R, Girolomoni G. Canine models of atopic dermatitis: a useful tool with untapped potential. J Invest Dermatol. 2009;129:2351–7.
7. Flicek P, Amode MR, Barrell D, Beal K, Brent S, Chen Y, et al. Ensembl 2011. Nucleic Acids Res. 2011;39:800–6.
8. Lindblad-Toh K, Wade CM, Mikkelsen TS, Karlsson EK, Jaffe DB, Kamal M, et al. Genome sequence, comparative analysis and haplotype structure of the domestic dog. Nature. 2005;438:803–19.
9. Kanda S, Sasaki T, Shiohama A, Nishifuji K, Amagai M, Kudoh J, et al. Characterization of canine filaggrin: gene structure and protein expression in dog skin. Vet Dermatol. 2013;24:25–31.e7.
10. Suriyaphol G, Suriyaphol P, Sarikaputi M, Theerawatanasirikul S, Sailasuta A. Association of filaggrin (*flg*) gene polymorphism with canine atopic dermatitis in small breed dogs. Thai J Vet Med. 2011;41:509–17.
11. Thompson JD, Higgins DG, Gibson TJ. CLUSTAL W: improving the sensitivity of progressive multiple sequence alignment through sequence weighting, position-specific gap penalties and weight matrix choice. Nucleic Acids Res. 1994;22:4673–80.
12. Cheng J, Randall AZ, Swerwdoski MJ, Baldi P. SCRATCH: a protein structure and structural feature prediction server. Nucleic Acid Res. 2005;33:W72–6.

13. Presland RB, Haydock PV, Fleckman P, Nirunsuksiri W, Dale BA. Characterization of the human epidermal profilaggrin gene. Genomic organization and identification of an S-100-like calcium binding domain at the amino terminus. J Biol Chem. 1992;267:23772–81.
14. Zhang D, Karunaratne S, Kessler M, Mahony D, Rothnagel JA. Characterization of mouse profilaggrin: evidence for nuclear engulfment and translocation of the profilaggrin B-domain during epidermal differentiation. J Invest Dermatol. 2002;119:905–12.
15. Rawlings AV, Harding CR. Moisturization and skin barrier function. Dermatol Ther. 2004;17 Suppl 1:43–8.
16. Gruber R, Janecke AR, Fauth C, Utermann G, Fritsch PO, Schmuth M. Filaggrin mutations p.R501X and c.2282del4 in ichthyosis vulgaris. Eur J Hum Genet. 2007;15:179–84.
17. Chervet L, Galichet A, McLean WH, Chen H, Suter MM, Roosje PJ, et al. Missing C-terminal filaggrin expression, NFkappaB activation and hyperproliferation identify the dog as a putative model to study epidermal dysfunction in atopic dermatitis. Exp Dermatol. 2010;19:343–6.
18. Theerawatanasirikul S, Suriyaphol G, Thanawongnuwech R, Sailasuta A. Histologic morphology and involucrin, filaggrin, and keratin expression in normal canine skin from dogs of different breeds and coat types. J Vet Sci. 2012;13:163–70.
19. Marsella R, Samuelson D, Harrington L. Immunohistochemical evaluation of filaggrin polyclonal antibody in atopic and normal beagles. Vet Dermatol. 2009;20:547–54.
20. Olivry T, Marsella R, Iwasaki T, Mueller R. Validation of CADESI-03, a severity scale for clinical trials enrolling dogs with atopic dermatitis. Vet Dermatol. 2007;18:78–86.
21. Plant JD, Gortel K, Kovalik M, Polissar NL, Neradilek MB. Development and validation of the Canine Atopic Dermatitis Lesion Index, a scale for the rapid scoring of lesion severity in canine atopic dermatitis. Vet Dermatol. 2012;23:515–e103.
22. Olivry T, Naydan DK, Moore PF. Characterization of the cutaneous inflammatory infiltrate in canine atopic dermatitis. Am J Dermatopathol. 1997;19:477–86.
23. Marsella R. Does filaggrin expression correlate with severity of clinical signs in dogs with atopic dermatitis? Vet Dermatol. 2013;24:266–e59.
24. Theerawatanasirikul S, Sailasuta A, Thanawongnuwech R, Nakbed T, Charngkaew K, Suriyaphol G. Differential expression patterns of proteins involved in epidermal proliferation and differentiation in canine atopic dermatitis. Thai J Vet Med. 2012;42:287–96.
25. Cork MJ, Danby SG, Vasilopoulos Y, Hadgraft J, Lane ME, Moustafa M, et al. Epidermal barrier dysfunction in atopic dermatitis. J Invest Dermatol. 2009;129:1892–908.
26. Magert HJ, Standker L, Kreutzmann P, Zucht HD, Reinecke M, Sommerhoff CP, et al. LEKTI, a novel 15-domain type of human serine proteinase inhibitor. J Biol Chem. 1999;274:21499–502.
27. Chavanas S, Bodemer C, Rochat A, Hamel-Teillac D, Ali M, Irvine AD, et al. Mutations in SPINK5, encoding a serine protease inhibitor, cause Netherton syndrome. Nat Genet. 2000;25:141–2.
28. Gertz EM. BLAST scoring parameters [Internet]. 2011 [updated 2005 Mar 16]. Available from: ftp://ftp.ncbi.nlm.nih.gov/blast/documents/developer/scoring.pdf.
29. Theerawatanasirikul S, Sailasuta A, Thanawongnuwech R, Suriyaphol G. Alterations of keratins, involucrin and filaggrin gene expression in canine atopic dermatitis. Res Vet Sci. 2012;93:1287–92.
30. Barros Roque J, O'Leary CA, Kyaw-Tanner M, Latter M, Mason K, Shipstone M, et al. Haplotype sharing excludes canine orthologous Filaggrin locus in atopy in West Highland White Terriers. Anim Genet. 2009;40:793–4.
31. Santoro D, Marsella R, Bunick D, Graves TK, Campbell KL, Ahrens K. Expression and distribution of canine filaggrin in the skin of healthy and atopic beagles [Abstracts]. Vet Dermatol. 2010;21:323.
32. Chen G, Gharib TG, Huang CC, Taylor JM, Misek DE, Kardia SL, et al. Discordant protein and mRNA expression in lung adenocarcinomas. Mol Cell Proteomics. 2002;1:304–13.
33. Fournier ML, Paulson A, Pavelka N, Mosley AL, Gaudenz K, Bradford WD, et al. Delayed correlation of mRNA and protein expression in rapamycin-treated cells and a role for Ggc1 in cellular sensitivity to rapamycin. Mol Cell Proteomics. 2010;9:271–84.
34. Wu Z, Hansmann B, Meyer-Hoffert U, Gläser R, Schröder JM. Molecular identification and expression analysis of filaggrin-2, a member of the S100 fused-type protein family. PLoS One. 2009;4:e5227.
35. Wood SH, Ollier WE, Nuttall T, McEwan NA, Carter SD. Despite identifying some shared gene associations with human atopic dermatitis the use of multiple dog breeds from various locations limits detection of gene associations in canine atopic dermatitis. Vet Immunol Immunopathol. 2010;138:193–7.
36. Salzmann CA, Olivry TJ, Nielsen DM, Paps JS, Harris TL, Olby NJ. Genome-wide linkage study of atopic dermatitis in West Highland White Terriers. BMC Genet. 2011;12:37.

Part V

Filaggrin and Cutaneous Disease

Ichthyosis Vulgaris and Other Disorders of Cornification

21

Sreya Talasila and Amy S. Paller

Contents

S. Talasila, BS • A.S. Paller, MD (✉)
Department of Dermatology, Feinberg School of Medicine, Northwestern University, 676 N. St. Claire Street, Suite 1600, Chicago, IL 60611-2997, USA
e-mail: stalasil@gmail.com; apaller@northwestern.edu

The ichthyoses (from Greek, *ichthys*, for fish) are a heterogeneous group of scaling dermatoses caused by disordered cornification. Ichthyosis vulgaris (IV; OMIM #146700) is by far the most common of these disorders of cornification and one of the mildest in severity. In fact, IV is one of the most frequent single-gene disorders found in humans [1–3].

21.1 Background

IV was first clearly documented in 1806 as "ichthyosis nacrée" by Alibert, who compared the scales of IV to mother-of-pearl [3, 4]. It took another 200 years before the discovery of the underlying gene defect in the filaggrin gene (*FLG*) was elucidated in 2006 [1, 5].

21.2 Pathogenesis

Loss-of-function mutations in *FLG*, which decrease the expression of filaggrin protein by approximately 50 % on each affected allele, cause IV [1, 6]. Mutations in the gene encoding pro-filaggrin had been suspected for many years to be the cause of IV based on the frequently observed reduction in the granular layer in IV, the site of filaggrin expression (see below) [1]. Genetic linkage studies of IV families first mapped the location of the *FLG* to chromosome 1q21.3 in 2002 [7, 8]. In 2006, Smith et al. and Palmer et al. simultaneously reported mutations

J.P. Thyssen, H. Maibach (eds.), *Filaggrin*,
DOI 10.1007/978-3-642-54379-1_21, © Springer-Verlag Berlin Heidelberg 2014

in *FLG* as the underlying cause of IV [1, 5]. One problem with verifying this suspicion was the impact of the environment on clinical manifestations. In particular, high humidity and warmer weather mask the phenotypic changes [8]. Similarly, many individuals with IV have "dry skin" that "runs in the family" and thus are not recognized to have IV. Also challenging for discovery are the multiple repeats and large size of *FLG* [1]. Detection of gene mutations required specific primers and use of long-range PCR for amplification within the highly homologous *FLG* sequence [9, 10].

Prior to the discovery of the underlying gene defect, IV was thought to be an autosomal dominant disorder with variable penetrance; recognition of the loss-of-function mutations in *FLG* allowed reclassification of IV as a semidominant disease with 83–96 % penetrance [1, 11–14]. Semidominant transmission means that the disease can manifest with either a single mutant allele [seen in a majority of individuals with IV] or with both alleles expressing a null mutation [12, 15, 16]. This semidominant pattern of inheritance also contributes to the reason why some patients have mild clinical signs such as dryness and scaling only in winter (heterozygotes with one mutant allele), whereas others have chronic, severe, generalized scaling and severe palmoplantar markings (homozygotes or compounds heterozygotes with two mutant alleles leading to a dose effect) [1].

The *FLG* gene is located in a cluster of proteins necessary for epidermal differentiation (epidermal differentiation cluster; EDC) [17, 18]. The EDC is a large sequence 2 Mb in length that includes genes that encode loricrin, involucrin, trichohyalin, and repetin [8]. *FLG* encodes pro-filaggrin, the primary protein of keratohyalin granules. The pathogenesis of IV has been linked to a decrease in both size and number and sometimes even a complete absence of epidermal (F-type) keratohyalin granules of the stratum granulosum [2, 6, 19]. After its synthesis, profilaggrin is phosphorylated and stored in the keratohyalin granules of the granular layer as 400 kDa pro-filaggrin polymers [1, 2]. The structure contains an N-terminal domain preceding 10–12 repeats of a 324-residue filaggrin sequence [20]. During the final processing steps of skin differentiation, pro-filaggrin is cleaved into 10–12 filaggrin monomers of approximately 37 kDa after dephosphorylation and cleavage by serine proteases [1]. Filaggrin then helps to aggregate keratin intermediate filaments (thus the name *filament-aggregating* prote*in*) and contributes to the cornified envelope by cross-linking proteins such as loricin and trichohyalin and providing a scaffold for ceramides secreted by the lamellar bodies [2, 21].

As a result of null mutations, pro-filaggrin mRNA in IV patients has a shortened half-life as compared to wild-type individuals [22] and loss-of-function mutations in the *FLG* gene, regardless of the location, result in unstable, truncated pro-filaggrin that cannot function normally [6, 10]. In addition to causing filament disorganization and a decrease in corneodesmosome density [19], *FLG* mutations are associated with irregular lamellar body secretion, maturation, organization, and distribution, with impaired lipid transport, further affecting the integrity of the stratum corneum [19, 23]. The result is an impaired epidermal barrier with increased transepidermal water loss (TEWL) leading to xerosis and a decrease in skin hydration as a function of the amount of filaggrin expression [9, 19]. Filaggrin proteolysis accelerates below 80 % humidity and increases as humidity declines in order to preserve skin hydration [24, 25]. In homozygous individuals, the decrease in stratum corneum hydration is 30 % [19]. This abnormal epidermal barrier function results in compensatory mechanisms such as epidermal hyperplasia which, coupled with the decrease in normal desquamation, leads to skin thickening and histological hyperkeratosis as a final product [21, 26].

Filaggrin is also degraded through a process requiring caspase-14 into hygroscopic amino acids (glutamine, arginine, and histidine) and their metabolites, such as urocanic acid (UCA) and pyrrolidone carboxylic acid [27]. These amino acids comprise the skin's "natural moisturizing factor" (NMF) along with urea and lactic acid [1, 14, 28] and are important in epidermal water retention [29]. Flaky tail mice with their null mutation in *FLG* show ichthyosis, whereas mice lacking caspase-14 show xerosis and UV sensitivity without ichthyosis; in these animals, pro-filaggrin to filaggrin cleavage is normal, but NMF production is impaired [30, 31]. NMFs (especially histidine and glutamine) also affect the stratum corneum pH, which is key for regulating protease activity and desquamation [32, 33]. Desquamation occurs at the outer stratum corneum when the attachments between corneocytes (corneodesmosomes) are weakened and requires a lower pH [19, 34, 35]. In contrast, serine proteases operate optimally in neutral or alkaline pH, and the elevated pH in the ichthyotic milieu created by filaggrin deficiency increases epidermal protein degradation but compromises hydrolases that normally organize and process important extracellular lipids in the epidermis [19, 36, 37]. Homozygous individuals display a greater increase in skin pH (the mechanism being a reduction in acidic filaggrin metabolites), displaying an inverse dose-dependent effect; the increased surface pH in individuals with filaggrin mutations likely contributes to scale retention of IV, as well as deactivation of enzymes responsible for processing ceramides [19, 24]. The elevated pH also impairs the acid mantle of the stratum corneum, thus impairing the inherent antimicrobial properties of the skin and increasing *Staphylococcus aureus* growth [33].

NMFs also protect keratinocytes from ultraviolet-B (UVB) light damage through the transformation of histidine to trans-urocanic acid, which absorbs UVB [6, 38]. The role of filaggrin in protecting against UVB cellular exposure has been hypothesized to explain differences in the frequency of finding *FLG* mutations based on latitude. Since UCA is protective against UVB radiation, which induces vitamin D_3 production, a reduction in UCA in individuals with IV has been correlated with increased levels of vitamin D when compared to controls [39]. German and Danish filaggrin mutation carriers displayed up to 10 % higher mean vitamin D levels when compared to controls, which may indicate a survival advantage against diseases resulting from hypovitaminosis D such as osteoporosis, rickets, and infections [21, 39]. On the other hand, the seemingly lower rate of *FLG* mutations in African-Americans may relate part, to the role of UCA in absorbing ultraviolet light, as an individual with a *FLG* mutation is theoretically sensitized to UVB-induced apoptosis [28, 38]. Although the more permeable barrier of IV has many drawbacks, another potential selective advantage to filaggrin deficiency could be the improved immunity from increased transcutaneous exposure to pathogens [28, 40].

21.3 Epidemiology

Most individuals affected by IV have "dry skin" and no knowledge of their underlying diagnosis. In fact, a 1966 study of English schoolchildren cited an incidence of up to 1 in 250 patients based on clinical features [3, 41], but more recent studies have shown that 4 % of Northern Europeans and 3 % of Asians show clinical evidence of IV [1, 5, 15, 41, 42]. Overall, the median prevalence of *FLG* mutations among Europeans is 7.7 % (range 2.7–14.2 %) and among Asians is 3.0 % (range 0–7.3 %) [5, 21].

FLG consists of three exons, and mutations are found in exon 3, which encodes the N-terminal domain and multiple 1 kb sequences that encode the filaggrin repeats [1, 9]. Although IV is a common disorder across a variety of populations, mutations tend to be population-specific and sometimes even vary by region [43]. For example, R501X, 2282del4, S2347X, and R2447X are the four most common *FLG* gene mutations in Northern Europeans (accounting for greater than 80 % of mutations in this population) but are uncommon in individuals of Southern European descent [19, 28, 44, 45]. Of these mutations, R501X and 2282del4 are most frequent [1, 5, 12, 14, 41]. Homozygosity for a null mutation or compound heterozygosity for both R501X and 2282del4 mutations has been shown to lead to more severe manifestations [1].

The following mutations have been discovered in population-specific studies of IV patients. S2554X and 3321delA mutations are most common in Japanese populations but less prevalent in the Korean population, where p.Y1767X has been reported [46]. In the Chinese population, K4671X and 3321delA are common, in addition to numerous other novel mutations that have been reported; 3321delA is also common in Taiwan [47–51]. 2282del4 and 3672del4 mutations were found in the Indian population [5, 52], while R501X and 2282del4 mutations were discovered in the Pakistani population; these mutations along with R826X and 2767insT have been found in the Bangladeshi population [15, 53, 54]. Individuals of African descent with clinical evidence of IV or atopic dermatitis (AD) have a low frequency of the *FLG* mutations that are seen in Northern European and Asian populations [5, 28]. A limitation of studies of populations outside of Northern Europe is that sequencing of the entire *FLG* gene has not been performed, but rather studies have sought known loss-of-function mutations in *FLG* or have looked in regions within the gene known to harbor mutations. As such, the prevalence frequencies listed here in non-European populations may be erroneously low. Explanations for lower prevalence in populations outside of Northern Europe, particularly in populations of African descent, may include novel *FLG* mutations outside of regions with previously found mutations in exon 3 or a mutation in an alternative gene with shared function, such as *FLG2*.

21.4 Clinical Features of Ichthyosis Vulgaris

The diagnosis of IV is clinical and made based on history and a thorough physical examination, including of the palms and soles. The family should be questioned about "dry skin" and the need for emollient administration, especially in colder and drier weather. If a parent or other family members are present, their lower legs and palms should be examined. Individuals with AD should be assessed for concomitant IV, especially if they display characteristic features of IV such as scaling on the lower extremities (Fig. 21.1), palmar hyperlinearity (Fig. 21.2), and, although less specific, keratosis pilaris (Fig. 21.3). If further confirmation is necessary, genetic testing or histological examination can be performed (see below).

IV does not tend to be present at birth, in contrast to many other forms of ichthyosis, but

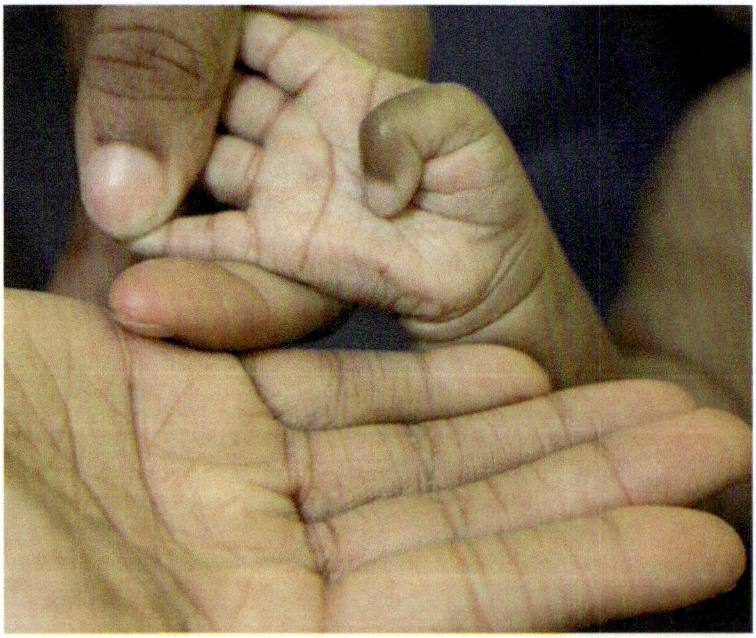

Fig. 21.1 Hyperlinear palms in a father and son with ichthyosis vulgaris (IV)

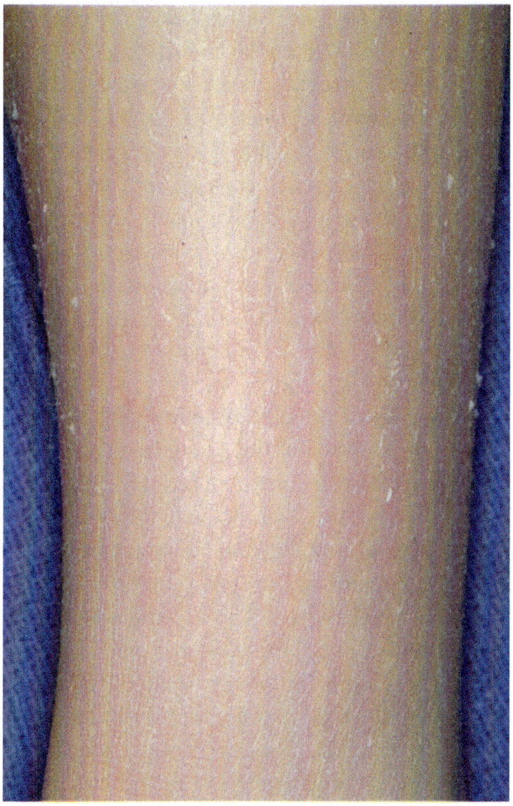

Fig. 21.2 Fine white desquamating scale with a larger polygonal scale on the lower extremity of an affected individual

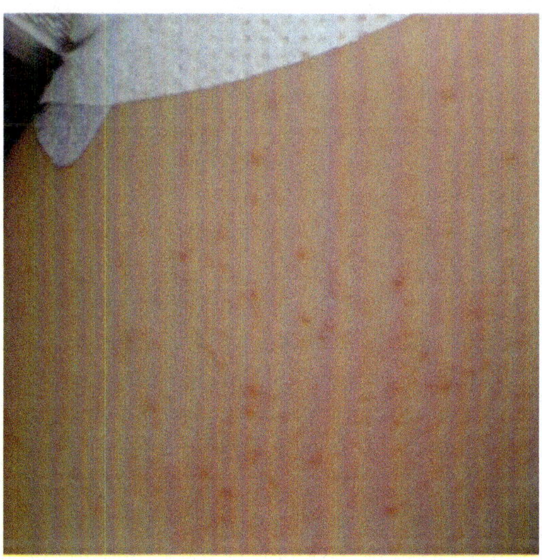

Fig. 21.3 Keratosis pilaris on the thigh of an infant with associated skin inflammation

rather tends to present at 3 months of age or later [3, 55]. On the other hand, concomitant AD may present as early as the first week of life with its xerosis and cutaneous inflammation. Heterozygotes show a very mild phenotype that can be masked by maintaining proper hydration with moisturizers, while homozygotes inherit a severe phenotype that tends to be more chronic and stable [15].

The *scaling* of IV occurs predominantly on the extensor surfaces and is exacerbated by low humidity (due to hastened proteolysis of remaining filaggrin) [12]. The more highly hydrated axillae and antecubital and popliteal fossae tend to show sparing [3, 8]. Scaling tends to range from fine and powdery to coarse and polygonal, with the latter morphology often seen on the lower extremities and to a lesser extent on the mid-face, scalp, and trunk [3]. NMF levels, as measured by Raman spectroscopy, are decreased on the lower legs, which may explain the increased incidence of dryness when compared to the upper extremities [33, 56]. NMF has also been shown to correlate with *FLG* null allele status, thus contributing to the phenotype of dry skin in IV and also AD [29]. Scaling on the scalp may be mistaken for seborrheic dermatitis. These scales are often attached centrally with the lateral edges peeling away from the skin surface [57]. In darker-skinned individuals, scales tend to be more hyperpigmented, given the retention of epidermal cells with melanin. Chapping, or painful fissures of the hands and heels, have been described in 81 % of affected school-age children in the UK, are seen more with severe disease, and are similarly exacerbated by decreased humidity [3]. There is notable improvement during summer months (reported by 80 % of IV patients), although some patients have reported hypohidrosis and heat intolerance as well [8, 55, 58].

Hyperlinear palms and soles show exaggerated skin markings. In a study of British schoolchildren, hyperlinear palms were strongly correlated with the *FLG* null genotype, particularly R501X, 2282del4, R2447X, and S3247X ($p<0.01$) [41]. The positive predictive value for hyperlinear palms is 71 %, and the negative predictive value is 90 % [41].

Keratosis pilaris is characterized by follicular-based keratotic papules, particularly on the lateral aspects of the face and extensor surfaces of the upper arms and thighs. Keratosis pilaris is noted in 100 % of homozygous *FLG* mutation carriers, 66 % of heterozygous individuals, and 30 % of wild-type individuals [41]. Keratosis pilaris is also significantly associated with R501X and 2282del4 mutations ($p<0.01$ for each) and with AD, which is also commonly associated with IV. The positive and negative predictive values for keratosis pilaris are 53 and 90 %, respectively [41]. The absence of keratosis pilaris and palmar hyperlinearity may yield a negative predictive value of *FLG* mutation as high as 92 % [41].

Although not yet validated, a scoring system for IV severity has been proposed to predict *FLG* mutations in the Caucasian population [59]. Five major clinical signs of IV (palmar hyperlinearity, diffuse xerosis, keratosis pilaris, leg scaling, and scalp desquamation) are graded on a 0–3 scale. A clinical severity score of <4 (out of 15) has a 90 % negative predictive value for having either an R501X or 2282del4 mutation [59]. Another proposed scale shows a high penetrance of summed skin signs known to be associated with *FLG* mutations (ichthyosis, keratosis pilaris, and palmar hyperlinearity): 100 % in homozygotes, 87.7 % in heterozygotes, and 46.5 % in wild-type individuals, suggesting highly penetrant haploinsufficiency [60].

21.5 Laboratory Testing

Laboratory testing is rarely necessary to make the diagnosis of IV, but features can be seen by studies of skin sections. Keratohyalin granules are a major component of the stratum granulosum and stain darkly with hematoxylin because they are mostly made up of electron-dense, histidine-rich insoluble pro-filaggrin aggregates. As such, on immunohistochemical staining, there is often "loss" of the granular layer as keratohyalin granules are reduced or absent in IV, regardless of anatomical location and time of sampling of the biopsy [1, 8, 61]. Nevertheless, a reduced granular layer is highly variable among individuals with IV, can be found in normal epidermis, and cannot be utilized for diagnostic confirmation [8]. The stratum corneum can show orthokeratosis or a basket weave pattern or may be compact and laminated [8, 38, 55]. The epidermis may be thicker than normal but tends to be thinner than in other proliferative disorders such as psoriasis. Studies have variably shown decreased mitotic counts and epidermal proliferation rates compared to control skin [57, 62] or increased numbers of Ki-67-positive keratinocytes, a marker of proliferation [19].

Immunohistochemical staining with a monoclonal antibody against the filaggrin repeat region and the N-terminal domain of pro-filaggrin shows a complete lack of staining in the suprabasal layers of individuals with null mutations in both alleles, compared to very strong staining of keratohyalin granules in normal individuals [2]. Interestingly, both heterozygotes and homozygotes with ichthyosis, but not AD, show increased expression of CD1a+ dendritic cells, although not to the extent seen with AD, suggesting sensitization in patients with IV, regardless of the concomitant presence of AD [14].

Although merely a research tool, electron microscopy can be used to distinguish heterozygous and homozygous carriers [14]. Homozygous carriers show a striking reduction in keratohyalin granules, while heterozygous carriers show minimal (but focally normal) keratohyalin granules [2, 14]. The amount of keratohyalin in samples correlates with filaggrin proteins in epidermal samples and with clinical severity [2]. As compared to control samples, lamellar bodies are smaller and abnormally shaped in patients with IV, but the number of lamellar bodies is not reduced [2, 38]. Keratohyalin granules may be "crumbly" or absent, and perinuclear keratin retractions may be seen in granular cells [2, 14, 19].

Immunoblot staining shows a reduction in or lack of the filaggrin protein, especially with greater severity [2, 8], but filaggrin is also reduced in the skin of individuals with AD (see below), even without a *FLG* mutation [63]. Filaggrin-deficient skin without AD shows normal mRNA expression of other proteins of differentiation, such as loricrin and keratin [19]. A slight increase in the expression of proline-rich molecule SPRR2A has been detected, but there is no difference in skin lipid composition [38].

21.6 Comorbidities

Atopic disease is found overall in 37–50 % of patients with IV, especially with two filaggrin mutations, and AD is strongly associated with IV [1, 3, 5, 12, 14, 64]. Approximately 8 % of individuals with AD have clinical features of IV [3], although 14–56 % of AD patients have a *FLG* mutation. Filaggrin mutations are the strongest genetic risk for developing AD [5, 65–69]. The odds ratio of developing AD with a *FLG* mutation has been estimated to be between 3.12 and 4.78 [64, 69]. The eczema subtype associated with *FLG* mutations is that of early-onset, severe, and persistent disease [41]. Copy number variation in number of filaggrin repeats also affects the chances of developing AD; an increased number of units (e.g., 12 repeats) may be protective against developing atopic dermatitis, while a lower number (e.g., 10 repeats) increases the risk [10, 20].

Filaggrin is the front line of defense in the cornified envelope that protects the skin from environmental substances that can trigger an immune response (which relates to the outside-in and inside-out model of AD) [1, 18]. The barrier abnormality produced by deficient filaggrin and the subsequent increase in permeability of allergens and haptens into the epidermis result in a Th2-weighted immune imbalance that typifies acute AD [23, 70]. This Th2 inflammation itself downregulates *FLG* expression, initiating a vicious cycle [19]. Interleukin 22 (IL-22), an important cytokine in chronic AD, and pro-inflammatory Th2 cytokines, such as IL-4 and IL-13, have also been shown to also downregulate filaggrin, further exacerbating the ichthyosis phenotype and contributing to the barrier impairment in genetically normal individuals with AD [71]. The role of filaggrin downregulation and barrier disruption as a primary effect or secondary to immune activation in the pathogenesis of AD has yet to be elucidated.

AD is often the first clinical manifestation of atopy during infancy or even the neonatal period. Since IV does not clinically manifest before about 3 months of age (and often later), AD with its cutaneous inflammation in association with dryness and scaling can be the first clinical manifestation of IV. Other atopic disorders, such as asthma and allergic rhinitis, generally appear subsequently through the "atopic march." Twenty to 25 % of patients with asthma have a *FLG* mutation, but in general, *FLG* mutations increase the risk of asthma only in patients with preceding or coexistent AD [5, 14, 72]. The strong association between *FLG* mutations and asthma is usually seen in individuals with initial AD, with AD typically presenting first [73]. Allergic rhinitis or high IgE levels have been noted in 42 % of patients (although only 23 % of these patients had a history of AD) [14]. Food sensitization and concomitant *FLG* mutations increase the positive predictive value of developing childhood asthma [74]. Double allele *FLG* mutations result in an earlier onset of AD, a more severe course, aeroallergen sensitization, and chronic disease; there is also an amplified risk of allergic rhinitis, food allergies, and severe asthma [75]. *FLG* mutations are associated with a higher risk of egg, milk, fish, and wheat allergy [76], as well as with clinically significant peanut allergy, even after controlling for confounding with AD (OR of 5.3) [77]. These findings suggest that the impaired epidermal barrier with *FLG* mutations allows penetration of food allergens which increase the risk of percutaneous sensitization, but does not necessarily lead to clinically relevant food allergies [78]. Filaggrin-deficient mice, with their homozygous *FLG* mutations, display increased allergen permeability and subsequent development of antigen-specific antibodies; these mice also develop AD at lower doses of hapten and have a reduced threshold for development of allergic and contact dermatitis when compared to wild-type mice [23, 70]. *FLG*-deficient mice are also more susceptible to dermatitis produced by the dust mite *Dermatophagoides pteronyssinus* [79].

Although predominantly linked to AD as a comorbidity, *FLG* mutations have also been associated with alopecia areata [80], nickel sensitization [42], irritant contact dermatitis [81], allergic contact dermatitis (including in mouse models) [42, 70, 82], eczema herpeticum (especially the R501X mutation) [83], and hand eczema [84]. One study has suggested a relationship between a p.K4022X *FLG* mutation and psoriasis vulgaris in a patient with a family history of concomitant psoriasis and IV [85].

21.7 Treatment and Prevention

Treatment goals include removal of excess scales and hydration of the skin. Emollients, especially those with ceramides, and keratolytic therapy utilizing urea, α-hydroxy acids such as lactic or glycolic acids, or salicylic acid (a β-hydroxy acid) are the mainstays of therapy in IV [21]. Daily application of 5–15 % lactic or glycolic acid containing emollients, 10–25 % urea, or propylene glycol has been suggested [86].

Urea uptake has been shown to have antimicrobial activity and also the ability to alter epidermal barrier function; it has been demonstrated that urea treatment may increase lipid synthesis [35]. Topical application of 20 % urea upregulates multiple proteins, including involucrin, filaggrin, loricrin, transglutaminase-1, and epidermal antimicrobial markers (β-defensin-2 and CAMP), while also minimizing TEWL [87]. Alpha hydroxy acids have been noted to promote desquamation, improve hydration, induce ceramide synthesis [24, 35], and also strengthen barrier function; these acids also increase keratinocyte growth and thus lead to a more even skin surface [35].

Patients should be followed closely for development of AD and/or the development of other atopic disorders [5, 41, 69, 73, 88]. Early results suggest a reduced incidence of developing AD when at-risk neonates are coated with an emollient cream at least once daily after bathing [89]. Given evidence that cat ownership increases the risk of developing AD, especially with a *FLG* mutation, families with IV may choose to consider a dog rather than a cat as a pet [90]. Exposure to tobacco smoke has been shown to lead to greater impairment of respiratory function and eventual asthma in individuals with *FLG* mutations (even though filaggrin is not expressed in the pulmonary airways) [91], suggesting that infants and children at risk of IV should be protected from exposure to tobacco smoke [90].

It has been suggested that individuals with IV be warned against nickel, metals, and other contact irritants in the home, given the possible increased risk of contact sensitization and hand eczema [42, 82]. Avoidance of wet work is also suggested for individuals with *FLG* mutations, and use of prophylactic emollient has been recommended to prevent fissuring of the hands [92]. Individuals with IV may also have increased sun sensitivity due to decreased UCA levels and subsequent UVB-induced cellular apoptosis; wearing sun-protective clothing prior to outdoor activities and sunscreen application should be highlighted [38].

21.8 The Role of Filaggrin in Other Ichthyoses and Disorders of Cornification

Individuals with X-linked recessive ichthyosis, caused by a mutation in the steroid sulfatase (arylsulfatase C) gene, show a more severe phenotype with a concomitant mutation in *FLG* when compared to affected family members with wild-type *FLG* [52, 93].

Acquired ichthyotic scaling (ichthyosis acquisita), a nonheritable form of ichthyosis, has many causes but is most commonly linked to Hodgkin's lymphoma [94, 95]. The role of filaggrin deficiency in acquired ichthyosis is unclear; however, onset is during adulthood, family history is usually lacking, and successful treatment of the underlying disease process tends to clear the scaling [94].

21.9 Future Considerations

The *FLG* promoter in humans has been found to contain retinoic acid and glucocorticoid response elements and is also regulated by PPARγ (peroxisome proliferator-activated receptor) [96]. In mice capable of expressing PPARγ, topical treatment with PPARγ agonists increases pro-filaggrin expression, suggesting that PPARγ activity modulates pro-filaggrin expression directly and promoting the use of topically applied PPARγ agonists as treatment [96]. Rosiglitazone, a PPARγ agonist, has been shown to lead to an improvement in AD, though *FLG* mutation status was not determined, and may be an appropriate topical treatment for IV with further testing [97]. Vigorous research is ongoing to identify compounds that can upregulate filaggrin expression in heterozygotes or even correct the nonsense *FLG* mutation for individuals with null mutations in both alleles [43].

References

1. Smith FJ, Irvine AD, Terron-Kwiatkowski A, Sandilands A, Campbell LE, Zhao Y, et al. Loss-of-function mutations in the gene encoding filaggrin cause ichthyosis vulgaris. Nat Genet. 2006;38(3):337–42.
2. Sybert VP, Dale BA, Holbrook KA. Ichthyosis vulgaris: identification of a defect in synthesis of filaggrin correlated with an absence of keratohyaline granules. J Invest Dermatol. 1985;84(3):191–4.
3. Wells RS, Kerr CB. Clinical features of autosomal dominant and sex-linked ichthyosis in an English population. Br Med J. 1966;1(5493):947–50.
4. Elaut L. The first illustrated treatise on dermatology, J. L. Alibert's Description des Maladies de la Peau. Ned Tijdschr Geneeskd. 1958;102(36):1749–53.
5. Palmer CN, Irvine AD, Terron-Kwiatkowski A, Zhao Y, Liao H, Lee SP, et al. Common loss-of-function variants of the epidermal barrier protein filaggrin are a major predisposing factor for atopic dermatitis. Nat Genet. 2006;38(4):441–6.
6. Sandilands A, Terron-Kwiatkowski A, Hull PR, O'Regan GM, Clayton TH, Watson RM, et al. Comprehensive analysis of the gene encoding filaggrin uncovers prevalent and rare mutations in ichthyosis vulgaris and atopic eczema. Nat Genet. 2007; 39(5):650–4.
7. Mischke D, Korge BP, Marenholz I, Volz A, Ziegler A. Genes encoding structural proteins of epidermal cornification and S100 calcium-binding proteins form a gene complex ("epidermal differentiation complex") on human chromosome 1q21. J Invest Dermatol. 1996;106(5):989–92.
8. Compton JG, DiGiovanna JJ, Johnston KA, Fleckman P, Bale SJ. Mapping of the associated phenotype of an absent granular layer in ichthyosis vulgaris to the epidermal differentiation complex on chromosome 1. Exp Dermatol. 2002;11(6):518–26.
9. Winge MC, Hoppe T, Berne B, Vahlquist A, Nordenskjold M, Bradley M, et al. Filaggrin genotype determines functional and molecular alterations in skin of patients with atopic dermatitis and ichthyosis vulgaris. PLoS One. 2011;6(12):e28254.
10. Brown SJ, McLean WH. One remarkable molecule: filaggrin. J Invest Dermatol. 2012;132(3 Pt 2):751–62.
11. Wells RS, Jennings MC. X-linked ichthyosis and ichthyosis vulgaris. Clinical and genetic distinctions in a second series of families. JAMA. 1967;202(6):485–8.
12. Sandilands A, O'Regan GM, Liao H, Zhao Y, Terron-Kwiatkowski A, Watson RM, et al. Prevalent and rare mutations in the gene encoding filaggrin cause ichthyosis vulgaris and predispose individuals to atopic dermatitis. J Invest Dermatol. 2006;126(8):1770–5.
13. Gruber R, Janecke AR, Fauth C, Utermann G, Fritsch PO, Schmuth M. Filaggrin mutations p.R501X and c.2282del4 in ichthyosis vulgaris. Eur J Hum Genet. 2007;15(2):179–84.
14. Oji V, Seller N, Sandilands A, Gruber R, Gerss J, Huffmeier U, et al. Ichthyosis vulgaris: novel FLG mutations in the German population and high presence of CD1a+cells in the epidermis of the atopic subgroup. Br J Dermatol. 2009;160(4):771–81.
15. Sinclair C, O'Toole EA, Paige D, El Bashir H, Robinson J, Dobson R, et al. Filaggrin mutations are associated with ichthyosis vulgaris in the Bangladeshi population. Br J Dermatol. 2009;160(5):1113–5.
16. McGrath JA, Uitto J. The filaggrin story: novel insights into skin-barrier function and disease. Trends Mol Med. 2008;14(1):20–7.
17. Marenholz I, Volz A, Ziegler A, Davies A, Ragoussis I, Korge BP, et al. Genetic analysis of the epidermal differentiation complex (EDC) on human chromosome 1q21: chromosomal orientation, new markers, and a 6-Mb YAC contig. Genomics. 1996;37(3):295–302.
18. Sandilands A, Sutherland C, Irvine AD, McLean WH. Filaggrin in the frontline: role in skin barrier function and disease. J Cell Sci. 2009;122(Pt 9):1285–94.
19. Gruber R, Elias PM, Crumrine D, Lin TK, Brandner JM, Hachem JP, et al. Filaggrin genotype in ichthyosis vulgaris predicts abnormalities in epidermal structure and function. Am J Pathol. 2011;178(5):2252–63.
20. Brown SJ, Kroboth K, Sandilands A, Campbell LE, Pohler E, Kezic S, et al. Intragenic copy number variation within filaggrin contributes to the risk of atopic dermatitis with a dose-dependent effect. J Invest Dermatol. 2012;132(1):98–104.
21. Thyssen JP, Godoy-Gijon E, Elias PM. Ichthyosis vulgaris – the filaggrin mutation disease. Br J Dermatol. 2013;168:1155–66.
22. Nirunsuksiri W, Zhang SH, Fleckman P. Reduced stability and bi-allelic, coequal expression of profilaggrin mRNA in keratinocytes cultured from subjects with ichthyosis vulgaris. J Invest Dermatol. 1998;110(6):854–61.
23. Scharschmidt TC, Man MQ, Hatano Y, Crumrine D, Gunathilake R, Sundberg JP, et al. Filaggrin deficiency confers a paracellular barrier abnormality that reduces inflammatory thresholds to irritants and haptens. J Allergy Clin Immunol. 2009;124(3):496–506, e1–6.
24. Rawlings AV, Harding CR. Moisturization and skin barrier function. Dermatol Ther. 2004;17 Suppl 1:43–8.
25. Scott IR, Harding CR. Filaggrin breakdown to water binding compounds during development of the rat

stratum corneum is controlled by the water activity of the environment. Dev Biol. 1986;115(1):84–92.
26. Schmuth M, Gruber R, Elias PM, Williams ML. Ichthyosis update: towards a function-driven model of pathogenesis of the disorders of cornification and the role of corneocyte proteins in these disorders. Adv Dermatol. 2007;23:231–56.
27. Hoste E, Kemperman P, Devos M, Denecker G, Kezic S, Yau N, et al. Caspase-14 is required for filaggrin degradation to natural moisturizing factors in the skin. J Invest Dermatol. 2011;131(11):2233–41.
28. Winge MC, Bilcha KD, Lieden A, Shibeshi D, Sandilands A, Wahlgren CF, et al. Novel filaggrin mutation but no other loss-of-function variants found in Ethiopian patients with atopic dermatitis. Br J Dermatol. 2011;165(5):1074–80.
29. Kezic S, Kemperman PM, Koster ES, de Jongh CM, Thio HB, Campbell LE, et al. Loss-of-function mutations in the filaggrin gene lead to reduced level of natural moisturizing factor in the stratum corneum. J Invest Dermatol. 2008;128(8):2117–9.
30. Presland RB, Boggess D, Lewis SP, Hull C, Fleckman P, Sundberg JP. Loss of normal pro-filaggrin and filaggrin in flaky tail (ft/ft) mice: an animal model for the filaggrin-deficient skin disease ichthyosis vulgaris. J Invest Dermatol. 2000;115(6):1072–81.
31. Denecker G, Hoste E, Gilbert B, Hochepied T, Ovaere P, Lippens S, et al. Caspase-14 protects against epidermal UVB photodamage and water loss. Nat Cell Biol. 2007;9(6):666–74.
32. Gibbs NK, Norval M. Urocanic acid in the skin: a mixed blessing? J Invest Dermatol. 2011;131(1):14.
33. Miajlovic H, Fallon PG, Irvine AD, Foster TJ, Miajlovic H, Fallon PG, Irvine AD, Foster TJ. Effect of filaggrin breakdown products on growth of and protein expression by Staphylococcus aureus. J Allergy Clin Immunol. 2010;126(6):1184–90 e3.
34. Egelrud T. Desquamation in the stratum corneum. Acta Derm Venereol Suppl. 2000;208:44–5.
35. Harding CR, Watkinson A, Rawlings AV, Scott IR. Dry skin, moisturization and corneodesmolysis. Int J Cosmet Sci. 2000;22(1):21–52.
36. Hachem JP, Man MQ, Crumrine D, Uchida Y, Brown BE, Rogiers V, et al. Sustained serine proteases activity by prolonged increase in pH leads to degradation of lipid processing enzymes and profound alterations of barrier function and stratum corneum integrity. J Invest Dermatol. 2005;125(3):510–20.
37. Demerjian M, Hachem JP, Tschachler E, Denecker G, Declercq W, Vandenabeele P, et al. Acute modulations in permeability barrier function regulate epidermal cornification: role of caspase-14 and the protease-activated receptor type 2. Am J Pathol. 2008;172(1):86–97.
38. Mildner M, Jin J, Eckhart L, Kezic S, Gruber F, Barresi C, et al. Knockdown of filaggrin impairs diffusion barrier function and increases UV sensitivity in a human skin model. J Invest Dermatol. 2010;130(9):2286–94.
39. Thyssen JP, Thuesen B, Huth C, Standl M, Carson CG, Heinrich J, et al. Skin barrier abnormality caused by filaggrin (FLG) mutations is associated with increased serum 25-hydroxyvitamin D concentrations. J Allergy Clin Immunol. 2012;130(5):1204–7 e2.
40. Irvine AD, McLean WH. Breaking the (un)sound barrier: filaggrin is a major gene for atopic dermatitis. J Invest Dermatol. 2006;126(6):1200–2.
41. Brown SJ, Relton CL, Liao H, Zhao Y, Sandilands A, Wilson IJ, et al. Filaggrin null mutations and childhood atopic eczema: a population-based case-control study. J Invest Dermatol. 2008;121(4):940–46 e3.
42. Thyssen JP, Johansen JD, Linneberg A, Menne T, Nielsen NH, Meldgaard M, et al. The association between null mutations in the filaggrin gene and contact sensitization to nickel and other chemicals in the general population. Br J Dermatol. 2010;162(6):1278–85.
43. Irvine AD, McLean WH, Leung DY. Filaggrin mutations associated with skin and allergic diseases. N Engl J Med. 2011;365(14):1315–27.
44. Giardina E, Paolillo N, Sinibaldi C, Novelli G. R501X and 2282del4 filaggrin mutations do not confer susceptibility to psoriasis and atopic dermatitis in Italian patients. Dermatology. 2008;216(1):83–4.
45. Cascella R, Foti Cuzzola V, Lepre T, Galli E, Moschese V, Chini L, et al. Full sequencing of the FLG gene in Italian patients with atopic eczema: evidence of new mutations, but lack of an association. J Invest Dermatol. 2011;131(4):982–4.
46. Lee DE, Park SY, Han JY, Ryu HM, Lee HC, Han YS. Association between filaggrin mutations and atopic dermatitis in Korean pregnant women. Int J Dermatol. 2013;52:772–3.
47. Xiong Z, Luo S, Xu X, Zhang L, Peng H, Li W, et al. Novel FLG mutations associated with ichthyosis vulgaris in the Chinese population. Clin Exp Dermatol. 2012;37(2):177–80.
48. Li M, Cheng R, Shi M, Liu J, Zhang G, Liu Q, et al. Analyses of FLG mutation frequency and filaggrin expression in isolated ichthyosis vulgaris (IV) and atopic dermatitis-associated IV. Br J Dermatol. 2013;168:1335–8.
49. Chen H, Ho JC, Sandilands A, Chan YC, Giam YC, Evans AT, et al. Unique and recurrent mutations in the filaggrin gene in Singaporean Chinese patients with ichthyosis vulgaris. J Invest Dermatol. 2008;128(7):1669–75.
50. Zhang X, Liu S, Chen X, Zhou B, Liu D, Lei G, et al. Novel and recurrent mutations in the filaggrin gene in Chinese patients with ichthyosis vulgaris. Br J Dermatol. 2010;163(1):63–9.
51. Hsu CK, Akiyama M, Nemoto-Hasebe I, Nomura T, Sandilands A, Chao SC, et al. Analysis of Taiwanese ichthyosis vulgaris families further demonstrates differences in FLG mutations between European and Asian populations. Br J Dermatol. 2009;161(2):448–51.
52. Ramesh R, Chen H, Kukula A, Wakeling EL, Rustin MH, McLean WH. Exacerbation of X-linked ichthyosis phenotype in a female by inheritance of filaggrin and steroid sulfatase mutations. J Dermatol Sci. 2011;64(3):159–62.

53. Naz N, Samdani AJ. Detection of filaggrin gene mutation (2282del4) in Pakistani Ichthyosis vulgaris families. J Coll Physicians Surg Pak. 2011;21(6):382–3.
54. Samdani AJ, Naz N, Ahmed N. Molecular studies of ichthyosis vulgaris in Pakistani families. J Coll Physicians Surg Pak. 2010;20(10):644–8.
55. Ziprkowski L, Feinstein A. A survey of ichthyosis vulgaris in Israel. Br J Dermatol. 1972;86(1):1–8.
56. Horii I, Nakayama Y, Obata M, Tagami H. Stratum corneum hydration and amino acid content in xerotic skin. Br J Dermatol. 1989;121(5):587–92.
57. Frost P, Van Scott EJ. Ichthyosiform dermatoses. Classification based on anatomic and biometric observations. Arch Dermatol. 1966;94(2):113–26.
58. Oji V, Traupe H. Ichthyosis: clinical manifestations and practical treatment options. Am J Clin Dermatol. 2009;10(6):351–64.
59. Ezzedine K, Droitcourt C, Ged C, Diallo A, Hubiche T, de Verneuil H, et al. Usefulness of a global clinical ichthyosis vulgaris scoring system for predicting common FLG null mutations in an adult caucasian population. Br J Dermatol. 2012;167(5):1165–9.
60. Brown SJ, Relton CL, Liao H, Zhao Y, Sandilands A, McLean WH, et al. Filaggrin haploinsufficiency is highly penetrant and is associated with increased severity of eczema: further delineation of the skin phenotype in a prospective epidemiological study of 792 school children. Br J Dermatol. 2009;161(4):884–9.
61. Nomura T, Akiyama M, Sandilands A, Nemoto-Hasebe I, Sakai K, Nagasaki A, et al. Prevalent and rare mutations in the gene encoding filaggrin in Japanese patients with ichthyosis vulgaris and atopic dermatitis. J Invest Dermatol. 2009;129(5):1302–5.
62. Kanitakis J, Hoyo E, Chouvet B, Thivolet J, Faure M, Claudy A. Keratinocyte proliferation in epidermal keratinocyte disorders evaluated through PCNA/cyclin immunolabelling and AgNOR counting. Acta Derm Venereol Suppl. 1993;73(5):370–5.
63. Gittler JK, Shemer A, Suarez-Farinas M, Fuentes-Duculan J, Gulewicz KJ, Wang CQ, et al. Progressive activation of T(H)2/T(H)22 cytokines and selective epidermal proteins characterizes acute and chronic atopic dermatitis. J Allergy Clin Immunol. 2012;130(6):1344–54.
64. Rodriguez E, Baurecht H, Herberich E, Wagenpfeil S, Brown SJ, Cordell HJ, et al. Meta-analysis of filaggrin polymorphisms in eczema and asthma: robust risk factors in atopic disease. J Allergy Clin Immunol. 2009;123(6):1361–70 e7.
65. Ruether A, Stoll M, Schwarz T, Schreiber S, Folster-Holst R. Filaggrin loss-of-function variant contributes to atopic dermatitis risk in the population of Northern Germany. Br J Dermatol. 2006;155(5):1093–4.
66. Irvine AD. Fleshing out filaggrin phenotypes. J Invest Dermatol. 2007;127(3):504–7.
67. Baurecht H, Irvine AD, Novak N, Illig T, Buhler B, Ring J, et al. Toward a major risk factor for atopic eczema: meta-analysis of filaggrin polymorphism data. J Allergy Clin Immunol. 2007;120(6):1406–12.
68. De D, Handa S. Filaggrin mutations and the skin. Indian J Dermatol Venereol Leprol. 2012;78(5):545–51.
69. van den Oord RA, Sheikh A. Filaggrin gene defects and risk of developing allergic sensitisation and allergic disorders: systematic review and meta-analysis. BMJ. 2009;339:b2433.
70. Fallon PG, Sasaki T, Sandilands A, Campbell LE, Saunders SP, Mangan NE, et al. A homozygous frameshift mutation in the mouse Flg gene facilitates enhanced percutaneous allergen priming. Nat Genet. 2009;41(5):602–8.
71. Howell MD, Kim BE, Gao P, Grant AV, Boguniewicz M, Debenedetto A, et al. Cytokine modulation of atopic dermatitis filaggrin skin expression. J Allergy Clin Immunol. 2007;120(1):150–5.
72. Rogers AJ, Celedon JC, Lasky-Su JA, Weiss ST, Raby BA. Filaggrin mutations confer susceptibility to atopic dermatitis but not to asthma. J Allergy Clin Immunol. 2007;120(6):1332–7.
73. Marenholz I, Nickel R, Ruschendorf F, Schulz F, Esparza-Gordillo J, Kerscher T, et al. Filaggrin loss-of-function mutations predispose to phenotypes involved in the atopic march. J Allergy Clin Immunol. 2006;118(4):866–71.
74. Marenholz I, Kerscher T, Bauerfeind A, Esparza-Gordillo J, Nickel R, Keil T, et al. An interaction between filaggrin mutations and early food sensitization improves the prediction of childhood asthma. J Allergy Clin Immunol. 2009;123(4):911–6.
75. Weidinger S, Rodriguez E, Stahl C, Wagenpfeil S, Klopp N, Illig T, et al. Filaggrin mutations strongly predispose to early-onset and extrinsic atopic dermatitis. J Invest Dermatol. 2007;127(3):724–6.
76. Linneberg A, Fenger RV, Husemoen LL, Thuesen BH, Skaaby T, Gonzalez-Quintela A, et al. Association between loss-of-function mutations in the filaggrin gene and self-reported food allergy and alcohol sensitivity. Int Arch Allergy Immunol. 2013;161(3):234–42.
77. Brown SJ, Asai Y, Cordell HJ, Campbell LE, Zhao Y, Liao H, et al. Loss-of-function variants in the filaggrin gene are a significant risk factor for peanut allergy. J Allergy Clin Immunol. 2011;127(3):661–7.
78. Tan HT, Ellis JA, Koplin JJ, Matheson MC, Gurrin LC, Lowe AJ, et al. Filaggrin loss-of-function mutations do not predict food allergy over and above the risk of food sensitization among infants. J Allergy Clin Immunol. 2012;130(5):1211–3 e3.
79. Moniaga CS, Egawa G, Kawasaki H, Hara-Chikuma M, Honda T, Tanizaki H, et al. Flaky tail mouse denotes human atopic dermatitis in the steady state and by topical application with Dermatophagoides pteronyssinus extract. Am J Pathol. 2010;176(5):2385–93.
80. Betz RC, Pforr J, Flaquer A, Redler S, Hanneken S, Eigelshoven S, et al. Loss-of-function mutations in the filaggrin gene and alopecia areata: strong risk factor for a severe course of disease in patients comorbid for atopic disease. J Invest Dermatol. 2007;127(11):2539–43.
81. Visser MJ, Landeck L, Campbell LE, McLean WH, Weidinger S, Calkoen F, et al. Impact of atopic dermatitis and loss-of-function mutations in the filaggrin

gene on the development of occupational irritant contact dermatitis. Br J Dermatol. 2013;168(2):326–32.
82. Novak N, Baurecht H, Schafer T, Rodriguez E, Wagenpfeil S, Klopp N, et al. Loss-of-function mutations in the filaggrin gene and allergic contact sensitization to nickel. J Invest Dermatol. 2008;128(6):1430–5.
83. Gao PS, Rafaels NM, Hand T, Murray T, Boguniewicz M, Hata T, et al. Filaggrin mutations that confer risk of atopic dermatitis confer greater risk for eczema herpeticum. J Allergy Clin Immunol. 2009;124(3):507–13, 13.e1–7.
84. Thyssen JP, Carlsen BC, Menne T, Linneberg A, Nielsen NH, Meldgaard M, et al. Filaggrin null mutations increase the risk and persistence of hand eczema in subjects with atopic dermatitis: results from a general population study. Br J Dermatol. 2010;163(1):115–20.
85. Hu Z, Xiong Z, Xu X, Li F, Lu L, Li W, et al. Loss-of-function mutations in filaggrin gene associate with psoriasis vulgaris in Chinese population. Hum Genet. 2012;131(7):1269–74.
86. Vahlquist A, Ganemo A, Virtanen M. Congenital ichthyosis: an overview of current and emerging therapies. Acta Derm Venereol. 2008;88(1):4–14.
87. Grether-Beck S, Felsner I, Brenden H, Kohne Z, Majora M, Marini A, et al. Urea uptake enhances barrier function and antimicrobial defense in humans by regulating epidermal gene expression. J Invest Dermatol. 2012;132(6):1561–72.
88. Weidinger S, O'Sullivan M, Illig T, Baurecht H, Depner M, Rodriguez E, et al. Filaggrin mutations, atopic eczema, hay fever, and asthma in children. J Allergy Clin Immunol. 2008;121(5):1203–9 e1.
89. Simpson EL, Berry TM, Brown PA, Hanifin JM. A pilot study of emollient therapy for the primary prevention of atopic dermatitis. J Am Acad Dermatol. 2010;63(4):587–93.
90. Bisgaard H, Simpson A, Palmer CN, Bonnelykke K, McLean I, Mukhopadhyay S, et al. Gene-environment interaction in the onset of eczema in infancy: filaggrin loss-of-function mutations enhanced by neonatal cat exposure. PLoS Med. 2008;5(6):e131.
91. Berg ND, Husemoen LL, Thuesen BH, Hersoug LG, Elberling J, Thyssen JP, et al. Interaction between filaggrin null mutations and tobacco smoking in relation to asthma. J Allergy Clin Immunol. 2012;129(2):374–80, 80.e1–2.
92. Thyssen JP, Ross-Hansen K, Johansen JD, Zachariae C, Carlsen BC, Linneberg A, et al. Filaggrin loss-of-function mutation R501X and 2282del4 carrier status is associated with fissured skin on the hands: results from a cross-sectional population study. Br J Dermatol. 2012;166(1):46–53.
93. Liao H, Waters AJ, Goudie DR, Aitken DA, Graham G, Smith FJ, et al. Filaggrin mutations are genetic modifying factors exacerbating X-linked ichthyosis. J Invest Dermatol. 2007;127(12):2795–8.
94. Patel N, Spencer LA, English 3rd JC, Zirwas MJ. Acquired ichthyosis. J Am Acad Dermatol. 2006;55(4):647–56.
95. Dykes PJ, Marks R. Acquired ichthyosis: multiple causes for an acquired generalized disturbance in desquamation. Br J Dermatol. 1977;97(3):327–34.
96. Mao-Qiang M, Fowler AJ, Schmuth M, Lau P, Chang S, Brown BE, et al. Peroxisome-proliferator-activated receptor (PPAR)-gamma activation stimulates keratinocyte differentiation. J Invest Dermatol. 2004;123(2):305–12.
97. Behshad R, Cooper KD, Korman NJ. A retrospective case series review of the peroxisome proliferator-activated receptor ligand rosiglitazone in the treatment of atopic dermatitis. Arch Dermatol. 2008;144(1):84–8.

Xerosis Means "Dry Skin": Mechanisms, Skin Conditions, and Its Management

22

Ana M. Gimenez-Arnau

Contents

A.M. Gimenez-Arnau, MD, PhD, MB BS
Department of Dermatology, Hospital del Mar
Universitat Autonoma de Barcelona,
Passeig MAritim 25-26, Barcelona,
Catalunya 08003, Spain
e-mail: 22505aga@comb.cat

Dry skin is common in the general population. Although there is no general consensus on a definition, it is generally agreed that the basic characteristic of the disorder is the presence of rough, flaky skin that has lost its normal mechanical properties. It can be the trigger factor and also the main symptom of certain dermatoses. The term "xerosis" is often used to refer to the concept of dry skin. Severe xerosis can lead to the onset of, for example, a type of eczema characterized by intensely itchy, fissured, and cracked skin called xerotic eczema or eczema craquele. Dry skin affects the patient's quality of life, and severe xerosis can interfere with work productivity, especially when the hands are affected.

22.1 Epidermal Homeostasis and Xerosis

The epidermis is the skin layer mostly involved in the pathogenic mechanisms that make the skin dry.

J.P. Thyssen, H. Maibach (eds.), *Filaggrin*,
DOI 10.1007/978-3-642-54379-1_22, © Springer-Verlag Berlin Heidelberg 2014

22.1.1 The Stratum Corneum

The epidermis is an avascular and stratified keratinized epithelium that undergoes constant renewal. The main physiologic epidermal process involved in maintaining epidermal homeostasis is keratinocyte differentiation. Disruption of the orderly and regulated stacking of corneocytes is one of the principal causes of dry skin. Acellular components of the stratum corneum include structural proteins, intercellular lipids, "natural moisturizing factor" (NMF), and enzyme systems. The metabolic and enzyme activity of the stratum corneum plays a key role in maintaining barrier functions and the skin's water content. The process involves the formation of lipids and envelope proteins (e.g., filaggrin) and the subsequent degradation of intercellular bonds (by way of corneodesmolysis) to produce physiologic desquamation. The stratum corneum response to certain external stimuli induces enzyme activity, promoting the synthesis of ceramides and NMF, both crucial in maintaining optimum skin hydration. Enzymes (particularly proteases, glucosidases, and phosphates) play roles in transforming glucoceramides into ceramides and breaking down corneocyte cohesion. The activity of these proteins is modulated by factors such as pH, temperature, and hydration. A decline in lipid content gives rise to an increase in insensible water loss, which in turn destabilizes the optimum environment for epidermal enzyme activity. This change affects the process of corneocyte maturation and inhibits cell desquamation. In this way, the alterations in enzyme activity initiate a self-perpetuating cycle that gives rise to persistent dry skin [1, 2].

22.1.2 Stratum Corneum Lipids

Lipid deficiency impairs stratum corneum hydration affecting the elasticity and flexibility of healthy skin. The intercellular lipid component is arranged in bilayers between the corneocytes. These lipids are formed in the lamellar or Odland bodies of the stratum granulosum and then migrate upward into the stratum corneum during the epidermal differentiation [3, 4]. The lipids secreted are the substrate of the enzymes that transform the glucosphingolipids into a mixture of nonpolar lipids, including ceramides (50 %), free sterols, essential and nonessential free fatty acids (10–2 0 %), and cholesterol (25 %) [2, 5]. Ceramides are the main source of essential fatty acids such as linoleic acid, a structural element that plays a key role in epidermal barrier function [5].

22.1.3 Stratum Corneum Peptides [6]

Filaggrin and involucrin are the major structural proteins in the stratum corneum. Filaggrin is produced as the precursor proprotein profilaggrin. Profilaggrin is expressed in terminally differentiating keratinocytes in the outmost layers of the human epidermis. It is a constituent of the keratohyalin granules in the stratum granulosum. During terminal differentiation at the granular to cornified layer, profilaggrin is dephosphorylated and cleaved by endoproteases (e.g., recombinant SASPase [7, 8], elastase-2 [9], serine proteases matriptase/MT-SP1 [10], or prostasin [11]) to generate functional filaggrin monomers. Filaggrin monomers bind to keratin 1 and 10 and other intermediate filament proteins within the keratinocyte cytoskeleton to form tight bundles contributing to mechanical strength and integrity of the stratum corneum, facilitating the collapse and flattening of cells in the outermost stratum corneum to produce squames. Filaggrin is a histidine-rich protein, and its metabolites include the organic acids trans-urocanic acid (trans-UCA) and pyrroliodone-5-carboxylic acid. Filaggrin breakdown products, together with chloride and sodium ions, lactate, and urea, form the NMF, which contributes to epidermal hydration and barrier functions. The filaggrin gene (*FLG*) is located in the epidermal differentiation complex on chromosome 1q21. Exon 3 of the *FLG* is a large exon encoding almost the entire profilaggrin protein. There are common size-variant *FLG* alleles in the general population with 10, 11, or 12 repeats. The number of filaggrin units in human subjects varies from 20 to 24, excluding null mutations. Since the discovery in 2006 that loss-of-function mutations in the *FLG* are the cause of ichthyosis vulgaris, a disorder of keratinization and also a genetic risk factor for atopic dermatitis (AD), *FLG* mutations acquired a special relevance in the study of eczema independent of sensitization [12]. Now we also know that heterozygous null alleles in the *FLG* contribute to clinical dry skin in young adults and the elderly [13].

22.1.4 Natural Moisturizing or Natural Hydration Factor

The physiologic water content of a healthy stratum corneum is 15–20 %. When the water content of the cornified layer falls below 10 %, the skin acquires a rough, dry appearance [2]. To prevent water loss, the epidermis contains hygroscopic substances such as the NMF, a compound composed of a mixture of amino acids, amino acid derivatives, and salts generated by filaggrin hydrolysis [2]. The water absorbed by the NMF from the environment and from inside the skin acts as an intracellular plasticizer in the stratum corneum ensuring that the corneocytes retain their turgidity and preventing abnormal skin cracking and desquamation [2]. Decreases in NMF levels and its content (lactate, potassium, sodium, and chlorine) in the stratum corneum induce decreased skin hydration and flexibility [14]. Dehydration impairs the skin's capacity to keep out irritants and potentially harmful pathogens [15]. The stratum corneum contains other moisturizing components as hyaluronic acid, glycerol, and lactate (independent from filaggrin or urea) that play also a role in maintaining the physical epidermal properties [16].

22.1.5 Corneocyte Binding

Specialized desmosomes, the corneodesmosomes, bind the corneocytes together. These transmembrane glycoprotein complexes are located in the corneocyte envelope [16]. Both desmoglein 1 and desmocollin 1, members of the cadherin family of molecules, are their basic components. These proteins, which span the corneocyte envelope and are embedded into the intercellular lipid cement, increase cohesion between adjacent cells. The keratin filaments bind through the corneodesmosomal plaques, which are composed of plakoglobin, desmoplakins, and plakophilins. The links are formed by calcium-dependent transglutaminases (TGase). TGase 1, 3, and 5 are involved in corneocyte differentiation. The strength and flexibility of the corneocyte envelopes are ensured by the connections between several proteins such as involucrin, the small proline-rich proteins, envoplakin, and periplakin [16]. The enzymatic hydrolysis of the corneodesmosomes promotes epithelial desquamation. The stratum corneum chymotryptic enzyme (SCCE), stratum corneum thiol protease (SCTP or cathepsin L2), and cathepsin D and E are enzymes that play an important role in this process. Most of these enzymes are synthesized inside the lamellar bodies as proenzymes and become activated when they are secreted into the intercellular space. Their proteolytic activity is regulated by the water content, lipid content, and pH of the intercellular microenvironment.

22.1.6 Influence of Environmental Humidity on Stratum Corneum Differentiation

The environment influences the structure and functions of the stratum corneum and can induce xerosis or dry skin. It is well known that a decreased amount of epidermal ceramides and fatty acids—mainly in winter—with low air humidity induces dry skin. Dryness activates the mechanisms required to restore epidermal barrier function. Also, epidermal filaggrin is hydrolyzed in a humidity-sensitive fashion to its constituent amino acids and their deiminated carboxylic acid derivates, the NMF. This process normally begins as environmental humidity declines below 80 % and accelerates as humidity continues to decline [17]. Exposure to low humidity conditions induces synthesis of epidermal DNA and interleukin 1 [16]. Dry skin can be improved or, on the contrary, exacerbated by environmental seasonal changes [16].

22.2 Dry or Xerotic Skin Shows an Impaired Barrier Function

Xerosis and the disruption of normal skin function can occur when the epidermal water level falls below 10 % [18]. The stratum corneum maintains a water gradient between its innermost layer (where the water content is the same as that of the deeper skin layers) and its outer surface, which is in direct contact with the atmosphere. The level of hydration is modified by water diffusion from the dermis to the epidermis, water diffusion within the stratum corneum, and water loss through superficial evaporation. The skin's water content is made up of transepidermal water and retained water. Transepidermal water originates from circulating blood, migrates through the dermis into the epidermis, and eventually evaporates from the surface of the skin. This movement of water plays a key role in the supply of nutrients to the epidermis. The water retained in the stratum corneum is located between the lipid bilayers and inside the corneocytes. This static water content maintains the mechanical properties of the cornified layer, increases the plasticity of the epidermis, and enhances the hydrophilic properties of keratin.

The stratum corneum should prevent the loss of fluids and electrolytes from the skin, through the structure and function of its lipids, proteins, and cells. The recovery of the barrier function after an acute depletion of the stratum corneum lipid content is based on the reconstitution of the lipid-enriched intercellular substance, a process involving two distinct stages: an initial short phase and a second longer phase. To start, the response is fast, and the lamellar bodies' content is released in the stratum granulosum [5]. The synthesis of new lipids (ceramides, cholesterol, and fatty acids) is slower, involving enzymes that govern lipid formation. HMG-CoA reductase and acetyl-CoA carboxylase are the enzymes involved in cholesterol synthesis. Fatty acid synthase and serine palmitoyltransferase play a role in ceramide synthesis [5]. The inhibition of the enzymes involved in the synthesis of cholesterol, fatty acids, ceramides, and glucosylceramides delays the recovery of barrier function by causing a decline in the production and excretion of lamellar bodies, which in turn disrupts the formation of intercellular lipid bilayers [5].

The mixture of the required lipids in the correct proportions is essential for the reconstitution of the epidermal lipid bilayers. Topical application of only one or two of the three lipids required to maintain the intercellular barrier (ceramides, cholesterol, and fatty acids) could be detrimental to the quality of the bilayers. However, the topical application of a mixture of the three lipids in the correct proportions accelerates the repair of the epidermal barrier. It has been reported that physiologic lipids applied to the skin pass through the stratum corneum and are taken up by the lamellar bodies in the stratum granulosum. This implies that topical application of more physiologic lipids would contribute to the repair of the epidermal barrier, as they would not only form an occlusive layer on the skin but would also deliver the raw materials required to create new lamellar bodies [5].

22.3 Molecular Signals Involved in the Recovery of the Epidermal Homeostasis

Corneocyte protein expression and the stratum corneum homeostasis are regulated by nuclear hormone receptors (NHRs) and their ligands. Type I NHRs, which are steroid hormones (glucocorticoids, androgens, and estrogens), regulate skin permeability in newborn infants and the homeostasis of the epidermal barrier in adults. Estrogens and glucocorticoids accelerate the formation of the epidermal barrier, while androgens delay this process. Type II NHRs bind to thyroid hormone, retinoic acid, and 1,25 (OH)2vitamin D_3. This group also includes another type of receptor (called lipid-sensing receptors) that binds to endogenous lipids (free fatty acids, leukotrienes, prostanoids, and oxygenated sterols) to induce development of fetal skin and keratinocyte differentiation. These lipid-sensing receptors also act on the intercellular matrix, regulating the synthesis of corneocyte proteins (involucrin, loricrin, and TGase 1) and epidermal lipids. Their function is not restricted to the regulation of the physiologic composition of the stratum corneum as they can also correct anomalies in the epidermal barrier and reduce epidermal hyperplasia and dermal inflammation. For this reason, these receptors may have some potential as therapeutic agents in a number of skin diseases [5].

Lipid synthesis in the stratum corneum is precisely regulated. The formation of cholesterol and fatty acid is modulated by a group of sterol regulatory element-binding transcription factors. Type 2 binding protein regulates the formation of cholesterol and fatty acids in the corneocytes, while type 1 protein regulates fatty acid synthesis [5]. Ceramide synthesis is mainly modulated by serine palmitoyltransferase and by the availability of its substrate—palmitic acid—a product of fatty acid synthesis. The expression of serine palmitoyltransferase increases by inflammatory stimuli, such as ultraviolet radiation, certain endotoxins, and cytokines (tumor necrosis factor and interleukin 1). Expression of certain cytokines, growth factors, and other inflammatory mediators also intervenes in the homeostasis of the epidermal barrier and can disrupt barrier function in some skin diseases. Certain cytokines induce the synthesis and proliferation of lipids in the keratinocytes [5].

22.4 Epidermal Desquamation and Xerosis

Desquamation is a specific consequence of the epidermal turnover. Correct desquamation is as important as the physiologic cornification process. The appearance of dry skin can be the consequence of an abnormal desquamation. The cells detach from the epidermis as a result of the degradation of the corneodesmosomes induced by proteases. The most important enzymes involved in the desquamation are the stratum corneum chymotryptic and tryptic enzymes (also called kallikrein 7 and kallikrein 5, respectively). The tryptic enzyme and the chymotryptic enzyme act directly on the corneodesmosomes. The activity

Table 22.1 Substrates and inhibitors of the proteases involved in epidermal desquamation

Protease	Substrate	Antiprotease
SCTE	Desmoglein 1, pro-SCCE	LEKTI-1
SCCE	Desmocollin 1, corneodesmosin	SDALP, SLPI, LEKTI-1
Stratum corneum cysteine protease	Desmocollin 1, corneodesmosin	Cystatin E/M, cystatin a, SLPI
Cathepsin G	Desmocollin 1, corneodesmosin	SLPI

Abbreviations: *LEKTI-1* serine proteinase inhibitors, *SCCE* stratum corneum chymotryptic enzyme, *SCTE* stratum corneum tryptic enzyme, *SDALP* skin-derived antileukocyte proteinase (elafin), *SLPI* secretory leukocyte protease inhibitor (antileukoprotease)

of the chymotryptic enzyme is regulated by the tryptic enzyme, the epidermal lipids (cholesterol sulfate and free fatty acids), and other lipids. Its activity is also regulated by certain stratum corneum conditions as acid pH, decreased hydration, and low calcium levels that inhibit chymotryptic enzyme activity. Disruption of the epithelial barrier eliminates the inhibiting interstitial lipids, thereby increasing its water and calcium content and pH. This in turn leads to enhanced chymotryptic enzyme activity and, consequently, an increase in desquamation. A number of antiproteases have been identified that inhibit the activity of the proteases that facilitate epidermal desquamation (Table 22.1). Secretory leukocyte protease inhibitor (antileukoprotease) and skin-derived antileukocyte proteinase (elafin) are complementary. Antileukoprotease mainly inhibits chymotryptic enzyme activity, while skin-derived antileukocyte proteinase inhibits the tryptic enzyme. Both antileukoprotease and antileukocyte proteinase have a low baseline expression, which increases in psoriasis, during tissue scar formation, and when the epidermis is damaged. A defect in protease inhibitor function can give rise to structural anomalies in the lamellar membrane similar to those found in Netherton syndrome [5]. The balance between protease and inhibitor activities controls desquamation and thereby the number of corneocyte cell layers in the stratum corneum. Excess of protease activity

Table 22.2 Etiologic factors involved in development of dry skin or xerosis

Inherited predisposition
Age
Comorbid diseases Atopic dermatitis Psoriasis Hypothyroidism Intestinal malabsorption
Related to environmental conditions Temperature Humidity Exposure to sunlight Air conditioning Heating
Related to chemical agents Soaps and bath gels Lotions and perfumes Detergents Pharmacotherapy
Related to physical insult Friction Abrasion Radiation

can lead to stratum corneum thinning, while reduced protease activity can lead to stratum corneum thickening. The accumulation of corneocytes on the surface of the stratum corneum leads to the condition termed "dry skin." Reductions in enzyme activities, together with retention of corneodesmosomes in the upper layers of the stratum corneum, induce "dry skin" that can be observed in pathologic conditions such as psoriasis or noneczematous atopic xerosis [19].

22.5 Etiologic Factors Involved in Xerosis

Xerosis or dry skin can be the consequence of internal and external trigger factors to which any individual can be exposed (Table 22.2) [16]. Specific mechanisms can impair or disrupt stratum corneum barrier function as defective lipids, impaired ceramide synthesis, or reduced NMF levels. Deterioration of ceramide synthesis in the stratum corneum tends to perpetuate the dry skin cycle by disrupting lipid bilayer structure, inducing dysfunctional keratinocyte differentiation and amplifying the inflammatory response [20]. A number of authors have observed ceramide deficiencies in the healthy skin of patients with atopic diathesis, a finding that suggests that this population may have a predisposing factor for constitutional xerosis [20]. In both xerosis skin and AD, besides alterations in lipid levels, changes also occur in the composition of NMF and in the free amino acid content of the stratum corneum. A relationship between certain genetic abnormalities and AD has recently been reported, and mutations in the *FLG* have been identified in atopic patients from a number of different geographical areas [21]. Low levels of filaggrin have also been described in aged skin from healthy subjects, which supports the role played by this protein in the genesis of increased dry skin suffered by old people [22].

Environmental factors can also influence the composition of NMF. For example, a decrease in air humidity has been shown to reduce the generation of free amino acids in the stratum corneum, increasing skin dryness [23].

The reduction in intercellular lipid content and NMF that characterizes xerosis is both a cause and a consequence of abnormal epidermal differentiation. In fact, disruption of epidermal differentiation perpetuates the phenomenon of dry skin. As mentioned above, disruption of the epidermal barrier activates a metabolic response directed toward recovering epithelial homeostasis and reestablishing normal corneocyte differentiation. The main response is an increase in the biosynthesis of lipids, such as cholesterol, ceramides, and fatty acids. Slight disturbances of barrier function usually only affect the superficial epidermis, but repeated or severe damage gives rise to an inflammatory response that involves the deeper epidermal layers and even the endothelium [24, 25]. These phenomena give rise to abnormal keratinization and close the cycle that perpetuates the lesions.

22.6 Diseases and Skin Conditions Characterized by Xerosis

Dry skin is often the result of a combination of etiologic factors, in particular genetic abnormalities, but also metabolic and environmental triggers. The following are the most important skin diseases in which dry skin may be a symptom.

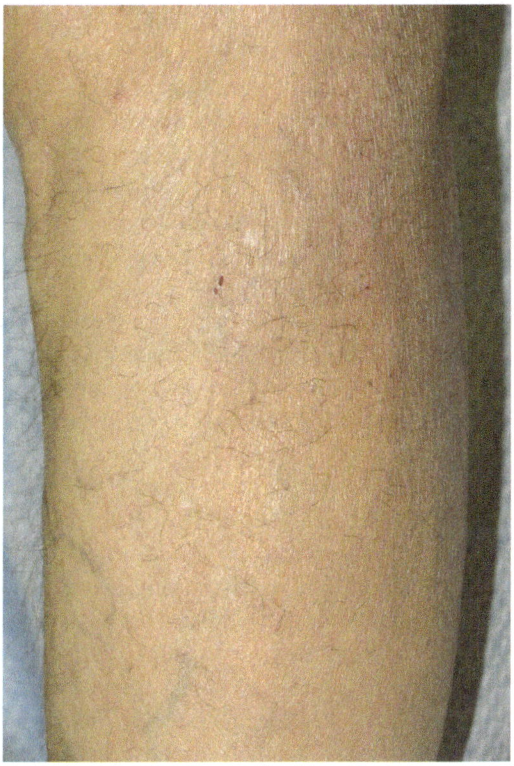

Fig. 22.1 Dry skin in a young adult during wintertime

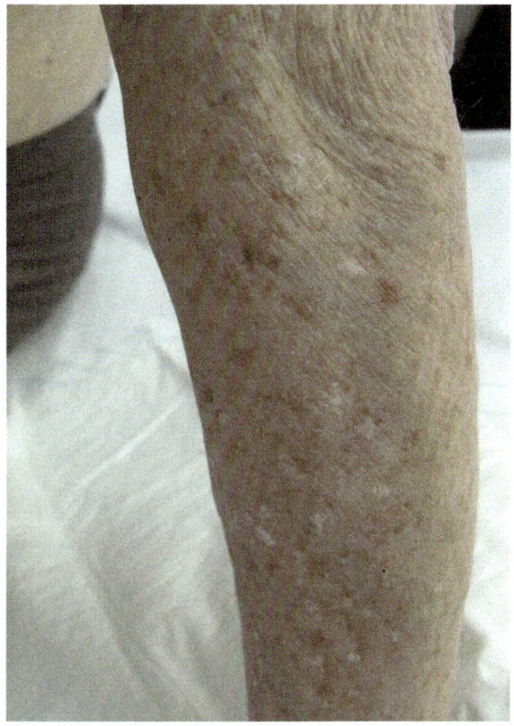

Fig. 22.2 Dry skin in older people is usually first noted on the lower limbs, and the disorder may spread to other areas of the body, such as the arms

22.6.1 Winter Xerosis

Cold weather and low ambient humidity during winter months are associated with decreased cutaneous hydration [26, 27]. Winter xerosis is aggravated by the warm, dry air produced by modern central heating systems [2] (Fig. 22.1).

22.6.2 Aged Skin

It is estimated that generalized or diffuse xerosis affects 75 % of individuals over 75 years of age and is the most common cause of itch in this age group. While dry skin in older people is usually first noted on the lower limbs, the disorder may spread to other areas of the body (Fig. 22.2). Itch is often more intense at night and after a hot bath. Changes in temperature, low air humidity, or exposure to solvents or detergents may worsen the dry skin. While there are many possible causes of dry skin in older people, the most common mechanism involved is a lower rate of epidermal proliferation compared with normal skin [28]. Skin aging involves other physiologic changes that may induce xerosis. Age-related changes in the collagen decrease skin elasticity and increase dryness sensation. The decline in gonadal and adrenal androgens is associated with decreased synthesis of sebum and cutaneous ceramides. Levels of filaggrin, the protein from which the components of NMF are derived, are also lower in aged skin [2].

22.6.3 Atopic Dermatitis

AD is a chronic inflammatory skin disorder characterized by recurrent outbreaks of symmetrical and bilateral eczema on the flexural surfaces of the body accompanied by severe pruritus. It is considered the first step of the atopic march that is associated with later development of allergic rhinitis and asthma. It is a multifactorial condition associated with genetic abnormalities that give rise, among other things, to immunological imbalance. Initial symptoms include dry skin and intense pruritus (Fig. 22.3). The isolated genetic component alone

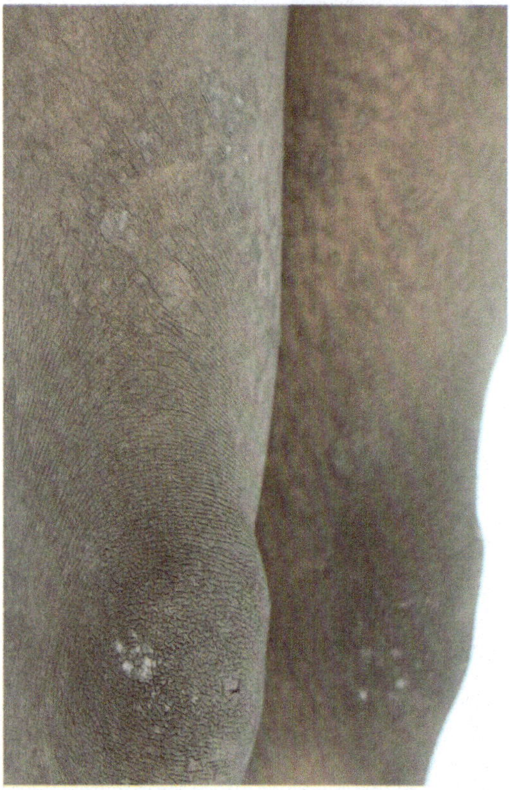

Fig. 22.3 Initial symptoms of atopic dermatitis include dry skin and intense pruritus

is not sufficient to cause the characteristic symptoms, and genetically predisposed individuals must be exposed to certain environmental antigens that promote disruption of the epidermal barrier and the manifestation of the disease [29]. These patients usually have a personal or family history of immunological processes related to immunoglobulin E. The skin of patients with AD is characterized by low ceramide levels [2], increased transepidermal water loss [30], and decreased water-binding capacity [31]. The unaffected skin of patients with AD has also been reported to have abnormally low levels of ceramides 1 and 3, molecules rich in polyunsaturated fatty acids, and particularly linoleic acid. It has recently been shown that the genetic predisposition to atopy also favors overexpression of stratum corneum chymotryptic enzyme and consequent disruption of the epidermal barrier as a result of premature corneodesmolysis. Detergents and topical corticosteroids increase the expression of this protease and contribute to the chronicity of the disease [29]. *FLG* null mutations (e.g., R501X and 2282del14) have been identified in as many 50 % of patients with moderate to severe AD in European population and 20 % in Asian population. These nonsense mutations significantly decrease filaggrin expression in the skin. The development of AD is common in *FLG* mutation carriers. Skin NMF is of significantly different composition in patients with AD with *FLG* mutations versus those without. Many patients suffering from severe AD do not carry mutations in the *FLG*, but the atopic response itself (Th2 cytokines) modulates decreasing keratinocyte expression of filaggrin. Patients with *FLG* null mutations show increased transepidermal water loss. This fact supports a barrier dysfunction induced by poor filaggrin function rather than inflammation causing barrier defects. Treating barrier dysfunction of AD proactively with effective emollients will benefit the prognosis of atopy [32–35].

22.6.4 Ichthyosis

Ichthyosis is a family of genetic skin disorders characterized by xerosis or dry skin. Different epidermal defects are responsible for the different types of ichthyosis (Figs. 22.4 and 22.5). Ichthyosis vulgaris is characterized by abnormalities in the formation of keratohyalin granules and, consequently, in filaggrin. Immunostaining of ichthyosis vulgaris skin biopsies showed a reduction in filaggrin protein expression and a reduction in profilaggrin mRNA within the keratinocytes [33]. The initial two loss-of-function mutations (R501X and 2282del14) were thus identified in the first repeat of *FLG* exon 3 [36]. Population-specific as well as recurrent *FLG* mutations were described (e.g., Q2147X, E2422X, R4307X in Singapore [37]). This abnormality leads to the formation of a stratum corneum deficient in many of the components of the NMF. Impaired corneodesmolysis caused by

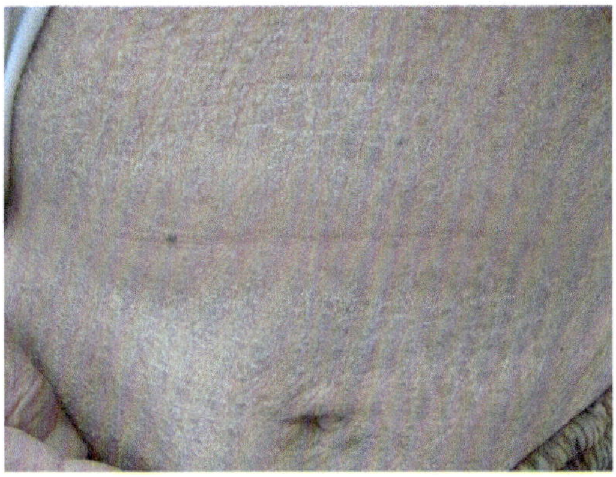

Fig. 22.4 Ichthyosis is a family of genetic skin disorders all characterized by xerosis or dry skin

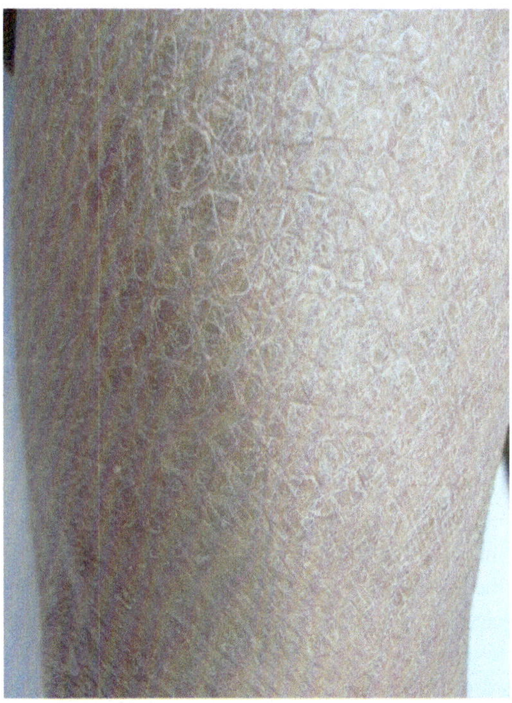

Fig. 22.5 Squamous dry skin characteristic of lamellar ichthyosis

defective water binding in the stratum corneum and alterations in skin pH give rise to visibly abnormal desquamation. X-linked recessive ichthyosis is characterized by the presence of large scales. This condition is caused by a steroid sulfatase deficiency that leads to an accumulation of cholesterol sulfate and a reduction in cholesterol levels in the stratum corneum [38]. These anomalies give rise to an abnormal intercellular lipid profile that increases intercellular cohesion. The accumulated cholesterol sulfate inhibits some of the proteases involved in desquamation [39] and contributes to the abnormal retention of corneodesmosomes. No inflammatory or repair mechanisms are triggered. Treatment with moisturizers does not have any major impact on the skin barrier properties [40].

22.6.5 Hand Eczema

Dermatitis of the hands can be caused by contact with active ingredients that irritate the skin. Although anyone can suffer from irritant contact dermatitis contacting with the irritant substance in a dose-dependent manner, it is most often found in individuals with constitutional impairment of the barrier function, such as patients with AD, especially those showing *FLG* null mutations [41]. Wet work is a risk factor by itself that can be responsible for epidermal barrier disturbance. People who suffer from AD show a reduced threshold of irritancy and, when exposed to water in an occupational setting or not, easily develop irritant contact dermatitis. Xerosis, desquamation, and cracked or fissured skin involve especially the palms [42]. I should specifically mention the significant association recently observed between *FLG* mutation and fissured skin on the hands and/or fingers in adults but with a nearly significant interaction with AD, suggesting that the effect occurred predominantly in subjects without AD [43, 44].

22.7 Topical Treatment of Dry Skin

The basis of the treatment of xerosis or dry skin is epidermal rehydration—repairing barrier function by applying lipids similar in composition and in similar concentrations to the physiologic lipids present in the skin.

22.7.1 General Care

Some recommendations should be followed to ensure appropriate care of the skin. In addition to proper hygiene, the condition of the skin is also related to nutrition and physical exercise. A balanced diet is recommended because this guarantees the nutritional intake required to maintain epidermal homeostasis. Most nutritional deficiencies are clearly manifested with dry skin (e.g., essential fatty acid deficiency syndrome). Physical exercise stimulates blood circulation, thereby increasing the transfer of nutrients and oxygen to the keratinocytes. It also favors epithelial regeneration, strengthens connective tissue, and increases collagen production. The skin barrier is susceptible also to the damaging effects of agents that accelerate skin aging, such as tobacco [2], alcohol, and solar radiation. Nicotine induces capillary constriction, which reduces blood flow and favors the storage of harmful substances in the skin. Table 22.3 shows basic skin care guidelines.

Table 22.3 Basic recommendations for skin care and hydration

Drink sufficient water
Eat a varied and balanced diet including plenty of fruit and vegetables
Avoid smoking, alcohol consumption, and direct exposure to sunlight
Take regular moderate exercise
For daily hygiene use soaps that have an acid pH and contain humectants

22.7.2 Specific Skin Care Regimens

Some additional measures can enhance the efficacy of the treatments. Patients should avoid rubbing or sponging the skin during baths. The use of smooth soaps without irritant ingredients is recommended. After bathing, oil or hydrating cream should be applied immediately, preventing transdermal water loss as a consequence of the immediate change of body temperature. Many moisturizing lotions can be applied over wet skin. Toweling the skin thoroughly could favor intercellular lipid loss and hamper the reconstitution of skin balance. Perfumed colognes, creams, and lotions are not recommended, especially if they contain alcohol. Topical solvents by themselves can eliminate skin's natural lipids from the stratum corneum, making the skin very dry. Soft fabrics, preferably natural materials such as cotton, are useful to minimize friction that could exacerbate the xerosis. Any detergents or fabric softeners used in laundry should be products specifically designed to be gentle on the skin and preferably those guaranteed to be suitable for sensitive skin [2]. Tight clothing may aggravate dry skin through a combination of physical aggression (rubbing) and constriction that could limit blood circulation and interfere with the proper delivery of oxygen and nutrients to the skin.

22.7.3 Topical Treatment

In addition to these measures aimed at maintaining the skin in optimum condition, dry skin will also benefit from the topical application of the components required by the epidermis to reestablish normal keratinocyte differentiation. Specialists currently advocate the usefulness of applying topical treatments containing active ingredients that rapidly penetrate the epidermis to stimulate the production pathways of intercellular lipids. This "inside out" approach, compared to the traditional "outside in" approach, appears to produce more effective therapeutic outcomes [2]. The topical preparations designed to treat dry skin are emollient or hydrating substances in preparations such as lotions or creams, that is, oil-in-water (O/W) emulsions (higher concentration of oil than water) or W/O emulsions (higher concentration of water than oil). I discuss in the following sections the mechanisms of action of different active principles. The main objective will be to restore epidermal homeostasis.

22.7.3.1 Repair of the Lipid Barrier

Lipids are the essential ingredients in formulations used to treat dry skin. The delivery of water alone will not repair the lipid barrier; natural physiologic lipids (cholesterol, ceramides, and fatty acids) must also be supplied [2]. Non-physiologic lipids are not recommended because they do not contribute to the reconstitution of the fatty bilayers [2, 43]. The main lipid components found in the epidermis are ceramides (50 %) and cholesterol derivatives (25 %). The physiologic lipids have several advantages over non-physiologic molecules: they are not occlusive, they penetrate the stratum corneum more easily, they gain better acceptance from patients because they are natural, and they restore proper epidermal differentiation. In short, physiologic lipids, such as the ceramides, act as structural elements in the epidermal barrier and mediate the stimuli that trigger epidermal repair [2, 45].

22.7.3.2 Supply and Retention of Water in the Stratum Corneum

We should differentiate between "humectant" and "hydrating" molecules. Humectants are substances that attract and retain water [2]; they play a passive role from the outside. A hydrating substance, however, is one that plays an active role in supplying and restoring water to the skin. Humectants are generally hygroscopic substances, such as glycerin or propylene glycol, while hydrating agents or moisturizers are complex mixtures of active ingredients or special combinations of amino acids. The inclusion of a humectant, such as glycerol or urea, as an active ingredient in a topical treatment for dry skin is based on the scientific evidence that humectants are capable of correcting defects in skin elasticity and barrier function when these deficiencies are not related to lipid loss [2, 46]. Glycerol plays a crucial role in keeping the stratum corneum hydrated: changes in aquaporin-3, an epidermal water/glycerol transporter, lead to decreased hydration and loss of skin elasticity that can be corrected by the topical application of glycerol [2]. It is therefore recommended that topical moisturizers include glycerol.

22.7.3.3 Alleviation of Pruritus

The sensation of itch induces scratching, and it represents a physical aggression that damages the epidermis. When the patient stops scratching, the epidermal restoration can then start [2]. Topical application of certain natural agents, such as glycine, blocks the release of histamine from the mast cells, thereby helping to break the self-perpetuating cycle of itching-scratching [2]. Glycine blocks the release of histamine by the mast cells [47]. Other products, particularly corticosteroids, are also used to treat the itch, but topical corticosteroids show an indirect effect on itch when they are used to cure cutaneous inflammation.

22.7.3.4 Restoring Normal Epidermal Turnover

As the skin and its different layers are structures that undergo continual renewal, dry skin can be treated by delivering components, such as dexpanthenol, that stimulate and accelerate the process of epidermal regeneration. Dexpanthenol promotes fibroblast proliferation and migration and stimulates intracellular protein synthesis [2], while hydroxy acids facilitate desquamation and improve lipid biosynthesis [16].

22.7.3.5 Active Ingredients for the Topical Treatment of Xerosis

Topical preparations for the treatment of dry skin should contain molecules that activate the epidermal regeneration process and restore the lipid content of the horny layer [2]. It is essential to choose the most suitable excipient for the area of the skin to be treated (Table 22.4) [2]. The ideal formulation would contain physiologic lipids (ceramides, cholesterol), a physiologic humectant (glycerol), an anti-itching agent (glycerol), and a component that enhances epidermal differentiation (dexpanthenol) [2].

Humectants are natural oily substances that do not intervene in the metabolic processes of the skin but rather act passively by preventing excessive

Table 22.4 Proposed hydrating preparations used to treat xerotic skin under different conditions

Facial area	Water-rich cream
Limbs and folds	Lipid-rich ointment
Acute eczema	Water-rich cream, solutions
Chronic eczema	Lipid-rich ointments and unguents
Summer	Water-rich cream
Winter	Lipid-rich ointment

Table 22.5 Substances included in topical preparations for the treatment of xerosis

Type	Compound	Characteristic
Humectants	Hydrocarbons	Mineral oils, such as paraffin and Vaseline
	Fatty oils and alcohols	
	Colloid substances	Some are hygroscopic, such as the cellulose derivatives (ethyl cellulose), natural polymers (xanthan gum), and synthetic polymers (carbopol)
	Silicones	No strong smell
		Not comedogenic
		Excellent tolerance
		Nongreasy formulations
Hydrating agents	Polyols	Highly hydrating
		Restore the flexibility of the cornified layer
		Prevent crystallization of lipids
		Promote corneodesmolysis
		Examples: glycerol, sorbitol, propylene glycol
	Urea	A component of natural moisturizing factor
		Highly hygroscopic and good exfoliating qualities
	Reconstituted natural moisturizing factor	Mixture of amino acids, sodium lactate, lactate acid, citrate, and others
		Repairs the upper layers of the stratum corneum with a hydrating action similar to that of natural moisturizing factor
	Hyaluronic acid	Creates a barrier layer
		High capacity to hydrate the stratum corneum
		Restores the flexibility and elasticity of the skin
		Well tolerated by skin
Relipidating active ingredients	Ceramides	Facilitate epidermal differentiation by reestablishing ingredients of the cellular lipids
	Cholesterol	Ensures the availability of this natural lipid in the stratum corneum to facilitate regeneration and epidermal differentiation
	Essential fatty acids	Provides consistency and cohesion in the stratum corneum
		Anti-inflammatory, immunogenic, and antimicrobial activity
		Principal fatty acids: linoleic, g-linoleic, and arachidonic acid
		Fatty acids are found in vegetable oils, such as evening primrose, shea, jojoba, borage, olive, wheat germ, and sunflower
Other active ingredients	Oats	Complex composition: very rich in water, proteins, glucides, lipids, mineral salts, and vitamins
		Hydrating, restructuring, antipruritic, and anti-inflammatory
		Improves the compatibility between the components in the preparation
	Allantoin	Conditioning, hydrating, and keratoplastic
	a-Bisabolol	Anti-inflammatory, emollient, and bactericidal
	Aloe vera	Soothing and emollient
	Glycyrrhetic acid	Anti-inflammatory and emollient

Source: From Barco et al. [51]; with permission of the author and the publisher

water loss. They can be classified according to general chemical categories: hydrocarbons, oils and fatty alcohols, colloid substances, and silicones.

Moisturizers or hydrating agents play an active role in the process of maintaining the water balance of the stratum corneum and are listed in Table 22.5.

Active relipidating agents supply the components the skin needs to balance the composition of the interlamellar lipid bilayers. In dry skin, the fatty acid content of these layers is low, and ceramide content is impaired and should be restored.

Some topical formulations for dry skin include an active ingredient that stimulates cell proliferation and lipid synthesis. One example of this type of component is dexpanthenol, a precursor of pantothenic acid and a constituent of coenzyme A [2]. It has been observed that pantothenic acid increases the proliferation and migration of fibroblasts [48] and stimulates intracellular protein synthesis [49]. Dexpanthenol has been used for some time to enhance skin barrier function in a number of different situations, including dermal regeneration following skin graft extraction and after x-ray irradiation [50].

Table 22.5 lists other active ingredients that play a role in restoring the xerotic stratum corneum [51].

Conclusion

Dry skin or xerosis management shows a global and multidisciplinary interest. It is a very common disorder in the general population showing a broad clinical spectrum between an uncomfortable skin status and disease. Xerosis can occur in healthy individuals at any age when several predisposing factors coincide. While the pathophysiology of this process is complex, disruption of normal epidermal differentiation is always present. This fact gives rise to an imbalance in water content in the stratum corneum and an impaired barrier function. Recent advances in our understanding of cutaneous homeostasis allow advocating an "inside out" approach to the treatment of xerosis complementary to the classic therapeutic approach. The development of clinical studies that could contribute to increase the level of evidence of the available treatments would be desirable. Still new and more effective topical treatments are needed, including active principles acting through concrete mechanisms with the objective of maintaining and restoring normal homeostasis of the epidermis.

References

1. Burdette-Taylor SR. Eczema, ichthyosis, psoriasis: conditions of cornification. Ostomy Wound Manage. 1995;41:36–8.
2. Proksch E, Lachapelle JM. The management of dry skin with topical emollients–recent perspectives. J Dtsch Dermatol Ges. 2005;3:768–74.
3. Feingold KR, Mao-Quian M, Menon GK, Cho SS, Brown BE, Elias PM. Cholesterol synthesis is required for cutaneous barrier function in mice. J Clin Invest. 1990;86:1738–45.
4. Grubauer G, Feingold KR, Harris RM, Elias PM. Lipid content and lipid type as determinants of the epidermal permeability barrier. J Lipid Res. 1989;30:89–96.
5. Elias PM. Stratum corneum defensive functions: an integrated view. J Invest Dermatol. 2005;125:183–200.
6. McAleer MA, Irvine AD. The multifunctional role of filaggrin in allergic skin disease. J Allergy Clin Immunol. 2013;131:280–91.
7. Matsui T, Miyamoto K, Kubo A, Kawasaki H, Ebihara T, Hata K, Tanahashi S, Ichinose S, Imoto I, Inazawa J, Kudoh J, Amagai M. SASPase regulates stratum corneum hydration through profilaggrin-to-filaggrin processing. EMBO Mol Med. 2011;3:320–33.
8. Sandilands A, Brown SJ, Goh CS, Pohler E, Wilson NJ, Campbell LE, et al. Mutations in the SASPase gene (ASPRV1) are not associated with atopic eczema or clinically dry skin. J Invest Dermatol. 2012;132:1507–10.
9. Bonnart C, Deraison C, Lacroix M, Uchida Y, Besson C, Robin A, et al. Elastase 2 is expressed in human and mouse epidermis and impairs skin barrier function in Netherton syndrome through filaggrin and lipid misprocessing. J Clin Invest. 2010;120: 871–82.
10. List K, Szabo R, Wertz PW, Segre J, Haudenschild CC, Kim SY, et al. Loss of proteolytically processed filaggrin caused by epidermal deletion of Matriptase/MT-SP1. J Cell Biol. 2003;163:901–10.
11. Leyvraz C, Charles RP, Rubera I, Guitard M, Rotman S, Breiden B, et al. The epidermal barrier function is dependent on the serine protease CAP1/Prss8. J Cell Biol. 2005;170:487–96.
12. Böhme M, Söderhäll C, Kull I, Bergström A, van Hage M, Wahlgren CF. Filaggrin mutations increase the risk for persistent dry skin and eczema independent of sensitization. J Allergy Clin Immunol. 2012;129:1153–5.
13. Sergeant A, Campbell LE, Hull PR, Porter M, Palmer CN, Smith FJ, et al. Heterozygous null alleles in filaggrin contribute to clinical dry skin in young adults and the elderly. J Invest Dermatol. 2009;129:1042–5.
14. Nakagawa N, Sakai S, Matsumoto M, Yamada K, Nagano M, Yuki T, et al. Relationship between NMF (lactate and potassium) content and the physical properties of the stratum corneum in healthy subjects. J Invest Dermatol. 2004;122:755–63.
15. Clark C. How to choose a suitable emollient. Pharm J. 2004;273:351–3.

16. Rawlings AV, Matts PJ. Stratum corneum moisturization at the molecular level: an update in relation to the dry skin cycle. J Invest Dermatol. 2005;124:1099–110.
17. Thyssen JP, Elias PM. Xerosis is latitude dependent and affects the propensity to develop atopic disease. J Allergy Clin Immunol. 2012;130:820.
18. Rycroft RJG. Low humidity and microtrauma. Am J Ind Med. 1985;8:371–3.
19. Rawlings AV, Voegeli R. Stratum corneum proteases and dry skin conditions. Cell Tissue Res. 2013;351:217–35.
20. Di Nardo A, Werz P, Giannetti A, Deidenari S. Ceramide and cholesterol composition of the skin of patients with atopic dermatitis. Acta Derm Venereol. 1998;78:27–30.
21. Morar N, Cookson WO, Harper JI, Moffatt MF. Filaggrin mutations in children with severe atopic dermatitis. J Invest Dermatol. 2007;127:1667–72.
22. Takahashi M, Tezuka T. The content of free amino acids in the stratum corneum is increased in senile xerosis. Arch Dermatol Res. 2004;295:448–52.
23. Katagiri C, Sato J, Nomura J, Denda M. Changes in environmental humidity affect the water-holding property of the stratum corneum and its free amino acid content, and the expression of the filaggrin in the epidermis of hairless mice. J Dermatol Sci. 2003;31:29–35.
24. Elias PM, Ansel JC, Woods LD, Feingold KR. Signaling networks in barrier homeostasis: the mystery widens. Arch Dermatol. 1996;132:1505–6.
25. Elias PM, Woods LD, Feingold KR. Epidermal pathogenesis of inflammatory dermatoses. Am J Contact Dermat. 1999;10:119–26.
26. Rycroft PJG, Smith WDL. Low humidity occupational dermatoses. Contact Dermatitis. 1980;6:488–92.
27. Agner T, Serup J. Seasonal variation of skin resistance to irritants. Br J Dermatol. 1989;121:323–8.
28. Engelke M, Jensen JM, Ekanayake-Mudiyanselage S, Proksch E. Effects of xerosis and ageing on epidermal proliferation and differentiation. Br J Dermatol. 1997; 137:219–25.
29. Cork MJ, Robinson DA, Vasilopoulos Y, Ferguson A, Moustafa M, MacGowan A, et al. New perspectives on epidermal barrier dysfunction in atopic dermatitis: gene environment interactions. J Allergy Clin Immunol. 2006;118:3–21.
30. Werner Y, Lindberg M. Transepidermal water loss in dry and clinically normal skin in patients with atopic dermatitis. Acta Derm Venereol. 1985;65:102–5.
31. Thune P. Evaluation of the hydration and water binding capacity in atopic skin and so-called dry skin. Acta Derm Venereol. 1989;144(Suppl):133–5.
32. Heimall J, Spergel JM. Filaggrin mutations and atopy: consequences for future therapeutics. Expert Rev Clin Immunol. 2012;8:189–97.
33. Brown SJ, McLean WHI. One remarkable molecule: filaggrin. J Invest Dermatol. 2012;132:751–62.
34. McGrath JA. Profilaggrin, dry skin, and atopic dermatitis risk: size matters. J Invest Dermatol. 2012;132: 10–1.
35. Kubo A, Nagao K, Amagai M. Epidermal barrier dysfunction and cutaneous sensitization in atopic diseases. J Clin Invest. 2012;122:440–7.
36. Smith FJ, Irvine AD, Terron-Kwiatkowski A, Sandilands A, Campdell LE, Zhao Y, Liao H, Evans AT, Goudie DR, Lewis-Jones S, Arseculeratne G, Munro CS, Sergeant A, O´Regan G, Bale SJ, Compton JG, DiGiovanna JJ, Presland RB, Fleckman P, McLean WH. Loss –of –function mutations in the gene encoding filaggrin cause ichthyosis vulgaris. Nat Genet. 2006;38:337–42.
37. Chen H, Ho JCC, Sandilands A, Chan YC, Giam YC, Evans AT, Lane EB, McLean I. Gene in Singaporean Chinese patients with ichthyosis vulgaris. J Invest Dermatol. 2008;128:1669–75.
38. Shapiro LJ, Weiss R, Frances JT, Webster D. X-linked ichthyosis due to steroid sulphatase deficiency. Lancet. 1978;1:70–2.
39. Sato J, Denda M, Nakinishi J, Nomura J, Koyama J. Cholesterol sulphate inhibits proteases that are involved in desquamation of stratum corneum. J Invest Dermatol. 1998;111:189–94.
40. Hoppe T, Winge MCG, Bradley M, Nordenskjöld M, Vahlquist A, Berne B, et al. X-linked recessive ichthyosis: an impaired barrier function evokes limited gene responses before and after moisturizing treatments. Br J Dermatol. 2012;167:514–22.
41. Thyssen JP. The association between filaggrin mutations hand eczema and contact dermatitis: a clear picture is emerging. Br J Dermatol. 2012;167:1197–9.
42. Ramírez C, Jacob SE. Dermatitis de manos. Actas Dermosifiliogr. 2006;97:363–73.
43. Thyssen JP, Ross-Hansen K, Johansen JD, Zachariae C, Carlsen BC, Linneberg A, et al. Filaggrin loss-of-function mutation R501X and 2282del4 carrier status is associated with fissured skin on the hands: results from a cross-sectional population study. Br J Dermatol. 2012;166:46–53.
44. Kaae J, Menné T, Carlsen BC, Zachariae C, Thyssen JP. The hands in health and disease of individuals with filaggrin loss-of-function mutations: clinical reflections on the hand eczema phenotype. Contact Dermatitis. 2012;67:119–24.
45. Chamlin SL, Kao J, Freiden IJ, Sheu MY, Fowler AJ, Fluhr JW, et al. Ceramide-dominant barrier repair lipids alleviate childhood atopic dermatitis: changes in barrier function provide a sensitive indicator of disease activity. J Am Acad Dermatol. 2002;47:198–208.

46. Hara M, Verkman AS. Glycerol replacement corrects defective skin hydration, elasticity and barrier function in aquaporin- 3-deficient mice. Proc Natl Acad Sci U S A. 2003;100:7360–5.
47. Paubert-Braquet M, Lefrançois G, Picquot S. Etude in vitrodupouvoirantiprurigineux du glycocolle: effetsur la dégranulation des mastocytes. Therapeutique. 1992;95:2–3.
48. Weimann BI, Hermann D. Studies on wound healing: effects of calcium pantothenate on the migration, proliferation and protein synthesis of human dermal fibroblasts in culture. Int J Vitam Nutr Res. 1999; 69:113–9.
49. Lacroix B, Didier E, Grenier JF. Effects of pantothenic acid on fibroblastic cell cultures. Res Exp Med. 1988;188:391–6.
50. Girard P, Beraud A, Goujon C. Effect of bepanthen ointment on the graft-donor site wound-healing model: double-blind biometrological and clinical study, with assessment by the patient versus the vehicle. Nouv Dermatol. 1998;17:559–70.
51. Barco D, Giménez-Arnau A. Xerosis: a dysfunction of the epidermal barrier. Actas Deramtosifilogr. 2008;99:671–82.

Allergic Contact Dermatitis

23

Wolfgang J.C. Uter

Contents

This chapter reviews evidence regarding the association between filaggrin gene (*FLG*) null mutations and allergic contact dermatitis (ACD) and contact allergy (CA), respectively. Hence, immediate-type hypersensitivity underlying contact urticaria, protein contact dermatitis, and extrinsic atopic eczema/dermatitis (AD) syndrome are not covered. After a brief definition of CA and ACD, the methods for retrieving evidence are described, an essential extract of available original data tabulated and discussed, finally arriving at a conclusion, which is considered valid at the time of writing. Obviously, new evidence in this emerging field of research will need to be considered in the future.

23.1 Definitions and Background

Contact allergy (CA, regarded here as synonymous with contact sensitization) is an acquired alteration of the immune system. CA occurs after an individual has been sufficiently exposed to a substance for sensitization to take place. While not all factors determining why a certain individual gets sensitized while another individual, exposed to a similar extent, does not get sensitized are fully understood, the general principle is well explored. Briefly, to be able to sensitize, the xenobiotic substance must be able (1) to penetrate the epidermal barrier; (2) to bind to skin (or, much more rarely, to mucous membrane) proteins, acting as a hapten, forming a full antigen with the protein; and (3) to mount a sufficient

W.J.C. Uter, MD
Department of Medical Informatics, Biometry and Epidemiology, University of Erlangen/Nürnberg, Waldstr. 4-6, Erlangen 91054, Germany
e-mail: wolfgang.uter@imbe.med.uni-erlangen.de

J.P. Thyssen, H. Maibach (eds.), *Filaggrin*,
DOI 10.1007/978-3-642-54379-1_23, © Springer-Verlag Berlin Heidelberg 2014

activation of the dendritic cells and lymphocyte subsets involved, including a probably nonspecific "danger signal." Components of both adaptive and innate immune systems are involved in this process [1, 2]. Recently, immunoregulation by regulatory T cells (Treg) has been identified as a major component of the sensitization process [3]. For a detailed description of this process, which is beyond the scope of this chapter, see, for example, the reviews cited above or [4]. Sensitization is not a "pass or fail" process, but a graded phenomenon, depending, among other factors, very much on the dose per area of exposure and on repetition of exposures: the higher the exposure dose is, not only the more likely the individual is to become sensitized but also the more intense the sensitization will be [5]. With repeated dosing, a super-additive effect has been noted [6]; that is, a few small doses that would not suffice to sensitize if given alone, and much less likely if given in one total dose, result in a rate and intensity of sensitization as achieved by one single dose massively exceeding the cumulated dose.

From this background, any alteration of the epidermal barrier facilitating allergen penetration, or protein binding, will have an impact on sensitization risk. Of interest, inflammatory cytokines may alter the metabolism of pro-filaggrin to filaggrin, and the degradation of filaggrin to the natural moisturizing factor (e.g., [7, 8]), thereby contributing to epidermal barrier disturbance. In a murine model, homozygous (*FLG-/-*) mice had been created and exhibited exaggerated responses both to irritant exposure to croton oil and to hapten exposure to DNFB (2,4-dinitrofluorobenzene) [9]. In these mice, increased penetration of Cr(III) ions to viable layers of the epidermis has been demonstrated, compared to wild type, apparently due to functional defects in the lower stratum corneum [10]. Also the commonly used flaky tail mouse (*ft/ft*), effectively lacking filaggrin, was more easily sensitized to the experimental hapten oxazolone [11].

After successful induction of a certain level of CA, ACD may develop after sufficient reexposure to the hapten (allergen) after weeks, months, or years. Sometimes, especially in case of massive accidental exposure (e.g., spills), the elicitation phase will follow induction within a few days. Generally, ACD is an inflammatory reaction of the epidermis and dermis [4]. In the acute stage, erythema, infiltration, vesicles, and possible oozing, with usually intense pruritus, are observed. Secondary infection, crusting, and erosion may occur. After chronic exposure or when acute ACD is starting to subside, the exudative reaction lessens, and secondary lesions, such as lichenification, hyperkeratosis, fissures, and scaling, are observed. Often, the site of ACD may give a substantial clue regarding the offending agent. Conversely, in chronic dermatitis, particularly hand eczema, which may be of mixed etiology, it may be very difficult to derive valuable clues from the patient's history and presentation.

The essential tool for diagnosing CA is patch testing [4]. Usually, patients suspected of having ACD, but also patients with long-standing dermatitis (eczema), including primarily irritant, atopic, or occupational eczema, or mixtures thereof, with the least suspicion of additional, secondary ACD, are eligible for patch testing. Guidelines for this diagnostic procedure exist [4], aiming at standardized application of allergen preparations and classification and interpretation of test results. Test series are under constant revision to achieve adaptation to changing allergen exposure patterns in the patients. Patch test results from single departments, national networks [12], or from international networks [13] are published to serve the purpose of CA surveillance, identifying targets of further in-depth research and preventive action, respectively.

23.2 Methods of Retrieval of Evidence

The Medline (PubMed) database was searched using the following search string: "filaggrin AND contact AND (allergy OR sensitization)"

(37 hits). An additional search in ISI Web of Science and reference lists of already identified publications yielded five additional publications eligible for screening. Screening of the full text excluded 30 publications with a different focus as well as reviews or comments. Therefore, altogether, 12 publications in terms of (clinical) epidemiological studies are the basis of the present review. Additional selected reviews, comments, and basic science publications are cited where these can provide useful perspectives.

23.3 Results

Overall, seven original studies, and one secondary analysis, were identified. Results from some of these studies have been presented in more than one publication. The studies are summarized in Table 23.1 and, regarding the Danish Glostrup III study, separately in Table 23.2. Specific or partly general comments to the studies' findings are found below the tables as annotations.

Another analysis of the Glostrup III study data not included in Table 23.2 identified a somewhat (if not significant) shorter time between piercing and onset of nickel dermatitis, according to a Kaplan-Meier analysis in participants with *FLG* null mutation (more pronounced in men than in women) in a sex-stratified analysis restricted to participants without ear piercing [26]. Furthermore, a nonsignificant association between *FLG* status and strength of patch test reaction to nickel (++/+++ vs. +) has been found: OR 2.2 (95 % CI: 0.8–6.4) [26].

A general criticism on the Glostrup III study is that the self-administered patch test was read by the study physicians only once, after 48 h (D2). This truncated reading frame is expected to miss overall 35 % of positive reactions observed after 72 h (D3), while, conversely, about 8 % of reactions considered allergic on D2 were graded as nonallergic at D3, according to a previous analysis of clinical patch test data [27]. Hence, the overall and the allergen-specific CA prevalences have been considerably underestimated in this study, the implications for risk estimation not being clear.

Notwithstanding this shortcoming, the analysis provides a detailed insight by addressing various outcomes in different subgroups (see Table 23.2). The pattern emerging from these results illustrates that:

- Nickel sensitization may occur quite independently from *FLG* mutation status, as the epidermal barrier is bypassed by the most common sensitizing exposure via nickel-releasing piercing posts ("bypass hypothesis") [26], and, therefore, an association is found preferentially in non-pierced persons. The results by Novak et al. [20], who found an overall association with nickel CA and particularly with nickel ACD, seem to be contradictory to this finding. However, the proportion of non-pierced persons in their sample is unclear; if higher, a higher overall OR could be expected.
- The role of thiomersal in this context appears less well established, albeit plausible, if sensitization occurs via vaccination.
- Some anatomical sites, such as the axilla (relevant for CA to fragrances), may offer increased skin penetration and sensitization risk even in the absence of a *FLG* defect [28], thus lessening the impact of *FLG* mutations. This may explain the lack of association with CA to the fragrance mix.
- In chronic AD/hand eczema, a strong association between *FLG* null mutations and CA risk had been observed in the latest series of analyses [16]. Notwithstanding likely increased exposure to therapeutic agents and skin care products in these conditions, and thus increased risk of CA, this finding may indicate that additional inflammation and epidermal barrier perturbation going along with chronic eczema considerably aggravate the otherwise limited (OR mostly <2) effect of *FLG* null mutation.

Table 23.1 Tabulation of original studies providing information on the association between *FLG* null allele status and contact allergy. Results of the Glostrup III study: see Table 23.2. OR: risk for CA (or outcome as indicated) associated with *FLG* mutations (either type listed), as given in original publication or recalculated from numerical results

Study group (effective sample size)	*FLG* mutations: FLG null (prevalence)	Outcome (overall %)	OR (95 % CI)	Comment	Reference
Twin sample with HE 8 years earlier ($n=181$)	R501X het (6.6 %) 2282del4 het (6.6 %)	Any CA (True Test) (25 %)	1.3 (0.5–3.4)	a,b	[17]
Secondary analysis of [17] and controls from [18] ($n=249$)	[17]: see above [18], controls: R501X het (2.4 %) 2282del4 het (3.6 %)	[17]: see above [18], controls: no PT	2.9 (1.1–7.5)	a,c	[19]
Population sample enriched in "allergic symptoms" (KORA-C) ($n=1,141$)	R501X het (2.2 %) 2282del4 het (5.7 %) Combined ($n=2$, 0.1 %)	Any CA (baseline ser.) (39.3 %) Nickel CA (13.1 %) Ni ACD (5.2 %)	1.1 (0.7–1.9) 1.9 (0.9–4.0) 4.0 (1.4–12.1)	d	[20]
Patients with chronic hand eczema ($n=122$)	R501X het (3/122) 2282del4 het (8/122)	"ACD" (44.3 %)	1.6 (0.5–5.5)	b,e	[21]
Patients with at least one CA ($n=430$)	R501X het (5.1 %) 2282del4 het (3.5 %)	–	–	f	[22, 23]
Patients with chronic occupational ICD ($n=296$)	R501X het (4.4 %) 2282del4 het (8.1 %)	Any CA (60 %)	0.5 (0.2–0.9)	g	[24]
Croatian students ($n=423$)	R501X het (0.2 %) 2282del4 het (2.4 %)	Any ACD (7 %)	1.4 (0.2–11.6)	h	[25]

[a]The inclusion of nickel may have biased the overall result toward unity, as nickel – usually the most common contact allergen overall – may sensitize bypassing the epidermis (with or without functional alterations due to *FLG* mutations) via piercing holes [26]

[b]Sample size was moderate, limiting the statistical power to detect differences as significant and the precision of risk estimates, respectively

[c]The controls from [18] are "hypernormal" in terms of excluding participants with AE or elevated IgE. Regarding the association between CA and *FLG*, (1) this may lead to some overrepresentation of AD-associated CA in the sample from [17] and (2) an underrepresentation of *FLG* (associated with AD) in the latter sample [18]. Moreover, in the latter sample [18], no patch testing had been performed to assess CA status. Hence, it is likely that the "control" group included some subjects with CA, which will have biased risk estimates in the secondary analysis toward unity. It is not clear what the resulting effects on risk estimation might be

[d]Only nickel sensitization, if relevant in terms of intolerance of jewelry, was significantly and strongly associated with *FLG* mutation. There had been no stratification or adjustment for piercing status and AD in the study by Novak et al., only adjustment for age and sex [20]. Hence, the association between nickel ACD and *FLG* may have theoretically been confounded by AD, even though in the Glostrup III study no confounding by AD had been observed (see Table 23.2, crude vs. adjusted ORs)

[e]Patch test methodology and results not given in detail; the final diagnoses "chronic allergic hand dermatitis" and "chronic allergic and irritant hand dermatitis" were aggregated as one compound outcome observed in 54/122 patients

[f]As the patient sample included only patients with at least one CA (to the baseline series), it is not possible to estimate risk for this event. However, within this study group, some, albeit nonsignificant, if tested for heterogeneity, variation of nickel CA according to AD and *FLG* status, resp., was noted: the prevalence was 30.9 % in patients without *FLG* null mutation and without AE and 43.8 % in those with AD and *FLG* null mutation

[g]Scope of patch testing not given. The comparison of CA prevalence in *FLG* null mutation versus wild-type carriers is based on a particular patient sample, namely, patients with chronic irritant contact dermatitis with occupational background. Assuming that (1) *FLG* mutation carriers are more susceptible to ICD due to irritant (occupational) exposure, which is reasonable in view of the relatively high carrier prevalences observed, (2) ICD may have occurred before CA developed, and (3) particularly in *FLG* mutation carriers, the observed lower prevalence of CA in *FLG* mutation carriers – a unique finding – can possibly be explained

[h]As outcome, the combination of a positive patch test and at least two of the following symptoms (redness, itching, vesicles, rash) was used

Table 23.2 Association between *FLG* null allele status and different contact allergy outcomes from the Glostrup III study, a Danish population sample, ages 18–69 (*n*=3,335 with successful *FLG* genotyping and patch test. Effective sample sizes may be slightly smaller for some outcomes due to missing/invalid data). The prevalence of *FLG* mutations was as follows: R501X het (3.3 %), 2282del4 het (4.6 %), R501Xhom (0.1 %), and 2282del4hom (0.1 %); these figures include two patients with mixed mutation

Outcome	(Sub)group	Prevalence (%)	OR (95 % CI)	Adj. OR (95 % CI)	Reference
Any CA (True Test Panel 1+2)	All	9.8	1.3 (0.9–1.9)	"Not significant"[a]	[14]
Ni CA	All	6.1	1.0 (0.6–1.7)	0.9 (0.5–1.6)[a]	
	Unpierced (*n*=1,507)	1.1	1.8 (0.4–8.0)	–	
	Unpierced females (*n*=317)	2.8	3.7 (0.7–19)	Not shown	
Ni ACD	All	4.4	1.1 (0.6–2.0)	1.0 (0.6–1.9)[b]	
	Females (*n*=1,844)	7.8	1.1 (0.6–2.0)	1.1 (0.6–2.0)[b]	
	Unpierced females (*n*=317)	1.8	6.8 (1.2–39)	–	
Fragrance mix I CA	All	1.6	0.9 (0.3–2.6)	–	
Neomycin CA	All	0.1	11 (0.7–182)	–	
Ethylenediamine CA	All	0.2	7 (1.7–30)	–	
Neomycin or ethylenediamine CA	All	–	–	4.9 (1.1–22)[c]	
Any CA (True Test Panel 1+2) except nickel	All	4.6	1.8 (1.1–2.9)	2.0 (1.2–3.5)[d]	[15]
Any CA (True Test Panel 1+2) except nickel and thiomersal	All	4.0	1.7 (1.03–2.9)	1.6 (0.9–2.7)[e]	[16]
As above	AD (*n*=324)	–	–	2.1 (0.9–5.0)	
As above	Hand eczema ever (*n*=716)	–	–	2.5 (1.1–5.6)	
As above	AD and hand eczema ever (*n*=159)	–	–	3.5 (1.1–11)	
As above	AD and frequent hand eczema (*n*=117)	–	–	5.7 (1.3–25)	

Nickel ACD has been defined as positive patch test to nickel and history of metal intolerance
Adjusted for:
[a]Age group, sex, piercing status
[b]Age group, sex, piercing status, atopic dermatitis (AD)
[c]Age group, sex, AD
[d]Age group, sex, AD, xerosis
[e]Age group, sex

Conclusion

Overall, available evidence at the time of writing points to an increased risk of CA associated with *FLG* null mutation, which is especially pronounced in persons with (longstanding) AD. In view of outcome definitions varying between studies, and also diverse subgroup analyses, it does not appear useful to summarize the findings quantitatively in terms of a meta-analysis.

Generally, confounding of the association between CA status and *FLG* mutation by AD is possible (if not taken into account by analyses stratifying or adjusting for AD), insofar as (1) a close relationship between AD and *FLG* mutation has been established (e.g., [18]); (2)

AD leads to, or increases, certain exposures; and (3) barrier perturbation and inflammation in AD likely have some additional potential of increasing overall sensitization risk.

CA risk may be generally diminished in AD in terms of TH2 skewing of the immune response, as recently shown using the model sensitizer dinitrochlorobenzene (DNCB), which penetrates the skin quantitatively in the doses used: persons with AD, independent of their *FLG* status, exhibited a reduced TH1 response and a specific propensity to TH2 programming [29]. However, in other, "real-world" allergens, in which skin penetration is affected by *FLG* status (and by AD with accompanying additional barrier perturbation), these latter factors may well override the lesser ability to mount a TH1 response in persons with AD.

References

1. Martin SF, Esser PR, Weber FC, Jakob T, Freudenberg MA, Schmidt M, et al. Mechanisms of chemical-induced innate immunity in allergic contact dermatitis. Allergy. 2011;66:1152–63.
2. Schmidt M, Raghavan B, Müller V, Vogl T, Fejer G, Tchaptchet S, et al. Crucial role for human Toll-like receptor 4 in the development of contact allergy to nickel. Nat Immunol. 2010;11:814–9.
3. Kimber I, Travis MA, Martin SF, Dearman RJ. Immunoregulation of skin sensitization and regulatory T cells. Contact Dermatitis. 2012;67:179–83.
4. Johansen JD, Frosch PJ, Lepoittevin J-P. Contact dermatitis. Heidelberg: Springer; 2011.
5. Moss C, Friedmann PS, Shuster S, Simpson JM. Susceptibility and amplification of sensitivity in contact dermatitis. Clin Exp Immunol. 1985;61:232–41.
6. Paramasivan P, Lai C, Pickard C, Ardern-Jones M, Healy E, Friedmann PS. Repeated low-dose skin exposure is an effective sensitizing stimulus, a factor to be taken into account in predicting sensitization risk. Br J Dermatol. 2010;162:594–7.
7. Gutowska-Owsiak D, Schaupp AL, Salimi M, Taylor S, Ogg GS. Interleukin-22 downregulates filaggrin expression and affects expression of profilaggrin processing enzymes. Br J Dermatol. 2011;165:492–8.
8. Hvid M, Johansen C, Deleuran B, Kemp K, Deleuran M, Vestergaard C. Regulation of caspase 14 expression in keratinocytes by inflammatory cytokines–a possible link between reduced skin barrier function and inflammation? Exp Dermatol. 2011;20:633–6.
9. Kawasaki H, Nagao K, Kubo A, Hata T, Shimizu A, Mizuno H, et al. Altered stratum corneum barrier and enhanced percutaneous immune responses in filaggrin-null mice. J Allergy Clin Immunol. 2012;129:1538–46.e1536.
10. Kubo A, Ishizaki I, Kubo A, Kawasaki H, Nagao K, Ohashi Y, Amagai M. The stratum corneum comprises three layers with distinct metal-ion barrier properties. Sci Rep. 2013;3:1731.
11. Scharschmidt TC, Man MQ, Hatano Y, Crumrine D, Gunathilake R, Sundberg JP, et al. Filaggrin deficiency confers a paracellular barrier abnormality that reduces inflammatory thresholds to irritants and haptens. J Allergy Clin Immunol. 2009;124:496–U164.
12. Warshaw EM, Belsito DV, Taylor JS, Sasseville D, Dekoven JG, Zirwas MJ, et al. North American Contact Dermatitis Group patch test results: 2009 to 2010. Dermatitis. 2013;24:50–9.
13. Uter W, Aberer W, Armario-Hita JC, Fernandez-Vozmediano JM, Ayala F, Balato A, et al. Current patch test results with the European baseline series and extensions to it from the 'European Surveillance System on Contact Allergy' network, 2007–2008. Contact Dermatitis. 2012;67:9–19.
14. Thyssen JP, Johansen JD, Linneberg A, Menné T, Nielsen NH, Meldgaard M, et al. The association between null mutations in the filaggrin gene and contact sensitization to nickel and other chemicals in the general population. Br J Dermatol. 2010;162:1278–85.
15. Thyssen JP, Johansen JD, Zachariae C, Menné T, Linneberg A. Xerosis is associated with atopic dermatitis, hand eczema and contact sensitization independent of filaggrin gene mutations. Acta Derm Venereol. 2013;93(4):406–10.
16. Thyssen JP, Linneberg A, Ross-Hansen K, Carlsen BC, Meldgaard M, Szecsi PB, et al. Filaggrin mutations are strongly associated with contact sensitization in individuals with dermatitis. Contact Dermatitis. 2013;68:273–6.
17. Lerbaek A, Bisgaard H, Agner T, Ohm Kyvik K, Palmer CN, Menné T. Filaggrin null alleles are not associated with hand eczema or contact allergy. Br J Dermatol. 2007;157:1199–204.
18. Weidinger S, Rodriguez E, Stahl C, Wagenpfeil S, Klopp N, Illig T, Novak N. Filaggrin mutations strongly predispose to early-onset and extrinsic atopic dermatitis. J Invest Dermatol. 2007;127:724–6.
19. Brown SJ, Cordell HJ. Are filaggrin mutations associated with hand eczema or contact allergy? We do not know. Br J Dermatol. 2008;158:1383–4.
20. Novak N, Baurecht H, Schäfer T, Rodriguez E, Wagenpfeil S, Klopp N, Heinrich J, et al. Loss-of-function mutations in the filaggrin gene and allergic contact sensitization to nickel. J Invest Dermatol. 2008;128:1430–5.
21. Molin S, Vollmer S, Weiss EH, Ruzicka T, Prinz JC. Filaggrin mutations may confer susceptibility to chronic hand eczema characterized by combined allergic and irritant contact dermatitis. Br J Dermatol. 2009;161:801–7.

22. Carlsen BC, Johansen JD, Menné T, Meldgaard M, Szecsi PB, Stender S, et al. Filaggrin null mutations and association with contact allergy and allergic contact dermatitis: results from a tertiary dermatology clinic. Contact Dermatitis. 2010;63:89–95.
23. Carlsen BC, Thyssen JP, Menné T, Meldgaard M, Linneberg A, Nielsen NH, et al. Association between filaggrin null mutations and concomitant atopic dermatitis and contact allergy. Clin Exp Dermatol. 2011;36:467–72.
24. de Jongh CM, Khrenova L, Verberk MM, Calkoen F, van Dijk FJ, Voss H, et al. Loss-of-function polymorphisms in the filaggrin gene are associated with an increased susceptibility to chronic irritant contact dermatitis: a case–control study. Br J Dermatol. 2008;159:621–7.
25. Sabolic Pipinic I, Varnai VM, Turk R, Breljak D, Kezic S, Macan J. Low frequency of filaggrin null mutations in Croatia and their relation with allergic diseases. Int J Immunogenet. 2013;40(3):192–8.
26. Ross-Hansen K, Menné T, Johansen JD, Carlsen BC, Linneberg A, Nielsen NH, et al. Nickel reactivity and filaggrin null mutations–evaluation of the filaggrin bypass theory in a general population. Contact Dermatitis. 2011;64:24–31.
27. Uter WJ, Geier J, Schnuch A. Good clinical practice in patch testing: readings beyond day 2 are necessary: a confirmatory analysis. Members of the Information Network of Departments of Dermatology. Am J Contact Dermat. 1996;7:231–7.
28. Hanifin JM. Filaggrin mutations and allergic contact sensitization. J Invest Dermatol. 2008;128:1362–4.
29. Newell L, Polak ME, Perera J, Owen C, Boyd P, Pickard C, et al. Sensitization via healthy skin programmes Th2 responses in individuals with atopic dermatitis. J Invest Dermatol. 2013;133(10):2372–80.

Irritant Contact Dermatitis

24

Sanja Kezic and Maaike J. Visser

Contents

S. Kezic, PhD (✉) • M.J. Visser, BS
Coronel Institute of Occupational Health, Academic Medical Centre, Meibergdreef 15, Amsterdam 1105 AZ, The Netherlands
e-mail: s.kezic@amc.uva.nl; m.j.visser@amc.nl

24.1 Background

Irritant contact dermatitis (ICD) may occur after direct contact of the skin with an irritating factor that leads to disruption of the skin barrier and a local inflammatory reaction that is largely mediated by innate immunity [1]. ICD can be caused by either chemicals (e.g. detergents, organic solvents, disinfectants, and water) or physical irritants (e.g. mechanical friction and cold and dry environment) [1, 2]. The mode of action and inflammatory response are dependent on the physicochemical properties of the irritating compound and the factors that influence its penetration across the skin, like magnitude and duration of the exposure. *Acute ICD* is caused by relatively major damage to the skin, usually caused by an accidental, short contact with a strong irritant. *Chronic ICD* may develop with repetitive skin exposure to one or multiple skin-irritating factors that cause only minor damage, but there is not sufficient time between subsequent exposures for the skin to completely recover, eventually leading to an eczematous skin reaction. Chronic ICD is one of the most prevalent work-related diseases, especially in so-called wet work occupations like nursing, cleaning, and hair dressing [2].

Experimental irritation studies and epidemiological data have convincingly shown that manifest or even a history of atopic dermatitis (AD) in childhood confers an increased risk for acute as well as chronic irritant dermatitis [3–5]. Enhanced susceptibility to ICD can at least partly

J.P. Thyssen, H. Maibach (eds.), *Filaggrin*,
DOI 10.1007/978-3-642-54379-1_24, © Springer-Verlag Berlin Heidelberg 2014

be explained by reduced skin barrier, which is a major hallmark of atopic skin. Jakasa et al. [6] showed that even the noninvolved skin in patients with AD has impaired skin barrier, allowing higher penetration of sodium lauryl sulphate (SLS), a model irritant commonly used in experimental studies. The same research group [7] showed that the barrier impairment and inflammation were dependent on the penetration rate of SLS. An impaired skin barrier function in AD can at least partly be explained by decreased levels of filaggrin protein. Filaggrin aggregates keratin filaments in the transition of keratinocytes into corneocytes, contributing to the mechanical strength of the SC. Filaggrin, as well as its precursor and degradation products, plays an important role in SC homeostasis and antimicrobial defence through different mechanisms [8]. As skin barrier condition determines the amount of irritant that penetrates the SC and reaches viable epidermis, it is likely that the gene encoding filaggrin (*FLG*) mutations also confer risk for ICD. Furthermore, as loss-of-function mutations in the *FLG* are a well-known predisposing factor for AD, an increased prevalence of *FLG* carriers in the ICD patients who have concomitant AD is to be expected.

24.2 Filaggrin and Irritant Contact Dermatitis

Several recent studies have investigated whether the presence of *FLG* loss-of-function mutations increases the risk of acute as well as chronic ICD.

24.2.1 Acute ICD

Scharschmidt et al. [9] showed that compared to wild-type mice, the filaggrin-deficient (flaky tail, *ft/ft*) mice had reduced inflammatory thresholds to irritants. Interestingly, this effect was only detectable when modest concentration of the irritant was applied, whereas high concentrations did not lead to a significant difference between the groups. Consistently, Kawasaki et al. [10] demonstrated that filaggrin null mice had a stronger irritation response than wild-type mice. However, studies in human subjects could not confirm *FLG* mutations as a risk factor for acute ICD. Angelova-Fisher et al. [11] studied inflammatory response following a 24-h single exposure to 1 % SLS in healthy subjects wild-type for *FLG* (CTRL) and AD patients with (AD-FLG) and without *FLG* mutations (AD-wt). Although AD-FLG group showed slightly higher irritation response as compared to AD-wt and controls, the difference between the groups was not significant. These results were consistent with a study of similar experimental design of Jungersted et al. [12] In an in vitro skin model of filaggrin deficiency, Küchler et al. [13] found elevated irritation response to SLS based on significantly higher lactate dehydrogenase leakage, and interleukin (IL)-6 and IL-8 levels, than in the control model.

24.2.2 Chronic ICD

In addition to studies on acute irritation, several epidemiological and clinical studies have recently been addressing chronic ICD. Molin et al. [14] investigated *FLG* mutations in 122 German nonatopic patients with different subtypes of chronic hand eczema and compared them to 95 healthy controls. Marginally significant associations with *FLG* mutations were reported for a subgroup of patients diagnosed with a combination of ICD and allergic contact dermatitis, but not in the subgroup with ICD alone. However, the small sample size and the choice of the control population limit the informative value of this study. De Jongh et al. [5] studied the prevalence of the R501X and 2282del4 *FLG* mutations in 296 patients with occupational ICD and a control group of 217 vocational school apprentices and showed that the carriers *of FLG* mutations are approximately twice as likely to acquire occupational ICD. However, due to limited group size, that study could not answer the question whether *FLG* mutations are a risk factor for ICD per se or modify the risk through AD. Namely, in the etiological relation between *FLG* mutations and ICD, AD can be an intermediate factor (as *FLG* mutations increase the risk of AD) as well as a codeterminant independent from *FLG*. As still approximately 60 % of *FLG* carriers do not develop AD [15, 16], a substantial part of individuals with potentially increased risk (i.e. *FLG* mutation carriers without AD) will not be recognized as susceptible in current prevention programmes. Thus, a follow-up study by the same research group sought to investigate relative contribution of *FLG* and AD in larger cohorts of patients and controls and included two additional *FLG* mutations (R2447X, and S3247X) [17]. The crude OR for the combined mutant allele was 2.1 (95 % CI 1.33–3.28), which was in a good agreement with the previous study by de Jongh et al. reporting an OR of 1.9 [5, 17]. A logistic regression model with AD and *FLG* mutations as independent risk factors revealed a significant association of ICD with *FLG* loss-of-function mutations, even when the analysis was adjusted for AD (OR 1.6; 95 % CI 1.01–2.58). A history of AD increased the risk of developing ICD approximately threefold (OR 2.9; 95 % CI 2.8–4.3), while the concomitant presence of AD and *FLG* mutations result in a 4.7-fold increased risk for ICD [17]. Interestingly, a recent study by Landeck et al. [18] revealed that *FLG* mutations in the presence of AD seem to be a modifier also of the severity of the clinical course in ICD, supporting the view that this group represents a special risk population for ICD.

It has been emphasized that all existing studies addressing the association between filaggrin and ICD are based on *FLG* loss-of-function mutations and not on the levels of filaggrin protein in the SC. This might be important because in addition to loss-of-function mutations, filaggrin expression can be reduced by other factors such as *FLG* intragenic copy number variations [19], downregulation by inflammation, or presence of inflammatory cytokines [20, 21] or by modulation of enzymatic processes responsible for the processing of pro-filaggrin into filaggrin or for the breakdown of filaggrin NMF [22]. Recently, it has been shown that altered lipid composition and organization in AD patients was not associated with *FLG* genotype but rather with the levels of filaggrin degradation products in the SC [23].

Conclusion

Present knowledge on the role of *FLG* mutations on acute skin irritation is limited. Two existing studies suggest that the presence of *FLG* mutations in AD patients does not lead to enhanced irritation response [11, 12]. So far, there are no studies that address acute irritation response in the individuals with *FLG* mutations but without a history of AD. Such studies might shed more light on the role of *FLG* mutations in the skin irritation response independently of AD.

In contrast to acute ICD, *FLG* mutations enhance individual susceptibility to chronic ICD. Although these results need confirmation from additional studies by preference prospective cohort investigations, *FLG* mutations have been shown to increase the risk of chronic persistent ICD even in the absence of AD [17]. The risk-modifying effect of *FLG* was the highest in the presence of AD. It can be speculated that the cutaneous cytokine milieu in AD might lead to further reduction of filaggrin in the SC. Furthermore, irritation response might be attenuated due to a distinct inflammatory status in *FLG*-related AD. Kezic et al. reported elevated levels of IL-1 cytokines in AD patients with *FLG* mutations and in a murine model of filaggrin deficiency, suggesting a pre-existing enhanced or proinflammatory status in the skin of patients with AD that relates to *FLG* mutations [24]. Future studies in well-defined subgroups will be required to elucidate the temporal and causal relationship between the levels of filaggrin in the SC, presence and severity of AD, skin barrier function, and development of ICD.

References

1. Smith HR, Basketter DA, McFadden JP. Irritant dermatitis, irritancy and its role in allergic contact dermatitis. Clin Exp Dermatol. 2002;27:138–46.

2. Diepgen TL. Occupational skin disease data in Europe. Int Arch Occup Environ Health. 2003;76:331–8.
3. Coenraads PJ, Diepgen TL. Risk for hand eczema in employees with past or present atopic dermatitis. Int Arch Occup Environ Health. 1998;71:7–13.
4. Dickel H, Bruckner TM, Schmidt A, Diepgen TL. Impact of atopic skin diathesis on occupational skin disease incidence in a working population. J Invest Dermatol. 2003;121:37–40.
5. de Jongh CM, Khrenova L, Verberk MM, Calkoen F, van Dijk FJ, Voss H, et al. Loss-of-function polymorphisms in the filaggrin gene are associated with an increased susceptibility to chronic irritant contact dermatitis: a case–control study. Br J Dermatol. 2008;159:621–7.
6. Jakasa I, de Jongh CM, Verberk MM, Bos JD, Kezic S. Percutaneous penetration of sodium lauryl sulphate is increased in uninvolved skin of patients with atopic dermatitis compared with control subjects. Br J Dermatol. 2006;155:104–9.
7. De Jongh CM, Jakasa I, Verberk MM, Kezic S. Variation in barrier impairment and inflammation of human skin as determined by sodium lauryl sulphate penetration rate. Br J Dermatol. 2005;154:651–7.
8. McAleer MA, Irvine AD. The multifunctional role of filaggrin in allergic skin disease. J Allergy Clin Immunol. 2013;131(2):280–91.
9. Scharschmidt TC, Man MQ, Hatano Y, Crumrine D, Gunathilake R, Sundberg JP, et al. Filaggrin deficiency confers a paracellular barrier abnormality that reduces inflammatory thresholds to irritants and haptens. J Allergy Clin Immunol. 2009;124:496–506.
10. Kawasaki H, Nagao K, Kubo A, Hata T, Shimizu A, Mizuno H, et al. Altered stratum corneum barrier and enhanced percutaneous immune responses in filaggrin null mice. J Allergy Clin Immunol. 2012;129:1538–46.e6.
11. Angelova-Fischer I, Mannheimer AC, Hinder A, Ruether A, Franke A, Neubert RH, et al. Distinct barrier integrity phenotypes in filaggrin-related atopic eczema following sequential tape stripping and lipid profiling. Exp Dermatol. 2011;20:351–6.
12. Jungersted JM, Scheer H, Mempel M, Baurecht H, Cifuentes L, Høgh JK, et al. Stratum corneum lipids, skin barrier function and filaggrin mutations in patients with atopic eczema. Allergy. 2010;65:911–8.
13. Küchler S, Henkes D, Eckl KM, Ackermann K, Plendl J, Korting HC, et al. Hallmarks of atopic skin mimicked in vitro by means of a skin disease model based on FLG knock-down. Altern Lab Anim. 2011;39:471–80.
14. Molin S, Vollmer S, Weiss EH, Ruzicka T, Prinz JC. Filaggrin mutations may confer susceptibility to chronic hand eczema characterized by combined allergic and irritant contact dermatitis. Br J Dermatol. 2009;161:801–7.
15. Irvine AD, McLean WH, Leung DY. Filaggrin mutations associated with skin and allergic diseases. N Engl J Med. 2011;365:1315–27.
16. O'Regan GM, Irvine AD. The role of filaggrin in the atopic diathesis. Clin Exp Allergy. 2010;40:965–72.
17. Visser MJ, Landeck L, Campbell LE, McLean WH, Weidinger S, Calkoen F, et al. Impact of loss-of-function mutations in the filaggrin gene and atopic dermatitis on the development of occupational irritant contact dermatitis. Br J Dermatol. 2013;168(2):326–32.
18. Landeck L, Visser M, Skudlik C, Brans R, Kezic S, John SM. Clinical course of occupational irritant contact eczema of the hands in relation to filaggrin genotype status and atopy. Br J Dermatol. 2012;167:1302–9.
19. Brown SJ, McLean WH. One remarkable molecule: filaggrin. J Invest Dermatol. 2011;132:751–62.
20. Howell MD, Kim BE, Gao P, Grant AV, Boguniewicz M, Debenedetto A, et al. Cytokine modulation of atopic dermatitis filaggrin skin expression. J Allergy Clin Immunol. 2007;120:150–5.
21. Kezic S, O'Regan GM, Yau N, Sandilands A, Chen H, Campbell LE, et al. Levels of filaggrin degradation products are influenced by both filaggrin genotype and atopic dermatitis severity. Allergy. 2011;66:934–40.
22. Pellerin L, Henry J, Hsu C-Y, Balica S, Jean-Decoster C, Méchin MC, et al. Defects of filaggrin-like proteins in both lesional and non-lesional atopic skin. J Allergy Clin Immunol. 2013;131(4):1094–102.
23. Janssens M, van Smeden SJ, Gooris GS, Bras W, Portale G, Caspers PJ, et al. Increase in short chain ceramides correlates with an altered lipid organization and decreased barrier function in atopic eczema patients. J Lipid Res. 2012;53:2755–66.
24. Kezic S, O'Regan GM, Lutter R, Jakasa I, Koster ES, Saunders S, et al. Filaggrin loss-of-function mutations are associated with enhanced expression of IL-1 cytokines in the stratum corneum of patients with atopic dermatitis and in a murine model of filaggrin deficiency. J Allergy Clin Immunol. 2012;129:1031–9.e1.

Hand Eczema

25

Richard Brans and Swen Malte John

Contents

R. Brans, MD (✉) • S.M. John, MD, PhD
Department of Dermatology,
Environmental Medicine, Health Theory,
University of Osnabrueck, Sedanstrasse 115 (D1),
Osnabrueck DE-49069, Germany
e-mail: rbrans@uni-osnabrueck.de;
sjohn@uni-osnabrueck.de

25.1 Background

Hand eczema is a common, noninfectious skin inflammation restricted to the hands. Although it can be transient or improved by adequate treatment, a chronic course with continual or intermittent symptoms over years is frequently seen [1]. The 1-year prevalence in middle and Northern Europe is approximately 10 % [1, 2]. Women are twice as often affected as men. Occupational hand eczema is one of the most common occupational diseases [3] resulting in substantial socioeconomic consequences for the affected individual and the society [1, 4]. Symptoms of hand eczema may vary from mild skin changes to severe lesions accompanied by itch and pain and can have a negative impact on quality of life [5–8]. Hand eczema can be found in different stages, i.e., acute dermatitis characterized by erythema and vesicles and chronic dermatitis defined by dry, lichenified, and fissured skin with scales. The cause, localization, and morphology of hand eczema are heterogeneous. Therefore, universally accepted classifications are difficult to establish [3, 9]. Subtypes include irritant and allergic contact dermatitis, atopic dermatitis (AD), mixed forms, and minor groups with vesicular and hyperkeratotic hand eczema. Multiple factors, both individual and environmental, are involved in the pathogenesis of hand eczema [10]. It is often difficult to identify the cause because mixed exposures to irritants and allergens, as well as different predispositions, interact

J.P. Thyssen, H. Maibach (eds.), *Filaggrint*,
DOI 10.1007/978-3-642-54379-1_25, © Springer-Verlag Berlin Heidelberg 2014

with each other. Factors impairing the skin barrier may facilitate the penetration of allergens and irritants through the epidermis and increase the likelihood of developing both allergic and irritant hand eczema [11–13]. Moreover, they may be responsible for an increased transepidermal loss of water causing dryness of the skin. Twin studies have shown that genetic factors increase the risk for hand eczema [14, 15]. Moreover, it has been established that the impairment of the skin barrier function is often based upon a genetic predisposition. In line with this, the main endogenous risk factor for hand eczema is atopic dermatitis, an inherent condition characterized by an impaired skin barrier and dry skin [15–17]. Involvement of the hands is frequently seen in patients with AD. Moreover, AD increases the likelihood of developing irritant contact dermatitis of the hands two to four times [18–20].

Several studies have demonstrated that the barrier function of the outermost layer of the epidermis (stratum corneum) is affected by mutations in the gene encoding the filament-aggregating protein filaggrin [21, 22]. Filaggrin and its metabolites are not only important for the mechanical strength of the skin but also for the hydration and homeostasis of the epidermis. Several loss-of-function mutations in the filaggrin gene (*FLG*) have been identified [23, 24]. The most common *FLG* mutations in European populations are R501X and 2282del4. They are present in 9–10 % of individuals of European descent [22], resulting in lower levels of filaggrin and its metabolites in the skin [25–27]. Reduced levels of filaggrin are associated with dry skin [28, 29] and may lead to increased penetration of irritants and allergens through the epidermis and subsequent inflammation [30, 31]. It has been demonstrated in multiple studies that *FLG* mutations are causative for ichthyosis vulgaris and strongly associated with AD. Approximately 20–30 % of patients with AD carry a *FLG* mutation [23, 24, 32–35]. Mutations in the *FLG* gene are more common in individuals with AD in combination with asthma and allergic rhinitis [33, 36], IgE sensitization [34, 37], early onset [34], and persistence into adulthood [32]. However, in some patients with AD, the skin barrier is reduced, although no *FLG* mutation is present [38] and the majority of heterozygous *FLG* mutation carriers never develop AD [35].

The levels of filaggrin in the skin are not only influenced by *FLG* loss-of-function mutations but also by other factors. It has been reported that skin inflammation decreases the synthesis of filaggrin, suggesting that filaggrin levels in already inflamed skin are lower, especially in *FLG* mutation carriers [39]. Currently, there is no way to differentiate between acquired filaggrin deficiency caused by skin inflammation and inherited filaggrin deficiency caused by *FLG* mutations. In addition, intragenic copy number variations within the *FLG* locus have been reported, with alleles encoding 10, 11, or 12 filaggrin monomers, affecting the amount of filaggrin expressed in the epidermis independently of *FLG* mutations. Excluding *FLG* mutation carriers, Brown et al. demonstrated that low copy numbers within the *FLG* locus are associated with AD [40]. Moreover, patients with AD and *FLG* mutations have elevated levels of proinflammatory IL-1 cytokines, which might influence inflammatory response after exposure to irritating chemicals [41]. Therefore, the lower threshold for irritants in patients with AD may reflect both barrier impairment and greater inflammatory reactivity [42].

It has to be noted that other endogenous risk factors may be involved in hand eczema. Several genetic polymorphisms have been associated with atopic, irritant, and allergic contact dermatitis. Relevant genes may also affect the skin barrier (e.g., *claudin-1*) [43], have an impact on metabolizing enzymes (e.g., *N-acetyltransferase*, *glutathion-S-transferase*) potentially influencing the detoxification or activation of irritants and allergens, or may alter individual immune responses (e.g., *IL-1*, *TNF-alpha*) [44–47].

25.2 Filaggrin and Hand Eczema

Several studies have investigated whether the presence of *FLG* loss-of-function mutations increases the risk of hand eczema. However, the results have been conflicting (Table 25.1). *FLG* mutations may only be of importance in distinct subtypes of hand eczema [61].

Table 25.1 Overview on publication addressing *FLG* loss-of-function mutations and hand eczema

	Association with *FLG* loss-of-function mutations		
	Yes	No	References
HE in twins		+	Lerbaek et al. [48], Brown and Cordell [49]
HE in patients with dermatitis	+		Thyssen et al. [50]
HE in patients with AD	+		Thyssen et al. [51]
Cause of HE:			
ICD of the hands	+		De Jong et al. [52], Visser et al. [20]
ICD + ACD of the hands	+		Molin et al. [53]
Worse prognosis of ICD of the hands in atopic patients	+		Landeck et al. [54]
Morphology of HE:			
Palmar hyperlinearity	+		Novak et al. [28], Brown et al. [55], Thyssen et al. [50], Landeck et al. [56]
Vesicular HE		+	Landeck et al. [56], Thyssen et al. [50], Thyssen et al. [57]
Involvement of the dorsal aspects of the hands	+		Thyssen et al. [58], Thyssen et al. [50], Kaae et al. [59], Carson et al. [60]

Abbreviations: *HE* hand eczema, *ICD* irritant contact dermatitis, *ACD* allergic contact dermatitis, *AD* atopic dermatitis

25.2.1 Cause of Hand Eczema

In 2007, 183 Danish adult twin individuals with self-reported or clinical signs of hand eczema were genotyped for the *FLG* loss-of-function mutations R501X and 2282del4. Twin individuals without hand eczema ($n = 50$) and children without AD ($n = 198$) served as controls. No association between the variant alleles and hand eczema or contact allergy was found [48]. However, the study was impaired by a small size and several confounders, resulting in a low statistical power [49]. In 2009, Molin et al. evaluated the prevalence of *FLG* mutations R501X and 2282del4 in 122 German nonatopic patients with chronic hand eczema compared to 95 control individuals of unknown origin. Patients with atopic hand eczema were excluded. In this study, *FLG* mutations were marginally associated with a distinct subtype of chronic hand eczema characterized by a combination of irritant and allergic contact dermatitis, not with irritant or allergic contact dermatitis alone [53]. In a pilot study, Thyssen et al. genotyped 128 patients for *FLG* mutations R501X and 2282del4. Mutation carriers were more likely to have hand eczema, although this association was not significant (OR 1.9, 95 % CI 0.47–7.76) [50]. This was followed by a cross-sectional study in 3,335 randomly selected adults of the Danish general population who were genotyped for R501X and 2282del4, patch tested, and asked for the presence of AD and hand eczema during the past 12 months. *FLG* mutations conferred an increased risk of hand eczema in subjects with AD (OR 2.89, 95 % CI 1.27–7.01), but not in subjects without AD (OR 0.82, 95 % CI 0.41–1.67). Moreover, *FLG* mutations were associated with early onset and persistence of hand eczema into adulthood [51].

In a preliminary study of de Jongh et al. in 296 patients of European descent from Germany with occupational irritant contact dermatitis of the hands and a control group of 217 vocational school apprentices, the presence of R501X and 2282del4 *FLG* mutations doubled the risk of occupational irritant dermatitis of the hands (OR =1.91, 95 % CI 1.02–3.59) [52]. In a follow-up study increasing the number of patients ($n = 634$) and controls ($n = 393$), the participants were genotyped for the *FLG* loss-of-function mutations R501X, 2282del4, R2447X, and S3247X. Together these four different mutations constitute more than 90 % of the *FLG* mutations found in European populations [22]. The study revealed that mutations in the *FLG* gene and AD are both independently associated with occupational irritant contact dermatitis of the hands. *FLG* mutations were found in 15.9 % of patients and in 8.3 % of controls, with a crude OR of 2.09 (95 % CI 1.33–3.28) for the combined genotype. The adjusted OR for *FLG* mutations, corrected for AD, was 1.62 (95 % CI 1.01–2.58). Having AD increased the risk of developing irritant contact dermatitis of the hands approximately threefold (OR = 2.89, 95 % CI 2.09–3.99). Individuals with AD and *FLG* mutations had the highest risk [20].

Landeck et al. performed a 3-year prospective observational study in 459 patients from Germany with occupational irritant contact dermatitis of the hands and evaluated the clinical course, recovery rate, periods of being absent from work, and job continuation [54]. The patients were examined for atopy and genotyped for the *FLG* loss-of-function mutations R501X, 2282del4, R2447X, and S3247X. Nonatopic patients with irritant contact dermatitis responded better to therapeutic approaches, while atopic patients had more resistant lesions, resulting in lower rates of recovery and job continuation. The presence of *FLG* loss-of-function mutations worsened the course of occupational irritant contact dermatitis, but only in combination with atopy, not as an independent risk factor. After 3 years, the rate of those abandoning the profession was significantly increased in atopic patients with irritant contact dermatitis of the hands and *FLG* mutations compared to nonatopic patients with irritant contact dermatitis and no *FLG* mutations (OR 3.1; $p < 0.05$).

It seems plausible that an impaired skin barrier facilitates the development of allergic contact dermatitis by enhanced penetration of allergens through the skin. Moreover, it has been suggested that inflammatory responses due to skin irritation may promote contact allergy [62, 63]. Therefore, it has been assumed that individuals with an impaired skin barrier may have a higher risk to develop allergic contact dermatitis. However, several studies have shown that AD is inversely associated with contact sensitization. This is possibly due to a reduced cell-mediated immunity in atopic dermatitis [64, 65]. Studies on evaluating *FLG* mutations as a risk factor for contact allergy have been contradictory. When genotyping and patch testing 430 Caucasians in Denmark, Carlsen et al. did not find a positive association between *FLG* mutations and contact sensitization [66]. Moreover, the subgroup of patients with AD and concomitant contact sensitization did not have a significantly higher prevalence of *FLG* mutations compared with patients with positive patch test without AD, or compared with the general population [67].

However, Novak et al. observed an association between *FLG* mutations (R501X and 2282del4) and contact sensitization to nickel combined with self-reported intolerance to fashion jewelry in a cross-sectional German population ($n = 1,537$) [28]. No association was found between the presence of *FLG* mutations and sensitization to other common haptens. This may indicate that probably not only the impaired skin barrier itself confers an increased risk of contact sensitization. Other, more complex mechanisms may be involved.

In a cross-sectional study in the general Danish population ($n = 3,335$), Thyssen et al. found a positive association between *FLG* mutations (R501X and 2282del4) and contact sensitization to nickel only in women without ear piercing [68]. This was explained by the bypass theory, suggesting that contact allergens like nickel bypass the upper skin layers by ear piercing with nickel-releasing jewelry [69]. It was speculated that filaggrin, a histidine-rich protein, could chelate nickel, preventing its penetration through the skin in cases of superficial exposure [70, 71]. Therefore, individuals with reduced levels of filaggrin, as seen in *FLG* mutation carriers [26], may be at higher risk to develop nickel sensitization when superficially exposed to nickel. However, when the skin is exposed to nickel by ear piercing, even individuals with an intact skin barrier may easily get sensitized to nickel independently of *FLG* mutations.

In addition, the prevalence of contact sensitization to chemicals used in topical products, such as ethylenediamine and neomycin, was higher in individuals with AD and *FLG* mutations compared to wild-type carriers without AD

[51]. This is probably related to a frequent application of emollients as well as topical corticosteroids and antibacterial creams in individuals with AD due to their dry and inflamed skin, increasing the likelihood of contact sensitization to ingredients of these products. In a subgroup of individuals with self-reported AD and frequent episodes of hand eczema, a strong association between *FLG* mutations and contact sensitization to allergens other than nickel (OR 5.71, 95 % CI 1.31–24.94) was found, whereas *FLG* mutations alone did not substantially increase the risk of sensitization [72].

In a study by Landeck et al., 496 patients from Germany with occupational irritant contact dermatitis of the hands were examined for atopy, patch tested, and genotyped for the *FLG* loss-of-function mutations R501X, 2282del4, R2447X, and S3247X. The sensitization rate in *FLG* mutation carriers was not increased. Only sensitizations to wool alcohols and para-tertiary-butylphenol formaldehyde resin were associated with *FLG* mutations. Wool alcohols are commonly present in topical products used for inflamed and dry skin. Therefore, the increased rate of sensitization to wool alcohol in *FLG* mutation carriers is probably due to an increased use of topical products by these individuals. The increased rate of sensitization to para-tertiary-butylphenol formaldehyde resin, which is mainly used in adhesives and in paints, was somewhat surprising. However, it was argued that the statistical power of the results was low and that larger cohorts are necessary to confirm these findings [73].

It might be speculated that sensitization to some contact allergens may be promoted by genetic predispositions, whereas sensitization to others may be mainly influenced by environmental factors, such as the nature and concentration of the hapten, time and frequency of exposure to the allergen, and the status of skin damage [74].

25.2.2 Morphology of Hand Eczema

Several attempts have been made to identify distinct morphological pattern that can differentiate between atopic hand eczema and other subtypes of hand eczema [75–78]. It has been demonstrated that involvement of the palmar aspects of the hand is not typical for atopic dermatitis [75], while palmar hyperlinearity [76, 78] and skin lesions on the dorsal aspect of the wrist [77] are associated with atopic hand eczema.

Novak et al. reported that palmar hyperlinearity is significantly more common in *FLG* mutations carriers among 1,502 German adults genotyped for R501X and 2282del4 [28]. In line with this, Brown et al. confirmed in a prospective study among 792 British schoolchildren that palmar hyperlinearity is associated with the presence of *FLG* mutations. The positive and negative predictive values for palmar hyperlinearity in FLG mutation carriers were 71 and 90 %, respectively [55]. In 459 patients with occupational irritant contact dermatitis of the hands, it was demonstrated by a mixture of clinical findings and self-reported clinical features that *FLG* mutations are significantly associated with pulpitis sicca and hyperlinear palms and not with recurring vesicular hand eczema [56].

In a case series by Thyssen et al., eight *FLG* mutation carriers with generalized AD were presented who suffered from a distinct type of hand eczema characterized by hyperkeratosis and dermatitis on the dorsal aspects of the hands and fingers, whereas the palmar involvement was sparse or not present at all [58]. Genotyping 128 patients in a pilot study for R501X and 2282del4, Thyssen et al. demonstrated that hand eczema in *FLG* mutation carriers was mainly located on the dorsal aspects of the hands. Moreover, the hands were often lichenified. Palmar hyperlinearity was common, whereas vesicular eruptions were rarely observed [50]. Moreover, apart from intermittent fissures in winter, the hands of *FLG* mutation carriers typically displayed thickened and folded skin on the dorsal aspects of the fingers [59]. In a general population study, 730 participants were genotyped for *FLG* mutations (R501X and 2282del4) and questioned about skin symptoms. *FLG* mutations were significantly associated with self-reported dry, fissured skin on the hands and/or fingers in adults with and without hand eczema (OR 1.93, 95 % CI 1.05–3.55). This association was predominantly seen in subjects without AD. It was suggested that in atopic individuals, dysfunction of skin barrier proteins other than filaggrin may be involved. Moreover, individuals with AD may apply moisturizers more frequently, resulting in an enhanced skin hydration and prevention of skin fissures. Another possibility is that skin inflammation in AD decreases the skin levels of filaggrin and its derivatives to such an extent that the effect of *FLG* mutations is diminished. No association was found between *FLG* mutations and vesicular hand eczema [57]. This was especially true for chronic hand eczema. It was speculated that in acute irritant and allergic contact dermatitis of the hands, the differences between *FLG* mutation carriers and non-mutation carriers were not as pronounced as in chronic hand eczema, possibly covered by the acute lesions [59]. Interestingly, skin inflammation may result in functional filaggrin deficiency [79]. Therefore, skin changes may not only be the results of *FLG* mutations but also of reduced levels of filaggrin due to chronic inflammation. It is likely that the skin of the dorsal aspects of the hands is more prone to drying, especially in winter, and more exposed to irritants. In *FLG* mutation carriers these effects may be more pronounced, therefore resulting in a preferential involvement of this part of the hand. Interestingly, in a prospective cohort study of children with AD and *FLG* mutations (R501X and 2282del4), skin lesions were predominantly seen on the backs of the hands and cheeks [60].

Conclusion

There is evidence that *FLG* mutations are associated with different subtypes of hand eczema and may influence the prognosis of the disease. Identification of this subgroup of individuals at risk may help to implement specific preventive and therapeutic measures at an early stage. This is particularly important in individuals who work in professions with a high prevalence of occupational hand eczema. In the future, individuals with filaggrin deficiency might be identifiable by minimal invasive tape-stripping methods detecting filaggrin breakdown products [80].

As *FLG* mutations are associated with an early onset and persistence of hand eczema, it would be important that mutation carriers are guided toward professions with as little as possible exposure to irritants and allergens. If high-risk professions have already been chosen, appropriate medical advice should be given before the start of training. Moisturizers and adequate personal protection equipment should be used consistently. In those who have already developed hand eczema, early, intensive, multidisciplinary intervention should be initiated to limit morbidity [81–83]. With regard to therapy of hand eczema, the use of topical anti-inflammatory agents is essential, particularly when considering that inflammation can reduce filaggrin expression [39]. However, the dermatologist should always question and, when possible, limit the use of topical corticosteroid therapy. If the skin barrier is already impaired by endogenous factors, such as *FLG* mutations, the use of topical corticosteroids may further increase the impairment [59]. Other treatments should be considered, if possible. At least corticosteroids with less atrophic effects on the skin should be favored [84]. In the future, skin barrier enhancement therapy that specifically targets filaggrin deficiency may be a promising approach to tackle hand eczema [22].

References

1. Thyssen JP, Johansen JD, Linneberg A, Menne T. The epidemiology of hand eczema in the general population–prevalence and main findings. Contact Dermatitis. 2010;62(2):75–87.
2. Meding B, Jarvholm B. Hand eczema in Swedish adults – changes in prevalence between 1983 and 1996. J Invest Dermatol. 2002;118(4):719–23.
3. Diepgen TL, Andersen KE, Brandao FM, Bruze M, Bruynzeel DP, Frosch P, et al. Hand eczema classification: a cross-sectional, multicentre study of the aetiology and morphology of hand eczema. Br J Dermatol. 2009;160(2):353–8.
4. Leino-Arjas P, Liira J, Mutanen P, Malmivaara A, Matikainen E. Predictors and consequences of unemployment among construction workers: prospective cohort study. BMJ. 1999;319(7210):600–5.

5. Meding B, Swanbeck G. Consequences of having hand eczema. Contact Dermatitis. 1990;23(1):6–14.
6. Cvetkovski RS, Zachariae R, Jensen H, Olsen J, Johansen JD, Agner T. Prognosis of occupational hand eczema: a follow-up study. Arch Dermatol. 2006;142(3):305–11.
7. Hald M, Berg ND, Elberling J, Johansen JD. Medical consultations in relation to severity of hand eczema in the general population. Br J Dermatol. 2008;158(4):773–7.
8. Meding B, Wrangsjo K, Jarvholm B. Hand eczema extent and morphology–association and influence on long-term prognosis. J Invest Dermatol. 2007;127(9): 2147–51.
9. Johansen JD, Hald M, Andersen BL, Laurberg G, Danielsen A, Avnstorp C, et al. Classification of hand eczema: clinical and aetiological types. Based on the guideline of the Danish Contact Dermatitis Group. Contact Dermatitis. 2011;65(1):13–21.
10. Coenraads PJ. Hand eczema is common and multifactorial. J Invest Dermatol. 2007;127(7):1568–70.
11. Fluhr JW, Darlenski R, Angelova-Fischer I, Tsankov N, Basketter D. Skin irritation and sensitization: mechanisms and new approaches for risk assessment. 1. Skin irritation. Skin Pharmacol Physiol. 2008;21(3): 124–35.
12. Smith HR, Basketter DA, McFadden JP. Irritant dermatitis, irritancy and its role in allergic contact dermatitis. Clin Exp Dermatol. 2002;27(2):138–46.
13. Proksch E, Folster-Holst R, Jensen JM. Skin barrier function, epidermal proliferation and differentiation in eczema. J Dermatol Sci. 2006;43(3):159–69.
14. Bryld LE, Agner T, Kyvik KO, Brondsted L, Hindsberger C, Menne T. Hand eczema in twins: a questionnaire investigation. Br J Dermatol. 2000; 142(2):298–305.
15. Lerbaek A, Kyvik KO, Ravn H, Menne T, Agner T. Incidence of hand eczema in a population-based twin cohort: genetic and environmental risk factors. Br J Dermatol. 2007;157(3):552–7.
16. Bryld LE, Hindsberger C, Kyvik KO, Agner T, Menne T. Risk factors influencing the development of hand eczema in a population-based twin sample. Br J Dermatol. 2003;149(6):1214–20.
17. Apfelbacher CJ, Funke U, Radulescu M, Diepgen TL. Determinants of current hand eczema: results from case–control studies nested in the PACO follow-up study (PACO II). Contact Dermatitis. 2010;62(6): 363–70.
18. Coenraads PJ, Diepgen TL. Risk for hand eczema in employees with past or present atopic dermatitis. Int Arch Occup Environ Health. 1998;71(1):7–13.
19. Dickel H, Kuss O, Schmidt A, Diepgen TL. Impact of preventive strategies on trend of occupational skin disease in hairdressers: population based register study. BMJ. 2002;324(7351):1422–3.
20. Visser MJ, Landeck L, Campbell LE, McLean WH, Weidinger S, Calkoen F, et al. Impact of loss-of-function mutations in the filaggrin gene and atopic dermatitis on the development of occupational irritant contact dermatitis. Br J Dermatol. 2013;168(2):326–32.
21. Brown SJ, McLean WH. One remarkable molecule: filaggrin. J Invest Dermatol. 2011;132(3 Pt 2): 751–62.
22. Irvine AD, McLean WH, Leung DY. Filaggrin mutations associated with skin and allergic diseases. N Engl J Med. 2011;365(14):1315–27.
23. Brown SJ, Sandilands A, Zhao Y, Liao H, Relton CL, Meggitt SJ, et al. Prevalent and low-frequency null mutations in the filaggrin gene are associated with early-onset and persistent atopic eczema. J Invest Dermatol. 2008;128(6):1591–4.
24. Sandilands A, Terron-Kwiatkowski A, Hull PR, O'Regan GM, Clayton TH, Watson RM, et al. Comprehensive analysis of the gene encoding filaggrin uncovers prevalent and rare mutations in ichthyosis vulgaris and atopic eczema. Nat Genet. 2007; 39(5):650–4.
25. Kezic S, Kemperman PM, Koster ES, de Jongh CM, Thio HB, Campbell LE, et al. Loss-of-function mutations in the filaggrin gene lead to reduced level of natural moisturizing factor in the stratum corneum. J Invest Dermatol. 2008;128(8):2117–9.
26. Kezic S, Kammeyer A, Calkoen F, Fluhr JW, Bos JD. Natural moisturizing factor components in the stratum corneum as biomarkers of filaggrin genotype: evaluation of minimally invasive methods. Br J Dermatol. 2009;161(5):1098–104.
27. Kezic S, O'Regan GM, Yau N, Sandilands A, Chen H, Campbell LE, et al. Levels of filaggrin degradation products are influenced by both filaggrin genotype and atopic dermatitis severity. Allergy. 2011;66(7): 934–40.
28. Novak N, Baurecht H, Schafer T, Rodriguez E, Wagenpfeil S, Klopp N, et al. Loss-of-function mutations in the filaggrin gene and allergic contact sensitization to nickel. J Invest Dermatol. 2008;128(6): 1430–5.
29. Flohr C, England K, Radulovic S, McLean WH, Campbel LE, Barker J, et al. Filaggrin loss-of-function mutations are associated with early-onset eczema, eczema severity and transepidermal water loss at 3 months of age. Br J Dermatol. 2010;163(6): 1333–6.
30. Scharschmidt TC, Man MQ, Hatano Y, Crumrine D, Gunathilake R, Sundberg JP, et al. Filaggrin deficiency confers a paracellular barrier abnormality that reduces inflammatory thresholds to irritants and haptens. J Allergy Clin Immunol. 2009;124(3):496–506, 506.e1–6.
31. Gruber R, Elias PM, Crumrine D, Lin TK, Brandner JM, Hachem JP, et al. Filaggrin genotype in ichthyosis vulgaris predicts abnormalities in epidermal structure and function. Am J Pathol. 2011;178(5):2252–63.
32. Barker JN, Palmer CN, Zhao Y, Liao H, Hull PR, Lee SP, et al. Null mutations in the filaggrin gene (FLG) determine major susceptibility to early-onset atopic dermatitis that persists into adulthood. J Invest Dermatol. 2007;127(3):564–7.
33. Palmer CN, Irvine AD, Terron-Kwiatkowski A, Zhao Y, Liao H, Lee SP, et al. Common loss-of-function

variants of the epidermal barrier protein filaggrin are a major predisposing factor for atopic dermatitis. Nat Genet. 2006;38(4):441–6.
34. Weidinger S, Rodriguez E, Stahl C, Wagenpfeil S, Klopp N, Illig T, et al. Filaggrin mutations strongly predispose to early-onset and extrinsic atopic dermatitis. J Invest Dermatol. 2007;127(3):724–6.
35. O'Regan GM, Irvine AD. The role of filaggrin in the atopic diathesis. Clin Exp Allergy. 2010;40(7):965–72.
36. Marenholz I, Nickel R, Ruschendorf F, Schulz F, Esparza-Gordillo J, Kerscher T, et al. Filaggrin loss-of-function mutations predispose to phenotypes involved in the atopic march. J Allergy Clin Immunol. 2006;118(4):866–71.
37. Weidinger S, Illig T, Baurecht H, Irvine AD, Rodriguez E, Diaz-Lacava A, et al. Loss-of-function variations within the filaggrin gene predispose for atopic dermatitis with allergic sensitizations. J Allergy Clin Immunol. 2006;118(1):214–9.
38. Jakasa I, Koster ES, Calkoen F, McLean WH, Campbell LE, Bos JD, et al. Skin barrier function in healthy subjects and patients with atopic dermatitis in relation to filaggrin loss-of-function mutations. J Invest Dermatol. 2011;131(2):540–2.
39. Gutowska-Owsiak D, Schaupp AL, Salimi M, Taylor S, Ogg GS. Interleukin-22 downregulates filaggrin expression and affects expression of profilaggrin processing enzymes. Br J Dermatol. 2011;165(3):492–8.
40. Brown SJ, Kroboth K, Sandilands A, Campbell LE, Pohler E, Kezic S, et al. Intragenic copy number variation within filaggrin contributes to the risk of atopic dermatitis with a dose-dependent effect. J Invest Dermatol. 2012;132(1):98–104.
41. Kezic S, O'Regan GM, Lutter R, Jakasa I, Koster ES, Saunders S, et al. Filaggrin loss-of-function mutations are associated with enhanced expression of IL-1 cytokines in the stratum corneum of patients with atopic dermatitis and in a murine model of filaggrin deficiency. J Allergy Clin Immunol. 2012;129(4):1031–9.e1.
42. Nassif A, Chan SC, Storrs FJ, Hanifin JM. Abnormal skin irritancy in atopic dermatitis and in atopy without dermatitis. Arch Dermatol. 1994;130(11):1402–7.
43. Ross-Hansen K, Linneberg A, Johansen JD, Hersoug LG, Brasch-Andersen C, Menne T, et al. The role of glutathione S-transferase and claudin-1 gene polymorphisms in contact sensitization: a cross-sectional study. Br J Dermatol. 2013;168(4):762–70.
44. de Jongh CM, John SM, Bruynzeel DP, Calkoen F, van Dijk FJ, Khrenova L, et al. Cytokine gene polymorphisms and susceptibility to chronic irritant contact dermatitis. Contact Dermatitis. 2008;58(5):269–77.
45. de Jongh CM, Khrenova L, Kezic S, Rustemeyer T, Verberk MM, John SM. Polymorphisms in the interleukin-1 gene influence the stratum corneum interleukin-1 alpha concentration in uninvolved skin of patients with chronic irritant contact dermatitis. Contact Dermatitis. 2008;58(5):263–8.
46. Schnuch A, Westphal G, Mossner R, Uter W, Reich K. Genetic factors in contact allergy–review and future goals. Contact Dermatitis. 2011;64(1):2–23.
47. Landeck L, Visser M, Kezic S, John SM. IL1A-889C/T gene polymorphism in irritant contact dermatitis. J Eur Acad Dermatol Venereol. 2013;27(8):1040–3.
48. Lerbaek A, Bisgaard H, Agner T, Ohm Kyvik K, Palmer CN, Menne T. Filaggrin null alleles are not associated with hand eczema or contact allergy. Br J Dermatol. 2007;157(6):1199–204.
49. Brown SJ, Cordell HJ. Are filaggrin mutations associated with hand eczema or contact allergy?–we do not know. Br J Dermatol. 2008;158(6):1383–4.
50. Thyssen JP, Carlsen BC, Johansen JD, Meldgaard M, Szecsi PB, Stender S, et al. Filaggrin haploinsufficiency among patients with dermatitis from a tertiary referral centre: early findings and possible phenotype. Contact Dermatitis. 2010;62(3):182–3.
51. Thyssen JP, Carlsen BC, Menne T, Linneberg A, Nielsen NH, Meldgaard M, et al. Filaggrin null mutations increase the risk and persistence of hand eczema in subjects with atopic dermatitis: results from a general population study. Br J Dermatol. 2010;163(1):115–20.
52. de Jongh CM, Khrenova L, Verberk MM, Calkoen F, van Dijk FJ, Voss H, et al. Loss-of-function polymorphisms in the filaggrin gene are associated with an increased susceptibility to chronic irritant contact dermatitis: a case–control study. Br J Dermatol. 2008;159(3):621–7.
53. Molin S, Vollmer S, Weiss EH, Ruzicka T, Prinz JC. Filaggrin mutations may confer susceptibility to chronic hand eczema characterized by combined allergic and irritant contact dermatitis. Br J Dermatol. 2009;161(4):801–7.
54. Landeck L, Visser M, Skudlik C, Brans R, Kezic S, John SM. Clinical course of occupational irritant contact dermatitis of the hands in relation to filaggrin genotype status and atopy. Br J Dermatol. 2012;167(6):1302–9.
55. Brown SJ, Relton CL, Liao H, Zhao Y, Sandilands A, McLean WH, et al. Filaggrin haploinsufficiency is highly penetrant and is associated with increased severity of eczema: further delineation of the skin phenotype in a prospective epidemiological study of 792 school children. Br J Dermatol. 2009;161(4):884–9.
56. Landeck L, Visser M, Kezic S, John SM. Genotype-phenotype associations in filaggrin loss-of-function mutation carriers. Contact Dermatitis. 2013;68(3):149–55.
57. Thyssen JP, Ross-Hansen K, Johansen JD, Zachariae C, Carlsen BC, Linneberg A, et al. Filaggrin loss-of-function mutation R501X and 2282del4 carrier status is associated with fissured skin on the hands: results from a cross-sectional population study. Br J Dermatol. 2012;166(1):46–53.
58. Thyssen JP, Carlsen BC, Johansen JD, Meldgaard M, Szecsi PB, Stender S, et al. Filaggrin null-mutations may be associated with a distinct subtype of atopic hand eczema. Acta Derm Venereol. 2010;90(5):528.

59. Kaae J, Menne T, Carlsen BC, Zachariae C, Thyssen JP. The hands in health and disease of individuals with filaggrin loss-of-function mutations: clinical reflections on the hand eczema phenotype. Contact Dermatitis. 2012;67(3):119–24.
60. Carson CG, Rasmussen MA, Thyssen JP, Menne T, Bisgaard H. Clinical presentation of atopic dermatitis by filaggrin gene mutation status during the first 7 years of life in a prospective cohort study. PLoS One. 2012;7(11):e48678.
61. Giwercman C, Lerbaek A, Bisgaard H, Menne T. Classification of atopic hand eczema and the filaggrin mutations. Contact Dermatitis. 2008;59(5):257–60.
62. McLelland J, Shuster S, Matthews JN. 'Irritants' increase the response to an allergen in allergic contact dermatitis. Arch Dermatol. 1991;127(7):1016–9.
63. Pedersen LK, Haslund P, Johansen JD, Held E, Volund A, Agner T. Influence of a detergent on skin response to methyldibromo glutaronitrile in sensitized individuals. Contact Dermatitis. 2004;50(1):1–5.
64. Jones HE, Lewis CW, McMarlin SL. Allergic contact sensitivity in atopic dermatitis. Arch Dermatol. 1973;107(2):217–22.
65. Uehara M, Sawai T. A longitudinal study of contact sensitivity in patients with atopic dermatitis. Arch Dermatol. 1989;125(3):366–8.
66. Carlsen BC, Johansen JD, Menne T, Meldgaard M, Szecsi PB, Stender S, et al. Filaggrin null mutations and association with contact allergy and allergic contact dermatitis: results from a tertiary dermatology clinic. Contact Dermatitis. 2010;63(2):89–95.
67. Carlsen BC, Thyssen JP, Menne T, Meldgaard M, Linneberg A, Nielsen NH, et al. Association between filaggrin null mutations and concomitant atopic dermatitis and contact allergy. Clin Exp Dermatol. 2011;36(5):467–72.
68. Thyssen JP, Johansen JD, Linneberg A, Menne T, Nielsen NH, Meldgaard M, et al. The association between null mutations in the filaggrin gene and contact sensitization to nickel and other chemicals in the general population. Br J Dermatol. 2010;162(6):1278–85.
69. Ross-Hansen K, Menne T, Johansen JD, Carlsen BC, Linneberg A, Nielsen NH, et al. Nickel reactivity and filaggrin null mutations–evaluation of the filaggrin bypass theory in a general population. Contact Dermatitis. 2011;64(1):24–31.
70. Fullerton A, Hoelgaard A. Binding of nickel to human epidermis in vitro. Br J Dermatol. 1988;119(5):675–82.
71. Thyssen JP, Carlsen BC, Menne T. Nickel sensitization, hand eczema, and loss-of-function mutations in the filaggrin gene. Dermatitis. 2008;19(6):303–7.
72. Thyssen JP, Linneberg A, Ross-Hansen K, Carlsen BC, Meldgaard M, Szecsi PB, et al. Filaggrin mutations are strongly associated with contact sensitization in individuals with dermatitis. Contact Dermatitis. 2013; 68(5):273–6.
73. Landeck L, Visser M, Skudlik C, Brans R, Kezic S, John SM. Contact Dermatitis. 2014;70(1):27–34.
74. Bryld LE, Hindsberger C, Kyvik KO, Agner T, Menne T. Genetic factors in nickel allergy evaluated in a population-based female twin sample. J Invest Dermatol. 2004;123(6):1025–9.
75. Bonven TF, Hansen L, Thestrup-Pedersen K. Palmar skin is not involved in severe atopic eczema. Dermatologica. 1991;183(4):237–8.
76. Fartasch M, Diepgen TL, Hornstein OP. Are hyperlinear palms and dry skin signs of a concomitant autosomal ichthyosis vulgaris in atopic dermatitis? Acta Derm Venereol Suppl (Stockh). 1989;144: 143–5.
77. Simpson EL, Thompson MM, Hanifin JM. Prevalence and morphology of hand eczema in patients with atopic dermatitis. Dermatitis. 2006;17(3):123–7.
78. Svensson A. Hand eczema: an evaluation of the frequency of atopic background and the difference in clinical pattern between patients with and without atopic dermatitis. Acta Derm Venereol. 1988;68(6): 509–13.
79. Howell MD, Kim BE, Gao P, Grant AV, Boguniewicz M, DeBenedetto A, et al. Cytokine modulation of atopic dermatitis filaggrin skin expression. J Allergy Clin Immunol. 2009;124(3 Suppl 2):R7–12.
80. Dapic I, Jakasaa I, Yau N, Kezic S, Kammeyer A. Evaluation of an HPLC method for the determination of natural moisturizing factors in the human stratum corneum. Anal Lett. 2013;46:2133–44.
81. Skudlik C, Wulfhorst B, Gediga G, Bock M, Allmers H, John SM. Tertiary individual prevention of occupational skin diseases: a decade's experience with recalcitrant occupational dermatitis. Int Arch Occup Environ Health. 2008;81(8):1059–64.
82. Skudlik C, Weisshaar E, Scheidt R, Elsner P, Wulfhorst B, Schonfeld M, et al. First results from the multicentre study rehabilitation of occupational skin diseases–optimization and quality assurance of inpatient management (ROQ). Contact Dermatitis. 2012; 66(3):140–7.
83. Weisshaar E, Skudlik C, Scheidt R, Matterne U, Wulfhorst B, Schonfeld M, et al. Multicentre study 'rehabilitation of occupational skin diseases -optimization and quality assurance of inpatient management (ROQ)'-results from 12-month follow-up. Contact Dermatitis. 2013;68(3):169–74.
84. Jensen JM, Pfeiffer S, Witt M, Brautigam M, Neumann C, Weichenthal M, et al. Different effects of pimecrolimus and betamethasone on the skin barrier in patients with atopic dermatitis. J Allergy Clin Immunol. 2009;124(3 Suppl 2):R19–28.

26 Filaggrin in Psoriasis

Peter Jensen and Lone Skov

Contents

26.1 Introduction

Psoriasis is a chronic inflammatory skin disease with a prevalence of 2–3 % [1]. Abnormal keratinocyte terminal differentiation is an important feature of psoriatic skin lesions, which are characterized by red patches or plaques on the skin with thick, silvery, and adherent scales. Barrier function in lesional psoriatic skin is reduced [2, 3]. Psoriasis is considered to be a disorder of multifactorial origin affecting patients who are genetically predisposed and who at the same time are exposed to some form of environmental trigger, such as tonsillitis caused by group A Streptococci. Linkage analyses of large families as well as genome-wide transmission studies have mapped a number of susceptibility loci known as psoriasis-associated chromosomal regions (PSORS) 1–10. Some of the psoriasis susceptibility loci are shared with susceptibility for atopic dermatitis (AD) [4]. This has led to speculations that filaggrin gene (*FLG*) mutations and changes in filaggrin expression may also be associated with psoriasis. This chapter presents the current knowledge on the possible association between psoriasis and *FLG* mutations as well as changes in filaggrin expression in psoriasis.

26.2 Filaggrin Gene Mutations in Psoriasis

Familial occurrence of psoriasis is well known, and the genetic predisposition to psoriasis is supported by twin studies showing high concordance

P. Jensen, MD, PhD (✉) • L. Skov, MD, PhD
Department of Dermato-Allergology,
Gentofte Hospital, Copenhagen University,
Niels Andersens Vej 65, Hellerup 2900, Denmark
e-mail: peter.jensen@regionh.dk;
lone.skov.02@regionh.dk

J.P. Thyssen, H. Maibach (eds.), *Filaggrin*,
DOI 10.1007/978-3-642-54379-1_26, © Springer-Verlag Berlin Heidelberg 2014

rates in monozygotic twins [5, 6]. Over the years, several loci and genes have been identified by linkage analyses and genome-wide association studies. It has been established that PSORS1 is the predominant locus for psoriasis vulgaris, the most common clinical variant of psoriasis. PSORS1 is located within the region of the major histocompatibility complex on chromosome 6p21.3, where half of patients with psoriasis vulgaris carry the human leukocyte antigen (HLA)-Cw6 allele [7]. Recently, HLA Cw6 was found to be the most likely PSORS1 candidate gene responsible for most of the genetic susceptibility in psoriasis [8]. Several other psoriasis susceptibility loci have since been identified including psoriasis-associated genes belonging to the interleukin 23/Th17 axis as well as nuclear factor-κβ [9–18]. The epidermal differentiation complex (EDC) on chromosome 1q21.3 is another pathway, which may be involved in the pathogenesis of psoriasis. The EDC is a cluster of genes (including *FLG*), which encode molecules in the superficial layers of the differentiating epidermis [4]. Psoriasis susceptibility locus 4 (PSORS4) has been linked to a region located at 1q21 that also contains the EDC. Filaggrin plays a key role in facilitating the terminal differentiation of the epidermis and formation of the skin barrier; therefore, *FLG* variants may in theory be a primary event in the pathogenesis of psoriasis and may influence the risk of psoriasis.

Several groups have studied the possible association between certain *FLG* variants and psoriasis [2, 19–25]. Table 26.1 provides an overview of these studies, which include both case-control studies and cross-sectional general population studies. Most studies investigated the association between psoriasis and the two major *FLG* variants associated with AD (R501X and 2282del4), and none have so far demonstrated a positive correlation. In addition, the study by Winge et al. including children with psoriasis failed to demonstrate an association between *FLG* mutations and onset of psoriasis in childhood [22]. However, two studies including individuals of Asian ethnicity have demonstrated an association between psoriasis and the two rare *FLG* variants p. K4022X and P478S [19, 23]. It therefore seems that the genetic background responsible for the epidermal barrier defect in psoriasis is different from that associated with AD, at least in Caucasian individuals. However, due to genetic heterogeneity, one cannot rule out the possibility that altered filaggrin structure and function caused by other *FLG* variants may influence psoriasis susceptibility in other non-Caucasian populations. In conclusion, available data indicate that the currently known *FLG* null genotypes can be

Table 26.1 Overview of studies that have investigated the possible association between psoriasis and filaggrin gene mutations

Author name	Filaggrin gene (*FLG*) mutation	Ethnicity	Number of patients with psoriasis (controls)	*FLG* variant associated with psoriasis
Hüffmeier et al. [2]	R501X 2282del4	Caucasian (German)	750 (376)	No association
Weichenthal et al. [20]	R501X 2282del4	Caucasian (German)	360 (276)	No association
Giardina et al. [21]	R501X 2282del4	Caucasian (Italian)	195 (210)	No association
Winge et al. [22]	R501X 2282del4 S3247X R2447X	Northern European (Finnish)	241 (314)	No association
Hu et al. [23]	p.K4022X	Asian (Chinese)	441 (500)	p.K4022X
Zhao et al. [24]	R501X, 2282del4	Caucasian (Irish, English)	691 (2117)	No association
Thyssen et al. [25]	R501X, 2282del4	Caucasian (Danish)	217 (3023)	No association

excluded as predisposing factors for psoriasis in Caucasian individuals, but certain other and rare *FLG* variants may play a role in psoriasis susceptibility in other populations.

26.3 Filaggrin Protein Expression in Psoriasis

Several histological studies have demonstrated that the expression of filaggrin is decreased in lesional psoriatic skin. Bernard et al. obtained skin biopsies from psoriatic lesions from 14 patients with psoriasis and compared them to skin biopsies from individuals without psoriasis. The authors showed that filaggrin was undetectable in stable psoriatic skin lesions [26]. This was confirmed in a study from 1991 by Watanabe et al. demonstrating that involved psoriatic skin revealed little or no reaction in the stratum corneum or in the granular layer with anti-filaggrin antibody [27]. Furthermore, Gerritsen et al. investigated epidermal and dermal aspects of the margin of the progressive psoriatic plaque, distant uninvolved skin, and normal healthy skin using immunohistochemistry for filaggrin and other markers for keratinization [28]. The authors found that in the margin of the spreading psoriatic plaque and in the plaque itself, the expression of filaggrin in the stratum corneum and stratum granulosum was discontinued or absent. The study also supported the notion that filaggrin is unaffected on the genetic level in patients with psoriasis by demonstrating that filaggrin expression in uninvolved skin was similar to that found in skin biopsies from individuals without psoriasis. In 2007, Hüffmeier studied the expression of filaggrin in biopsies obtained from lesional skin from 10 patients with psoriasis and showed that epidermal *FLG* expression was reduced in 8 out of 10 patients, and in 2 patients there was an almost complete lack of *FLG* expression [2]. In contrast, Kim et al. demonstrated deficiency of filaggrin expression in both lesional and non-lesional skin from patients with psoriasis; however, filaggrin expression was significantly reduced in lesional compared to non-lesional skin samples [29]. There are several possible explanations for the downregulation of filaggrin expression in lesional psoriatic skin. Expression of filaggrin is a calcium-dependent process that is downregulated after blocking the N-methyl-D-ASPARTATE (NMDA) receptor [30]. In 2004, Fischer et al. demonstrated that the expression of NMDA in the superficial epidermis was reduced in psoriasis. The NMDA receptor participates in the regulation of keratinocyte intracellular calcium, and, therefore, NMDA receptor deficiency associated with psoriasis may result in concomitant reductions of filaggrin expression [31]. Another possible explanation for the downregulation of filaggrin expression in psoriasis may be the overexpression of TNF-α as seen in psoriasis [32, 33]. Kim et al. showed that filaggrin expression was significantly inhibited in keratinocytes treated with TNF-α and that TNF-α-neutralizing antibody blocked the in vitro TNF-α-mediated inhibition [29]. Another mechanism was proposed by Hvid et al. investigating the effect of caspase 14, a proteinase that probably plays a role in the maturation process of the epidermis [34]. Caspase 14 deficiency results in abnormal accumulation of filaggrin fragments in the stratum corneum due to decreased breakdown of filaggrin into hygroscopic amino acids resulting in decreased barrier function. Hvid et al. demonstrated that protein levels of caspase 14 were significantly reduced in keratinocytes following stimulation with the Th1-associated cytokine IFN-γ, which is elevated in lesional psoriatic skin. In addition, the authors showed that levels of filaggrin mRNA in keratinocytes decreased significantly in response to stimulation with IFN-γ. Downregulation of filaggrin expression in psoriatic skin lesions might also be due to elevated levels of circulating interleukin (IL)-17. IL-17 is an important cytokine in psoriasis, which is elevated in both serum and skin in these patients [35]. Little is known about how IL-17 exerts its effect on filaggrin expression and function; however, the study showed that profilaggrin mRNA levels decreased significantly in cell cultures exposed to IL-17. Taken together, current knowledge indicates that inflammatory mechanisms in psoriatic as in atopic skin can lead to downregulation of filaggrin expression.

In summary, there is little evidence to support a correlation between *FLG* gene mutations and psoriasis except in Asian individuals, in which two rare *FLG* gene variants have been shown to be associated with psoriasis. Several studies have shown a decreased filaggrin expression in lesional psoriatic skin probably due to secondary downregulation of filaggrin expression by inflammatory mechanisms.

References

1. Smith CH, Barker JN. Psoriasis and its management. BMJ. 2006;333(7564):380–4. Epub 2006/08/19.
2. Huffmeier U, Traupe H, Oji V, Lascorz J, Stander M, Lohmann J, et al. Loss-of-function variants of the filaggrin gene are not major susceptibility factors for psoriasis vulgaris or psoriatic arthritis in German patients. J Invest Dermatol. 2007;127(6):1367–70. Epub 2007/01/27.
3. Proksch E, Brandner JM, Jensen JM. The skin: an indispensable barrier. Exp Dermatol. 2008;17(12):1063–72. Epub 2008/12/02.
4. Bowcock AM, Cookson WO. The genetics of psoriasis, psoriatic arthritis and atopic dermatitis. Hum Mol Genet. 2004;13(Spec No 1):R43–55. Epub 2004/03/05.
5. Brandrup F, Hauge M, Henningsen K, Eriksen B. Psoriasis in an unselected series of twins. Arch Dermatol. 1978;114(6):874–8. Epub 1978/06/01.
6. Lonnberg AS, Skov L, Skytthe A, Kyvik KO, Pedersen OB, Thomsen SF. Heritability of psoriasis in a large twin sample. Br J Dermatol. 2013;169:412–6.
7. Trembath RC, Clough RL, Rosbotham JL, Jones AB, Camp RD, Frodsham A, et al. Identification of a major susceptibility locus on chromosome 6p and evidence for further disease loci revealed by a two stage genome-wide search in psoriasis. Hum Mol Genet. 1997;6(5):813–20. Epub 1997/05/01.
8. Nair RP, Stuart PE, Nistor I, Hiremagalore R, Chia NV, Jenisch S, et al. Sequence and haplotype analysis supports HLA-C as the psoriasis susceptibility 1 gene. Am J Hum Genet. 2006;78(5):827–51. Epub 2006/04/28.
9. Nair RP, Stuart P, Henseler T, Jenisch S, Chia NV, Westphal E, et al. Localization of psoriasis-susceptibility locus PSORS1 to a 60-kb interval telomeric to HLA-C. Am J Hum Genet. 2000;66(6):1833–44. Epub 2000/05/10.
10. Elder JT, Nair RP, Henseler T, Jenisch S, Stuart P, Chia N, et al. The genetics of psoriasis 2001: the odyssey continues. Arch Dermatol. 2001;137(11):1447–54. Epub 2001/11/16.
11. International Psoriasis Genetics Consortium. The International Psoriasis Genetics Study: assessing linkage to 14 candidate susceptibility loci in a cohort of 942 affected sib pairs. American journal of human genetics. 2003;73(2):430–7. Epub 2003/07/10.
12. Capon F, Trembath RC, Barker JN. An update on the genetics of psoriasis. Dermatol Clin. 2004;22(4):339–47, vii. Epub 2004/09/29.
13. Capon F, Helms C, Veal CD, Tillman D, Burden AD, Barker JN, et al. Genetic analysis of PSORS2 markers in a UK dataset supports the association between RAPTOR SNPs and familial psoriasis. J Med Genet. 2004;41(6):459–60. Epub 2004/06/03.
14. Morar N, Bowcock AM, Harper JI, Cookson WO, Moffatt MF. Investigation of the chromosome 17q25 PSORS2 locus in atopic dermatitis. J Invest Dermatol. 2006;126(3):603–6. Epub 2005/12/24.
15. Bowcock AM, Krueger JG. Getting under the skin: the immunogenetics of psoriasis. Nat Rev Immunol. 2005;5(9):699–711. Epub 2005/09/03.
16. Kere J. Mapping and identifying genes for asthma and psoriasis. Philos Trans R Soc Lond B Biol Sci. 2005;360(1460):1551–61. Epub 2005/08/13.
17. Goldminz AM, Au SC, Kim N, Gottlieb AB, Lizzul PF. NF-kappaB: an essential transcription factor in psoriasis. J Dermatol Sci. 2013;69(2):89–94. Epub 2012/12/12.
18. Lowes MA, Russell CB, Martin DA, Towne JE, Krueger JG. The IL-23/T17 pathogenic axis in psoriasis is amplified by keratinocyte responses. Trends Immunol. 2013;34(4):174–81. Epub 2013/01/08.
19. Chang YC, Wu WM, Chen CH, Hu CF, Hsu LA. Association between P478S polymorphism of the filaggrin gene and risk of psoriasis in a Chinese population in Taiwan. Arch Dermatol Res. 2008;300(3):133–7. Epub 2008/01/15.
20. Weichenthal M, Ruether A, Schreiber S, Nair R, Voorhees JJ, Schwarz T, et al. Filaggrin R501X and 2282del4 mutations are not associated with chronic plaque-type psoriasis in a German cohort. J Invest Dermatol. 2007;127(6):1535–7. Epub 2007/03/24.
21. Giardina E, Paolillo N, Sinibaldi C, Novelli G. R501X and 2282del4 filaggrin mutations do not confer susceptibility to psoriasis and atopic dermatitis in Italian patients. Dermatology. 2008;216(1):83–4. Epub 2007/11/23.
22. Winge MC, Suneson J, Lysell J, Nikamo P, Lieden A, Nordenskjold M, et al. Lack of association between filaggrin gene mutations and onset of psoriasis in childhood. J Eur Acad Dermatol Venereol. 2013;27(1):e124–7. Epub 2011/12/21.
23. Hu Z, Xiong Z, Xu X, Li F, Lu L, Li W, et al. Loss-of-function mutations in filaggrin gene associate with psoriasis vulgaris in Chinese population. Hum Genet. 2012;131(7):1269–74. Epub 2012/03/13.
24. Zhao Y, Terron-Kwiatkowski A, Liao H, Lee SP, Allen MH, Hull PR, et al. Filaggrin null alleles are not associated with psoriasis. J Invest Dermatol. 2007;127(8):1878–82. Epub 2007/04/06.
25. Thyssen JP, Johansen JD, Carlsen BC, Linneberg A, Meldgaard M, Szecsi PB, et al. The filaggrin null

genotypes R501X and 2282del4 seem not to be associated with psoriasis: results from general population study and meta-analysis. J Eur Acad Dermatol Venereol. 2012;26(6):782–4. Epub 2011/05/14.

26. Bernard BA, Asselineau D, Schaffar-Deshayes L, Darmon MY. Abnormal sequence of expression of differentiation markers in psoriatic epidermis: inversion of two steps in the differentiation program? J Invest Dermatol. 1988;90(6):801–5. Epub 1988/06/01.
27. Watanabe S, Wagatsuma K, Ichikawa E, Takahashi H. Abnormal distribution of epidermal protein antigens in psoriatic epidermis. J Dermatol. 1991;18(3):143–51. Epub 1991/03/01.
28. Gerritsen MJ, Elbers ME, de Jong EM, van de Kerkhof PC. Recruitment of cycling epidermal cells and expression of filaggrin, involucrin and tenascin in the margin of the active psoriatic plaque, in the uninvolved skin of psoriatic patients and in the normal healthy skin. J Dermatol Sci. 1997;14(3):179–88. Epub 1997/03/01.
29. Kim BE, Howell MD, Guttman-Yassky E, Gilleaudeau PM, Cardinale IR, Boguniewicz M, et al. TNF-alpha downregulates filaggrin and loricrin through c-Jun N-terminal kinase: role for TNF-alpha antagonists to improve skin barrier. J Invest Dermatol. 2011;131(6):1272–9. Epub 2011/02/25.
30. Elias PM, Ahn SK, Denda M, Brown BE, Crumrine D, Kimutai LK, et al. Modulations in epidermal calcium regulate the expression of differentiation-specific markers. J Invest Dermatol. 2002;119(5):1128–36. Epub 2002/11/26.
31. Fischer M, William T, Helmbold P, Wohlrab J, Marsch W. Expression of epidermal N-methyl-D-aspartate receptors (NMDAR1) depends on formation of the granular layer–analysis in diseases with parakeratotic cornification. Arch Dermatol Res. 2004;296(4):157–62. Epub 2004/09/01.
32. Kristensen M, Chu CQ, Eedy DJ, Feldmann M, Brennan FM, Breathnach SM. Localization of tumour necrosis factor-alpha (TNF-alpha) and its receptors in normal and psoriatic skin: epidermal cells express the 55-kD but not the 75-kD TNF receptor. Clin Exp Immunol. 1993;94(2):354–62. Epub 1993/11/01.
33. Ettehadi P, Greaves MW, Wallach D, Aderka D, Camp RD. Elevated tumour necrosis factor-alpha (TNF-alpha) biological activity in psoriatic skin lesions. Clin Exp Immunol. 1994;96(1):146–51. Epub 1994/04/01.
34. Hvid M, Johansen C, Deleuran B, Kemp K, Deleuran M, Vestergaard C. Regulation of caspase 14 expression in keratinocytes by inflammatory cytokines–a possible link between reduced skin barrier function and inflammation? Exp Dermatol. 2011;20(8):633–6. Epub 2011/05/05.
35. Gutowska-Owsiak D, Schaupp AL, Salimi M, Selvakumar TA, McPherson T, Taylor S, et al. IL-17 downregulates filaggrin and affects keratinocyte expression of genes associated with cellular adhesion. Exp Dermatol. 2012;21(2):104–10. Epub 2012/01/11.

Filaggrin and Skin Infections

27

Emily Y. Gan, Sophie C.S. Cai, and Mark B.Y. Tang

Contents

E.Y. Gan, MBBS(Hons), MRCP(UK), MMed(IntMed) • M.B.Y. Tang, MBBS, MMED, FRCP (UK), FAMS (✉)
Department of Dermatology, National Skin Centre, 1 Mandalay Road, Singapore 308205, Singapore
e-mail: emily.ganyp@gmail.com; marktang@nsc.gov.sg

S.C.S. Cai, BA (Hons), MS
Duke-National University of Singapore Graduate Medical School, 8 College Road, Singapore 169857, Singapore
e-mail: scai@nus.edu.sg

27.1 Introduction

One of the key functions of the skin is to act as a barrier against microbes and infections. The skin barrier is composed of both the physical epidermal barrier and the innate immune barrier, which act together to protect and maintain microbial balance of the skin.

Filaggrin is one of the major components of the stratum corneum, the tough, outermost layer of the physical skin barrier. Together with other structural proteins such as loricrin, involucrin, and small proline-rich proteins, filaggrin is heavily cross-linked with the lipid constituents, including ceramides, cholesterol, and free fatty acids, to form the hydrophobic lipid-protein extracellular matrix. This extracellular matrix seals and envelops the multiple layers of corneocytes, which comprise the cellular component of this complex stratum corneum barrier [1].

Once thought to contribute only to the protective water-proofing function of the stratum corneum through its keratin-binding properties and production of "natural moisturizing factors" from proteolytic degradation, filaggrin is now increasingly known to be the driving force in a wide range of metabolic, immune, and inflammatory processes of the skin barrier.

J.P. Thyssen, H. Maibach (eds.), *Filaggrin*,
DOI 10.1007/978-3-642-54379-1_27, © Springer-Verlag Berlin Heidelberg 2014

This chapter discusses the crucial role that filaggrin plays in maintaining stratum corneum function and homeostasis, focusing especially on skin infections in patients with atopic dermatitis (AD). It also reviews the vicious cycle where skin infections affect filaggrin production and degradation, leading to an acquired filaggrin deficiency and barrier abnormality, which further perpetuates the existing infection.

27.2 Filaggrin Deficiency in Atopic Dermatitis: Both Genetic and Acquired Causes

Mutations in the human filaggrin gene (*FLG*) have been shown to be the most significant risk factor to date for AD [2], a highly prevalent, chronic relapsing inflammatory skin condition. AD patients with *FLG* null mutations have been shown to have early-onset disease, increased disease severity, more persistent disease, and a greater risk of allergic sensitization and asthma [3]. *FLG* mutations have also been found to be population specific and heterogeneous, with up to 50 % of Irish AD patients having at least one null mutation, compared to only 20 % in a Chinese AD cohort [4].

It is noteworthy that even in patients without *FLG* null mutations, acquired filaggrin deficiency has been demonstrated due to the predominant Th2 cytokine milieu in AD [5]. Specifically, it has been shown that Th2 cytokines, such as interleukin (IL)-4 and IL-13, significantly downregulate keratinocyte filaggrin expression. Furthermore, the upregulation of IL-25 [6], IL-22 [7], and IL-17 [8] in AD has been shown to inhibit the synthesis of filaggrin. These findings reinforce the intricate role of inflammation and cytokines in inhibiting filaggrin expression in AD, regardless of *FLG* status. Recently, a dose-dependent effect on AD risk attributed to intragenic copy number variation within the *FLG* has been reported, highlighting the importance of filaggrin dose in disease pathogenesis [9]. Taken together, these findings highlight the ubiquitous presence of filaggrin deficiency in AD, whether secondary to inherent genetic causes or to acquired immune responses, reinforcing its crucial role in the disease's pathogenesis.

It is well established that AD patients have a unique propensity to be colonized or infected by microbial organisms. This is fundamentally attributed to the underlying physical skin barrier dysfunction and abnormalities in both innate and adaptive immune responses in AD [10, 11]. Essentially, the skin barrier dysfunction readily allows penetration of multiple allergens or haptens, which enhance allergic inflammation. The allergic inflammation derived from underlying immunological abnormalities in turn further damages the skin barrier. This incessant sequence cycle therefore explains why AD patients are particularly vulnerable to surface infections by pathogenic bacteria, fungi, and viruses.

Recent evidence is emerging to support the central role that filaggrin deficiency has in contributing to the increased risk of cutaneous infections in AD.

27.3 Filaggrin Deficiency and the Increased Risk of Bacterial Infections in AD

Recurrent bacterial infections are a major complication and therapeutic challenge in AD patients. In particular, up to 90 % of AD patients are colonized with *Staphylococcus aureus* (*S. aureus*) [12], compared to a third of the general community [13], and recolonization after a course of systemic antimicrobial therapy occurs rapidly, often with the same toxin-producing strains [14]. Patients infected with *S. aureus* can present with widespread impetiginized eczematous plaques, which typically have a golden honey yellow crust or may be weepy (Fig. 27.1).

S. aureus colonization plays an important role in disease pathogenesis and triggering of recurrent flares in AD. The density of *S. aureus* on AD skin has also been associated with an increase in clinical severity [15].

Multiple *S. aureus*-associated virulence factors, including pro-inflammatory, superantigenic enterotoxins, such as staphylococcal enterotoxin-b (SEB), can directly activate the inflammatory cascade in AD. *S. aureus* adhesins, part of the family of microbial surface components recognizing adhesive matrix molecules (MSCRAMMs),

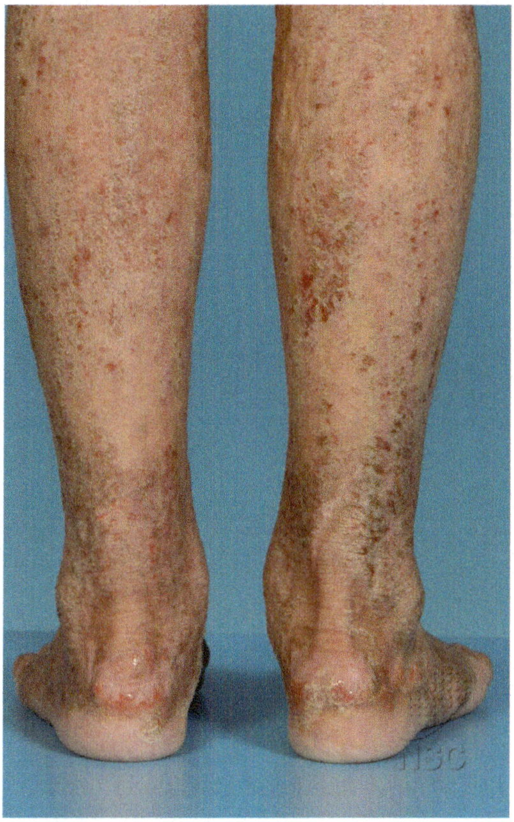

Fig. 27.1 Infected and fissured eczematous plaques on the legs of an AD patient with a known R4307X mutation in the *FLG* gene

function as bridging or anchoring molecules between the bacterium and host cells and may play a role in *S. aureus* adhesion to AD skin [16]. *S. aureus* induces scratching by stimulating the release of the highly pruritogenic cytokine, IL-31 [17]. It can also directly disrupt the skin barrier by the production of bacterial proteases, which break down epidermal proteins such as filaggrin.

Host factors are similarly responsible for the increased *S. aureus* colonization found in AD, most importantly, the dysfunctional physical and immune skin barrier that facilitates *S. aureus* adhesion and invasion. This inherently vulnerable skin barrier is further disrupted by the incessant itch-scratch cycle [18], a hallmark of AD that results in recurrent traumatic breaks in the skin that predispose to skin sepsis. In addition, a defective immune system, with an innate or acquired deficiency of antimicrobial peptides including human beta-defensins 2 and 3 and reduced Toll-like receptor-2 [19, 20], has also increased susceptibility to recurrent infections in AD.

The role of filaggrin and recurrent bacterial infection was reported in a recent landmark study that highlighted the association of *FLG* null mutations and the risk of recurrent skin infection in a cohort of Singaporean Chinese patients with AD [21]. The study found that AD patients with *FLG* mutations had approximately seven times higher risk of having more than four episodes of skin infection in the past year compared to those without *FLG* mutations (odds ratio 6.74; 95 % confidence interval 2.29–19.79). Patients with *FLG* null mutations also had significantly more episodes (median five episodes) compared to those without the mutation (median one episode). This risk was independent of disease severity and previous use of oral corticosteroids. In fact, the susceptibility to recurrent infections was greater in those with mild or moderate disease, suggesting that *FLG* status was the key contributory factor for the increased risk, even in patients with mild AD.

Further evidence for the role of filaggrin in recurrent bacterial infection can be gathered from "Netherton syndrome" (NS), a genodermatosis with severe AD-like features, elevated IgE levels, underlying Th2 predominance, and a propensity to recurrent skin infections, especially with *S. aureus* [22]. NS is caused by a loss-of-function mutation in the serine protease inhibitor *SPINK5*, leading to loss of inhibition of serine protease activity, causing severe skin barrier degradation, including the uncontrolled degradation of filaggrin [23]. It is possible that the relative filaggrin deficiency may contribute in part to the recurrent infections in NS.

27.4 Filaggrin Deficiency and the Increased Risk of Viral Infections in AD

27.4.1 Herpes Simplex Virus

AD patients are at increased risk of various viral infections such as herpes simplex infection and molluscum contagiosum. Of greatest clinical

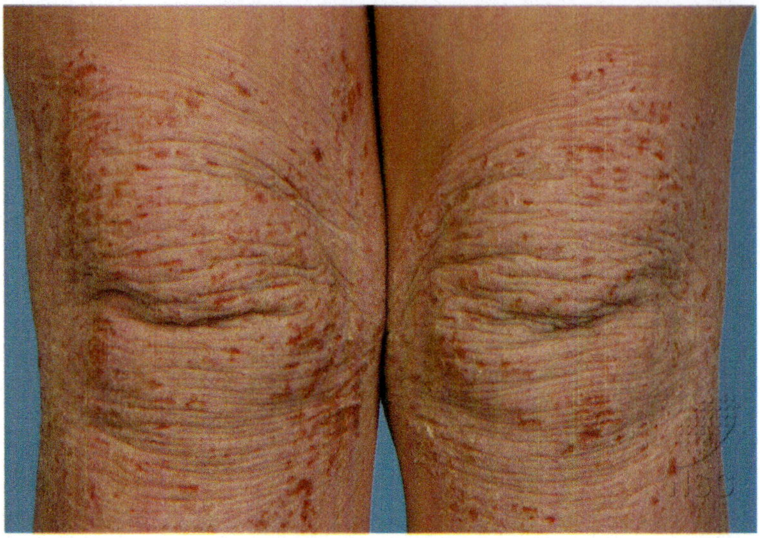

Fig. 27.2 Eczema herpeticum. Numerous monomorphic erosions on a background of severe hyperkeratosis and lichenification of the knees of an AD patient with a known S406X mutation in the *FLG* gene

significance is eczema herpeticum (EH), a serious infective complication of AD, resulting from disseminated cutaneous herpes simplex virus infection. EH is associated with high patient morbidity. The typical clinical presentation of EH is an acute, sudden worsening of rashes and skin pain, with the pathognomonic findings of widespread monomorphic vesicles and crusted erosions on an erythematous base (Fig. 27.2). Secondary complications such as keratoconjunctivitis, meningitis, encephalitis, viremia, and secondary bacterial sepsis can occur.

Epidemiological evidence suggests that AD patients with increased disease severity, asthma, increased sensitization to external allergens, and higher levels of circulating eosinophils and serum chemokine (C-C motif) ligand 17 (CCL17) are at greater risk of developing EH [24]. Recently, the link between filaggrin deficiency and the risk of EH has also been reported. Gao et al. showed that loss-of-function null mutations R501X and 2282del4 in the *FLG* were significantly associated with EH in a cohort of white American AD patients [25]. Similarly, the frequency of the R501X mutation was higher in African-American AD patients with EH compared to those without.

A reduction in claudin-1, the key protein of the tight junctions located at the stratum granulosum, has been reported to enhance susceptibility to herpes simplex viral infection [26] in AD patients. Interestingly, tight junction disruption of the stratum corneum has also been reported in filaggrin-deficient AD patients, due to reduced tight junction protein and corneodesmosin expression [27], and this may lend support to the role of the dysfunctional skin barrier, including filaggrin deficiency, in raising the risk of herpes infections.

27.4.2 Molluscum Contagiosum (MC)

Some studies have shown that AD patients are at increased risk of infection with the *molluscum contagiosum virus*, which is a DNA poxvirus. The typical lesion is a skin-colored or yellowish papule with central umbilication and can occur singly or present in a disseminated pattern (Fig. 27.3).

This increased risk has been attributed to the disturbed skin barrier and impaired cellular immunity [28]. More recent studies show that an impaired recruitment of plasmacytoid dendritic cells in AD patients is a relevant risk factor for viral infections such as MC and EH [29]. A reduced level of antimicrobial peptides, such as cathelicidin LL37, also increases the risk of AD patients developing MC infection [30]. Compared to non-AD patients, MC infection may be more therapeutically challenging with a

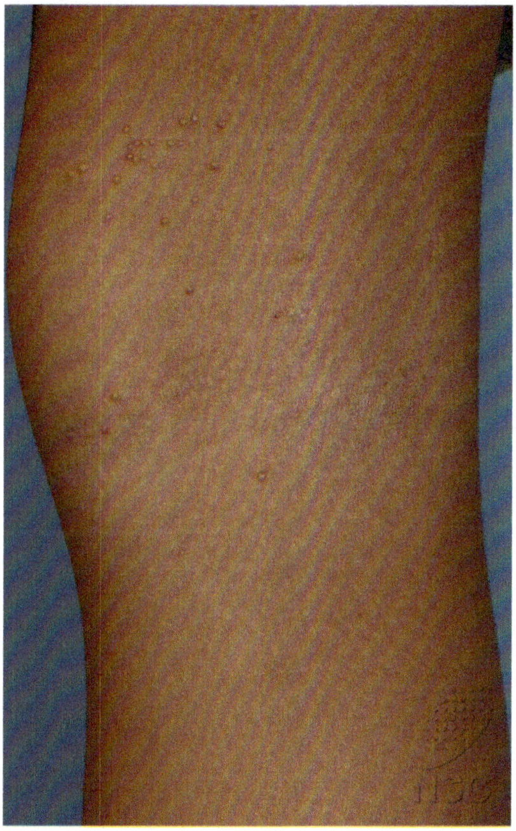

Fig. 27.3 Molluscum contagiosum. Skin-colored, umbilicated papules clustered around the popliteal fossa in a young child with flexural AD

higher risk of persistence and dissemination in AD patients [31].

Altered filaggrin expression has been reported in MC infection with findings of early expression of involucrin and filaggrin secondary to virally induced, altered keratinocyte differentiation and maturation [32, 33]. The altered filaggrin expression may therefore potentiate any preexisting filaggrin deficiency in AD patients and further facilitate the dissemination and persistence of the virus.

27.4.3 Human Papillomavirus

Much less commonly, AD patients have increased human papillomavirus (HPV) infections or generalized verrucosis [34–36]. This observation still remains controversial, as different population-based studies have reported contradictory data [37–40]. However, it is the association between the presence or absence of active eczema in AD patients and the number of warts, which is interesting, where Beltrani et al. showed that AD patients with active eczema demonstrated a significantly higher prevalence of multiple warts compared to those with inactive disease (39 % versus 20 %) [41].

Lower levels of cytokine production, specifically interferons (IFNs) such as IFN-γ, have been found in AD [42]. Further downregulation of IFN production and major histocompatibility class I expression by certain HPV subtypes (e.g., HPV 38) allows the viral warts to escape recognition by the immune system and persist [35].

An acquired filaggrin deficiency occurs in HPV infection arising from abnormal keratinization affecting the expression of profilaggrin. Modifications of filaggrin expression have been reported in patients diagnosed with Bowen's disease, genital bowenoid papulosis [43], and epidermodysplasia verruciformis plane wart-type lesions [44]. These changes are related to the abnormal keratinization process rather than to the presence of HPV DNA [43]. The pattern of filaggrin distribution in samples from patients with cervical HPV infections [45] varies with oncogenic HPV strains and the presence or absence of cervical intraepithelial neoplasia, suggesting that the severity of the disturbance in keratinocyte differentiation and, hence, filaggrin expression may be influenced by the degree and virulence of HPV infection.

Although the above findings were not reported specifically in AD or other diseases with an inherited deficiency of filaggrin, it is possible that the acquired filaggrin deficiency in HPV infection may have a role in perpetuation of infection by contributing to barrier dysfunction.

27.4.4 Vaccinia Virus

Finally, it has been recommended that smallpox vaccination be avoided in AD patients and their household contacts because they can have a severe, widespread, and potentially fatal cutaneous infection called eczema vaccinatum.

In ovalbumin-sensitized BALB/c (wild-type) mice with similar allergic skin inflammation as in AD, vaccinia virus inoculation led to a more severe dissemination of the virus with increased IL-17 expression compared to unsensitized controls [46]. In a filaggrin-deficient flaky tail mouse AD model subject to epicutaneous sensitization with protein antigens, there was also increased IL-17 expression with eczematous inflammation [47]. Increased IL-17 is known to reduce filaggrin synthesis [8] and to suppress natural killer cell activity against the vaccinia virus [48].

Similar to MC and HPV infections, the clinical applicability or significance of how infection with vaccinia virus leads to an acquired filaggrin deficiency and the extent to which this can account for the increased severity of infection in AD patients remain to be elucidated.

27.5 Filaggrin Deficiency and the Increased Risk of Fungal Infections in AD

Fungi such as *Malassezia* [49], *Candida albicans* [50], and possibly *Trichophyton rubrum* [51] have been reported to play a role in the pathogenesis of AD and have been associated with increased disease severity and disease flares.

Malassezia colonization, which is more prevalent in AD patients compared to healthy individuals [52], shows a different pattern on AD skin compared to healthy skin [53] and may aggravate AD due to an allergic reaction, especially on the head and neck of adults [54]. A significant proportion of AD patients have demonstrated positive results with skin prick tests, intradermal tests, and atopy patch tests to *Malassezia* antigens [52, 55]. An increased reaction of specific IgE to *Malassezia furfur* in AD patients is also known to occur, when compared to healthy controls [56], suggesting that IgE-mediated sensitization to *Malassezia* has a role in the pathogenesis of AD [57]. Notably, *Candida albicans* is also a potent intrinsic factor in inducing AD skin lesions, attributed to the IgE-mediated hypersensitivity of the *Candida albicans* antigen [58]. In therapeutic proof of concept, antifungal therapy has been reported to be beneficial in the treatment of some AD patients [59].

It is known that tinea corporis, a superficial mycotic infection caused by dermatophytes such as *Trichophyton rubrum*, can lead to skin barrier disruption and altered epidermal proliferation and differentiation with reduced expressed of keratin K10 and cornified envelope proteins including involucrin, loricrin, and filaggrin [60]. Reduced filaggrin expression correlated with the reduced skin hydration present in these patients, which may explain how tinea infections can aggravate eczematous lesions in AD.

To our knowledge, there has been no published data that seals the link between fungal infection, filaggrin deficiency, and the greater risk of these infections in AD patients. However, there have been reports of localized [61] and widespread dermatophyte infection [62, 63] occurring in patients with ichthyosis vulgaris (IV), the most common inherited disorder of keratinization, attributed to semidominant loss-of-function mutations in *FLG*. In an Indian study of chronic dermatophytosis, IV was present in 25 % of patients, and atopy was the most common systemic manifestation [64]. It has been postulated that as a result of altered keratinization, ichthyotic skin provides a more favorable habitat for fungi growth and survival compared to normal skin. Notwithstanding, since IV is commonly associated with atopy, the inherited filaggrin deficiency may not be the sole contributing factor toward more fungal infections in such patients.

27.6 Filaggrin Deficiency and Recurrent Skin Infections: Pathogenic Mechanisms

The mechanisms by which filaggrin deficiency contributes to increased susceptibility to infections are multifold and include structural, biophysical, and functional changes in the epidermal barrier as well as modified immunomodulatory responses that perpetuate the skin barrier dysfunction [65] (Fig. 27.4).

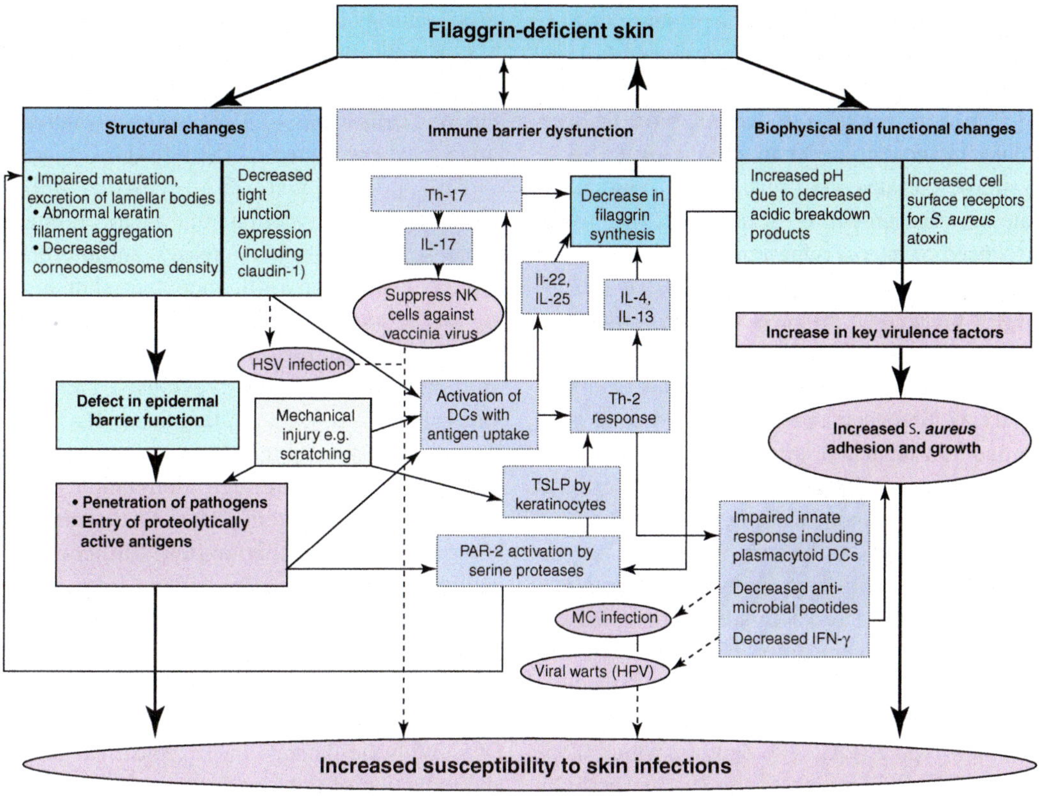

Fig. 27.4 Possible mechanisms of filaggrin deficiency associated with increased susceptibility to infections. *DCs* dendritic cells, *HSV* herpes simplex virus, *HPV* human papillomavirus, *IL* interleukin, *MC* molluscum contagiosum, *NK* natural killer, *PAR-2* protease-activated receptor-2, *S. aureus Staphylococcus aureus*, *TSLP* thymic stromal lymphopoietin

27.6.1 Structural Changes

The presence of *FLG* mutations with resultant filaggrin deficiency is associated with dose-dependent alterations in epidermal structure and permeability. These structural changes are wide ranging, especially in the extracellular protein-lipid matrix component of the stratum corneum, with abnormalities in keratin filament organization, lamellar body function, lamellar bilayer architecture, corneodesmosome density, and tight junction protein expression [27]. As a consequence, the defective epidermal barrier loses its protective function and predisposes to greater pathogen penetration as demonstrated by increased transepidermal water loss (TEWL) and enhanced percutaneous penetration of allergens in the flaky tail mouse model with murine *FLG* mutation [66, 67] as well as a disturbed diffusion barrier in the dye penetration assay [68]. Increased disease severity of AD patients with *FLG* mutations may also be associated with increased intensity of pruritus, leading to more frequent scratching and breaks within the skin, which ease the entry of pathogens, leading to a higher risk of infection or koebnerization of viral infections such as MC.

27.6.2 Biophysical and Functional Changes

In the final stages of epidermal differentiation, filaggrin is ultimately degraded into component contributing to the "natural moisturizing factor," in particular pyrrolidone carboxylic acid and urocanic acid [69]. These acidic breakdown products play a crucial role in maintaining the protective acidic skin mantle [70], and the presence of *FLG* mutations has been linked to an increased skin-surface pH [71, 72].

Pathogens, such as *S. aureus*, have to overcome this acidic barrier to survive on the skin. Notably, elevated skin-surface pH may facilitate bacterial growth of *S. aureus* [73], through enhanced *S. aureus* adhesion and multiplication [74]. This may be attributable to the lack of filaggrin breakdown products that have also been shown to reduce the expression of key virulence factors produced by *S aureus*, such as clumping factor B, fibronectin-binding protein A, protein A, and iron-regulated surface determinant protein A, which are involved in the colonization of keratinocytes and immune evasion [74, 75].

The acidic pH of the skin barrier is also critical for the regulation of epidermal serine proteases, and an increased skin-surface pH results in serine protease overactivity. This can lead to further filaggrin degradation [76]. The overactivity of serine proteases also activates plasminogen activator type 2 (PAR-2) receptors, which block lamellar body secretion in the stratum granulosum [77, 78] leading to further disruption of the skin barrier in AD.

Finally, filaggrin has been found to play a protective role against *S. aureus* α-toxin by mediating the secretion of sphingomyelinase, an enzyme that reduces the expression of keratinocyte cell-surface receptors for α-toxin [79]. Thus, filaggrin-deficient keratinocytes are preferentially targeted and destroyed by *S. aureus* α-toxin.

27.6.3 Immune Barrier Dysfunction

Both the physical skin barrier and the cutaneous immune skin barrier are inextricably linked. Defects in epidermal barrier secondary to filaggrin deficiency have been shown to impact the cutaneous immune barrier, and vice versa. For instance, PAR-2 activation by serine proteases, as a result of filaggrin deficiency or barrier perturbation, leads to a cascade of innate inflammatory responses, including the activation of thymic stromal lymphopoietin (TSLP), the master switch that initiates Th2 inflammatory responses [80], central in the immunopathogenesis of AD. Barrier disruption inflicted by persistent scratching in AD patients also polarizes the skin dendritic cells to elicit a Th2 response by upregulating local expression of TSLP [81].

This predominant Th2 cytokine milieu decreases the induction of innate immune response genes [82], decreases antimicrobial peptide expression [30], increases skin binding of *S. aureus* [16], and induces a defective IFN-γ response to bacterial toxins [83]. Moreover, as highlighted previously, Th2 cytokines such as IL-4, IL-13, IL-25, and IL-17 can also indirectly inhibit keratinocyte expression of filaggrin [5, 7], through modulation of the calcium-sensing protein S100A11 that impairs keratinocyte differentiation [84], further aggravating the physical barrier defect.

Another key factor is the close interaction of dendritic cell (DC) dendrites and keratinocyte tight junctions [85], where antigens need to penetrate leaky tight junctions to be taken up by DCs. With the damage to the epidermal barrier and reduction of tight junctions in filaggrin deficiency as highlighted earlier, DC function is disorganized, and this may promote their uptake of antigens from the skin surface. This subsequently leads to the release of epithelial-derived adjuvants that can favor Th2 and Th17 immune responses [26]. DCs produce IL-25, which not only stimulates Th2 response [86] but also downregulates filaggrin synthesis in vitro [6]. There is also an increased expression of IL-22 by CD4+ T cells, which are induced by epidermal Langerhans cells and dermal DCs [87], resulting in a decreased expression of filaggrin [88]. These cytokine products therefore establish an autocrine feedback loop that exacerbates the skin barrier dysfunction.

27.7 Clinical Implications

While personalized medicine is the ultimate goal for informed, targeted therapy in AD, to our knowledge, there are no current guidelines that differentiate the treatment or prophylaxis of skin infections in filaggrin-deficient AD patients compared to the general AD population.

As such, a multipronged approach targeting the key pathogenic factors in AD is critical in the management of this chronic disease in order to reduce recurrent skin infections. Active skin barrier restoration with lipid-enriched emollients has been shown to decrease disease severity [89] and reduce TEWL and is fundamental for long-term disease control. Ceramide-dominant moisturizers are thought to permeate the stratum corneum and actively restore the lipid barrier defects [90]. The use of low-pH, soap-free cleansers, instead of harsh alkaline soaps, to restore and maintain the acid mantle of the skin barrier is often recommended for AD patients.

Anti-inflammatory therapy with topical corticosteroids and calcineurin inhibitors is critical as part of AD treatment, and they reduce bacterial skin colonization [91–94]. Although topical tacrolimus is not associated with an increased risk of infections [95], there have been reports of herpes simplex infection, MC [96], and extensive varicella zoster infection [97] so caution should be exercised with its use. Interestingly however, the use of systemic immunosuppressives such as ciclosporin has not been associated with an increased risk of infections in a cohort of Korean AD patients [98]. Suppressing the Th2 inflammatory response in AD with corticosteroids [99] and calcineurin inhibitors is the cornerstone in AD treatment, and theoretically this may decrease the Th2 cytokine milieu-mediated suppression of filaggrin synthesis and reduce the allergic inflammation triggered by *S. aureus*, thereby concomitantly improving the dysfunctional barrier and decreasing bacteria-induced damage. However, studies have paradoxically shown that topical corticosteroids cause thinning of the epidermal barrier and affect epidermal lipid synthesis [2], so a balance needs to be struck when assessing the need for continuation of therapy.

At present, there is no clear consensus on the best way to eradicate *S. aureus* or reduce the recolonization of AD patients. Although several studies have demonstrated that *S. aureus* decolonization may improve AD outcomes, a recent Cochrane review concluded that further long-term, randomized controlled antimicrobial intervention trials are urgently required in this area [100]. Unfortunately, this review did not stratify patients according to disease severity or whether *FLG* mutations were present. It is clear that AD patients require long-term maintenance decolonization since short-term, acute interventions cannot be effectively sustained [14]. Some of the factors responsible for recolonization of *S. aureus* in AD include nasal carriage, drug resistance, environmental contamination, and the presence of colonized close contacts [101]. Specific prophylactic measures that have been used to reduce bacterial infections and recolonization in AD include the regular use of soap-free, alcohol-free antiseptic washes, bleach baths, and decolonization protocols such as the nasal application of nasal mupirocin ointment, oral decontamination with 0.2 % chlorhexidine gargle, and topical application of mupirocin ointment to major skin folds and clinically infected eczematous lesions. However, the judicious use of both oral and topical antibiotics, especially fusidic acid and mupirocin ointment, is highly recommended and essential to minimize the development of antibiotic resistance, which is increasing.

A high index of suspicion is needed for the early recognition of EH to ensure timely treatment and to reduce complications. Often, EH is misdiagnosed as a bacterial infection or as multiple excoriations in the context of an AD flare. The use of long-term antiviral drugs such as acyclovir or valaciclovir to suppress herpes reactivation has been reported in AD patients with recurrent or recalcitrant EH attacks [102].

Finally, given the dose-dependent effect of filaggrin on disease pathogenesis, it has been postulated that a modest upregulation of epidermal filaggrin might have critical protective and therapeutic effects [9], including the prevention of skin infections in AD.

Conclusion

Filaggrin is an important component of the epidermal barrier, and its deficiency may be associated with increased infections. The evidence for this is strongest in filaggrin-deficient AD patients who are at significantly higher risk of both bacterial and viral skin infection. Skin infections, especially secondary to *S. aureus*, can also result in changes in filaggrin expression, either directly or through immune-mediated mechanisms. Interestingly, patients with only IV, a genodermatosis due to loss-of-function *FLG* mutations, do not appear to be at greatly increased risk of bacterial or viral skin infections, despite having absent or reduced filaggrin expression in their epidermal barrier. This highlights the need for further research into the key pathogenic mechanisms of host-pathogen interactions in filaggrin deficiency, to identify other compensatory mechanisms that may have a crucial role in this area. Further research into the upregulation or replacement of filaggrin in the skin as a means of protection against infection is also needed, as it may prove to be an exciting, novel treatment option for AD patients.

References

1. Candi E, Schmidt R, Melino G. The cornified envelope: a model of cell death in the skin. Nat Rev Mol Cell Biol. 2005;6:328–40.
2. Cork MJ, Danby SG, Vasilopoulos Y, Hadgraft J, Lane ME, Moustafa M, et al. Epidermal barrier dysfunction in atopic dermatitis. J Invest Dermatol. 2009;129:1892–908.
3. Henderson J, Northstone K, Lee SP, Liao H, Zhao Y, Pembrey M, et al. The burden of disease associated with filaggrin mutations: a population-based, longitudinal birth cohort study. J Allergy Clin Immunol. 2008;121:872–7.e9.
4. Chen H, Common JEA, Haines RL, Balakrishnan A, Brown SJ, Goh CSM, et al. Wide spectrum of filaggrin-null mutations in atopic dermatitis highlights differences between Singaporean Chinese and European populations. Br J Dermatol. 2011;165:106–14.
5. Howell MD, Kim BE, Gao P, Grant AV, Boguniewicz M, DeBenedetto A, et al. Cytokine modulation of atopic dermatitis filaggrin skin expression. J Allergy Clin Immunol. 2007;120:150–5.
6. Hvid M, Vestergaard C, Kemp K, Christensen GB, Deleuran B, Deleuran M. IL-25 in atopic dermatitis: a possible link between inflammation and skin barrier dysfunction? J Invest Dermatol. 2011;131:150–7.
7. Gutowska-Owsiak D, Schaupp AL, Salimi M, Taylor S, Ogg GS. Interleukin-22 downregulates filaggrin expression and affects expression of profilaggrin processing enzymes. Br J Dermatol. 2011;165:492–8.
8. Gutowska-Owsiak D, Schaupp AL, Salimi M, Selvakumar TA, McPherson T, Taylor S, et al. IL-17 downregulates filaggrin and affects keratinocyte expression of genes associated with cellular adhesion. Exp Dermatol. 2012;21:104–10.
9. Brown SJ, Kroboth K, Sandilands A, Campbell LE, Pohler E, Kezic S, et al. Intragenic copy number variation within filaggrin contributes to the risk of atopic dermatitis with a dose-dependent effect. J Invest Dermatol. 2012;132:98–104.
10. Boguniewicz M, Schmid-Grendelmeier P, Leung DYM. Clinical pearls: atopic dermatitis. J Allergy Clin Immunol. 2006;118:40–3.
11. Boguniewicz M, Leung DYM. Recent insights into atopic dermatitis and implications for management of infectious complications. J Allergy Clin Immunol. 2010;125:4–13.
12. Verhagen J, Akdis M, Traidl-Hoffman C, Schmid-Grendelmeier P, Hijnen D, Knol EF, et al. Absence of T-regulatory cell expression and function in atopic dermatitis skin. J Allergy Clin Immunol. 2006;117:176–83.
13. Kluytmans J, van Belkum A, Verbrugh H. Nasal carriage of *Staphylococcus aureus:* epidemiology, underlying mechanisms, and associated risks. Clin Microbiol Rev. 1997;10:505–20.
14. Boguniewicz M, Sampson H, Leung SB, Harbeck R, Leung DY. Effects of cefuroxime axetil on *Staphylococcus aureus* colonization and superantigen production in atopic dermatitis. J Allergy Clin Immunol. 2001;108:651–2.
15. Williams RE, Gibson AG, Aitchison TC, Lever R, Mackie RM. Assessment of a contact-plate sampling technique and subsequent quantitative bacterial studies in atopic dermatitis. Br J Dermatol. 1990;123:493–501.
16. Cho SH, Strickland I, Boguniewicz M, Leung DY. Fibrinogen and fibronectin contribute to the enhanced binding of *Staphylococcus aureus* to atopic skin. J Allergy Clin Immunol. 2001;108:269–74.
17. Sonkoly E, Muller A, Lauerma AI, Pivarcsi A, Soto H, Kemeny L, et al. IL-31, a new link between T cells and pruritus in atopic skin inflammation. J Allergy Clin Immunol. 2006;117:411–7.
18. Yosipovitch G, Papolu AD. What causes itch in atopic dermatitis? Curr Allergy Asthma Rep. 2008;8:306–11.
19. Hasannejad H, Takahashi R, Kimishima M, Hayakawa K, Shiohara T. Selective impairment of Toll-like receptor 2-mediated proinflammatory cytokine production by monocytes from patients with atopic dermatitis. J Allergy Clin Immunol. 2007;120:69–75.
20. Niebuhr M, Lutat C, Sigel S, Werfel T. Impaired TLR-2 expression and TLR-2-mediated cytokine secretion in macrophages from patients with atopic dermatitis. Allergy. 2009;64:1580–7.
21. Cai SCS, Chen H, Koh WP, Common JEA, van Bever HP, McLean WHI, et al. Filaggrin mutations are associated with recurrent skin infection in Singaporean Chinese patients with atopic dermatitis. Br J Dermatol. 2012;166:200–3.

22. Renner ED, Harti D, Rylaarsdam S, Young ML, Monaco-Shawver L, Kleiner G, et al. Comel-Netherton syndrome defined as primary immunodeficiency. J Allergy Clin Immunol. 2009;124:536–43.
23. Heimall J, Spergel JM. Filaggrin mutations and atopy: consequences for future therapeutics. Expert Rev Clin Immunol. 2012;8:189–97.
24. Beck LA, Boguniewicz M, Hata T, Schneider LC, Hanifin J, Gallo R, et al. Phenotype of atopic dermatitis subjects with a history of eczema herpeticum. J Allergy Clin Immunol. 2009;124:260–9.
25. Gao PS, Rafaels NM, Hand T, Murray T, Boguniewicz M, Hata T, et al. Filaggrin mutations that confer risk of atopic dermatitis confer greater risk for eczema herpeticum. J Allergy Clin Immunol. 2009;124:507–13.
26. De Benedetto A, Rafaels NM, McGirt LY, Ivanov AI, Georas SN, Cheadle C, et al. Tight junction defects in patients with atopic dermatitis. J Allergy Clin Immunol. 2011;127:773–86.
27. Gruber R, Elias PM, Crumrine D, Lin TK, Brandner JM, Hachem JP, et al. Filaggrin genotype in ichthyosis vulgaris predicts abnormalities in epidermal structure and function. Am J Pathol. 2011;178:2252–63.
28. Wetzel S, Wollenberg A. Eczema molluscatum in tacrolimus treated atopic dermatitis. Eur J Dermatol. 2004;14:73–4.
29. Wollenberg A, Wagner M, Gunther S, Towarowski A, Tuma E, Moderer M, et al. Plasmacytoid dendritic cells: a new cutaneous dendritic cell subset with distinct role in inflammatory skin diseases. J Invest Dermatol. 2002;119:1096–102.
30. Ong PY, Ohtake T, Brandt C, Strickland I, Boguniewicz M, Ganz T, et al. Endogenous antimicrobial peptides and skin infections in atopic dermatitis. N Engl J Med. 2002;347:1151–60.
31. Seize MB, Ianhez M, Cestari Sda C. A study of the correlation between molluscum contagiosum and atopic dermatitis in children. An Bras Dermatol. 2011;86:663–8.
32. Takahashi M, Izutani A, Tezuka T. An immunohistochemical study of abnormal keratinocyte differentiation in molluscum contagiosum. Br J Dermatol. 1999;141:116–8.
33. Callegaro CF, Sotto MN. Molluscum contagiosum: immunomorphological aspects of keratinocytes markers of differentiation and adhesion. J Cutan Pathol. 2009;36:1279–85.
34. Lipman BL. Atopic dermatitis in an infant complicated by generalized verrucae vulgares. Ann Allergy. 1983;51:33–4.
35. Slodkowska EA, Schwartz J, Rady PL, Tyring SK, Carlson JA. Atopic dermatitis associated with HPV 38 lower leg verrucosis. Am J Dermatopathol. 2010;32:610–3.
36. Sri JC, Dubina MI, Kao GF, Rady PL, Tying SK, Gaspari AA. Generalized verrucosis: a review of the associated diseases, evaluation, and treatments. J Am Acad Dermatol. 2012;66:292–311.
37. Williams H, Pottier A, Strachan D. Are viral warts seen more commonly in children with eczema? Arch Dermatol. 1993;129:717–20.
38. Larsson PA, Liden S. Prevalence of skin diseases among adolescents 12-16 years of age. Acta Derm Venereol. 1980;60:415–23.
39. Currie JM, Wright RC, Miller OG. The frequency of warts in atopic dermatitis. Cutis. 1971;8:243–4.
40. Bonifazi E, Garofalo L, Pisani V, Meneghini CL. Role of some infectious agents in atopic dermatitis. Acta Derm Venereol Suppl (Stockh). 1985;114: 98–100.
41. Beltrani VS. Warts and the atopic. Arch Dermatol. 1994;130:388–9.
42. Nakagawa S, Aiba S, Tagami H. Decreased frequency of interferon-gamma producing CD4+ cells in the peripheral blood of patients with atopic dermatitis. Exp Dermatol. 1998;7:112–8.
43. Guerin-Reverchon I, Chardonnet Y, Viac J, Chouvet B, Chignol MC, Thivolet J. Human papillomavirus infection and filaggrin expression in paraffin-embedded biopsy specimens of extragenital Bowen's disease and genital bowenoid papulosis. J Cancer Res Clin Oncol. 1990;116:295–300.
44. Barcelos ACN, Sotto MN. Comparative analysis of the expression of cytokeratins (1, 10, 14, 16, 4), involucrin, filaggrin and e-cadherin in plane warts and epidermodysplasia verruciformis plane wart-type lesions. J Cutan Pathol. 2009;36:647–54.
45. Cintorino M, Syrjanen S, Leoncini P, et al. Altered expression of filaggrin in human papillomavirus (HPV) lesions of the uterine cervix. Arch Gynecol Obstet. 1988;241:235–48.
46. Oyoshi MK, Elkhal A, Kumar L, Scott JE, Koduru S, He R, et al. Vaccinia virus inoculation in sites of allergic skin inflammation elicits a vigorous cutaneous IL-17 response. Proc Natl Acad Sci U S A. 2009;106:14954–9.
47. Oyoshi MK, Murphy GF, Geha RS. Filaggrin-deficient mice exhibit TH17-dominated skin inflammation and permissiveness to epicutaneous sensitization with protein antigen. J Allergy Clin Immunol. 2009;124:485–93.
48. Kawakami Y, Tomimori Y, Yumoto K, Hasegawa S, Ando T, Tagaya Y, et al. Inhibition of NK cell activity by IL-17 allows vaccinia virus to induce severe skin lesions in a mouse model of eczema vaccinatum. J Exp Med. 2009;206:1219–25.
49. Gueho E, Boekhout T, Ashbee HR, Guillot J, van Belkum A, Faergemann J. The role of *Malassezia* species in the ecology of human skin and as pathogens. Med Mycol. 1998;36 Suppl 1:220–9.
50. Savolainen J, Lammintausta K, Kalimo K, Viander M. *Candida albicans* and atopic dermatitis. Clin Exp Allergy. 1993;23:332–9.
51. Klein PA, Clark RA, Nicol NH. Acute infection with *Trichophyton rubrum* associated with flares of atopic dermatitis. Cutis. 1999;63:171–2.
52. Nikkels AF, Pierard GE. Framing the future of antifungals in atopic dermatitis. Dermatology. 2003;206:398–400.
53. Faergemann J. Atopic dermatitis and fungi. Clin Microbiol Rev. 2002;15:545–63.

54. Brehler S, Luger T. Atopic dermatitis: the role of *Pityrosporum ovale*. J Eur Acad Dermatol Venereol. 2001;15:5–6.
55. Johansson C, Sandstrom MH, Bartosik J, Sarnhult T, Christiansen J, Zargari A, et al. Atopy patch test reactions to *Malassezia* antigens differentiate subgroups of atopic dermatitis patients. Br J Dermatol. 2003;148:479–88.
56. Scalabrin M, Bavbek S, Perzanowski S, Wilson B, Platts-Mills A, Wheatley M. Use of specific IgE in assessing the relevance of fungal and dust mite allergens to atopic dermatitis: a comparison with asthmatic and nonasthmatic control subjects. J Allergy Clin Immunol. 1999;104:1273–9.
57. Chang FY, Lee JH, Yang YH, Yu HH, Wang LC, Lin YT, et al. Analysis of the serum levels of fungi-specific immunoglobulin E in patients with allergic diseases. Int Arch Allergy Immunol. 2011;154:49–56.
58. Morita E, Hide M, Yoneya Y, Kannbe M, Tanaka A, Yamamoto S. An assessment of the role of *Candida albicans* antigen in atopic dermatitis. J Dermatol. 1999;26:282–7.
59. Lintu P, Savolainen J, Kortekangas-Savolainen O, Kalimo K. Systemic ketoconazole is an effective treatment of atopic dermatitis with IgE-mediated hypersensitivity to yeasts. Allergy. 2001;56:512–7.
60. Jensen JM, Pfeiffer S, Akaki T, et al. Barrier function, epidermal differentiation and human β-defensin 2 expression in tinea corporis. J Invest Dermatol. 2007;127:1720–7.
61. Agostini G, Geti V, Difonzo EM, Giannotti B. Dermatophyte infection in ichthyosis vulgaris. Mycoses. 1992;35(7–8):197–9.
62. Hoetzenecker W, Schanz S, Schaller M, Fierlbeck G. Generalized tinea corporis due to Trichophyton rubrum in ichthyosis vulgaris. J Eur Acad Dermatol Venereol. 2007;21:1129–31.
63. Grahovac M, Budimcic D. Unrecognized dermatophyte infection in ichthyosis vulgaris. Acta Dermatovenerol Croat. 2009;17:127–30.
64. Sentamilselvi G, Kamalam A, Ajithadas K, Janaki C, Thambiah AS. Scenario of chronic dermatophytosis: an Indian study. Mycopathologia 1997–1998; 140:129–35.
65. Irvine AD, McLean WHI, Leung DYM. Filaggrin mutations associated with skin and allergic diseases. N Engl J Med. 2011;365:1315–27.
66. Fallon PG, Sasaki T, Sandilands A, Campbell LE, Saunders SP, Mangan NE, et al. A homozygous frameshift mutation in the mouse *flg* gene facilitates enhanced percutaneous allergen priming. Nat Genet. 2009;41:602–8.
67. Scharschmidt TC, Man MQ, Hatano Y, Crumrine D, Gunathilake R, Sundberg JP, et al. Filaggrin deficiency confers a paracellular barrier abnormality that reduces inflammatory thresholds to irritants and haptens. J Allergy Clin Immunol. 2009;124:496–506.
68. Mildner M, Jin J, Eckhart L. Knockdown of filaggrin impairs diffusion barrier function and increases UV sensitivity in a human skin model. J Invest Dermatol. 2010;130:2286–94.
69. Sandilands A, Terron-Kwiatkowski A, Hull PR, O'Regan GM, Clayton TH, Watson RM, et al. Comprehensive analysis of the gene encoding filaggrin uncovers prevalent and rare mutations in ichthyosis vulgaris and atopic eczema. Nat Genet. 2007;39:650–4.
70. Krien PM, Kermici M. Evidence for the existence of a self-regulated enzymatic process within the human stratum corneum -an unexpected role for urocanic acid. J Invest Dermatol. 2000;115:414–20.
71. Kezic S, O'Regan GM, Yau N, Sandilands A, Chen H, Campbell LE, et al. Levels of filaggrin degradation products are influenced by both filaggrin genotype and atopic dermatitis severity. Allergy. 2011;66:934–40.
72. Jungersted JM, Scheer H, Mempel M, Baurecht H, Cifuentes L, Hogh JK, et al. Stratum corneum lipids, skin barrier function and filaggrin mutations in patients with atopic eczema. Allergy. 2010;65:911–8.
73. O'Reagan GM, Irvine AD. The role of filaggrin loss-of-function mutations in atopic dermatitis. Curr Opin Allergy Clin Immunol. 2008;8:406–10.
74. Miajlovic H, Fallon PG, Irvine AD, Foster TJ. Effect of filaggrin breakdown products on growth of and protein expression by *Staphylococcus aureus*. J Allergy Clin Immunol. 2010;126(6):1184–90.
75. Clarke SR, Andre G, Walsh EJ, Dufrene YF, Foster TJ, Foster SJ. Iron-regulated surface determinant protein A mediates adhesion of *Staphylococcus aureus* to human corneocyte envelope proteins. Infect Immun. 2009;77:2408–16.
76. Brattsand M, Stefansson K, Lundh C, Haasum Y, Egelrud T. A proteolytic cascade of kallikreins in the stratum corneum. J Invest Dermatol. 2005;124:198–203.
77. Hachem JP, Man MQ, Crumrine D, Uchida Y, Brown BE, Rogiers V, et al. Sustained serine proteases activity by prolonged increase in pH leads to degradation of lipid processing enzymes and profound alterations of barrier function and stratum corneum integrity. J Invest Dermatol. 2005;125:510–20.
78. Demerjian M, Hachem JP, Tschachler E, Denecker G, Declercq W, Vandenabeele P, et al. Acute modulations in permeability barrier function regulate epidermal cornification: role of caspase-14 and the protease-activated receptor type 2. Am J Pathol. 2008;172:86–97.
79. Brauweiler AM, Bin L, Kim BE, Oyoshi MK, Geha RS, Goleva E, et al. Filaggrin-dependent secretion of sphingomyelinase protects against staphylococcal α-toxin-induced keratinocyte death. J Allergy Clin Immunol. 2013;131:421–7.
80. Paller AS. Latest approaches to treating atopic dermatitis. Chem Immunol Allergy. 2012;96:132–40.
81. Oyoshi MK, Larson RP, Ziegler SF, Geha RS. Mechanical injury polarizes skin dendritic cells to elicit a Th2 response by inducing cutaneous TSLP expression. J Allergy Clin Immunol. 2010;126:976–84.
82. Nomura I, Goleva E, Howell MD, Hamid QA, Ong PY, Hall CF, et al. Cytokine milieu of atopic dermatitis, as compared to psoriasis, skin prevents induction of innate immune response genes. J Immunol. 2003;171(6):3262–9.

83. Habu Y, Seki S, Takayama E, Ohkawa T, Koike Y, Ami K, et al. The mechanism of a defective IFN-gamma response to bacterial toxins in an atopic dermatitis model, NC/Nga mice, and the therapeutic effect of IFN-gamma, IL-12, or IL-18 on dermatitis. J Immunol. 2001;166:5439–47.
84. Howell MD, Fairchild HR, Kim BE, Bin L, Boguniewicz M, Redzic JS, et al. Th2 cytokines act on S100/A11 to downregulate keratinocyte differentiation. J Invest Dermatol. 2008;128:2248–58.
85. Kubo A, Nagao K, Yokouchi M, Sasaki H, Amagai M. External antigen uptake by Langerhans cells with reorganization of epidermal tight junction barriers. J Exp Med. 2009;206:2937–46.
86. Fort MM, Cheung J, Yen D, Li J, Zurawski SM, Lo S, et al. IL-25 induces IL-4, IL-5, and IL-13 and Th2-associated pathologies in vivo. Immunity. 2001;15:985–95.
87. Fujita H, Nograles KE, Kikuchi T, Gonzalez J, Carucci JA, Krueger JG. Human Langerhans cells induce distinct IL-22-producing CD41 T cells lacking IL-17 production. Proc Natl Acad Sci U S A. 2009;106:21795–800.
88. Boniface K, Bernard FX, Garcia M, Gurney AL, Lecron JC, Morel F. IL-22 inhibits epidermal differentiation and induces proinflammatory gene expression and migration of human keratinocytes. J Immunol. 2005;174:3695–702.
89. Chamlin SL, Kao J, Frieden IJ, Sheu MY, Fowler AJ, Fluhr JW, et al. Ceramide-dominant barrier repair lipids alleviate childhood atopic dermatitis: changes in barrier function provide a sensitive indicator of disease activity. J Am Acad Dermatol. 2002;47:198–208.
90. Sajic D, Asiniwasis R, Skotnicki-Grant S. A look at epidermal barrier function in atopic dermatitis: physiologic lipid replacement and the role of ceramides. Skin Therapy Lett. 2012;17:6–9.
91. Remitz A, Kyllonen H, Granlund H, Reitamo S. Reduction of staphylococcal colonization of skin in atopic dermatitis with 0.1 % tacrolimus ointment therapy. J Allergy Clin Immunol. 2001;107:196–7.
92. Gong JQ, Lin L, Lin T, Hao F, Zeng FQ, Bi ZG, et al. Skin colonization by *Staphylococcus aureus* in patients with eczema and atopic dermatitis and relevant combined topical therapy: a double-blind multicentre randomized controlled trial. Br J Dermatol. 2006;155:680–7.
93. Hung SH, Lin YT, Chu CY, Lee CC, Liang TC, Yang YH, et al. *Staphylococcus* colonization in atopic dermatitis treated with fluticasone or tacrolimus with or without antibiotics. Ann Allergy Asthma Immunol. 2007;98:51–6.
94. Svensson A, Chambers C, Ganemo A, Mitchell SA. A systematic review of tacrolimus ointment compared with corticosteroids in the treatment of atopic dermatitis. Curr Med Res Opin. 2011;27:1395–406.
95. Fleischer Jr AB, Ling M, Eichenfield L, Satoi Y, Jaracz E, Rico MJ, et al. Tacrolimus ointment for the treatment of atopic dermatitis is not associated with an increase in cutaneous infections. J Am Acad Dermatol. 2002;47:562–70.
96. Soter NA, Fleischer Jr AB, Webster GF, Monroe E, Lawrence I. Tacrolimus ointment for the treatment of atopic dermatitis in adult patients: Part II, Safety. J Am Acad Dermatol. 2001;44:S39–46.
97. Bovenschen HJ, Groeneveld-Haenen CP. Topical tacrolimus induced extensive varicella zoster infection. Dermatol Online J. 2011;15:5.
98. Kim SW, Park YW, Kwon IH, Kim KH. Cyclosporin treatment of atopic dermatitis: is it really associated with infectious diseases? Ann Dermatol. 2010;22:170–2.
99. Norris DA. Mechanisms of action of topical therapies and the rationale for combination therapy. J Am Acad Dermatol. 2005;53:S17–25.
100. Bath-Hextall FJ, Birnie AJ, Ravenscroft JC, Williams HC. Interventions to reduce *Staphylococcus aureus* in the management of atopic eczema: an updated Cochrane review. Br J Dermatol. 2010;163:12–26.
101. Chiu LS, Ho MS, Hsu LY, Tang MB. Prevalence and molecular characteristics of *Staphylococcus aureus* isolates colonizing patients with atopic dermatitis and their close contacts in Singapore. Br J Dermatol. 2009;160:965–71.
102. Frisch S, Siegfried EC. The clinical spectrum and therapeutic challenge of eczema herpeticum. Pediatr Dermatol. 2011;28:46–52.

Atopic Dermatitis and Skin Cancer 28

Lennart Emtestam, Lena Hagströmer, and Karin Sartorius

Contents

L. Emtestam, MD, PhD (✉)
Department of Dermatology, Karolinska Institutet, Stockholm SE-141 86, Sweden
e-mail: lennart.emtestam@ki.se

L. Hagströmer, MD, PhD • K. Sartorius, MD, PhD
Department of Dermatology and Venereolgy, Södersjukhuset, Stockholm SE-118 83, Sweden

Section of Dermatology, Department of Clinical Science and Education, Karolinska Institutet, Södersjukhuset, Stockholm, Sweden
e-mail: lena.hagstromer@sodersjukhuset.se; karin.sartorius@sodersjukhuset.se

Filaggrin gene (*FLG*) mutation carriage increases the risk of atopic dermatitis (AD), contact allergy, asthma, hay fever, and peanut allergy. These genetic variants also influence the severity of asthma and alopecia areata and susceptibility to herpetic infection [1]. Squamous cell carcinoma (SSC) in the skin is perhaps the most rapidly increasing malignancy arising through mutations in epidermal cells in sun-exposed areas. UVB radiation damages the DNA and its repair system and causes mutations in tumor-suppressing genes. The risk of metastasis increases if the mutated cells penetrate the dermis. Other risk factors for SCC include chronic sun exposure to UVB and UVA, a suppressed immune system due to treatment effects, exposure to cyclic aromatic hydrocarbons in tar, or long-standing inflammation. Basal cell carcinoma (BCC) arises from basal keratinocytes in epidermis. In this kind of tumor, UVB is also important for the induction of damaging the DNA and its repair system and also causing mutations in tumor-suppressing genes. BCC grows by direct extension and almost never metastasizes. Malignant melanoma (MM) is a malignancy of melanocytes located mostly in the skin but also in the eyes, ears, central nervous system, and gastrointestinal (GI) tract. Malignant melanoma accounts for approximately 4 % of skin cancers but is responsible for more than 74 % skin cancer deaths [2]. The melanocyte transformation is poorly understood. It involves a series of steps, several unknown, with progressive genetic mutation that alters cell proliferation, differentiation, and death, and has an impact from

J.P. Thyssen, H. Maibach (eds.), *Filaggrin*,
DOI 10.1007/978-3-642-54379-1_28, © Springer-Verlag Berlin Heidelberg 2014

the carcinogenic effects of ultraviolet radiation [3]. So far, not much published material exists, but this chapter is an attempt to summarize the recent literature regarding relations between *FLG* mutations, AD, and skin cancer.

28.1 Filaggrin Expression in Skin Tumors

In the pre-*FLG* genotyping era, filaggrin expression in SCC was evaluated by indirect immunofluorescence in two studies. Klein-Szanto et al. investigated 27 keratoacanthomas and SCC and found that filaggrin was absent in all SCCs except one. It was suggested that the absence of filaggrin in SCCs could be useful in diagnostic procedures of skin tumors [4].

Kvedar et al. examined three markers of keratinocyte differentiation in invasive SCCs—SCCs in situ, actinic keratoses, and seborrheic keratoses—by staining with three antibodies, one of which was directed against filaggrin. They concluded that these antibodies were not able to distinguish between the various types of lesions and therefore would not be reliable tools for distinguishing malignant from benign lesions [5].

28.2 Epidemiological Studies

We previously studied the cancer incidence in a cohort of patients with AD [6]. To our knowledge, this was the first large-scale follow-up study of patients with AD with cancer as the outcome. A total of 15,666 patients hospitalized for AD were identified through a Swedish national inpatient database, and their records were linked to a national cancer registry. A list of registers and numbers used in the study is shown in Table 28.1. On average, patients with AD were followed up for 15.4 years, yielding 241,867 accumulated patient-years at risk, 15,471 of which were during the first year. After excluding the first year of observation, during which 1 case of SCC of the skin occurred, we found a total of 331 cases of cancer (190 in women, 141 in men). The average age at diagnosis of cancer was 53.0 years for women and 54.9 years for men. The incidence of any cancer (all sites) was increased by 13 % (95 % CI, 1–25 %), compared with the age- and sex-matched general Swedish population. The estimated relative risks for all cancer sites and cancer of the skin are listed in Table 28.2. There was a statistically nonsignificant risk elevation for nonmelanoma skin cancer (SIR, 1.5; 95 % CI, 0.8–2.6; 12 patients) but a decreased risk for melanoma (SIR, 0.6; 95 % CI, 0.3–1.2; 10 patients) [6].

The findings were in contrast with those of a case-control study [7] that used a mailed survey and found that 254 patients with a history of AD did not seem to develop nonmelanoma skin cancers more often than patients with other

Table 28.1 Registers and numbers used in cohort study by Hagströmer et al. [6]

National Inpatient Register [I]	A register of hospitalizations and discharge diagnoses
National Registration Number [I]	A unique personal identifier
ICD code [I]	International Identification of Disease
National Swedish Cancer Register [I]	Codes close to 98 % of all malignant neoplasms
Total Population Register [I]	Living persons in Sweden
Death Register [I]	IDs of dead persons and cause of death
Emigrations Register [I]	Persons who have moved from Sweden

Table 28.2 Standardized incidence ratio (SIR) and 95 % confidence interval (CI) for all cancer types and cancer of the skin among patients hospitalized for atopic dermatitis from 1965 to 1999 in Sweden [6]

Cancer type	No. of patients observed	SIR (95 % CI)
All site	331	1.13 (1.01–1.25)
Male	141	1.30 (1.10–1.54)
Female	190	1.02 (0.88–1.88)
Cancer of the skin		
Squamous cell carcinoma	12	1.5 (0.8–2.6)
Male	8	2.0 (0.9–4.0)
Female	3	1.0 (0.3–2.5)
Melanoma	10	0.6 (0.3–1.2)

dermatologic conditions. Skin tumors, in contrast to tumors of internal organs, are readily observable. For example, prevalent but undiagnosed skin tumors may be detected in connection with hospital care for AD. However, neither patients with BCC nor those with actinic keratosis were at the time of study reported in the National Swedish Cancer Register. Thus, the nonmelanoma skin cancer category included only patients with SCC and not those with squamous intraepidermal neoplasia or Bowen disease. If a skin tumor was diagnosed during the first hospitalization and recorded in the Inpatient Register, the affected patient would not be included in the cohort. Consequently, the cohort members had, in effect, been screened for skin tumors before inclusion. Therefore, the rate of skin tumors diagnosed during the ensuing years was probably somewhat lower than in an unscreened population, at least in the older age groups (10 years or older) in whom skin tumors are not exceedingly rare. This hypothetical incidence deficit is expected to be balanced by a bias toward increased detection during follow-up as a result of closer dermatologic surveillance linked to the presence of any chronic skin disorder. As the initial screening effect wears off, the presumed detection bias will dominate the findings. The net effect on overall skin cancer risk is difficult to predict, but because the cohort, as it is aging, is slowly moving from lower to higher absolute risk, any bias will have greater impact during the last years of follow-up.

It is likely that overestimation of the incidence resulting from detection bias will dominate. The shift from underestimation (resulting from initial screening) to overestimation (resulting from detection bias during follow-up) is expected to lead to impressions of an increasing relative risk over the follow-up period. Although the small numbers of patients observed hamper the interpretation of data, the observed decrease in relative risk with increasing follow-up time is probably not the effect of screening or detection bias. Instead, it creates questions about the biological relevance of the nonsignificant 50 % overall excess risk of nonmelanoma skin cancer in our Swedish cohort, seemingly confined to the first 9 years of follow-up [6].

A Danish study on 6,275 hospitalized AD patients showed similar results as ours, regarding risks for all cancer among the 2030 adults included (standard morbidity ratio [SMR] 1.5) (95 % CI, 1.2–1.9) [8]. They also showed, among the adults patients, an increased risk of "keratinocyte cancer" (SCC and BCC) (SMR 2.4) (95 % CI, 1.2–5.4) [8]. The prevalence of melanomas was not increased. A very recent Danish cohort study [9], on more than 30,000 AD patients, also showed results in the same direction. The overall observed number of MM cases among AD patients was 12, with 21 expected, yielding an SIR of 0.59 (95 % CI, 0.30, 1.02), with the most pronounced protective effect among AD patients with more than 5 years of follow-up (SIR=0.46; 95 % CI, 0.19, 0.95). The SIRs for BCC and SCC were increased among AD patients (1.41 [95 % CI, 1.07, 1.83] and 2.48 [95 % CI, 1.00, 5.11], respectively) [9]. All comparisons with our Swedish cohort are somewhat hampered by the fact that BCCs were not registered at the time in the Swedish Cancer Register. Synnerstad et al. proposed that patients with AD have a decreased risk to develop MM [10], which was not the case in our cohort [6].

On the other hand, an American case-control study by Ming et al. of 1,533 cases of NMSC and 1,378 unmatched controls found that patients with NMSC were less likely to have had eczema than individuals with other benign skin conditions (OR=0.78 [95 % CI, 0.61–0.98]), and the remaining work did not demonstrate any statistically significantly increased risk [7]. Another American study that was recently published evaluated prospectively predictors of BCC in 1,131 high-risk patients, with sun-damaged skin. Ninety-seven percent were men, and the mean age was 72 years. The most important predictor was the number of BCCs in the prior 5 years, but among the other independent risk factors was a history of eczema. None of those who reported eczema had used topical calcineurin inhibitors [11].

For MM, a population-based cohort study by Arana et al. reported a significant risk increase in patients with AD (IRR=1.74 [95 % CI, 1.25–2.41]) in contrast to the other publications, including ours (three cohort and two case-control studies) [9, 10, 12–14].

In our previous study, a slight excess of malignant neoplasms was noted among patients with AD. The greatest relative excesses were not really for skin cancers but for cancers of the esophagus, lung, brain, and pancreas and lymphoma [6]. Confounding by smoking and alcohol abuse, however, could not be excluded. The risk elevations, all of which were of borderline statistical significance, should be interpreted with caution. The combination of multiple significance testing and few observed patients may have generated chance findings. However, because (1) the cohort had a high proportion of young patients, who were not yet in the age groups most at risk for developing cancer, and (2) the frequency of occurrence of most types of cancer increases with age, future follow-ups of our cohort would be interesting.

Also, eczema and cancer risks have recently been critically reviewed by Wedgeworth et al. [15].

28.3 Skin Cancer and Filaggrin

To date few data have been published on the subject filaggrin (*FLG*) mutations and skin cancer. In a case-control study, a total of 159 patients with BCC and 3,346 adult controls from the general population were genotyped for the three most prevalent *FLG* mutations (R501X, 2282del4, and R2447X). No association between *FLG* mutations and BCC was found in a gender- and age-adjusted analysis (OR=0.81; 95 % CI, 0.43–1.54) [16].

The pathogenesis of AD is not fully understood, but several immunological aberrations are found, including filaggrin defects, impaired cellular-mediated immunity, elevated serum IgE and eosinophil levels, and IgE-bearing Langerhans cells. Furthermore, colonization of the more or less chronically inflamed lesions by microbes—most important, the yeast *Malassezia* (formerly *Pityrosporum*) *orbiculare* and the bacterium *Staphylococcus aureus*—may contribute to the perpetuation of the lesions. In a study from Singapore, patients with AD who had *FLG* mutations had approximately a seven times increased risk of more than four episodes of skin infection requiring antibiotics in the past year (odds ratio 6.74; 95 % confidence interval, 2.29–19.79). Patients without the *FLG*-null mutations were controls. The risk was much greater in those with mild or moderate disease and was seen in both users and nonusers of oral steroids [17].

Chronic inflammation and microbial colonization or infection in combination with the immune impairments (primary infection or adverse effects of treatment) may lead to proliferative epidermal changes, hence the suspicion of a link to cancer development.

Both sexes are affected by AD; among adults, more women than men have the disease [18]. However, among children 12 years or younger, more boys have it than girls. The reason for this sex difference is unclear. In Sweden, due to, on average, earlier onset of AD among men than among women, the duration of the chronic disease at any given age has generally been longer in men. In children, it also seems as if boys have a more severe disease than girls [18]. Therefore, it seems reasonable that there might be sex differences in risks of adverse long-term consequences of the disease. The atopic disease triad consists of hay fever, asthma, and AD. Over the life span, an individual with atopic disease may suffer from one, two, or all three of the manifestations. The connection between AD and lung disease has also been manifested previously, where lung disease (mainly asthma) was the most commonly found non-AD diagnosis, followed by other skin diseases and infectious diseases [6]. Although little is known about associations between cancer and hay fever, patients with asthma seem to be at increased risk for lung cancer [19–23], although contradicting results have been reported [24–26]. Among other allergic manifestations investigated for cancer risks, positive epicutaneous tests were reportedly linked to an overall increased risk of cancer within 20 years of follow-up in men but not in women [27]. No risk elevations were found among 1,155 patients with chronic urticaria observed for up to 27 years [27]. Thyssen et al. suggested that *FLG* mutation carriers with self-reported dermatitis have an increased risk of contact sensitization, whereas *FLG* mutations alone may not, or may only slightly, increase this risk [28].

28.4 UV Light and Caspase-14

Several experimental data indicate that AD patients with *FLG* mutations would be more sensitive to UV light. This would perhaps explain the increased incidence of skin cancer in this group of patients, as shown in epidemiological studies. Mildner et al. simply demonstrated that filaggrin deficiency by itself is sufficient to impair epidermal barrier formation and that absence of filaggrin results in enhanced UV sensitivity, most likely due to the reduction of epidermal urocanic acid [29]. Normal filaggrin levels and correct functioning of its degrading machinery, which involves caspase-14, are of great importance for optimal protection of the skin against harmful UVB radiation [30]. Interestingly, Denecker et al. showed that the skin of caspase-14-deficient mice was highly sensitive to the formation of cyclobutane pyrimidine dimers after UVB irradiation, leading to increased levels of UVB-induced apoptosis. Removal of the stratum corneum indicates that caspase-14 controls the UVB scavenging capacity of the stratum corneum [31]. Further, it has been indicated that accumulation of filaggrin fragments in caspase-$14^{-/-}$ mice is due to a defect in the terminal filaggrin degradation pathway. Therefore, the defective filaggrin degradation in caspase-14-deficient skin results in substantial reduction in the amount of natural moisturizing factors, such as urocanic acid and pyrrolidone carboxylic acid [32].

In a study by Uddin et al., Skh-1 mice were given oral β-damascenone followed by irradiation with UV radiation. It was shown that filaggrin and caspase-14 were increased in enlarged sebaceous glands. β-Damascenone protected against sunburn by activating a sebaceous gland-based pathway that fortified and thickened the cornified envelope plus sebum layer in a way that previously has been observed to occur only in keratinocytes, thus indicating a previously not described possible role of filaggrin [33].

28.5 Bringing Science to the Clinical Reality

None of the epidemiological cancer studies on AD so far has looked at the effect of treatment, such as phototherapy or immunosuppressive therapy, on skin cancer risk. New treatments for AD include one of the latest calcineurin inhibitors. The topical calcineurin inhibitors (TCIs) tacrolimus and pimecrolimus were approved in the United States for the treatment of AD in 2000 and 2001, respectively, and in Sweden shortly thereafter. In 2005, the Pediatric Advisory Committee of the US FDA implemented a "black box" warning for topical calcineurin inhibitors due to the lack of long-term safety data and the potential risk of the development of malignancies. Although the elevated cancer risks after topical treatment with calcineurin inhibitors really are suspected, they have not yet been proven. Tennis et al. reviewed the published literature and found inconsistent data on lymphomas following TCI use, insufficient to conclude on their causal role. They found no evidence that would indicate that nonmelanoma or melanoma skin cancer is associated with the use of TCI [34].

Therefore, repeated large cohort cancer incidence studies after the introduction of calcineurin inhibitors should be performed.

Improved knowledge of the mechanisms that drive the inflammation in AD may lead to a better understanding of this disease and shed light on the critical role of the epidermal barrier function and the immune system as well as the potential link between lack of filaggrin and the development of skin cancer. The first step is always immunomodulation. Different types of topical anti-inflammatory agents have been shown to normalize the reduced filaggrin expression found in lesional AD, a finding that is consistent with the concept that inflammation can reduce filaggrin expression [35]. The second step is moisturizers. Grether-Beck et al. studied various differentiation markers in the human skin, and the filaggrin expression showed the strongest upregulation after urea treatment [36]. Their findings indicate that filaggrin expression, which is reduced in a substantial proportion of AD patients, is improved by topical applications of urea to the skin of heterozygous null allele carriers [36]. A long time ago, we treated 22 patients with AD and found that a moisturizer containing urea and sodium chloride was somewhat more effective than the same moisturizer without sodium chloride, at least concerning the ability to reverse impedance indices of atopic skin toward normal, an effect ascribed mainly to changes in hydration of the stratum corneum [37]. The third step is to avoid excessive sun exposure.

Future studies should be aimed at characterizing the molecular mechanisms involved and to provide insight on how to improve treatment for patients with AD and to find better ways to reduce their risk of getting skin cancer.

References

1. Irvine AD, McLean WH, Leung DY. Filaggrin mutations associated with skin and allergic diseases. N Engl J Med. 2009;365(14):1315–27.
2. Linos E, Swetter SM, Cockburn MG, Colditz GA, Clarke CA. Increasing burden of melanoma in the United States. J Invest Dermatol. 2009;129(7): 1666–74.
3. Demierre MF, Nathanson L. Chemoprevention of melanoma: an unexplored strategy. J Clin Oncol. 2003;21(1):158–65.
4. Klein-Szanto AJ, Barr RJ, Reiners Jr JJ, Mamrack MD. Filaggrin distribution in keratoacanthomas and squamous cell carcinoma. Arch Pathol Lab Med. 1984;108(11):888–90.
5. Kvedar JC, Fewkes J, Baden HP. Immunologic detection of markers of keratinocyte differentiation. Its use in neoplastic and preneoplastic lesions of skin. Arch Pathol Lab Med. 1986;110(3):183–8.
6. Hagströmer L, Ye W, Nyrén O, Emtestam L. Incidence of cancer among patients with atopic dermatitis. Arch Dermatol. 2005;141(9):1123–7.
7. Ming ME, Levy R, Hoffstad O, Filip J, Abrams BB, Fernandez C, et al. The lack of a relationship between atopic dermatitis and nonmelanoma skin cancers. J Am Acad Dermatol. 2004;50(3):357–62.
8. Olesen AB, Engholm G, Storm HH, Thestrup-Pedersen K. The risk of cancer among patients previously hospitalized for atopic dermatitis. J Invest Dermatol. 2005;125(3):445–9.
9. Jensen AO, Svaerke C, Kormendine Farkas D, Olesen AB, Kragballe K, Sorensen HT. Atopic dermatitis and risk of skin cancer: a Danish nationwide cohort study (1977-2006). Am J Clin Dermatol. 2012;13(1): 29–36.
10. Synnerstad I, Fredrikson M, Ternesten-Bratel A, Rosdahl I. Low risk of melanoma in patients with atopic dermatitis. J Eur Acad Dermatol Venereol. 2008;22(12):1423–8.
11. Dyer RK, Weinstock MA, Cohen TS, Rizzo AE, Bingham SF. Predictors of basal cell carcinoma in high-risk patients in the VATTC (VA Topical Tretinoin Chemoprevention) trial. J Invest Dermatol. 2012; 132(11):2544–51.
12. Arana A, Wentworth CE, Fernandez-Vidaurre C, Schlienger RG, Conde E, Arellano FM. Incidence of cancer in the general population and in patients with or without atopic dermatitis in the U.K. Br J Dermatol. 2010;163(5):1036–43.
13. El-Zein M, Parent ME, Ka K, Siemiatycki J, St-Pierre Y, Rousseau MC. History of asthma or eczema and cancer risk among men: a population-based case-control study in Montreal, Quebec, Canada. Ann Allergy Asthma Immunol. 2010;104(5): 378–84.
14. Beral V, Evans S, Shaw H, Milton G. Cutaneous factors related to the risk of malignant melanoma. Br J Dermatol. 1983;109(2):165–72.
15. Wedgeworth E, Powell AM, Flohr C. Eczema and cancer risk: a critical appraisal and review of the literature. Br J Dermatol. 2011;165(3):457–62.
16. Kaae J, Thyssen JP, Johansen JD, Meldgaard M, Linneberg A, Allen M, Skov L. Filaggrin gene mutations and risk of basal cell carcinoma. Br J Dermatol. 2013;169(5):1162–4.
17. Cai SC, Chen H, Koh WP, Common JE, van Bever HP, McLean WH, et al. Filaggrin mutations are associated with recurrent skin infection in Singaporean Chinese patients with atopic dermatitis. Br J Dermatol. 2012;166(1):200–3.
18. Rajka G. Essential aspects of atopic dermatitis. Berlin/Heidelberg: Springer; 1989.
19. Boffetta P, Ye W, Adami HO, Mucci LA, Nyren O. Risk of cancers of the lung, head and neck in patients hospitalized for alcoholism in Sweden. Br J Cancer. 2001;85(5):678–82.
20. Reynolds P, Kaplan GA. Asthma and cancer. Am J Epidemiol. 1987;125(3):539–40.
21. Robinette CD, Fraumeni Jr JF. Asthma and subsequent mortality in World War II veterans. J Chronic Dis. 1978;31(9–10):619–24.
22. Vesterinen E, Pukkala E, Timonen T, Aromaa A. Cancer incidence among 78,000 asthmatic patients. Int J Epidemiol. 1993;22(6):976–82.
23. Alderson M. Mortality from malignant disease in patients with asthma. Lancet. 1974;2(7895): 1475–7.
24. Markowe HL, Bulpitt CJ, Shipley MJ, Rose G, Crombie DL, Fleming DM. Prognosis in adult asthma: a national study. BMJ (Clin Res Ed). 1987;295(6604): 949–52.
25. Mills PK, Beeson WL, Fraser GE, Phillips RL. Allergy and cancer: organ site-specific results from the Adventist Health Study. Am J Epidemiol. 1992; 136(3):287–95.

26. Sigurgeirsson B, Lindelöf B. Positive patch tests and cancer: an epidemiological study. Am J Contact Dermat. 1991;2:161–4.
27. Lindelöf B, Sigurgeirsson B, Wahlgren CF, Eklund G. Chronic urticaria and cancer: an epidemiological study of 1155 patients. Br J Dermatol. 1990;123(4): 453–6.
28. Thyssen JP, Linneberg A, Ross-Hansen K, Carlsen BC, Meldgaard M, Szecsi PB, et al. Filaggrin mutations are strongly associated with contact sensitization in individuals with dermatitis. Contact Dermatitis. 2013;68(5):273–6.
29. Mildner M, Jin J, Eckhart L, Kezic S, Gruber F, Barresi C, et al. Knockdown of filaggrin impairs diffusion barrier function and increases UV sensitivity in a human skin model. J Invest Dermatol. 2010; 130(9):2286–94.
30. Devos M, Prawitt J, Staumont-Salle D, Hoste E, Fleury S, Bouchaert E, et al. Filaggrin degradation by caspase-14 is required for UVB photoprotection but does not influence allergic sensitization in a mouse model of atopic dermatitis. J Invest Dermatol. 2012; 132(12):2857–60.
31. Denecker G, Hoste E, Gilbert B, Hochepied T, Ovaere P, Lippens S, et al. Caspase-14 protects against epidermal UVB photodamage and water loss. Nat Cell Biol. 2007;9(6):666–74.
32. Hoste E, Kemperman P, Devos M, Denecker G, Kezic S, Yau N, et al. Caspase-14 is required for filaggrin degradation to natural moisturizing factors in the skin. J Invest Dermatol. 2011;131(11):2233–41.
33. Uddin AN, Labuda I, Burns FJ. A novel mechanism of filaggrin induction and sunburn prevention by beta-damascenone in Skh-1 mice. Toxicol Appl Pharmacol. 2012;265(3):335–41.
34. Tennis P, Gelfand JM, Rothman KJ. Evaluation of cancer risk related to atopic dermatitis and use of topical calcineurin inhibitors. Br J Dermatol. 2011; 165(3):465–73.
35. Jensen JM, Pfeiffer S, Witt M, Brautigam M, Neumann C, Weichenthal M, et al. Different effects of pimecrolimus and betamethasone on the skin barrier in patients with atopic dermatitis. J Allergy Clin Immunol. 2009;124(3 Suppl 2):R19–28.
36. Grether-Beck S, Felsner I, Brenden H, Kohne Z, Majora M, Marini A, et al. Urea uptake enhances barrier function and antimicrobial defense in humans by regulating epidermal gene expression. J Invest Dermatol. 2012;132(6):1561–72.
37. Hagströmer L, Nyren M, Emtestam L. Do urea and sodium chloride together increase the efficacy of moisturisers for atopic dermatitis Skin? A comparative, double-blind and randomised study. Skin Pharmacol Appl Skin Physiol. 2001;14(1):27–33.

Filaggrin Gene Mutations: A Clinician's Perspective

29

David Bregnbak, Claus O.C. Zachariae, and Torkil Menné

Contents

D. Bregnbak, MD (✉)
C.O.C. Zachariae, MD, DMSci
Department of Dermato-Allergology, Copenhagen University, Gentofte Hospital, Niels Andersens Vej 65, Hellerup DK-2900, Denmark
e-mail: david.kaspar.bregnbak@regionh.dk; claus.zachariae@regionh.dk

T. Menné, MD
Department of Dermatology KA-1502, Gentofte Hospital, University of Copenhagen, Niels Andersensvej 65, Hellerup 2900 Denmark
e-mail: torkil.menne@regionh.dk

29.1 Introduction

Filaggrin gene (*FLG*) mutations have a significant association with a number of common diseases such as atopic dermatitis (AD), asthma, and hand eczema [1, 2]. The primary effect of the *FLG* mutations is through the effects on the skin barrier leading to xerosis [3].

29.2 Atopic Dermatitis

Historically, AD was first described as an entity in the early 1930s. Nexmand, in his thesis from 1948 [4], noted an increase in the number of AD patients hospitalized in the Copenhagen Dermatology Departments during 1930–1946 (Fig. 29.1).

Later, population-based studies at the end of the twentieth century illustrated a significant increase in the prevalence of AD [5, 6].

Recently, the question was raised whether AD was associated with *FLG* mutations and as a specific variant of the disease. We analyzed this question in a retrospective cohort and found that the frequency of AD increased slightly more in those with the *FLG* mutations, arguing in favor of a disease modified by gene-environment interaction [7].

J.P. Thyssen, H. Maibach (eds.), *Filaggrin*,
DOI 10.1007/978-3-642-54379-1_29, © Springer-Verlag Berlin Heidelberg 2014

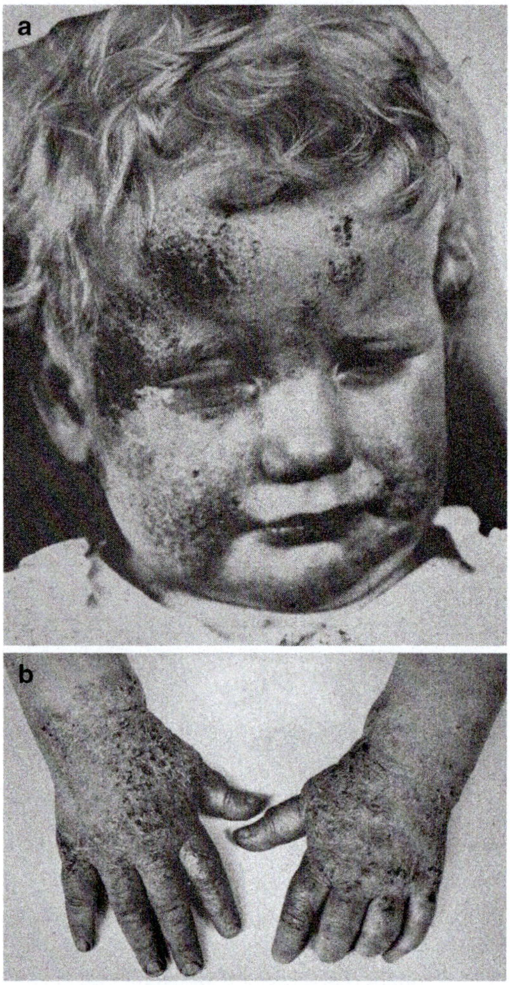

Fig. 29.1 (a, b) Historical pictures from Nexmand's thesis of severe AD in a child 2 years of age

29.3 The Primary Clinical Effects of the Filaggrin Gene Mutation, Xerosis, and Early Onset of Atopic Dermatitis

In a prospective study on children born by mothers with asthma, we found that the *FLG* mutations were associated with a specific endotype of AD with a dominance of dermatitis on the cheeks and the dorsal part of the hands. Interestingly, this is the part of the skin that is mainly in prolonged contact with the climate (leading to aggravation of xerosis) and endures friction from clothes and bed linen [8]. These early signs of dermatitis may be the first step to a lifelong disease with AD, asthma, rhinitis, food allergy, and irritant allergic contact dermatitis (Fig. 29.2).

AD is today one of the most frequent childhood diseases, often continuing into adult life. AD, particularly combined with hand eczema in childhood, is also the most significant risk factor for hand eczema in adults [9]. Females are at particular risk due to both domestic and occupational irritant exposures, leading to onset of hand eczema in the early twenties that often persists as a chronic disease [10, 11]. Adult-onset hand eczema may provoke the first outbreak of AD in patients without AD in childhood [12, 13].

Even if AD is common and in some cases difficult to treat, the historical perspective tells a different story. The individuals with the *FLG* mutations represent a special subgroup that often requires more intensive treatment and closer follow-ups, particularly the homozygotic, which constitutes 0.3 % of the population [14], with the tendency to early-onset AD and more severe and prolonged problems.

AD in patients with *FLG* mutations still represents a major challenge to clinicians who treat patients with chronic widespread dermatitis and severe fissures, particularly on the hands and feet (Fig. 29.3).

29.4 Allergic Hand Eczema

Contact allergy is more common in individuals with both dermatitis and the *FLG* mutations compared to individuals without the mutation [15]. This has been known for nickel allergy for some years but seems to be a general trend [16, 17]. Probably because of the high histidine content in filaggrin and its degradation products, it is functioning as a specific skin barrier for nickel and probably other divalent metal ions [18, 19]. The *FLG* mutations not only increase the risk of contact allergy but also seem to modify the cause of the clinical disease. We recently reported a young female who developed severe vesicular bullous acute dermatitis on the hands, imitating rubber glove dermatitis. She had been contact sensitized to sorbitan sesquioleate and had applied a steroid ointment containing this emulsifier and had worn a pair of rubber gloves for housekeeping, at the same time. The point is that a weak contact allergen can simulate a reaction to a more potent allergen in individuals with *FLG* mutations [20].

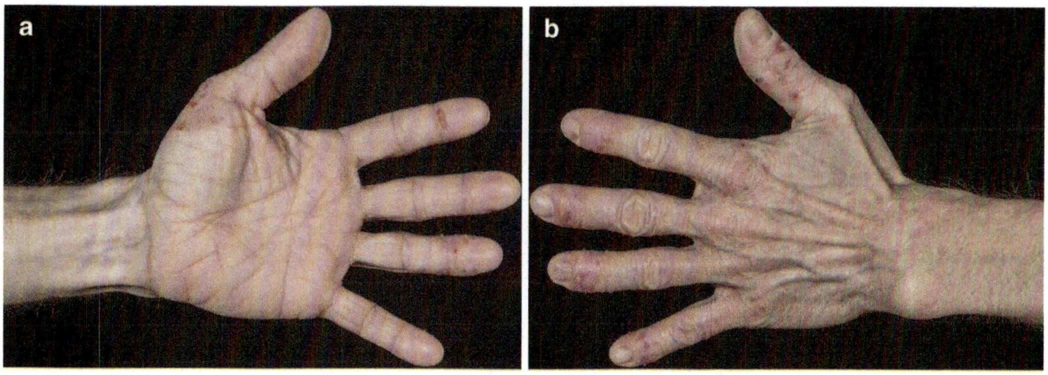

Fig. 29.2 (**a**, **b**) A 51-year-old male patient who was a homozygous *FLG* mutation carrier with a lifelong history of AD. As long as he remembered, he had prominent skin folds on the palmar and dorsal parts of the hands and fingers independent of his AD. This symptom is a sign seen in many individuals with AD with the *FLG* mutation starting early in life and probably a sign of subclinical inflammation due to a defect skin barrier. The patients often tell us that they have been teased during childhood for having "elephant fingers"

Figure 29.4 depicts a patient with an *FLG* mutation and contact allergy to thiurams who has been exposed to rubber gloves. It is our clinical impression that the patients with allergic contact dermatitis and the *FLG* mutation more frequently have problems with long-lasting dermatitis and fissures.

Figure 29.5 depicts a 66-year-old female who is heterozygotic for *FLG* mutation; she has had xerotic skin and slight ichthyosis on her lower legs all her life, but she never had AD or hand eczema in childhood. At 40 years of age, she developed acute vesicular hand eczema due to contact allergy to thiuram and exposure to rubber gloves. The vesicular component of the hand eczema disappeared gradually, and then the dermatitis transformed into lichenified, slightly hyperkeratotic hand eczema located on both the dorsal and the volar part of the hands. Her ichthyosis on the lower legs at the same time changed into a lichenified chronic dermatitis. Most dermatologists, including us, would until a few years ago have called such a case "adult-onset AD," but these types of cases [12] are probably not uncommon in individuals with an inborn defect of the skin barrier who, because of environmental skin exposures to allergens and irritants, develop chronic skin inflammation that leads to degeneration of an already weak skin barrier [21].

The progress in the understanding of disease mechanisms will improve treatment and classification of such cases.

29.5 Protein Contact Dermatitis

Protein contact dermatitis is an inflammatory skin disease that can be a result of the protein exposure acting either as an allergen or as an irritant. An increased risk of type I sensitization to food items has been described in individuals with the *FLG* mutations, specifically for peanuts [22] and more generally in a large population-based study [23].

We have recently described a heterozygous carrier of an *FLG* mutation who most likely developed primary occupational type I skin allergy to salmon [24]. She later developed asthma and anaphylaxis when she was working in a room where salmon were handled. Protein contact dermatitis is a difficult condition to classify. In the future, further subclassifications by genomic studies might be possible in those cases that have a skin barrier defect and thereby are exposed to the protein through the skin.

29.6 Irritant Contact Dermatitis

Irritant contact dermatitis (ICD) depends on the interplay between the quality of the skin barrier and the magnitude of the irritant exposure. Irritant exposure qualifying for *occupational* ICD is today defined as 2 h of wet work or 2 h of glove use daily [25]. The definition probably only holds true for a fraction of individuals. In individuals with a defect in the skin barrier, the trauma

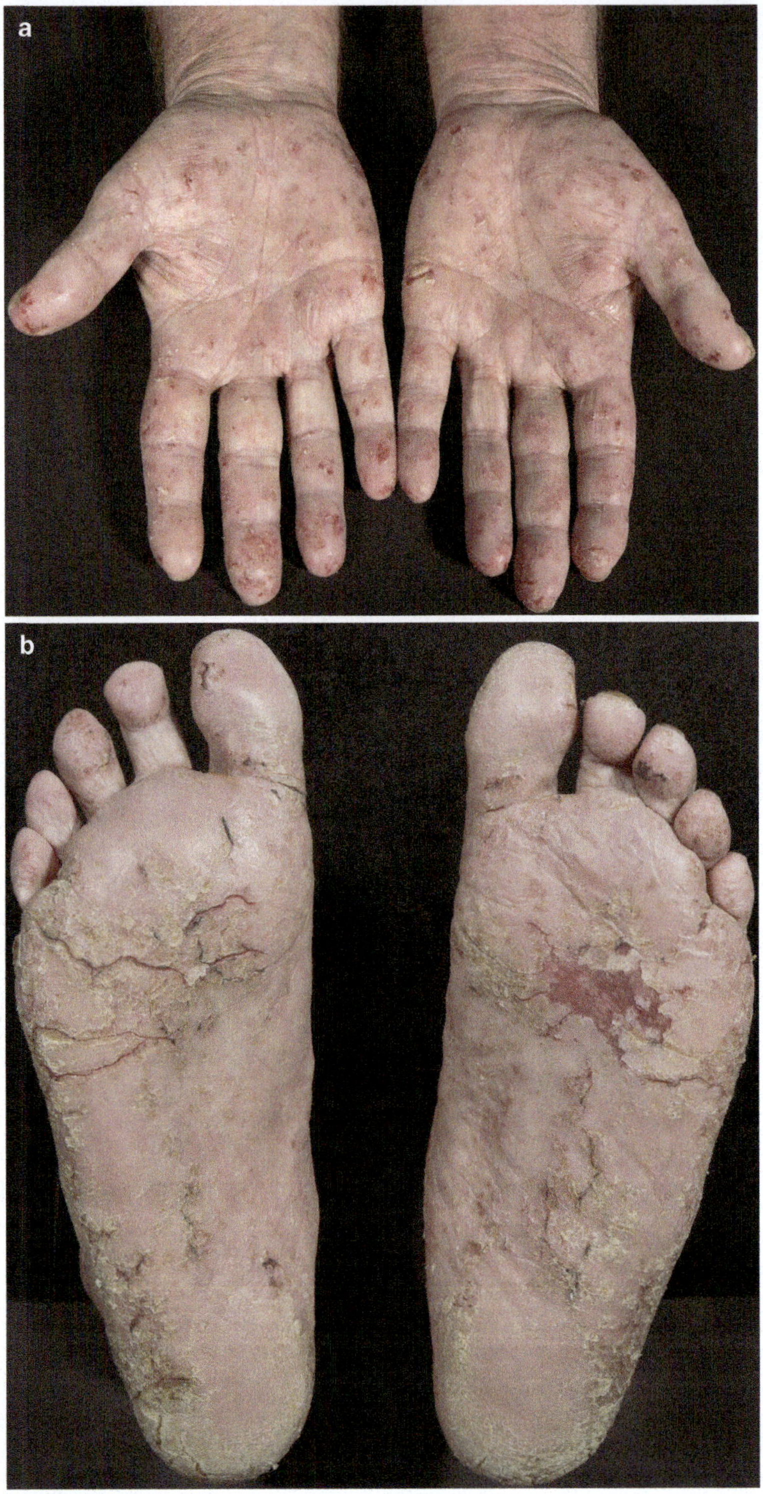

Fig. 29.3 (**a**, **b**) Severe fissured hands and feet on a 45-year-old male patient with generalized AD since early childhood and an *FLG* mutation

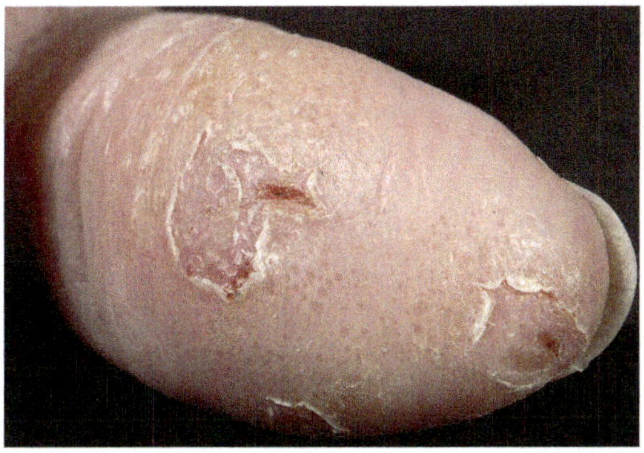

Fig. 29.4 A 48-year-old female's thumb of the right hand, with allergic contact dermatitis to thiurams from rubber gloves. She was heterozygous to a common *FLG* mutation. Note the vesicular inflammation with secondary fissures

required to make an irritant skin exposure is much less, and in another group of individuals, much larger. Individuals with *FLG* mutations have a risk of developing dorsal hand eczema within the first few months of life simply from exposure to air and friction from clothes [8]. We have recently shown that individuals with the *FLG* mutations, AD, and hand eczema in childhood to a large extent avoid wet work and occupational irritant exposure in adulthood [26]. These types of jobs account for 40 % of all available jobs. This observation is in agreement with the fact that individuals with AD are underrepresented among hairdresser trainees/apprentices [27]. Early-onset occupational ICD has been a repeated observation, particularly in young females with childhood AD and with or without childhood hand eczema [10, 11].

Notwithstanding this observation, it is clear from the epidemiological studies that a large group of individuals with the *FLG* mutations and only dry skin in childhood and teenage years, but without AD and hand eczema, will enter into wet work jobs. In a recent study based on selected occupational ICD, it was clearly shown that *FLG* mutations were individual risk factors for this disease independent of AD and childhood eczema [28]. In this context, it needs to be remembered

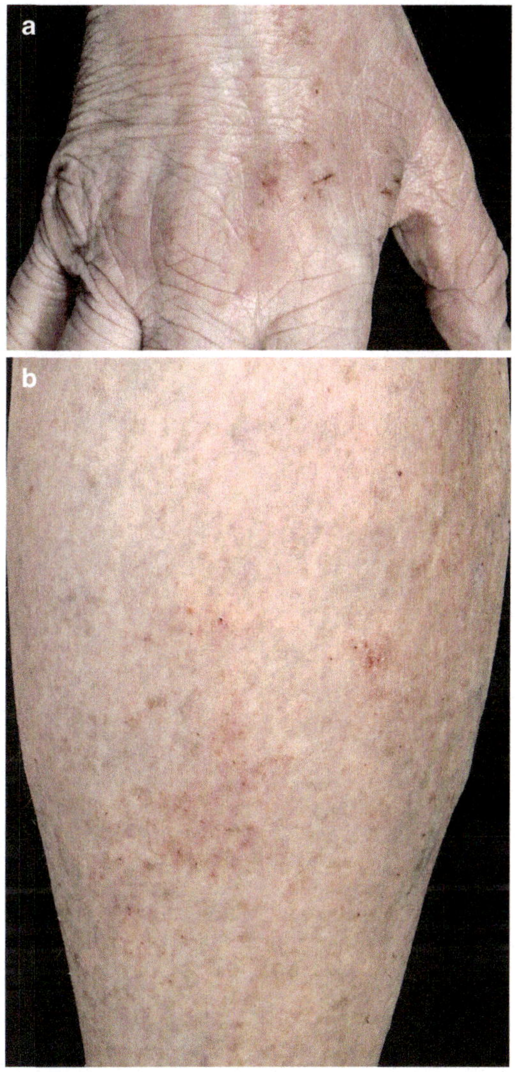

Fig. 29.5 (**a**, **b**) A 66-year-old female patient heterozygous for an *FLG* mutation. She has never had AD. She developed lasting skin changes similar to AD after an acute episode of allergic contact dermatitis due to thiurams. Such a patient should not be classified as having AD but as sequela from allergic contact dermatitis in a patient with the *FLG* mutation. This may have significant medicolegal consequences

that in large population studies the *FLG* mutation only acts as a risk factor for hand eczema in conjunction with AD and individuals with only the *FLG* mutation and no AD have a risk for hand

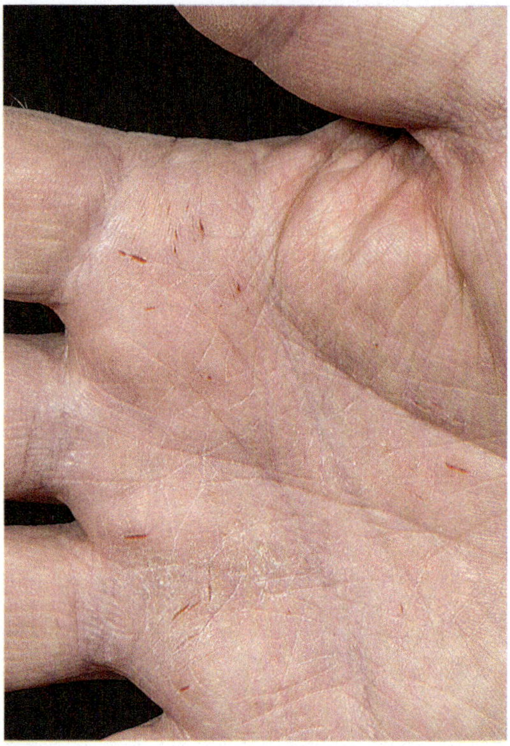

Fig. 29.6 A moderate case of a steatotic ICD based on professional oil contact in a patient with the *FLG* mutation. The case is notable because of the multiple knifelike cut fissures following the fine lines in the palms. The patients have many expressions for the vulnerability of their hands. One patient expressed to us "when the winter comes with its dry air, my hands sing like crystal and then they crack"

eczema in the same magnitude as the population in general [29]. So this is an area still open to further research to understand and identify the group at particular risk. First of all, the *FLG* mutation gives more severe and persistent disease, and painful skin fissures are a typical symptom [29, 30] (Fig. 29.6).

Conclusion

The discovery of the association between xerosis, AD, and the *FLG* mutations [1, 3] has been an eye opener for the clinical dermatologists who work with eczema disease. In our clinic, we have used this knowledge for the last 5 years. Even though information about whether a patient has a mutation may not result in immediate help for the patient, it is a major step in their understanding of the disease and of their whole family history.

It is our impression that this information and understanding lead to more focus on the rationale for using emollients and anti-inflammatory topical drugs. It also leads to less speculation on all other reasons for the background of their illness. The understanding of the *FLG* mutation for the skin barrier has opened up a completely new area for understanding the eczema mechanism and prognosis and, further, for development of specific topical and systemic treatments.

References

1. McLean WH, Irvine AD. Heritable filaggrin disorders: the paradigm of atopic dermatitis. J Invest Dermatol. 2012;132(E1):E20–1.
2. Palmer CN, Irvine AD, Terron-Kwiatkowski A, Zhao Y, Liao H, Lee SP, et al. Common loss-of-function variants of the epidermal barrier protein filaggrin are a major predisposing factor for atopic dermatitis. Nat Genet. 2006;38(4):441–6.
3. Thyssen JP, Johansen JD, Zachariae C, Menne T, Linneberg A. Xerosis is associated with atopic dermatitis, hand eczema and contact sensitization independent of filaggrin gene mutations. Acta Derm Venereol. 2013;93(4):406–10.
4. Nexmand P-H. Clinical studies of Besnier's prurigo. Thesis, Faculty of Health and Medical Science. Copenhagen: Roskilde and Bagger Publishers; 1948.
5. Asher MI, Montefort S, Bjorksten B, Lai CK, Strachan DP, Weiland SK, et al. Worldwide time trends in the prevalence of symptoms of asthma, allergic rhinoconjunctivitis, and eczema in childhood: ISAAC Phases One and Three repeat multicountry cross-sectional surveys. Lancet. 2006;368(9537):733–43.
6. Taylor B, Wadsworth J, Wadsworth M, Peckham C. Changes in the reported prevalence of childhood eczema since the 1939-45 war. Lancet. 1984;2(8414):1255–7.
7. Thyssen JP, Johansen JD, Linneberg A, Menne T, Engkilde K. The association between contact sensitization and atopic disease by linkage of a clinical database and a nationwide patient registry. Allergy. 2012;67(9):1157–64.
8. Carson CG, Rasmussen MA, Thyssen JP, Menne T, Bisgaard H. Clinical presentation of atopic dermatitis by filaggrin gene mutation status during the first 7 years of life in a prospective cohort study. PLoS One. 2012;7(11):e48678.
9. Rystedt I. Factors influencing the occurrence of hand eczema in adults with a history of atopic dermatitis in childhood. Contact Dermatitis. 1985;12(4):185–91.

10. Dickel H, Bruckner TM, Schmidt A, Diepgen TL. Impact of atopic skin diathesis on occupational skin disease incidence in a working population. J Invest Dermatol. 2003;121(1):37–40.
11. Schwensen JF, Friis UF, Menne T, Johansen JD. One thousand cases of severe occupational contact dermatitis. Contact Dermatitis. 2013;68(5):259–68.
12. McFadden JP, White JM, Basketter DA, Kimber I. Does hapten exposure predispose to atopic disease? The hapten-atopy hypothesis. Trends Immunol. 2009;30(2):67–74.
13. Puangpet P, Lai-Cheong J, McFadden JP. Chemical atopy. Contact Dermatitis. 2013;68(4):208–13.
14. Thyssen JP, Carlsen BC, Bisgaard H, Giwercman C, Johansen JD, Linneberg A, et al. Individuals who are homozygous for the 2282del4 and R501X filaggrin null mutations do not always develop dermatitis and complete long-term remission is possible. J Eur Acad Dermatol Venereol. 2012;26(3):386–9.
15. Thyssen JP, Linneberg A, Ross-Hansen K, Carlsen BC, Meldgaard M, Szecsi PB, et al. Filaggrin mutations are strongly associated with contact sensitization in individuals with dermatitis. Contact Dermatitis. 2013;68(5):273–6.
16. Novak N, Baurecht H, Schafer T, Rodriguez E, Wagenpfeil S, Klopp N, et al. Loss-of-function mutations in the filaggrin gene and allergic contact sensitization to nickel. J Invest Dermatol. 2008;128(6):1430–5.
17. Weidinger S, Illig T, Baurecht H, Irvine AD, Rodriguez E, Diaz-Lacava A, et al. Loss-of-function variations within the filaggrin gene predispose for atopic dermatitis with allergic sensitizations. J Allergy Clin Immunol. 2006;118(1):214–9.
18. Ross-Hansen K, Menne T, Johansen JD, Carlsen BC, Linneberg A, Nielsen NH, et al. Nickel reactivity and filaggrin null mutations–evaluation of the filaggrin bypass theory in a general population. Contact Dermatitis. 2011;64(1):24–31.
19. Ross-Hansen K, Østergaard O, Tanassi JT, Thyssen JP, Johansen JD, Menne T, et al. Filaggrin is a predominant member of the nickel binding proteome of human epidermis. J Invest Dermatol. 2013. doi:10.1038/jid.2013.445.
20. Hald M, Menne T, Johansen JD, Zachariae C. Allergic contact dermatitis to sorbitan sesquioleate imitating severe glove dermatitis in a patient with filaggrin mutation. Contact Dermatitis. 2013;69(5):313–5.
21. Irvine AD, McLean WH, Leung DY. Filaggrin mutations associated with skin and allergic diseases. N Engl J Med. 2011;365(14):1315–27.
22. Brown SJ, Asai Y, Cordell HJ, Campbell LE, Zhao Y, Liao H, et al. Loss-of-function variants in the filaggrin gene are a significant risk factor for peanut allergy. J Allergy Clin Immunol. 2011;127(3):661–7.
23. Linneberg A, Fenger RV, Husemoen L, Carlsen BC, Johansen JD, Menne T, et al. Association between loss-of-function mutations in the filaggrin gene and self-reported food allergy and alcohol sensitivity. Int Arch Allergy Immunol. 2013;161(3):234–42.
24. Kaae J, Menne T, Thyssen JP. Severe occupational protein contact dermatitis caused by fish in 2 patients with filaggrin mutations. Contact Dermatitis. 2013; 68(5):319–20.
25. Uter W, Pfahlberg A, Gefeller O, Schwanitz HJ. Hand dermatitis in a prospectively-followed cohort of hairdressing apprentices: final results of the POSH study. Prevention of occupational skin disease in hairdressers. Contact Dermatitis. 1999;41(5):280–6.
26. Bandier J, Ross-Hansen K, Carlsen BC, Menne T, Linneberg A, Stender S, et al. Carriers of filaggrin gene (FLG) mutations avoid professional exposure to irritants in adulthood. Contact Dermatitis. 2013;69(6):355–62.
27. Bregnhoj A, Sosted H, Menne T, Johansen JD. Validation of self-reporting of hand eczema among Danish hairdressing apprentices. Contact Dermatitis. 2011;65(3):146–50.
28. Landeck L, Visser M, Skudlik C, Brans R, Kezic S, John SM. Clinical course of occupational ICD of the hands in relation to filaggrin genotype status and atopy. Br J Dermatol. 2012;167(6):1302–9.
29. Thyssen JP, Carlsen BC, Menne T, Linneberg A, Nielsen NH, Meldgaard M, et al. Filaggrin null mutations increase the risk and persistence of hand eczema in subjects with atopic dermatitis: results from a general population study. Br J Dermatol. 2010;163(1):115–20.
30. Thyssen JP, Ross-Hansen K, Johansen JD, Zachariae C, Carlsen BC, Linneberg A, et al. Filaggrin loss-of-function mutation R501X and 2282del4 carrier status is associated with fissured skin on the hands: results from a cross-sectional population study. Br J Dermatol. 2012;166(1):46–53.

Part VI

Filaggrin and Non-cutaneous Disease

Filaggrin, the Skin Barrier, and Diabetes

30

Elena Godoy-Gijón

Contents

E. Godoy-Gijón, MD
Department of Dermatology, Hospital De Cabueñes, Secretaría del Servicio de Dermatología, Policlínicas de la Casa del Mar, Avda. Príncipe de Asturias, 76, Gijón, Asturias 33212, Spain
e-mail: e.godoy.gijon@gmail.com

Epidermal barrier abnormalities are present in diabetic patients. A significant influence of insulin and hyperglycemia on keratinocyte homeostasis has been demonstrated. Skin barrier abnormalities in diabetics are not well defined, but several researchers agree that diabetic skin models have a decreased stratum corneum hydration.

Recent studies have suggested an epidemiological and genetic linkage between filaggrin gene mutations and type 2 diabetes mellitus. While additional studies are necessary in this area, it is possible that filaggrin molecules could play an important role for diabetics' skin as well as the development of diabetes. This chapter provides an overview of studies that have investigated skin barrier functions in diabetics and discusses the potential role of filaggrin.

30.1 Introduction

Diabetes mellitus (DM) is a prevalent and chronic disease with a tendency toward multiple organ involvement. During the last decades, the incidence of type 1 and 2 diabetes has increased remarkably [1]. At the same time, the filaggrin gene (*FLG*) mutation-related atopic disorders such as atopic dermatitis (AD), rhinitis, asthma, and food allergy have continued to become more widespread in industrialized countries [2].

Skin problems are common in diabetic patients, affecting up to 30 % [3–5]. Pertinently, they can be the first sign of DM. Patients generally have impaired skin homeostasis, not only caused

J.P. Thyssen, H. Maibach (eds.), *Filaggrin*,
DOI 10.1007/978-3-642-54379-1_30, © Springer-Verlag Berlin Heidelberg 2014

by vasculopathy and neuropathy (which has been thought to be the only cause in the past) but also due to abnormalities of skin metabolism. Furthermore, delayed wound healing [6, 7], reduced antimicrobial peptide (AMP) function [8, 9], and epidermal barrier structure alterations [3, 8, 10–12] are involved in the etiology and pathogenesis of diabetes skin abnormalities. Nonetheless, there are still several controversies regarding the diabetic skin model.

A hypothesis about an inherited skin barrier abnormality in diabetic patients has recently emerged. Genetic studies have suggested a linkage between *FLG* mutations and diabetes [13, 14]. This could be explained either by a common origin of genetic alterations due to mutations in the same locus or by early low-grade inflammation status caused by a skin barrier deficiency due to *FLG* mutations.

While pathological skin disorders in diabetic patients seem to be clearly associated with metabolic, neurological, and vascular dysregulation promoted by diabetes, future studies need to clarify whether a common genetic background with *FLG* mutations exists.

30.2 Diabetes and Skin

Diabetes induces different forms of dermatological disorders such as necrobiosis lipoidica diabeticorum, scleredema diabeticorum, bullosis diabeticorum, diabetic dermopathy, and acanthosis nigricans [15]. Pruritus, xerosis, delayed wound healing, and skin infection [5, 16–18] are also major skin factors in these patients, affecting their quality of life and leading to skin diseases.

Xerosis is one of the earliest and most common skin abnormalities in diabetic patients [5, 19–21], with a prevalence ranging between 6 and 80 % [5, 17, 19, 22, 23]. Xerosis is believed to act as a major factor promoting skin fissures, hyperkeratosis, and ulceration in the diabetic foot [20]. Yosipovitch et al. noticed that 48 % of patients with type 1 DM (T1DM) in their series had clinical ichthyosiform shin changes [24]. While dysfunction of sympathetic nerves (including sudomotor dysfunction) seems to contribute, resulting in hypohidrosis and xerosis [18, 25], epidermal lipid abnormalities and *FLG* mutations are being studied because of their possible significant contribution in this entity.

Delayed wound healing and dermal abnormalities are also a major problem in diabetic patients. Dermal collagen accumulation of toxic substances such as advanced glycosylation end products (AGEs) [12, 26–28] is induced in diabetic skin. This accumulation is thought to promote oxidative stress [29], develop microvascular complications [30], and produce dermal thickness [31] and stiffness [32]. Receptors for AGEs are also increased in epidermal keratinocytes from diabetic skin [8, 33]. Delayed wound healing in these patients is linked with AGEs [6, 7], and AGE-RAGE interactions in epidermal keratinocytes have been presumed to cause impaired skin barrier homeostasis [8]. Histological subcutaneous fat atrophy and dermal vessel abnormalities have also been noticed [12].

30.3 Diabetes and Skin Barrier Function

Insulin is a key factor for epidermal homeostasis, acting as an important growth factor in cultured keratinocytes. Insulin receptors are expressed constitutively in these cells, having their own glucose uptake system [34]. Hence, insulin has an essential role on keratinocyte differentiation [35, 36], proliferation [37], and migration [38–40], which are main factors for stratum corneum (SC) functions. Exposure of keratinocytes to high-glucose conditions inhibits the expression and autophosphorylation of insulin-like growth factor [34, 41].

Epidermal barrier function can be quantified by different biophysical measurements like SC hydration, transepidermal water loss (TEWL), sebum lipids, and pH values. Several studies have tried to establish epidermal barrier functional properties in patients with DM; however, results are conflicting.

When defining an animal model for studying the underlying skin abnormalities, it is also critical to standardize studies in diabetic skin. Sakai and colleagues established a model of T1DM skin using hairless mice treated with streptozotocin or alloxan to destroy pancreatic β cells [10]. This reproduced the T1DM animal model [42] with the clear advantage of having hairless mice for skin measurements. As for type 2 DM (T2DM), Otsuka Long-Evans Tokushima Fatty (OLETF) rats have been used as a model of chronic hyperglycemia in one diabetic skin research report [8], although this is not a usual T2DM model used in endocrinologic diabetes studies. Leptin-deficient and leptin receptor-deficient mice are probably the most commonly used models for T2DM research, but no skin barrier studies have yet been published using these mice. Anyway, none of these three models are hairless rodent, so they need to be shaved, and this process could modify the measurements.

30.3.1 Stratum Corneum Hydration

Hydration in the SC is regulated basically by three factors: natural moisturizing factors (NMFs) (amino acids derived from filaggrin cleavage), sebum lipids, and intercellular lipids (mainly ceramides [43, 44]).

Hitherto, the only study that has been performed on epidermal barrier changes in a T2DM animal model showed decreased SC hydration with a weakened SC integrity and delayed barrier recovery rate. Decreased total epidermal lipid content, with lower lamellar body production and lipid-related enzymes, was also found in this study, providing an explanation for the impaired barrier recovery. Filaggrin and other differentiation marker proteins (involucrin and loricrin) were evaluated, but no quantitative differences were observed [8].

In T1DM mice models, studies have shown a decreased water content of the SC together with an abnormal lipid content (see later) as well as unaltered epidermal concentrations of profilaggrin, filaggrin, loricrin, and keratins 1, 5, and 10 [10].

It should be noted that all these findings reflect (1) a relationship between decreased SC hydration in diabetic skin and lipid abnormalities and (2) no relationship between hyperglycemia or low insulin levels and epidermal concentrations of filaggrin or profilaggrin in animal hyperglycemic models.

Human studies have shown a significantly decreased SC hydration in a group of type 1 ($n=38$) and type 2 ($n=11$) DM patients with high fasting plasma glucose levels but not in patients with high glycosylated hemoglobin (HbA1c) [11]. Similar results have been obtained in different studies [17]. One study with a mixed diabetic population (type 1 $n=34$ and type 2 $n=4$) showed no significant differences [3].

These findings suggest that the current hyperglycemic state is more related to SC hydration than the past state, but it is more severe in aged patients with diabetes than in younger ones [11]. That could be the reason why some studies with younger diabetic patients have shown no differences when compared with controls [3].

30.3.2 Transepidermal Water Loss

Transepidermal water loss is a measure of the water barrier function of the SC. It reflects the quantity of water that passes, by diffusion or evaporation, through the epidermal barrier.

Epidermal barrier measurements in both T1DM [10] and T2DM [8] diabetes animal models have demonstrated normal TEWL compared with controls.

Human studies in T1DM ($n=11$) and T2DM ($n=38$) patients have shown significantly lower TEWL levels in forearm skin related with high HbA1c levels but not with high fasting plasma glucose levels [11]. One study from Iran found no differences between diabetic patients (T1DM $n=4$; T2DM $n=34$) and controls, even adjusting for fasting plasma glucose and HbA1c levels, but their measurements obtain higher TEWL values, which the authors attributed to a racial and climate difference [3].

30.3.3 Lipids

Sebum and intercellular lipids play an important role for skin barrier function and SC hydration [43–45]. Sebaceous gland can bind insulin, but this ability seems to be impaired in DM [46].

In animal models, T1DM rats show decreased sebum secretion from the sebaceous glands [47], while mice have abnormalities in sebum components [10]. The latter study found a significant decrease of the triglyceride content in the SC, with higher ratios of cholesterol-wax ester and fatty acids, while ceramides and cholesterol remained unchanged [10].

In humans, diabetic patients (T1DM and T2DM) have a lower skin surface lipid content associated with high fasting plasma glucose (significantly reduced) and high HbA1c (not significantly but reduced) [11], demonstrating an influence not only by long-term previous hyperglycemia but also by current hyperglycemic state. Some studies, on the other hand, have obtained different results, depending on the anatomical locations, and have found a lower sebum content on the forehead, but not on the forearm and lower leg [3].

Both decreased sebum secretion and altered sebum components could have a major influence on the decreased SC hydration observed in diabetic patients.

30.3.4 Skin Elasticity

Skin elasticity function is measured by evaluation of the time needed for the propagation of an acoustical shock wave between two sensors placed on the skin surface. This time is inversely related to skin elasticity [48].

It has been used to evaluate diabetic skin in a mixed group of T1DM ($n=4$) and T2DM ($n=34$) patients, showing lower skin elasticity in some locations (forehead and forearm) compared with controls [3]. This has been also demonstrated in facial skin from diabetes patients [49]. The diminished skin elasticity could be due to a dermal change in diabetic skin, perhaps due to dermal collagen accumulation of AGEs, but there are no studies clearly defining this relationship.

30.3.5 Corneocyte Surface Area

The surface area of the corneocytes has been directly related to the epidermal turnover rate. Diseases with high epidermal turnover rate, like psoriasis or AD, have shown a small corneocyte surface area [50, 51]. By contrast, the aged skin, which is characterized by a low epidermal turnover, showed a large corneocyte surface area [52, 53].

Some studies have shown an increased corneocyte surface area in T1DM mice [10]. In humans, patients with DM have a larger surface than normal individuals [54]. Studies measuring epidermal thickness and proliferating cell nuclear antigen (PCNA) have shown a lower rate among basal cells [10] and the absence of the multilayer epithelium with decrease of epidermal cell number [12] in T1DM animal models. Also a mitotic dysfunction in epidermal cells has been observed [12]. The increased surface area, the decreased PCNA levels, the mitotic dysfunction, and the thinner epidermis all suggest decreased epidermal cell turnover, which could be related to insulin differentiation and proliferation influence.

No studies in T2DM animal models focusing on the corneocyte surface area have yet been performed. However, epidermal and dermal thickness and expression of PCNA seem not to be modified [8]. Perhaps these characteristics are due to hyperinsulinemia and could partially compensate for insulin resistance.

30.3.6 Skin pH Values

The skin pH in normal adult skin is acidic, and it is considered mainly a product of amino acids, lactic acid, and sebum of the stratum corneum [55].

Elevated pH levels have been associated with a high rate of skin infections [56]. Reduced production of acidic metabolites of filaggrin may result in increased skin surface pH [57].

Yosipovitch et al. [58] showed that while skin pH from forearm was similar between T2DM patients and controls, the intertriginous areas from the diabetic group revealed a higher pH. This could explain the higher rate of candidal infection in the intertriginous areas of diabetic patients [4, 5, 59].

30.4 Diabetes and Filaggrin

In general, filaggrin proteins play a major role for normal epidermal barrier functions. Only a few studies have yet investigated whether filaggrin and *FLG* mutations may be involved in diabetes. However, several studies have associated diabetes with filaggrin-related diseases such as AD, rhinoconjunctivitis, and asthma.

30.4.1 Type 1 Diabetes

T1DM is defined by decreased pancreatic insulin production with a secondary hyperglycemia.

Rodent models of T1DM showed no quantitative abnormalities of filaggrin [10]. However, remember that these mice had experimentally induced DM.

Saleh et al. have investigated a genetic relationship between FLG mutations and T1DM. They found no association between T1DM and the most common *FLG* mutations in Northern European population (R501X and 2282del4) [60], proinflammatory cytokines such as tumor necrosis factor (TNF)-α, interleukin (IL)-1β, IL-18, and interferon γ in 6,743 T1DM cases and 7,864 controls [61]. However, a much smaller Danish case-control study found a markedly lower prevalence of *FLG* mutations in T1DM patients when compared to controls, although the difference did not reach statistical significance [13].

An inverse association between atopic conditions and T1DM has previously been suggested. While AD has been mostly associated with a lower risk of T1DM [62–67], some authors have not been able to confirm this finding [68]. Patients with asthma also seem to have a lower risk of developing T1DM [62, 63, 65, 69], but other studies have not confirmed this [63, 64, 66–68, 70], and some have even found a higher risk [70–72]. Moreover, studies on rhinitis and rhinoconjunctivitis have shown a lower risk [63, 65], no association [64–68], or even an increased risk [73].

All these epidemiological studies but one [68] have been conducted in a Northern European population [63–67, 70, 71]. They were mainly case-control studies [63–67, 70], with data obtained from questionnaires [64, 65, 67] or general population databases [63, 66, 70, 71].

Among the studies that showed a higher risk of atopic diseases, there are some important weaknesses. Stene and colleagues [72] did not use well-defined criteria for asthma in their study, which was assigned by an affirmative response to "ever asthma" or "any wheeze in the past year" questions. Simpson et al. [71] only found a positive association with Th1 diseases, including in this group not only T1DM but also other diseases as psoriasis and rheumatoid arthritis.

The inverse association between AD and DM may be explained by the T-helper (Th) cell 1 and 2 dichotomy [74]. T1DM is a Th1-skewed immune response, whereas AD and allergic IgE responses mainly display a Th2 pattern. It is known that Th1 responses tend to inhibit Th2 cell proliferation, and vice versa [74]. This paradigm, in its simplest form, could explain the lower risk of T1DM in patients with AD. But the disease development and Th role are quite complex, and that explains why AD and type 1 diabetes can exist in the same patient.

30.4.2 Type 2 Diabetes

T2DM is defined by insulin resistance in peripheral tissue and lack of insulin production in the pancreas. It is considered the most common form of DM, accounting for more than 80 % of all DM patients [75, 76].

The only animal study published about epidermal barrier abnormalities in T2DM showed no quantitative abnormalities in filaggrin expression [8]. This animal model is characterized by late onset of hyperglycemia and mild obesity.

Surprisingly, a recent Danish study showed a significant association between *FLG* null mutations (R501X and 2282del4) and patients with established T2DM. Also *FLG* mutations were associated with subjects from a Danish general population database who reported diabetes (Table 30.1) [13]. Interestingly, *FLG* mutations mainly constituted a risk factor for T2DM in those with a low body mass index.

The so-called epidermal differentiation complex is located on human chromosome 1q21.

Table 30.1 Logistic regression analyses with the outcome *FLG* null-mutation status and DM (adjusted for variables shown in the table as well as age)

–	General population, patients with type 1 and 2 diabetes ($n=4{,}213$)		–	General population and patients with type 2 diabetes ($n=4{,}109$)	
	FLG null mutation			Diabetes[b]	
Explanatory variables	% (n/n_{total})	Adjusted OR[a] with 95 % CI	Explanatory variables	% (n/n_{total})	Adjusted OR[a] with 95 % CI
Sex	–	–	*Sex*	–	–
Men	7.8 (159/2,029)	1 (reference)	Men	29.8 (586/1,967)	1 (reference)
Women	8.8 (193/2,184)	1.18 (0.94–1.47)	Women	18.1 (387/2,142)	0.56 (0.46–0.67)[c], $p=0.001$
Group	–	–	Filaggrin	–	–
Nondiabetic group	7.8 (246/3,136)	1 (reference)	Wild type	23.2 (874/3,764)	1 (reference)
Screen-detected diabetes group	9.1 (6/66)	1.23 (0.52–2.88)	Null mutation	28.2 (99/345)	1.50 (1.10–2.06)[c], $p=0.011$
Self-reported diabetes group	12.8 (17/133)	1.78 (1.05–3.04)[c], $p=0.032$	*BMI (kg/m²)*	–	–
Type 1 diabetes group	6.7 (7/104)	0.86 (0.39–1.87)	<25	9.7 (167/1,713)	1 (reference)
Type 2 diabetes group	9.8 (76/774)	1.37 (1.003–1.89)[c], $p=0.048$	25–30	23.2 (341/1,469)	1.97 (1.56–2.47)[c], $p=0.001$
–	–	–	>30	49.1 (461/905)	7.36 (5.79–9.36)[c], $p=0.001$

Key: *OR* odds ratio, *CI* confidence interval; nondiabetic group, healthy controls from the general population in Copenhagen; screen-detected diabetes group, diabetes screening group defined as subjects who did not report diabetes but who had fasting blood glucose ≥7 and/or HbA1c ≥6.5 %; self-reported diabetes group, diabetes group defined as subjects who gave an affirmative answer to the question, "Have you ever been told that you suffered from diabetes?"

[a]Mutually adjusted for variables shown in table and age

[b]Diabetes was defined as belonging to either the "screen-detected diabetes group," the "self-reported diabetes group," or the "type 2 diabetes group." The "type 1 diabetes group" was regarded as missing data

[c]Statistically significant

This complex comprises many epidermal crucial genes, including genes encoding profilaggrin and filaggrin-2 as a part of the S100 fused-type protein family [77]. In recent years, identification of genetic linkage of polymorphisms in the chromosome 1q21-25 to T2DM has been demonstrated [14]. This coincidence at the exact same chromosome locus suggests a common genetic background in both diseases. However, it is interesting that *FLG* mutation carriers also have 10 % higher serum levels of 25-hydroxy vitamin D [78]. Since vitamin D deficiency has been repeatedly associated with T2DM, one would expect *FLG* mutations instead to lower the risk of T2DM. This conundrum needs to be addressed in future large-scale studies, as it seems to be unclear at this point whether *FLG* mutations affect the risk of T2DM.

30.4.3 Gestational Diabetes

Maternal gestational diabetes is defined as glucose intolerance that is first detected during pregnancy [79]. It is the third of the four groups included in the diabetes classification. Women with gestational diabetes remain at high risk of development of T2DM after delivery [80–82], with a variable incidence rate from 2.6 to 70 % among different studies [80].

Significant association between maternal gestational diabetes and the development of AD and allergen sensitization in term infants has also been shown [83]. The consequence of this finding needs to be studied in the future.

Conclusion

The above findings suggest that T1DM skin displays decreased SC hydration, mainly due to alterations in lipid content, with a normal barrier function and a decreased epidermal turnover. Keratinocyte proliferation, differentiation, and migration are clearly downregulated in T1DM skin. Therefore, this could be a major factor in delayed barrier recovery and wound healing.

T2DM skin displays decreased SC hydration in both human and rodent models. The latter also demonstrates decreased epidermal lipid content. Delayed skin barrier recovery with normal keratinocyte proliferation and differentiation was shown in the only study conducted in diabetic rats. Skin elasticity seems to be decreased also in both T1DM and T2DM models.

FLG mutations have been epidemiologically associated with T2DM in two populations, and there seems to be an inverse relationship between, respectively, AD or asthma and development of T1DM, which could be explained by the Th1/Th2 dichotomy. Although filaggrin is not affected by hyperglycemia or low insulin levels according to both T1DM and T2DM animal models, remember that these animals only share alterations in glucose and insulin with diabetic humans, but not environmental exposures, neither immune nor genetic background.

Still it is unclear why the *FLG* null genotype was associated with the development of T2DM in two populations and whether this finding can be replicated. Filaggrin-deficient skin could increase inflammation, leading in concert with other factors to diabetes development. Another nascent hypothesis is that *FLG* mutations are randomly associated with T2DM as a marker of polymorphisms in a common locus of chromosome 1q. Development of diabetes may be also related to secondary effects of cutaneous disease treatments, e.g., topical and oral corticosteroids used to treat ichthyosis vulgaris or AD.

More studies are clearly required to clarify the origin of the relationship between skin barrier problems due to *FLG* mutations and diabetes, also taking vitamin D levels into account. Future research should differentiate between T1DM and T2DM, as both groups show different metabolic and skin changes. Special attention should be focused on T2DM, not only because it is the most frequent type of DM but also due to the high incidence of epidermal barrier abnormalities in these patients, as well as the possible genetic linkage with *FLG* mutations.

Acknowledgment Elena Godoy-Gijón was funded by an unrestricted research grant from ISDIN as a part of a national Spanish award.

References

1. Patterson CC, Dahlquist GG, Gyurus E, Green A, Soltész G, EURODIAB Study Group. Incidence trends for childhood type 1 diabetes in Europe during 1989-2003 and predicted new cases 2005-20: a multicentre prospective registration study. Lancet. 2009; 373:2027–33.
2. von Mutius E, Vercelli D. Farm living: effects on childhood asthma and allergy. Nat Rev Immunol. 2010;10:861–8.
3. Seirafi H, Farsinejad K, Firooz A, Davoudi SM, Robati RM, Hoseini MS, et al. Biophysical characteristics of skin in diabetes: a controlled study. J Eur Acad Dermatol Venereol. 2009;23:146–9.
4. Wahid Z, Kanjee A. Cutaneous manifestations of diabetes mellitus. J Pak Med Assoc. 1998;48:304–5.
5. Behm B, Schreml S, Landthaler M, Babilas P. Skin signs in diabetes mellitus. J Eur Acad Dermatol Venereol. 2012;26:1203–11.
6. Franzen LE, Roberg K. Impaired connective tissue repair in streptozotocin-induced diabetes shows ultrastructural signs of impaired contraction. J Surg Res. 1995;58:407–14.

7. Bitar MS. Glucocorticoid dynamics and impaired wound healing in diabetes mellitus. Am J Pathol. 1998;152:547–54.
8. Park HY, Kim JH, Jung M, Chung CH, Hasham R, Park CS, et al. A long-standing hyperglycaemic condition impairs skin barrier by accelerating skin ageing process. Exp Dermatol. 2011;20:969–74.
9. Rivas-Santiago B, Trujillo V, Montoya A, Gonzalez-Curiel I, Castañeda-Delgado J, Cardenas A, et al. Expression of antimicrobial peptides in diabetic foot ulcer. J Dermatol Sci. 2012;65:19–26.
10. Sakai S, Endo Y, Ozawa N, Sugawara T, Kusaka A, Sayo T, et al. Characteristics of the epidermis and stratum corneum of hairless mice with experimentally induced diabetes mellitus. J Invest Dermatol. 2003;120:79–85.
11. Sakai S, Kikuchi K, Satoh J, Tagami H, Inoue S. Functional properties of the stratum corneum in patients with diabetes mellitus: similarities to senile xerosis. Br J Dermatol. 2005;153:319–23.
12. Chen XF, Lin WD, Lu SL, Xie T, Ge K, Shi YQ, et al. Mechanistic study of endogenous skin lesions in diabetic rats. Exp Dermatol. 2010;19:1088–95.
13. Thyssen JP, Linneberg A, Carlsen BC, Johansen JD, Engkilde K, Hansen T, et al. A possible association between a dysfunctional skin barrier (filaggrin null-mutation status) and diabetes: a cross-sectional study. BMJ Open. 2011;1:e000062.
14. Prokopenko I, Zeggini E, Hanson RL, Mitchell BD, Rayner NW, Akan P, et al. Linkage disequilibrium mapping of the replicated type 2 diabetes linkage signal on chromosome 1q. Diabetes. 2009;58: 1704–9.
15. Jelinek JE. The skin in diabetes. Diabet Med. 1993;10:201–13.
16. Mahajan S, Koranne RV, Sharma SK. Cutaneous manifestation of diabetes mellitus. Indian J Dermatol Venereol Leprol. 2003;69:105–8.
17. Seite S, Khemis A, Rougier A, Ortonne JP. Importance of treatment of skin xerosis in diabetes. J Eur Acad Dermatol Venereol. 2011;25:607–9.
18. Yamaoka H, Sasaki H, Yamasaki H, Ogawa K, Ohta T, Furuta H. Truncal pruritus of unknown origin may be a symptom of diabetic polyneuropathy. Diabetes Care. 2010;33:150–5.
19. Pavlovic MD, Milenkovic T, Dinic M, et al. The prevalence of cutaneous manifestations in young patients with type 1 diabetes. Diabetes Care. 2007;30:1964–7.
20. Garrigue E, Martini J, Cousty-Pech F, et al. Evaluation of the moisturizer Pedimed® in the foot care of diabetic patients. Diabetes Metab. 2011;37:330–5.
21. Sehgal VN, Bhattacharya SN, Verma P. Juvenile, insulin-dependent diabetes mellitus, type 1-related dermatoses. J Eur Acad Dermatol Venereol. 2011;25:625–36.
22. Litzelman DK, Marriott DJ, Vinicor F. The role of footwear in the prevention of foot lesions in patients with NIDDM. Conventional wisdom or evidence-based practice? Diabetes Care. 1997;20:156–62.
23. Romano G, Moretti G, Di BA, Giofrè C, Di Cesare E, Russo G, et al. Skin lesions in diabetes mellitus: prevalence and clinical correlations. Diabetes Res Clin Pract. 1998;39:101–6.
24. Yosipovitch G, Hodak E, Vardi P, Shraga I, Karp M, Sprecher E, et al. The prevalence of cutaneous manifestations in IDDM patients and their association with diabetes risk factors and microvascular complications. Diabetes Care. 1998;21:506–9.
25. Pavicic T, Korting HC. Xerosis and callus formation as a key to the diabetic foot syndrome: dermatologic view of the problem and its management. J Dtsch Dermatol Ges. 2006;4:935–41.
26. Kennedy L, Baynes JW. Non-enzymatic glycosylation and the chronic complications of diabetes: an overview. Diabetologia. 1984;26:93–8.
27. Sternberg M, Cohen-Forterre L, Peyroux J. Connective tissue in diabetes mellitus: biochemical alterations of the intercellular matrix with special reference to proteoglycans, collagens and basement membranes. Diabete Metab. 1985;11:27–50.
28. Liao H, Zakhaleva J, Chen W. Cells and tissue interactions with glycated collagen and their relevance to delayed diabetic wound healing. Biomaterials. 2009;30:1689–96.
29. Stern DM, Yan SD, Yan SF, Schmidt AM. Receptor for advanced glycation endproducts (RAGE) and the complications of diabetes. Ageing Res Rev. 2002;1:1–15.
30. Nakai K, Yoneda K, Ishihara Y, Ohmori K, Moriue T, Igarashi J, et al. Lipid peroxidation-induced VEGF expression in the skin of KKAy obese mice. Exp Dermatol. 2011;20:388–93.
31. Forst T, Kann P, Pfutzner A, Lobmann R, Schäfer H, Beyer J. Association between "diabetic thick skin syndrome" and neurological disorders in diabetes mellitus. Acta Diabetol. 1994;31:73–7.
32. Aoki Y, Yazaki K, Shirotori K, Yanagisawa Y, Oguchi H, Kiyosawa K, et al. Stiffening of connective tissue in elderly diabetic patients: relevance to diabetic nephropathy and oxidative stress. Diabetologia. 1993;36: 79–83.
33. Wertheimer E, Spravchikov N, Trebicz M, Gartsbein M, Accili D, Avinoah I, et al. The regulation of skin proliferation and differentiation in the IR null mouse: implications for skin complications of diabetes. Endocrinology. 2001;142:1234–41.
34. Spravchikov N, Sizyakov G, Gartsbein M, Accili D, Tennenbaum T, Wertheimer E. Glucose effects on skin keratinocytes: implications for diabetes skin complications. Diabetes. 2001;50:1627–35.
35. Sadagurski M, Yakar S, Weingarten G, Holzenberger M, Rhodes CJ, Breitkreutz D, et al. Insulin-like growth factor 1 receptor signaling regulates skin development and inhibits skin keratinocyte differentiation. Mol Cell Biol. 2006;26:2675–87.
36. Wertheimer E, Trebicz M, Eldar T, Gartsbein M, Nofeh-Moses S, Tennenbaum T. Differential roles of insulin receptor and insulin-like growth factor-1 receptor in differentiation of murine skin keratinocytes. J Invest Dermatol. 2000;115:24–9.
37. Yoo H, Kim SJ, Kim Y, Lee H, Kim TY. Insulin-like growth factor-II regulates the 12-lipoxygenase gene expression and promotes cell proliferation in human keratinocytes via the extracellular regulatory kinase

and phosphatidylinositol 3-kinase pathways. Int J Biochem Cell Biol. 2007;39:1248–59.
38. Li Y, Fan J, Chen M, Li W, Woodley DT. Transforming growth factor-alpha: a major human serum factor that promotes human keratinocyte migration. J Invest Dermatol. 2006;126:2096–105.
39. Benoliel AM, Kahn-Perles B, Imbert J, Verrando P. Insulin stimulates haptotactic migration of human epidermal keratinocytes through activation of NF-kappa B transcription factor. J Cell Sci. 1997;110(Pt 17):2089–97.
40. Ando Y, Jensen PJ. Epidermal growth factor and insulin-like growth factor I enhance keratinocyte migration. J Invest Dermatol. 1993;100:633–9.
41. Sertznig P, Seifert M, Tilgen W, Reichrath J. Peroxisome proliferator-activated receptors (PPARs) and the human skin: importance of PPARs in skin physiology and dermatologic diseases. Am J Clin Dermatol. 2008;9:15–31.
42. Cheta D. Animal models of type I (insulin-dependent) diabetes mellitus. J Pediatr Endocrinol Metab. 1998; 11:11–9.
43. Imokawa G, Kuno H, Kawai M. Stratum corneum lipids serve as a bound-water modulator. J Invest Dermatol. 1991;96:845–51.
44. Akimoto K, Yoshikawa N, Higaki Y, Kawashima M, Imokawa G. Quantitative analysis of stratum corneum lipids in xerosis and asteatotic eczema. J Dermatol. 1993;20:1–6.
45. Denda M, Hori J, Koyama J, Yoshida S, Nanba R, Takahashi M, et al. Stratum corneum sphingolipids and free amino acids in experimentally-induced scaly skin. Arch Dermatol Res. 1992;284:363–7.
46. Jo N, Watanabe M, Kiyokane K, Shimada M. In vivo microradioautographic study of insulin binding in the skin of normal and NIDDM mice: with special reference to acanthosis nigricans. Cell Mol Biol (Noisy-le-grand). 1997;43:157–64.
47. Toh YC. Effect of streptozotocin-induced diabetes on the activity of the sebaceous glands in rats. Endokrinologie. 1982;80:56–9.
48. Paye M, Mac-Mary S, Elkhyat A, Tarrit C, Mermet P, Humbert PH. Use of the Reviscometer for measuring cosmetics-induced skin surface effects. Skin Res Technol. 2007;13:343–9.
49. Yoon HS, Baik SH, Oh CH. Quantitative measurement of desquamation and skin elasticity in diabetic patients. Skin Res Technol. 2002;8:250–4.
50. Goldschmidt H. Surface area measurements of psoriatic corneocytes: effects of intralesional steroid therapy. J Invest Dermatol. 1979;73:558–60.
51. Watanabe M, Tagami H, Horii I, Takahashi M, Kligman AM. Functional analyses of the superficial stratum corneum in atopic xerosis. Arch Dermatol. 1991;127:1689–92.
52. Corcuff P, Leveque J-L. Size and shape of corneocytes at various body sites: influence of age. In: Leveque J-L, editor. Aging skin. New York: Marcel Dekker; 1993. p. 199–216.
53. Hara M, Kikuchi K, Watanabe M, et al. Senile xerosis. Functional, morphological and biochemical studies. J Geriatr Dermatol. 1993;1:111–20.
54. Yajima Y, Sueki H, Fujisawa R. Increased corneocyte surface area in the diabetic skin. Nihon Hifuka Gakkai Zasshi. 1991;101:129–34.
55. Jolly Jr HW, Hailey CW, Netick J. pH determinations of the skin. Readings under normal and abnormal conditions. J Invest Dermatol. 1961;36:305–8.
56. Korting HC, Kober M, Mueller M, Braun-Falco O. Influence of repeated washings with soap and synthetic detergents on pH and resident flora of the skin of forehead and forearm. Results of a cross-over trial in health probationers. Acta Derm Venereol. 1987; 67:41–7.
57. De D, Handa S. Filaggrin mutations and the skin. Indian J Dermatol Venereol Leprol. 2012;78:545–51.
58. Yosipovitch G, Tur E, Cohen O, Rusecki Y. Skin surface pH in intertriginous areas in NIDDM patients. Possible correlation to candidal intertrigo. Diabetes Care. 1993;16:560–3.
59. Huntley AC. Cutaneous manifestations of diabetes mellitus. Dermatol Clin. 1989;7:531–46.
60. Thyssen JP, Godoy-Gijon E, Elias PM. Ichthyosis vulgaris – the filaggrin mutation disease. Br J Dermatol. 2013;168(6):1155–66.
61. Saleh NM, Raj SM, Smyth DJ, Wallace C, Howson JM, Bell L, et al. Genetic association analyses of atopic illness and proinflammatory cytokine genes with type 1 diabetes. Diabetes Metab Res Rev. 2011; 27:838–43.
62. Douek IF, Leech NJ, Gillmor HA, Bingley PJ, Gale EA. Children with type-1 diabetes and their unaffected siblings have fewer symptoms of asthma. Lancet. 1999;353:1850.
63. Decreased prevalence of atopic diseases in children with diabetes. The EURODIAB Substudy 2 Study Group. J Pediatr. 2000;137:470–4.
64. Olesen AB, Juul S, Birkebaek N, Thestrup-Pedersen K. Association between atopic dermatitis and insulin-dependent diabetes mellitus: a case-control study. Lancet. 2001;357:1749–52.
65. Meerwaldt R, Odink RJ, Landaeta R, Aarts F, Brunekreef B, Gerritsen J, et al. A lower prevalence of atopy symptoms in children with type 1 diabetes mellitus. Clin Exp Allergy. 2002;32:254–5.
66. Rosenbauer J, Herzig P, Giani G. Atopic eczema in early childhood could be protective against type 1 diabetes. Diabetologia. 2003;46:784–8.
67. Stene LC, Joner G. Atopic disorders and risk of childhood-onset type 1 diabetes in individuals. Clin Exp Allergy. 2004;34:201–6.
68. Siemiatycki J, Colle E, Campbell S, Dewar RA, Belmonte MM. Case-control study of IDDM. Diabetes Care. 1989;12:209–16.
69. Mattila PS, Tarkkanen J, Saxen H, Pitkäniemi J, Karvonen M, Tuomilehto J. Predisposition to atopic symptoms to inhaled antigens may protect from childhood type 1 diabetes. Diabetes Care. 2002;25:865–8.
70. Kero J, Gissler M, Hemminki E, Isolauri E. Could TH1 and TH2 diseases coexist? Evaluation of asthma incidence in children with coeliac disease, type 1 diabetes, or rheumatoid arthritis: a register study. J Allergy Clin Immunol. 2001;108:781–3.

71. Simpson CR, Anderson WJ, Helms PJ, Taylor MW, Watson L, Prescott GJ, et al. Coincidence of immune-mediated diseases driven by Th1 and Th2 subsets suggests a common aetiology. A population-based study using computerized general practice data. Clin Exp Allergy. 2002;32:37–42.
72. Stene LC, Nafstad P. Relation between occurrence of type 1 diabetes and asthma. Lancet. 2001;357:607–8.
73. Stromberg LG, Ludvigsson GJ, Bjorksten B. Atopic allergy and delayed hypersensitivity in children with diabetes. J Allergy Clin Immunol. 1995;96:188–92.
74. Abbas AK, Murphy KM, Sher A. Functional diversity of helper T lymphocytes. Nature. 1996;383:787–93.
75. Simon D. Epidemiological features of type 2 diabetes. Rev Prat. 2010;60:469–73.
76. Ahmed I, Goldstein B. Diabetes mellitus. Clin Dermatol. 2006;24:237–46.
77. Henry J, Toulza E, Hsu CY, Pellerin L, Balica S, Mazereeuw-Hautier J, et al. Update on the epidermal differentiation complex. Front Biosci. 2012;17: 1517–32.
78. Thyssen JP, Thuesen B, Huth C, Standl M, Carson CG, Heinrich J, et al. Skin barrier abnormality caused by filaggrin (FLG) mutations is associated with increased serum 25-hydroxyvitamin D concentrations. J Allergy Clin Immunol. 2012;130:1204–7.
79. Proceedings of the 4th international workshop-conference on gestational diabetes mellitus, Chicago, 14–16 Mar 1997. Diabetes Care. 1998;21 Suppl 2:B1-167.
80. Kim C, Newton KM, Knopp RH. Gestational diabetes and the incidence of type 2 diabetes: a systematic review. Diabetes Care. 2002;25:1862–8.
81. Jarvela IY, Juutinen J, Koskela P, Hartikainen AL, Kulmala P, Knip M, et al. Gestational diabetes identifies women at risk for permanent type 1 and type 2 diabetes in fertile age: predictive role of autoantibodies. Diabetes Care. 2006;29:607–12.
82. Bellamy L, Casas JP, Hingorani AD, Williams D. Type 2 diabetes mellitus after gestational diabetes: a systematic review and meta-analysis. Lancet. 2009;373:1773–9.
83. Kumar R, Ouyang F, Story RE, Pongracic JA, Hong X, Wang G, et al. Gestational diabetes, atopic dermatitis, and allergen sensitization in early childhood. J Allergy Clin Immunol. 2009;124:1031–8.

The Eye and Filaggrin

31

Anne Sofie Lund Bateman and Steffen Heegaard

Contents

Filaggrin is an important epidermal barrier protein. Since the eye has barrier functions, it is interesting to look at the role of filaggrin in the eye. The eye region can be divided into extraocular and intraocular structures. The extraocular surface structures comprise the eyelid, the conjunctiva, and the cornea. The intraocular structures with important barrier functions are the corneal endothelium, the ciliary body, and the retinal pigment epithelium (RPE). The blood-aqueous barrier is located in the ciliary body and the blood-retinal barrier in the retina.

31.1 Extraocular Structures and Filaggrin

31.1.1 Eyelid

The eyelids are modified skin with an inner lining of conjunctival epithelium. They form the anterior part of a moist tear-filled chamber that is essential for keeping the cornea moist and thereby transparent. Without eyelids, the corneal epithelium will keratinize. The eyelids form a protective barrier between the cornea and the outer surroundings. In one study, immunostainings were carried out on epithelium from the skin of the eyelid using impression cytology [1]. The study showed filaggrin staining in the squamous epithelium of the skin. This may also be demonstrated using routine immunohistology (Fig. 31.1).

A.S.L. Bateman, MD
Department of Neuroscience and Pharmacology, Eye Pathology Institute, University of Copenhagen, Frederik V's Vej 11, Copenhagen 2100, Denmark
e-mail: kragegaardlund@gmail.com

S. Heegaard, PhD (✉)
Department of Neuroscience and Pharmacology, Eye Pathology Institute, University of Copenhagen, Frederik V's Vej 11, Copenhagen 2100, Denmark

Department of Ophthalmology, Glostrup Hospital, University of Copenhagen, Frederik V's Vej 11, Copenhagen 2100, Denmark
e-mail: sthe@sund.ku.dk

J.P. Thyssen, H. Maibach (eds.), *Filaggrin*,
DOI 10.1007/978-3-642-54379-1_31, © Springer-Verlag Berlin Heidelberg 2014

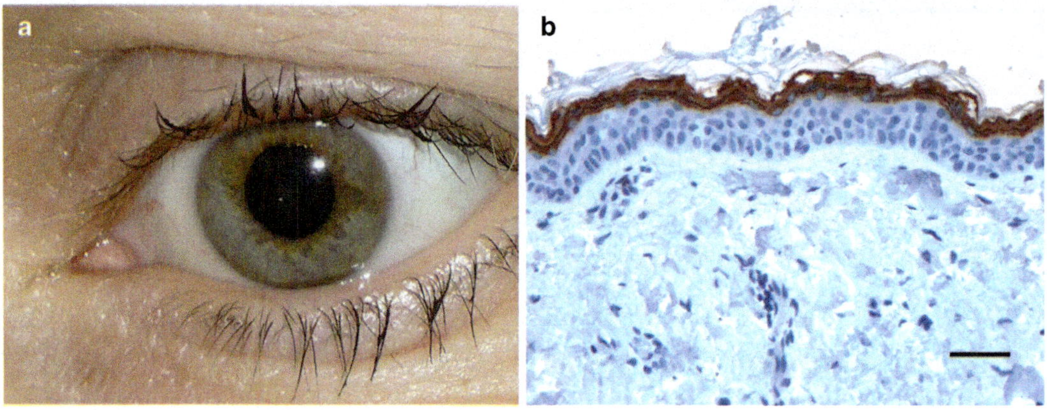

Fig. 31.1 (**a**) The eye with normal skin on the upper and lower eyelids. (**b**) The eyelid shows marked expression (brown staining) of filaggrin in the keratinized layers of the epithelium (bar = 50 μm)

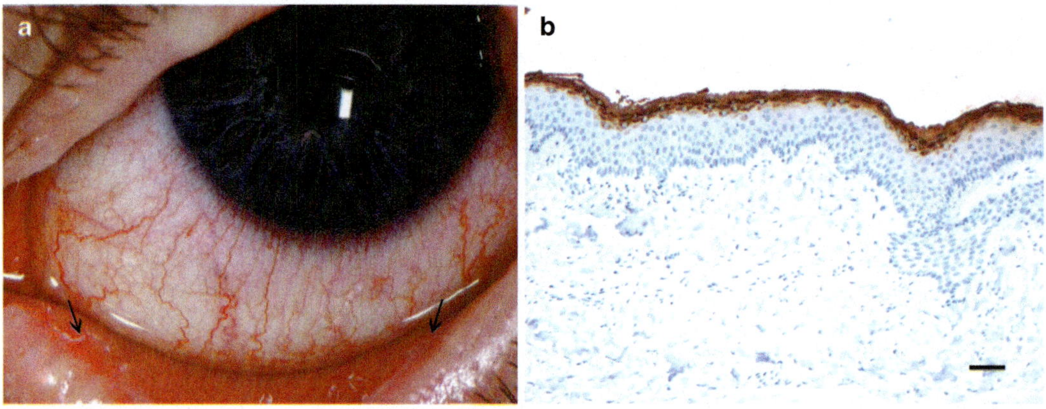

Fig. 31.2 (**a**) Late stage of Stevens-Johnson syndrome with metaplasia of the eyelid margin (*arrows*). (**b**) A marked binding of filaggrin antibodies to the keratinized superficial layers of the epithelium is seen (bar = 50 μm)

31.1.2 Conjunctiva

The conjunctiva covers the anterior surface of the eyeball and the posterior surface of the eyelids. It is a mucous membrane consisting of a two-to-five-cell-thick nonkeratinized stratified columnar epithelium with goblet cells and an underlying stroma. Lymphocytes and plasma cells are part of the defense mechanism of the eye and are normally found in the stroma. The conjunctiva is an important barrier to outside surroundings.

Most studies show no expression of filaggrin in normal conjunctival epithelium [1–4]. One study showed filaggrin in the mucocutaneous junction at the eyelid margin [1], and another study showed expression of filaggrin in the entire conjunctiva by using immunohistochemical procedures without verifying this by other methods [5]. They found filaggrin in the superficial cells of the normal bulbar conjunctival epithelium, but not in the suprabasal and basal cells [5]. The same study showed no expression of filaggrin in one case of conjunctival intraepithelial neoplasia. The authors concluded that the absence of filaggrin indicates an abnormal dedifferentiation of dysplastic cells. In contrast to this study, another study found a marked expression of filaggrin in a parakeratinized conjunctiva with moderate dysplasia [4].

The expression of filaggrin in other diseases of the conjunctiva has been investigated. It has been shown in Stevens-Johnson syndrome (SJS) (Fig. 31.2), in ocular cicatricial pemphigoid, in alkali injuries in the chronic cicatricial phase, and in one case of moderate dysplasia of the conjunctiva [2, 4]. These conditions are

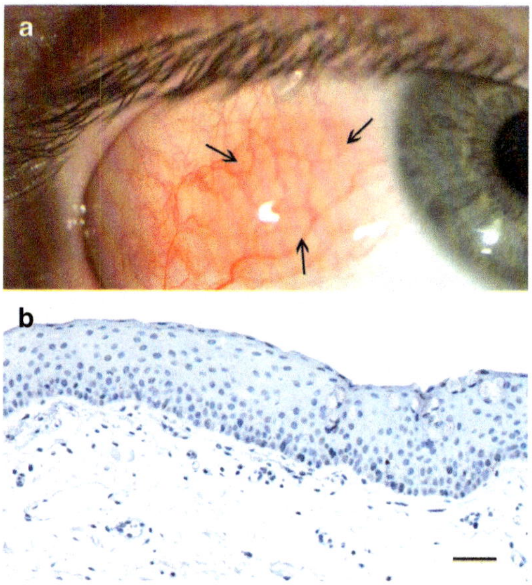

Fig. 31.3 (**a**) A pinguecula (*arrows*), an elastoid degeneration of the stroma of the conjunctiva. (**b**) The similar micrograph stained with anti-filaggrin. Filaggrin is not expressed (bar = 50 μm)

characterized by severe damage to the conjunctiva with different degrees of keratinization in the surface epithelia. When the conjunctival epithelium keratinizes, it changes with a loss of goblet cells and a conversion from stratified columnar to stratified squamous configuration [6].

We found no expression of filaggrin in the conjunctiva of patients with pinguecula and pterygium, both with milder degrees of parakeratinization (Fig. 31.3). Therefore, it has been concluded that filaggrin is only expressed in conjunctivas with severe forms of parakeratinization [4].

The reason why the expression of filaggrin is upregulated in the conjunctiva is still unknown. It may simply be a result of the metaplastic changes of the epithelium.

Chronic allergic keratoconjunctivitis (CAK) is also associated with atopic diseases such as atopic dermatitis (AD); however, no association between filaggrin gene (*FLG*) mutations and CAK has been found [7].

31.1.3 Cornea

The anterior surface of the cornea consists of a five-cell-thick nonkeratinized epithelium.

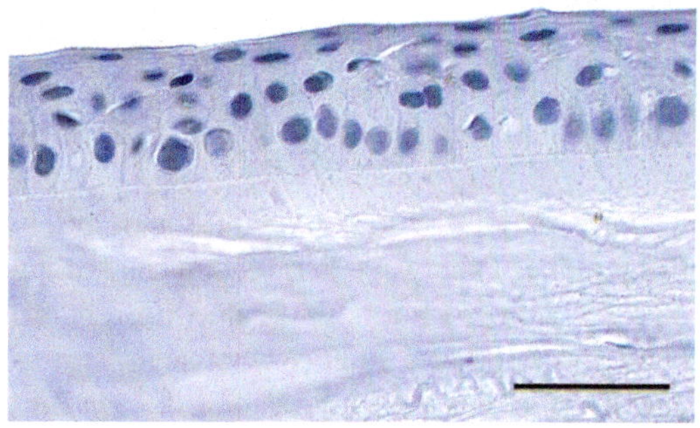

Fig. 31.4 The normal cornea does not express filaggrin (bar = 50 μm)

Bowman's layer separates the epithelium from the stroma. The posterior part of the cornea is composed of a monolayer of endothelial cells resting on Descemet's membrane. Since the anterior part of the cornea is facing the outer surroundings, it has a barrier function.

Two studies have shown a weak expression of filaggrin in the normal corneal epithelium by using Western blot [8, 9]. The first study also showed filaggrin in the corneal epithelium by immunofluorescence staining, whereas the latter found no staining. The latter study is supported by another study [5] and more stainings (Fig. 31.4).

One study found filaggrin expression in squamous metaplasia of a vascularized pannus removed from the corneal surface inflicted with SJS, aniridia, chemical burn, and recurrent pterygium [10]. This is consistent with findings in keratinized conjunctiva [2, 4].

Another study investigated the expression of filaggrin in the corneal filaments in filamentary keratitis. No expression of filaggrin was found [11].

Atopic diseases have been associated with keratoconus in several studies [12–17]. These studies found that between 20 and 47 % of patients with keratoconus had AD depending on the study design and patient recruitment. In keratoconus, the cornea degenerates and becomes thinner, leading to protrusion of the thinned cornea. This causes severe myopia and irregular astigmatism, affecting quality of vision. In severe cases, it may lead to acute stromal edema—known as hydrops—significant pain, and vision loss. One study hypothesized that the skin and cornea had a common dysfunction linked to *FLG* mutations [9]. This could explain the weakening that causes the thinning of the cornea as well as the impaired vision in patients with keratoconus and thereby the association between keratoconus and atopic disease. However, no significant difference in the expression of filaggrin between control and keratoconus cornea was found [9]. Both showed a very weak expression of filaggrin. Whether the association between atopic disease and keratoconus is caused by environmental or underlying genetic factors such as *FLG* mutations or excessive eye rubbing is still unknown.

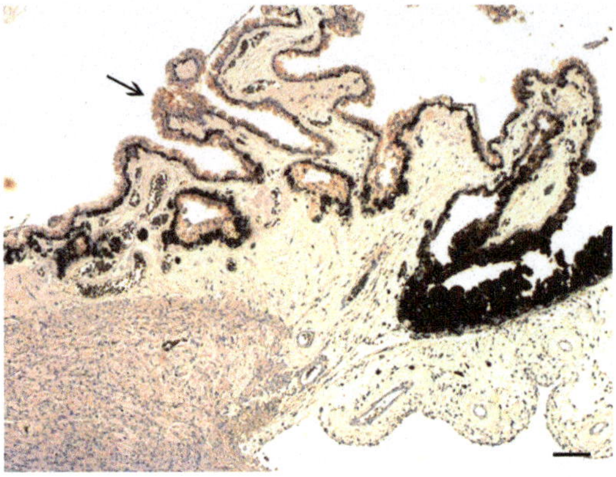

Fig. 31.5 The normal ciliary body. Note the expression of anti-filaggrin in the nonpigmented epithelium (*arrows*; bar = 100 μm)

31.2 Intraocular Structures and Filaggrin

The ciliary body and the retinal pigment epithelium make up the two most important intraocular barriers. The blood-aqueous barrier is mainly controlled by a ciliary epithelial bilayer consisting of the pigmented ciliary epithelial layer facing the ciliary stroma and the nonpigmented ciliary epithelial layer facing the aqueous humor.

The blood-retinal barrier is located at two levels, forming an outer barrier in the retinal pigment epithelium and an inner barrier in the endothelial membrane of the retinal vessels [18]. Formation and maintenance of the blood-aqueous barrier and the blood-retinal barrier are required for proper vision, and loss of these barriers contributes to the pathology of a wide number of ophthalmological diseases such as diabetes and systemic hypertension [19, 20].

Immunofluorescence staining shows a presence of filaggrin in the nonpigmented epithelium of the ciliary body (Fig. 31.5). It is an interesting finding that needs further investigation. Filaggrin is not found in the RPE or intraocularly.

Filaggrin in ophthalmology has only been investigated relatively recently. Some studies have focused on whether a skin disease—associated with *FLG* mutations—also has ocular surface lesions associated with *FLG* mutations. But this has so far not been the case. In all previous studies, only a little filaggrin has been found in the healthy cornea and conjunctiva compared to the amounts found in the skin. A reason why no or very little filaggrin is found could be that the antibodies used do not bind the type of filaggrin found in the eye because the antigenicity of the filaggrin in the eye might be different from filaggrin in the skin. Two studies have found filaggrin in the cornea by Western blot [8, 9], but only one was supported by immunostainings [8]. Only one study has shown the presence of filaggrin in the conjunctiva by using only immunostainings [5]. Both studies found a very weak expression. Therefore, an *FLG* mutation would probably be almost unnoticeable in immunohistochemical stainings of the healthy conjunctiva and cornea. A severe parakeratinization is therefore necessary before the expression of filaggrin becomes clearly visible. Therefore, it is questionable if *FLG* mutations could have any effect on the development of ocular diseases.

If an association between *FLG* mutations and an ocular disease is found, could it then be compared to the association between *FLG* mutations and allergic asthma? In allergic asthma, *FLG* mutations together with AD are risk factors for developing asthma [21]. Similar to the eye region, where only little filaggrin is found, no filaggrin is found in the respiratory epithelium [22]. The hypothesis is that AD causes systemic allergen sensitization, which, together with *FLG* mutations, is a risk factor of developing asthma [21]. This hypothesis may be comparable to diseases involving the conjunctiva and *FLG* mutations since the conjunctiva is part of the conjunctival and lacrimal gland-associated lymphoid tissue (CALT) system. CALT is part of the widespread mucosa-associated lymphoid tissue (MALT), which also includes the bronchus-associated lymphoid tissue (BALT). The possible pathomechanism is that allergens penetrate the impaired skin barrier, resulting in primary sensitization, and then secondary allergen exposure in the conjunctiva causes allergic conjunctivitis.

References

1. Jalbert I, Madigan MC, Shao M, Ng J, Cheng J, Wong D, et al. Assessing the human lid margin epithelium using impression cytology. Acta Ophthalmol. 2012;90(7):e547–52.
2. Nakamura T, Nishida K, Dota A, Matsuki M, Yamanishi K, Kinoshita S. Elevated expression of transglutaminase 1 and keratinization-related proteins in conjunctiva in severe ocular surface disease. Invest Ophthalmol Vis Sci. 2001;42(3):549–56.
3. Krenzer KL, Freddo TF. Cytokeratin expression in normal human bulbar conjunctiva obtained by impression cytology. Invest Ophthalmol Vis Sci. 1997;38(1):142–52.
4. Lund AS, Heegaard S, Prause JU, Toft PB, Skov L. Expression of filaggrin in normal and keratinized conjunctiva. Open Ophthalmol J. 2012;6:137–40.
5. Saika S, Minamide A, Tanaka T, Miyamoto T, Okada Y, Kawashima Y, et al. Expression of involucrin by ocular surface epithelia of patients with benign and malignant disorders. Curr Eye Res. 2000;21(5):877–85.
6. Spencer WH. Ophthalmic pathology, an atlas and textbook. 3rd ed. Philadelphia: W.B. Saunders; 1985.
7. Iwamoto S, Ebihara N, Hori K, Funaki T, Asada Y, Yokoi N, et al. Filaggrin mutations are not associated with chronic allergic keratoconjunctivitis. Br J Ophthalmol. 2012;96(9):1272–3.
8. Tong L, Corrales RM, Chen Z, Villarreal AL, De Paiva CS, Beuerman R, et al. Expression and regulation of cornified envelope proteins in human corneal epithelium. Invest Ophthalmol Vis Sci. 2006;47(5):1938–46.
9. Droitcourt C, Touboul D, Ged C, Ezzedine K, Cario-Andre M, de Verneuil H, et al. A prospective study of filaggrin null mutations in keratoconus patients with or without atopic disorders. Dermatology. 2011; 222(4):336–41.
10. Li W, Chen YT, Hayashida Y, Blanco G, Kheirkah A, He H, et al. Down-regulation of Pax6 is associated with abnormal differentiation of corneal epithelial cells in severe ocular surface diseases. J Pathol. 2008;214(1):114–22.
11. Tanioka H, Yokoi N, Komuro A, Shimamoto T, Kawasaki S, Matsuda A, et al. Investigation of the corneal filament in filamentary keratitis. Invest Ophthalmol Vis Sci. 2009;50(8):3696–702.
12. Copeman PW. Eczema and keratoconus. Br Med J. 1965;2(5468):977–9.
13. Gasset AR, Hinson WA, Frias JL. Keratoconus and atopic diseases. Ann Ophthalmol. 1978;10(8): 991–4.
14. Harrison RJ, Klouda PT, Easty DL, Manku M, Charles J, Stewart CM. Association between keratoconus and atopy. Br J Ophthalmol. 1989;73(10):816–22.
15. Ohmachi N, Sasabe T, Kojima M, Adachi J, Endo K, Fukuzumi T, et al. Eye complications in atopic dermatitis. Arerugi. 1994;43(7):796–9.
16. Bawazeer AM, Hodge WG, Lorimer B. Atopy and keratoconus: a multivariate analysis. Br J Ophthalmol. 2000;84(8):834–6.
17. Kaya V, Karakaya M, Utine CA, Albayrak S, Oge OF, Yilmaz OF. Evaluation of the corneal topographic characteristics of keratoconus with orbscan II in patients with and without atopy. Cornea. 2007; 26(8):945–8.
18. Tomi M, Hosoya K. The role of blood-ocular barrier transporters in retinal drug disposition: an overview. Expert Opin Drug Metab Toxicol. 2010;6(9): 1111–24.
19. Krogsaa B, Lund-Andersen H, Mehlsen J, Sestoft L. Blood-retinal barrier permeability versus diabetes duration and retinal morphology in insulin dependent diabetic patients. Acta Ophthalmol (Copenh). 1987; 65(6):686–92.
20. Krogsaa B, Lund-Andersen H, Parving HH, Bjaeldager P. The blood-retinal barrier permeability in essential hypertension. Acta Ophthalmol (Copenh). 1983;61(4):541–4.
21. McLean WH, Palmer CN, Henderson J, Kabesch M, Weidinger S, Irvine AD. Filaggrin variants confer susceptibility to asthma. J Allergy Clin Immunol. 2008;121(5):1294–5; author reply 5–6.
22. Ying S, Meng Q, Corrigan CJ, Lee TH. Lack of filaggrin expression in the human bronchial mucosa. J Allergy Clin Immunol. 2006;118(6):1386–8.

32 Filaggrin Dysfunction and Its Association with Inflammatory Conditions of the GI Tract

Andreas P. Thyssen and Erik P. Thyssen

Contents

A.P. Thyssen, BS, MD (✉)
Internal Medicine/Pediatrics, Maine Medical Center,
22 Bramhall Street, Portland, ME 4102, USA
e-mail: andreas.thyssen@gmail.com

E.P. Thyssen, MD
Specialists in Gastroenterology,
11525 Olde Cabin Rd., St. Louis, MO 63141, USA
e-mail: et@gidoctor.net

32.1 Introduction

The gastrointestinal (GI) tract plays many vital roles in maintaining host homeostasis. Beyond its digestive role of taking up water and nutrients, the GI tract also serves as a highly specialized and regulated immune organ with critical barrier functions. New investigation into the mechanics of the gut barrier and its dysfunction is helping to explain the pathogenesis of previously unexplained diseases. Two such pathologies that investigators are gaining a better understanding of through a new working knowledge of the gut barrier are inflammatory bowel disease (IBD) and eosinophilic esophagitis (EoE).

32.2 Inflammatory Bowel Disease

IBD encompasses two chronic, idiopathic intestinal disorders: Crohn's disease (CD) and ulcerative colitis (UC). Although these diseases affect people across the entire globe, there is a higher recognized burden of disease in Northern Europe and North America, where it is estimated that 2.2 million and 1.4 million people are affected, respectively [1, 2]. The clinical course and phenotypes are different in these two disorders, though the pathophysiology is thought to be similar. CD is an inflammatory process that can affect the intestinal wall of any part of the GI tract. The terminal ileum and colon are usually involved, though not always, and the inflammation occurs along the GI tract

J.P. Thyssen, H. Maibach (eds.), *Filaggrin*,
DOI 10.1007/978-3-642-54379-1_32, © Springer-Verlag Berlin Heidelberg 2014

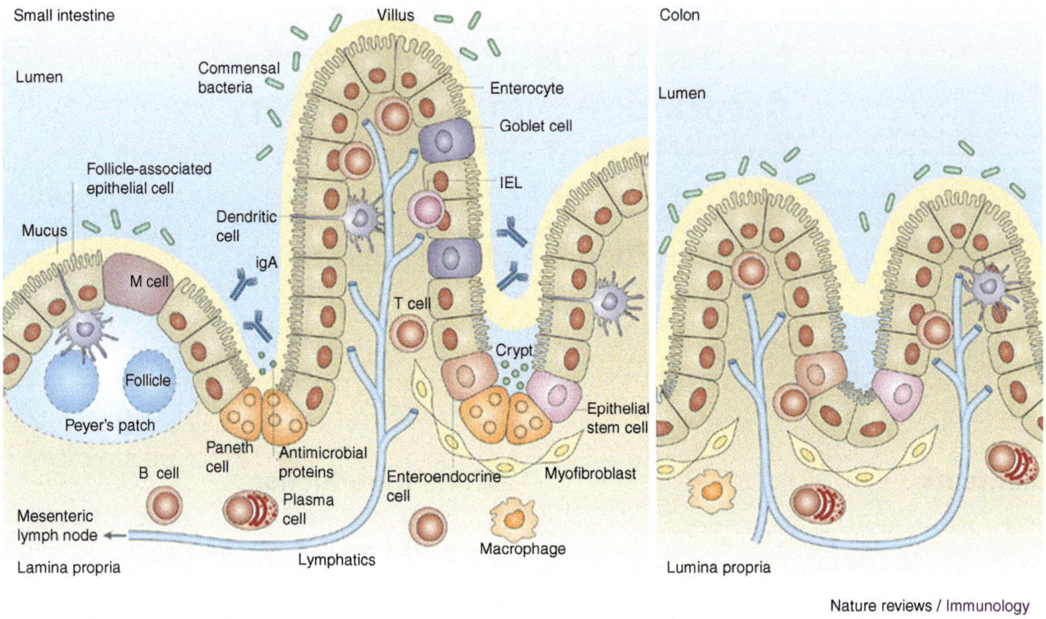

Fig. 32.1 The gastrointestinal tract is a highly active immune organ with critical barrier function. In healthy individuals, an intact gastrointestinal epithelium serves to limit the transit of pathogens across gut barrier. In inflammatory bowel disease, this barrier is defective and inflammation results from the increased permeability to gut bacteria (Reprinted by permission from Abreu [24], Macmillan Publisher Ltd, copyright 2010)

in discontinuous segments, leading to "skip lesions" [2, 3]. The inflammation in CD is typically transmural through the gut wall, which can lead to intestinal complications such as strictures, abscesses, and fistula formation [2–4]. This is in contrast to UC, which affects the GI tract from the rectum to the colon, in a continuous, uninterrupted fashion with gut wall inflammation being limited to the mucosa [2–4]. Stricture, abscess, and fistula formation are rarely seen in UC [2]. Both UC and CD can have extraintestinal disease manifestations, which can include musculoskeletal, oral, hepatobiliary, or dermal manifestations [2].

Although the exact mechanism underlying inflammatory bowel disease is unknown, new research is pointing toward an increased permeability of the GI tract toward gut flora [5]. This barrier dysfunction leads to a dysregulation between the host immune system and gut microbiome and a subsequent pro-inflammatory state [2, 5]. The resulting chronic and unchecked inflammation is thought to lead to the complications of CD and UC. The intestinal epithelium constitutes the main component of the gut barrier and serves as a physical barrier between intraluminal organisms and the circulation [2]. An intact GI barrier is mainly dependent on the proper functioning of tight junctions, which create a seal between epithelial cells and limit the transit of various compounds across the paracellular space (Fig. 32.1) [2]. In IBD, this seal becomes defective, which leads to an increase in the epithelial permeability to the gut flora. There are various hypotheses as to what causes the defect in the gut barrier, including environmental and genetic causes [2]. Some mouse models of colitis are illuminating how structural protein breakdown could be the trigger for barrier collapse and subsequent increase in barrier permeability and inflammation [5, 6]. One such study by Haptezion et al. showed that mice deficient in keratin-8, the major intermediate filament in the intestinal epithelium, developed a chronic, spontaneous, nonlethal colitis, similar in character to CD [7]. These discoveries have led to a wider genomic search for other barrier proteins that might be dysfunctional in cases of IBD.

32.3 Filaggrin and Inflammatory Bowel Disease

Like the gut barrier, the crucial role of an intact skin barrier has become apparent in recent years. The dysfunction and breakdown of the skin barrier has been linked to a host of dermal disorders [8]. The proper functioning of one particular protein, filaggrin, and its metabolites seems to be of particular relevance for normal skin barrier function. Because of its role in dermatologic pathology, common loss-of-function mutations in the filaggrin gene (*FLG*) have been hypothesized to play a role in the breakdown of barrier function observed in IBD. However, there is some controversy as to whether or not filaggrin is even expressed in gastrointestinal epithelium, and in reality, its expression likely varies throughout the tract. One study by De Bendetto et al. showed that filaggrin is not expressed in nasal or esophageal mucosa but is expressed in oral mucosa [9]. Other reports from online genomic databases show that filaggrin is moderately expressed in oral and esophageal epithelium and mildly in the lower stomach and colon endothelium [10, 11]. To date, only two studies have investigated the association of *FLG* mutations and IBD, neither of which has found an association. The first study was conducted on a cohort of 681 patients with CD and 427 with UC from Northern Germany [12, 13]. There was no association found between disease and *FLG* mutations [13]. A second study of a cohort of 378 Scottish children with IBD found that 40 of the children (10.6 %) were found to be carriers of one or more *FLG* null alleles, compared to 103/944 (10.9 %) of controls [14]. Although no link has yet been established between *FLG* mutations and the development of IBD, it remains possible that filaggrin, if expressed, may affect gut barrier functions and alter the risk of various GI disorders. In the skin, a connection between impaired tight junctions and dysfunctional filaggrin has been shown [15, 16] and, if present in the GI tract, could be a potential mechanism by which *FLG* mutation contributes to the development of IBD.

32.4 Eosinophilic Esophagitis

IBD is not the only GI pathology linked to barrier dysfunction. Indeed, there is evidence to suggest that *FLG* mutations are associated with the development of eosinophilic esophagitis (EoE). EoE, once thought to be a complication of gastroesophageal reflux disease (GERD), is now recognized as its own clinical entity [17, 18]. An expert panel defined EoE as "a chronic, immune/antigen-mediated, esophageal disease characterized clinically by symptoms related to esophageal dysfunction and histologically by eosinophil-predominant inflammation" [18]. A systematic review of the clinical signs and symptoms of EoE reported that the predominant presenting features are dysphagia to solid foods (93 %), food impaction (61.9 %), and heartburn (23.6 %) [19]. In the same review, the most common endoscopic findings were mucosal fragility/edema (59.3 %), rings or corrugated esophagus (49.2 %), and strictures (39.7 %) [19]. Due to the relatively new recognition of EoE, there remains a paucity of data on the natural progression of the disease. However, in one study of 30 untreated patients, dysphagia persisted in 97 % at 7 years of follow-up [20]. The pathophysiology of EoE remains unknown and is likely the result of both environmental and genetic factors.

32.5 Filaggrin and Eosinophilic Esophagitis

A study by Carine Blanchard et al. found that in a cohort of 144 patients with active EoE, there was a 16-fold down regulation of filaggrin mRNA in esophageal biopsies compared to normal controls [21]. In the same study, 6.1 % of patients with EoE had a *FLG* loss-of-function mutation versus only 1.3 % of normal controls ($p=.0172$) [21]. A more recent study on a cohort of 47 EoE patients found that filaggrin was greatly underexpressed compared to controls [22]. Specifically, filaggrin expression was positive in 88 % of controls, in 100 % of GERD patients, and in 0 % of patients with EoE ($p<.001$) [22]. While *FLG* mutation likely does not explain the entire disease process of EoE, it shows how GI pathology is indeed linked to this barrier protein and its dysfunction.

Conclusion

The importance of a properly functioning barrier between the internal organs and outside world is becoming more and more apparent, as researchers continue to find evidence that chronic, systemic diseases are associated with barrier breakdown. Continued research efforts must be placed on learning more about barrier proteins and their role in maintaining proper structure. In particular, further studies are needed to determine the extent to which filaggrin is expressed in the GI tract, with special attention focused on the esophagus and colon. To date, studies have only looked at genetic mutations of the *FLG*; however, future studies are needed to investigate the possibility of post-translational changes of filaggrin or even related proteins, such as filaggrin-2 [23]. Gleaning more information about filaggrin and its related proteins will help give us a better understanding of the integrity of the GI epithelium and the disease processes related to its breakdown.

References

1. Loftus Jr EV, Sandborn WJ. Epidemiology of inflammatory bowel disease. Gastroenterol Clin North Am. 2002;31(1):1–20.
2. Abraham C, Cho JH. Inflammatory bowel disease. N Engl J Med. 2009;361(21):2066–78.
3. Waye JD. The role of colonoscopy in the differential diagnosis of inflammatory bowel disease. Gastrointest Endosc. 1977;23(3):150–4.
4. Schwartz DA, Loftus EV, Tremaine WJ, Panaccione R, Harmsen WS, Zinsmeister AR, et al. The natural history of fistulizing Crohn's disease in Olmsted County, Minnesota. Gastroenterology. 2002;122(4):875–80.
5. Turner JR. Molecular basis of epithelial barrier regulation: from basic mechanisms to clinical application. Am J Pathol. 2006;169(6):1901–9.
6. Kaser A, Lee AH, Franke A, Glickman JN, Zeissig S, Tilg H, et al. XBP1 links ER stress to intestinal inflammation and confers genetic risk for human inflammatory bowel disease. Cell. 2008;134(5):743–56.
7. Habtezion A, Toivola DM, Butcher EC, Omary MB. Keratin-8-deficient mice develop chronic spontaneous Th2 colitis amenable to antibiotic treatment. J Cell Sci. 2005;118(Pt 9):1971–80.
8. de Cid R, Riveira-Munoz E, Zeeuwen PL, Robarge J, Liao W, Dannhauser EN, et al. Deletion of the late cornified envelope LCE3B and LCE3C genes as a susceptibility factor for psoriasis. Nat Genet. 2009;41(2):211–15.
9. De Benedetto A, Qualia CM, Baroody FM, Beck LA. Filaggrin expression in oral, nasal, and esophageal mucosa. J Invest Dermatol. 2008;128(6):1594–7.
10. Available from: The human protein atlas. Web site: http://www.proteinatlas.org/ENSG00000143631/normal. Cited 1 Apr 2013.
11. Available from: neXtProt. Web site: http://www.nextprot.org/db/entry/NX_P20930/expression. Cited 1 Apr 2013.
12. Stoll M, Corneliussen B, Costello CM, Waetzig GH, Mellgard B, Koch WA, et al. Genetic variation in DLG5 is associated with inflammatory bowel disease. Nat Genet. 2004;36(5):476–80.
13. Ruether A, Stoll M, Schwarz T, Schreiber S, Fölster-Holst R. Filaggrin loss-of-function variant contributes to atopic dermatitis risk in the population of Northern Germany. Br J Dermatol. 2006;155(5):1093–4.
14. Van Limbergen J, Russell RK, Nimmo ER, Zhao Y, Liao H, Drummond HE, et al. Filaggrin loss-of-function variants are associated with atopic comorbidity in pediatric inflammatory bowel disease. Inflamm Bowel Dis. 2009;15(10):1492–8.
15. Yuki T, Komiya A, Kusaka A, Kuze T, Sugiyama Y, Inoue S. Impaired tight junctions obstruct stratum corneum formation by altering polar lipid and profilaggrin processing. J Dermatol Sci. 2013;69(2):148–58.
16. Sugawara T, Iwamoto N, Akashi M, Kojima T, Hisatsune J, Sugai M, et al. Tight junction dysfunction in the stratum granulosum leads to aberrant stratum corneum barrier function in claudin-1-deficient mice. J Dermatol Sci. 2013;70(1):12–8.
17. Winter HS, Madara JL, Stafford RJ, Grand RJ, Quinlan JE, Goldman H. Intraepithelial eosinophils: a new diagnostic criterion for reflux esophagitis. Gastroenterology. 1982;83(4):818–23.
18. Liacouras CA, Furuta GT, Hirano I, Atkins D, Attwood SE, Bonis PA, et al. Eosinophilic esophagitis: updated consensus recommendations for children and adults. J Allergy Clin Immunol. 2011;128(1):3–20.e6; quiz 1–2.

19. Sgouros SN, Bergele C, Mantides A. Eosinophilic esophagitis in adults: a systematic review. Eur J Gastroenterol Hepatol. 2006;18(2):211–17.
20. Straumann A, Spichtin HP, Grize L, Bucher KA, Beglinger C, Simon HU. Natural history of primary eosinophilic esophagitis: a follow-up of 30 adult patients for up to 11.5 years. Gastroenterology. 2003;125(6):1660–9.
21. Blanchard C, Stucke EM, Burwinkel K, Caldwell JM, Collins MH, Ahrens A, et al. Coordinate interaction between IL-13 and epithelial differentiation cluster genes in eosinophilic esophagitis. J Immunol. 2010;184(7):4033–41.
22. Matoso A, Mukkada VA, Lu S, Monahan R, Cleveland K, Noble L, et al. Expression microarray analysis identifies novel epithelial-derived protein markers in eosinophilic esophagitis. Mod Pathol. 2013;26(5):665–76.
23. Irvine AD, McLean WHI, Leung DYM. Filaggrin mutations associated with skin and allergic diseases. N Engl J Med. 2011;365(14):1315–27.
24. Abreu MT. Toll-like receptor signaling in the intestinal epithelium: how bacterial recognition shapes intestinal function. Nat Rev Immunol. 2010;10(2):131–44.

33 Filaggrin, Human Papillomavirus, and the Cervix

Simon W. Fage, Thomas Iftner, Jacob P. Thyssen, and Susanne K. Kjaer

Contents

S.W. Fage, MD (✉)
Department of Dermato-Allergology,
National Allergy Research Centre, Gentofte Hospital,
University of Copenhagen, Niels Andersens Vej 65,
Hellerup 2900, Denmark
e-mail: simon.w.fage@gmail.com

T. Iftner, Dr rer nat (PhD)
Division of Experimental Virology,
Institute for Medical Virology and Epidemiology
of Viral Diseases, University Hospital,
Eberhard-Karls-Universität Tübingen,
Elfriede-Aulhorn-Str. 6, Tuebingen 72076, Germany
e-mail: thomas.iftner@med.uni-tuebingen.de

J.P. Thyssen, MD, PhD
Department of Dermato-Allergology,
National Allergy Research Centre,
Gentofte Hospital, University of Copenhagen,
Niels Andersens Vej 65, Hellerup 2900, Denmark
e-mail: jacob.p.thyssen@regionh.dk

S.K. Kjaer, MD, DMSc
Unit of Virus, Lifestyle and Genes,
Gynecologic Clinic, Juliane Marie Centre,
Rigshospitalet, and Danish Cancer Research Center,
Strandboulevarden 49, Copenhagen
DK-2100, Denmark
e-mail: susanne@cancer.dk

33.1 Introduction

While little is known about the association between filaggrin expression and diseases of the human cervix uteri, a few studies have demonstrated an altered expression of filaggrin and the precursor pro-filaggrin in HPV-infected cervical tissue, in different stages of cervical intraepithelial neoplasia (CIN), and in cervical cancer [2, 6, 7, 9, 10].

Filaggrin is a 37 kD cytoplasmatic protein that is synthesized by maturating squamous epithelial cells and derived from the much larger precursor pro-filaggrin, which is the main constituent of the keratohyalin granules in the granular cell layers [27, 28]. In the skin, the essential role of filaggrin is partially to enable the aggregation of keratin filaments into macrofibrils to form the cellular cytoskeleton as well as the generation of natural moisturizing factors (NMFs). Skin pH is also influenced by filaggrin expression, since some of the metabolites, which are released by degradation of filaggrin on the skin surface, are acidic [1, 12, 29, 30]. Hence, filaggrin is important for terminal differentiation and homeostasis of the epidermis and is known to play an important role for skin barrier functions [1, 2, 9, 12, 27, 31, 32].

Several studies have been made on the association between filaggrin gene (*FLG*) mutations, resulting in lower levels of filaggrin and its metabolites, and cutaneous and non-cutaneous disorders [1, 4, 33, 34]. Genotyping of the *FLG* has shown that loss-of-function mutations are common in Europeans and Asians with a median

J.P. Thyssen, H. Maibach (eds.), *Filaggrin*,
DOI 10.1007/978-3-642-54379-1_33, © Springer-Verlag Berlin Heidelberg 2014

prevalence of 7.7 % (2.7–14.2) and 3.0 % (0–7.3), respectively [1]. Whether the high prevalence of *FLG* mutations and thus potential insufficient barrier function of the squamous epithelium of the human cervix also play a role for cervical disorders such as HPV infection, CIN, and cancer is currently unknown. This chapter provides an overview of selected cervical disorders and discusses the possible role of filaggrin.

33.2 The Cervical Barrier

The vaginal and cervical mucosa provides the first line of physical and immunological defense against sexually transmitted pathogens [14]. The cervical mucosa is covered with stratified squamous epithelium adherent to the vagina (exocervix) and with columnar epithelium adherent to corpus uteri (endocervix). In between is the transformation zone (T-zone), where the columnar epithelium undergoes metaplasia to squamous epithelium. The position of the T-zone varies with age, hormone exposure, and parity [35, 36]. The squamous epithelium is similar to the epidermis covering most of the body's surface, with a single layer of basal cells (stratum basale), overlying layers of spinous cells (stratum spinosum), and granular cells (stratum granulosum) [36]. Proteins on the surface of adjacent cells can interact to form intercellular junctions (ICJs), which make the cells able to communicate with each other and prevent toxins and pathogens from entering the underlying tissue, vasculature, or immune system [37].

A final layer of cornified cells and keratin (stratum corneum) is found on most stratified epithelium and is thought to play an important role for epithelial barrier functions [37–40]. Filaggrin is an essential factor in the formation of the superficial layers as shown in several studies. However, in a study by Minh H. Dinh et al. on cervical, vaginal, and penile tissue from adult human donors, they found that a true cornified layer and filaggrin expression were absent or decreased in female tissue, but clearly visible in male tissue [3]. This could in part be relevant to the observation that women are more susceptible to sexually transmittable infections (STIs), for example to HPV, than men [41, 42] and may emphasize the potential importance of filaggrin in barrier functions of the cervical mucosa. While the single layer of columnar cells adherent to the uterus suggests increased vulnerability to pathogens, several antimicrobial factors, along with less exposure and a thick mucus containing immune cells and immunoglobulins overlying the epithelium, are secreted at higher levels to protect this surface [3, 43].

The role of keratinization and the thickness of epithelia in vaginal transmission have been explored in both animals and humans. Thus, progesterone treatment of rhesus macaques yielded a thinner vaginal epithelium and greater rates of simian immunodeficiency virus (SIV) infection [44], whereas the opposite effect was seen in estrogen-treated ovariectomized rhesus macaques [45, 46]. Similar results have been observed in mice treated with hormones and challenged with HSV-2 [47], while studies on human subjects and progesterone have failed to demonstrate significant changes [48–50]. The effects of hormonal therapy and mucosal keratin layers are therefore largely unknown with regard to the risk of STIs.

Some infectious agents will not get through an intact and normal mucosa and are believed to get access to the underlying tissue through wounds and microabrasions. This is thought to be the case for HPV infections [15, 16, 51, 52]. Studies have demonstrated a decreased resistance of filaggrin-deficient skin to stress, for example stretch, resulting in keratinocyte disruption and reduced integrity [12, 19, 20]. In relation to the cervix, it is thus possible, at least in theory, that filaggrin deficiency due to *FLG* mutations could increase the risk of HPV infection through microwounds and ultimately the development of neoplasia. Furthermore, it has been shown that inflammation in the skin and exposure of keratinocytes to certain cytokines, among others IL-17, IL-25, and IL-22, reduce filaggrin and profilaggrin expression and mRNA along with other factors of differentiation and barrier function [21–26]. The latter includes downregulation of genes involved in processing filaggrin into amino acids and the generation of NMFs, adhesion molecules, and tight junction proteins. These cytokines have also been suggested to exert their effect in other organs and surfaces of the body [21] and are not necessarily produced by cells at the main site of inflammation in the skin or in the skin at all in atopic dermatitis [23, 25]. Similar effects could thus be present in cervical epithelium.

A vaginal pH range of 4.0–4.5 is considered normal for premenopausal women, whereas an increase occurs after menopause. It is primarily determined by lactic acid production of the epithelium and from *Lactobacillus* species that dominate the vaginal microflora [53]. An elevated vaginal pH has been shown to be associated with a loss of natural epithelial defense, increased rate of colonization with pathogens in the vagina and the urinary tract, and increased susceptibility to STIs. Thus, it could possibly be associated with increased risk of HPV acquisition [13, 14, 54]. As mentioned earlier, the reduction in filaggrin metabolites causes an increase in skin pH. Whether this is also the case in vaginal and cervical mucosa is unknown.

33.3 Human Papillomavirus

HPV is a small DNA virus that infects keratinocytes of the skin and mucosa. More than 100 types have been identified. Around 40 types have a special affinity for the epithelium in the anogenital area, and around 15 types have been characterized as being carcinogenic (high risk); however, not all of these types have the same carcinogenic potential [83]. HPV-16 has a unique carcinogenic potential, and together with another high-risk HPV type, HPV-18, it is associated with around 70 % of all cervical cancers. Among the noncarcinogenic (low-risk) HPV types, HPV-6 and HPV-11 are associated with more than 90 % of all cases of genital warts [5].

HPV is almost exclusively sexually transmitted and has increased in prevalence during the past decades in many parts of the world [5]. It is the most common STI worldwide, and a high proportion of sexually active men and women will acquire it at some point during their life. The prevalence has been estimated to be 27.4 % worldwide [84], with type 16 detected in about 24 % and type 18 in about 9 % of women with HPV infection. Among women, the highest prevalence is observed in those younger than 25 years [55–57].

HPV is inoculated in the basal cells of the epithelium and depends on the terminal differentiation of these to complete its life cycle and produce new viral particles. The HPV life cycle takes approximately 2–3 weeks, which is the time for the cell to migrate from the basal cell layer to the superficial layers, mature, undergo senescence, die, and release new viral particles [5, 16, 59–61]. Among others, HPV encodes two proteins, E6 and E7, which together can promote cellular proliferation, prolong cell cycle progression, and prevent apoptosis. These two proteins are both major contributive factors to the carcinogenesis of persistent infection with a high-risk HPV type [62, 63]. E6 and E7 interact with, respectively, p53 and retinoblastoma proteins, and for some high-risk HPV infections, they block these negative regulators of the cell cycle so efficiently that the infected cells never mature and cease to apoptose [5, 11]. Due to this, viral structural proteins are less frequently expressed in severe lesions, and HPV infections can be demonstrated only by using HPV DNA probes. The resulting genomic instability enables genetic alterations to accumulate, eventually causing malignant transformation. In addition, the development of cancer also depends on immune evasion that enables the virus to remain undetected for long periods. High-risk HPV and E6 and E7 exhibit several mechanisms to do so. Progressing preinvasive neoplasia accumulates genetic alterations that further assists with immune evasion, so-called immunoediting [5, 64, 65].

33.4 CIN and Cervical Cancer

Cervical cancer is one of the most common female malignancies worldwide. A definite causal role of HPV in the development of cervical cancer has been demonstrated [85]. HPV is a necessary but not sufficient factor in the

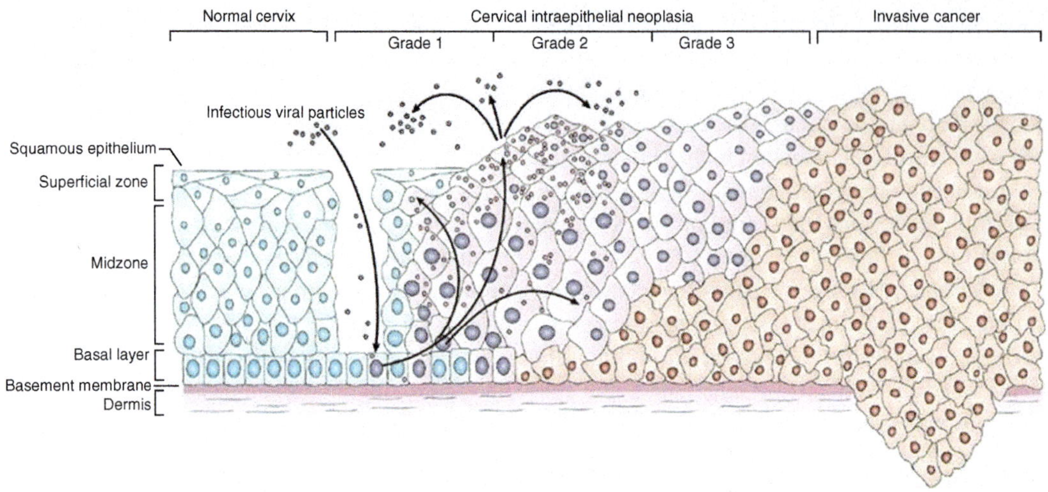

Fig. 33.1 HPV infection through microabrasion, life cycle, and development of CIN and cervical cancer (Reprinted by permission from Woodman et al. [60], Macmillan Publishers Ltd, copyright 2007)

development of cervical cancer, and HPV is also associated with a proportion of other anogenital and oropharyngeal cancers. Persistent infection with a high-risk HPV type is a prerequisite for progression to CIN and cancer [5]. HPV DNA is found in >99 % of cervical cancers [66]. Most HPV infections, irrespective of the type, are often harmless and clear spontaneously without symptoms, but persistent infection can disrupt the normal cell cycle control, promoting uncontrolled cell division and the accumulation of genetic damage [5] as described above. The main site of origin of all premalignant and malignant lesions of the cervical epithelium is considered to be the T-zone [8, 77]. CIN-1 is considered a low-grade lesion or a morphological manifestation of an HPV infection and refers to mildly atypical cellular changes in the lower third of the epithelium (formerly called mild dysplasia). HPV viral cytopathic effect (koilocytotic atypia) is often present. CIN-2 is considered a high-grade lesion. It refers to moderately atypical cellular changes confined to the basal two-thirds of the epithelium (formerly called moderate dysplasia) with preservation of epithelial maturation. CIN-3 is also considered a high-grade lesion. It refers to severely atypical cellular changes encompassing greater than two-thirds of the epithelial thickness and includes full-thickness lesions (formerly called severe dysplasia or carcinoma in situ) [82]. HPV types 16 and 18 account for roughly 70 % of all cervical cancer [57], and type 16 is by far the likeliest to persist and cause CIN-3 and cervical cancer, followed by type 18 [78–80, 83]. Several prophylactic vaccines against especially high-risk types have been developed and are in use, and studies on therapeutic vaccines are ongoing [5] (Fig. 33.1).

The HPV type and immunodeficiency (e.g., HIV positivity and organ transplant recipients) are the only clear nonbehavioral risk factors for the acquisition, persistence, and progression of infection, although others have been explored [58, 67–70]. The latter includes alterations in the vaginal microenvironment, where, for example, an acidic vaginal pH is known to be a key component of defense [13, 71]. In a study by Clarke et al. [13], they investigated the relationship between vaginal pH and the risk of HPV infection and HPV-related cytological abnormalities in cervical specimens. In accordance with other studies [14], a significant relationship was found between elevated vaginal pH and the risk of HPV infection, which was particularly pronounced in premenopausal women (<25 and 25–34 years). Elevated vaginal pH was also associated with a 30 % greater risk of infection with multiple HPV types and with low-grade squamous intraepithelial lesions, predominantly in women younger than 35 and older than 65 years of age. However, a sub-analysis in HPV-positive women revealed no association between elevated pH and high-grade squamous intraepithelial lesions.

Bacterial vaginosis is a very common complaint, where the vaginal pH is elevated due to alterations in vaginal microflora. It is recognized that bacterial vaginosis has an influence on acquisition of certain genital infections, for example, Neisseria gonorrhoeae, Chlamydia trachomatis, HSV-1 and HPV-2, and HIV [72–74]. In a meta-analysis from 2011 by Gillet et al. [54] including a total of 6,372 women, a positive association between bacterial vaginosis and HPV was found with an odds ratio 1.43 (95 % CI, 1.11–1.84). It should be emphasized that bacterial vaginosis exerts several changes in the vaginal milieu besides elevated pH, including changes in cytokine production [75] and decreased signs of inflammation [76], which may play a role in this possible association.

33.5 Filaggrin and the Cervix

While keratinization and filaggrin expression were found to be markedly decreased in vaginal and cervical tissue compared to penile tissue from human donors without known disease in one study [3], most other studies have demonstrated a positive homogeneous stain in the upper parabasal, intermediate, and superficial cell layers of normal squamous cervical epithelium with increasing intensity toward the surface of the epithelium and filaggrin intracellularly distributed diffusely throughout the cytoplasm. The conflicting results in terms of vaginal and cervical tissue may be due to difference in staining methods. Lara et al. [7] studied 87 cervical biopsies, classified as either normal cervical epithelium, non-neoplastic lesions, preneoplastic lesions, or neoplastic lesions, and performed filaggrin protein immunostaining. Non-squamous epithelium, either benign or malignant, showed lack of filaggrin expression as also demonstrated elsewhere [8]. As for normal squamous epithelium, they found the aforementioned homogeneous suprabasal pattern. In cervical condylomas, a predominantly diffuse cytoplasmatic homogeneous pattern similar to that found in normal tissue was present in 85.7 % of cases. In CIN, the extent of filaggrin immunoreactivity was inversely related to the severity of the lesion ($P<0.001$) as it has been shown in other studies [2, 9]. While all CIN-1 lesions showed a homogeneous distribution of filaggrin, most CIN-2 and CIN-3 lesions exhibited an irregular staining pattern, with a mixture of reactive and unreactive cells scattered at all levels of the epithelium. Two CIN-3 specimens showed complete absence of filaggrin expression. These findings are in line with several other reports [2, 7–9]. With respect to the invasive squamous cell carcinomas of the cervix, regular filaggrin expression was associated with the grade of tumor differentiation ($P<0.05$). Hence, poorly and moderately differentiated carcinomas were predominantly associated with focal or negative immunostainings, while well-differentiated carcinomas were associated with a homogenous or irregular distribution of filaggrin [7]. With regard to the results in this report, it must be emphasized that in 35.7 % of condylomatous lesion and in most cases of CIN-1 to CIN-3, filaggrin was found to be distributed throughout the whole thickness of the epithelium, including the basal cell layer. Studying the figures in the article, especially comparing Figures 2 and 3, this staining of the basal cell layers could possibly be due to low specificity of the antibody used. In conjunction with other experiments dealing with HPV infection, epithelial differentiation, and filaggrin [86], including the ones described in this text, filaggrin staining of basal cells seems at least unlikely.

Cintorino et al. [2] studied the association between filaggrin expression, HPV type, and CIN in follow-up biopsies and healthy controls. HPV infection was detected using an in situ hybridization technique with DNA probes for HPV-6, HPV-11, HPV-16, HPV-18, and HPV-31, and distinction was made between HPV-CIN, noHPV-CIN, and HPV-noCIN. Three patterns of filaggrin expression were used: pattern 1, where all layers above the basal cells stained positive regularly; pattern 2, where all layers above the basal cells stained irregularly; and pattern 3, where only scattered superficial cells stained positive. Similar to others, they found that filaggrin distribution in lesions infected with HPV-6 or HPV-11 resembled that of normal exocervix, with pattern 1 being present in 36.7 % and pattern 2 in 34.7 %. This contrasted HPV-16- and HPV-18-infected cases, where 46.7 % showed pattern 3. Also in this study, there was an inverse relationship between both HPV-CIN and noHPV-CIN and filaggrin expression according to the patterns. There was a significant difference between HPV-noCIN and HPV-CIN lesions, in the manner that 77.7 % in the former expressed pattern 1 and 43.5 % expressed pattern 3 in the latter. Moreover, a significant difference was found between noHPV-CIN and HPV-CIN, since pattern 3 was expressed in 4.3 and 43.5 %, respectively ($P<0.001$). Among these pattern 1 was expressed in 30.4 and 21.7 %. This study suggests that filaggrin expression is unaffected by HPV infection in the vast majority of lesions when not accompanied by CIN. Bear in mind that only HPV types 6, 11, 16, 18, and 31 were detected in these trials, whereby the neoplasia in some noHPV-CIN could be due to other HPV types. Still it is indicated that CIN lesions of the same severity can differ in filaggrin expression depending on the type of HPV infection, HPV-16 and HPV-18, causing the greatest alterations from normal. Contrasted to this, in a study on Bowen's disease and genital Bowen's papulosis [6], no correlation could be established between the presence of HPV DNA or its type and the expression of filaggrin/pro-filaggrin. Changes in filaggrin/pro-filaggrin expression were related to abnormal keratinization rather than to the presence of HPV DNA or type. With regard to the T-zone, Serra et al. [8] showed that filaggrin was completely absent from the endocervical columnar epithelium, but with diffuse and intense immunoreactivity even in the most immature metaplastic epithelium. In the area of very incipient squamous metaplasia, unnoticed histologically, filaggrin expression was found to be restricted to the reserve subcolumnar cells, which is thought to play a role in the histogenesis of the T-zone [8, 81].

The prospective value of filaggrin immunoreactivity in the differential diagnosis of cervical lesions was investigated by Lara et al. [7]. Nonhomogeneous filaggrin immunostaining patterns (irregular, focal, or absent) identified premalignant/malignant squamous cervical lesions (CIN or invasive carcinomas) with a sensitivity of 77.4 %, a specificity of 95.2 %, a positive predictive value of 92.3 %, and a negative predictive value of 85.1 %. Ronald C. McGlennen [9] found that immunohistochemical staining could qualitatively distinguish between normal and CIN-1 to CIN-2 lesions and between CIN-3 and squamous cell carcinoma, though not quantitatively separate further between histologic groups. It should though be taken into consideration that filaggrin could be a marker of squamous epithelium differentiation in both normal and pathological cervical tissue, and thus even neoplastic lesion can exhibit a regular homogeneous immunostaining pattern if well differentiated [7].

33.6 Discussion

No studies have yet explored the association between mutations in the *FLG* and the risk of HPV infection, CIN, and cervical cancer. Several studies have looked at filaggrin expression along with HPV type in these disorders [2, 7–10] and found changes in comparison to normal cervical epithelium as described above. Expression of filaggrin in CIN has thus been found to be inversely related to the severity of the lesion with an irregular pattern present in most high-grade CIN, and it was closely related to the degree of differentiation ($P<0.05$) in squamous cell carcinomas of the cervix [7, 9]. Furthermore, a difference was seen in filaggrin immunostaining in tissues, histologically equivalent, infected with HPV types 16 or 18 compared to low-risk types, the former showing the greatest alterations from normal cervical tissue [2].

As mentioned earlier, filaggrin expression and hence barrier function is affected by inflammation and different cytokines. In continuation of this, Kim et al. [22] demonstrated that IL-25 enhances herpes simplex virus (HSV)-1 and vaccinia virus replication through inhibition of filaggrin expression. Filaggrin was found to be a critical protein to inhibit HSV-1 replication in that filaggrin RNA knockdown enhances HSV-1 replication in vitro. We also mentioned that cytokines secreted in response to inflammation of different origin in the skin is believed to exert their effect in other organs and surfaces of the body [21]. Thus, it is conceivable that besides impaired barrier function of the cervical mucosa due to *FLG* mutations, additional impairment could be caused by secretion of cytokines from cells in inflamed skin and in the peripheral blood in response to, for example, severe atopic dermatitis. With impaired barrier function, susceptibility to STIs increases, leading to further inflammation, which again leads to further impaired barrier function. A vicious circle.

FLG mutations are significantly associated with atopic dermatitis; however, they are not a prerequisite for the development of this condition. Despite a 13 % increase in the risk of any cancer among atopic dermatitis patients, Hagströmer et al. [17] could not demonstrate an increased risk of cervical cancer in a cohort of 15,666 hospitalized patients with a discharge diagnosis of atopic dermatitis between 1965 and 1999. Neither could Olesen et al. [18] in a 2005 Danish follow-up study. It should be emphasized that only a small number of patients with cervical and female genital cancers was included in these studies and that only a 20–30 % of atopic dermatitis patients have *FLG* mutations.

References

1. Thyssen JP, Godoy-Gijon E, Elias PM. Ichthyosis vulgaris – the filaggrin mutation disease. Br J Dermatol. 2013;168(6):1155–66.
2. Cintorino M, Syrjänen S, Leoncini P, Bellizzi De Marco E, Petracca R, Pallini V, et al. Altered expression of filaggrin in human papillomavirus (HPV) lesions of the uterine cervix. Arch Gynecol Obstet. 1988;241:235–48.
3. Dinh MH, Okocha EA, Koons A, Veazey RS, Hope TJ. Expression of structural proteins in human female and male genital epithelia and implications for sexually transmitted infections. Biol Reprod. 2012; 86(2):32.
4. Gao PS, Rafaels NM, Hand T, Murray T, Boguniewicz M, Hata T, et al. Filaggrin mutations that confer risk of atopic dermatitis confer greater risk for eczema herpeticum. J Allergy Clin Immunol. 2009;124(3): 507–13, 513.e1–7.
5. Crosbie EJ, Einstein MH, Franceschi S, Kitchener HC. Human papillomavirus and cervical cancer. http://dx.doi.org/10.1016/S0140-6736(13)60022-7. Published Online 23 Apr 2013.
6. Guerin-Reverchon I, Chardonnet Y, Viac J, Chouvet B, Chignol MC, Thivolet J. Human papillomavirus infection and filaggrin expression in paraffin-embedded biopsy specimens of extragenital Bowen's disease and genital bowenoid papulosis. J Cancer Res Clin Oncol. 1990;116:295–300.
7. Lara C, Serra V, Marzo C, Valcuende F, Castells A, Bonilla-Musoles FM. Immunohistochemical localization of filaggrin in benign, premalignant and malignant cervical tissue. Arch Gynecol Obstet. 1994;255:73–9.
8. Serra V, Lara C, Ramirez AA, Marzo MC, Valcuende F, Castells A, et al. Precocious appearance of markers of squamous differentiation in metaplastic cells of human endocervix. Arch Gynecol Obstet. 1989;246:233–42.
9. McGlennen RC, Ostrow RS, Carson LF, Stanley MS, Faras AJ. Expression of cytokine receptors and markers of differentiation in human papillomavirus-infected cervical tissues. Am J Obstet Gynecol. 1991;165(3):696–705.
10. Viac J, Guérin-Reverchon I, Chignol MC, Chardonnet Y. Filaggrin expression in cutaneous and mucosal human papillomavirus induced lesions. Pathol Res Pract. 1989;185(3):342–6.
11. Münger K, Phelps WC, Bubb V, Howley PM, Schlegel R. The E6 and E7 genes of the human papillomavirus type 16 together are necessary and sufficient for transformation of primary human keratinocytes. J Virol. 1989;63(10):4417–21.
12. Gruber R, Elias PM, Crumrine D, Lin TK, Brandner JM, Hachem JP, et al. Filaggrin genotype in ichthyosis vulgaris predicts abnormalities in epidermal structure and function. Am J Pathol. 2011;178(5): 2252–63.
13. Clarke MA, Rodriguez AC, Gage JC, Herrero R, Hildesheim A, Wacholder S, et al. A large, population-based study of age-related associations between vaginal pH and human papillomavirus infection. BMC Infect Dis. 2012;12:33.
14. Campos ACC, Murta EFC, Michelin MA, Reis C. Evaluation of cytokines in endocervical secretion and vaginal pH from women with bacterial vaginosis or human papillomavirus. ISRN Obstetr Gynecol. 2012;2012, Article ID 342075.
15. Longworth MS, Laimins LA. Pathogenesis of human papillomaviruses in differentiating epithelia. Microbiol Mol Biol Rev. 2004;68(2):362.
16. Pyeon D, Pearce SM, Lank SM, Ahlquist P, Lambert PF. Establishment of human papillomavirus infection requires cell cycle progression. PLoS Pathog. 2009; 5(2):e1000318.
17. Lena Hagströmer L, Ye W, Nyrén O, Emtestam L. Incidence of cancer among patients with atopic dermatitis. Arch Dermatol. 2005;141:1123–7.

18. Olesen AB, Engholm G, Storm HH, Thestrup-Pedersen K. The risk of cancer among patients previously hospitalized for atopic dermatitis. J Invest Dermatol. 2005;125(3):445–9.
19. Sandilands A, Sutherland C, Irvine AD, McLean WH. Filaggrin in the frontline: role in skin barrier function and disease. J Cell Sci. 2009;122(Pt 9):1285–94.
20. Scharschmidt TC, Man MQ, Hatano Y, Crumrine D, Gunathilake R, Sundberg JP, et al. Filaggrin deficiency confers a paracellular barrier abnormality that reduces inflammatory thresholds to irritants and haptens. J Allergy Clin Immunol. 2009;124(3):496–506.e6.
21. Gutowska-Owsiak D, Schaupp AL, Salimi M, Selvakumar TA, McPherson T, Taylor S, et al. IL-17 downregulates filaggrin and affects keratinocyte expression of genes associated with cellular adhesion. Exp Dermatol. 2012;21(2):104–10.
22. Kim BE, Bin L, Ye YM, Ramamoorthy P, Leung DY. IL-25 enhances HSV-1 replication by inhibiting filaggrin expression, and acts synergistically with TH2 cytokines to enhance HSV-1 replication. J Invest Dermatol. 2013;133:2678–85.
23. Hvid M, Vestergaard C, Kemp K, Christensen GB, Deleuran B, Deleuran M. IL-25 in atopic dermatitis: a possible link between inflammation and skin barrier dysfunction? J Invest Dermatol. 2011;131(1):150–7.
24. Deleuran M, Hvid M, Kemp K, Christensen GB, Deleuran B, Vestergaard C. IL-25 induces both inflammation and skin barrier dysfunction in atopic dermatitis. Chem Immunol Allergy. 2012;96:45–9.
25. Gutowska-Owsiak D, Schaupp AL, Salimi M, Taylor S, Ogg GS. Interleukin-22 downregulates filaggrin expression and affects expression of profilaggrin processing enzymes. Br J Dermatol. 2011;165(3):492–8.
26. Brown SJ, McLean WH. One remarkable molecule: filaggrin. J Invest Dermatol. 2012;132(3 Pt 2): 751–62.
27. Ford MJ. Filaggrin. Int J Dermatol. 1986;25(9): 547–51.
28. Manabe M, O'Guin WM. Keratohyalin, trichohyalin and keratohyalin-trichohyalin hybrid granules: an overview. J Dermatol. 1992;19(11):749–55.
29. Manabe M, Sanchez M, Sun TT, Dale BA. Interaction of filaggrin with keratin filaments during advanced stages of normal epidermal differentiation and in ichthyosis vulgaris. Differentiation. 1991;48(1):43–50.
30. Angelova-Fischer I, Mannheimer AC, Hinder A, Ruether A, Franke A, Neubert RH, et al. Distinct barrier integrity phenotypes in filaggrin-related atopic eczema following sequential tape stripping and lipid profiling. Exp Dermatol. 2011;20(4):351–6.
31. Kanitakis J, Ramirez-Bosca A, Reano A, Viac J, Roche P, Thivolet J. Filaggrin expression in normal and pathological skin. A marker of keratinocyte differentiation. Virchows Arch A Pathol Anat Histopathol. 1988;412:375–82.
32. Kvedar JC, Fewkes J, Baden HP. Immunologic detection of markers of keratinocyte differentiation. Its use in neoplastic and preneoplastic lesions of skin. Arch Pathol Lab Med. 1986;110:183–8.
33. Irvine AD, McLean WH, Leung DY. Filaggrin mutations associated with skin and allergic diseases. N Engl J Med. 2011;365:1315–27.
34. van den Oord RA, Sheikh A. Filaggrin gene defects and risk of developing allergic sensitisation and allergic disorders: systematic review and meta-analysis. BMJ. 2009;339:b2433.
35. Fenger C, Baandrup U, Clausen PP, Graem N, Jacobsen GK. Klinisk Patologi, 1. udgave, 3. oplag, s:409–10.
36. Nestle FO, Di Meglio P, Qin JZ, Nickoloff BJ. Skin immune sentinels in health and disease. Nat Rev Immunol. 2009;9:679–91.
37. Niessen CM. Tight junctions/adherens junctions: basic structure and function. J Invest Dermatol. 2007; 127:2525–32.
38. Green KJ, Simpson CL. Desmosomes: new perspectives on a classic. J Invest Dermatol. 2007;127: 2499–515.
39. Candi E, Schmidt R, Melino G. The cornified envelope: a model of cell death in the skin. Nat Rev Mol Cell Biol. 2005;6:328–40.
40. Madison KC. Barrier function of the skin: "la raison d'être" of the epidermis. J Invest Dermatol. 2003;121: 231–41.
41. Centers for Disease Control and Prevention. Sexually transmitted disease surveillance 2009. Atlanta: US Department of Health and Human Services; 2010. p. 1–123.
42. Markowitz LE, Sternberg M, Dunne EF, McQuillan G, Unger ER. Seroprevalence of human papillomavirus types 6, 11, 16, and 18 in the United States: National Health and Nutrition Examination Survey 2003–2004. J Infect Dis. 2009;200:1059–67.
43. Wira CR, Patel MV, Ghosh M, Mukura L, Fahey JV. Innate immunity in the human female reproductive tract: endocrine regulation of endogenous antimicrobial protection against HIV and other sexually transmitted infections. Am J Reprod Immunol. 2011;65: 196–211.
44. Marx PA, Spira AI, Gettie A, Dailey PJ, Veazey RS, Lackner AA, et al. Progesterone implants enhance SIV vaginal transmission and early virus load. Nat Med. 1996;2:1084–9.
45. Smith SM, Baskin GB, Marx PA. Estrogen protects against vaginal transmission of simian immunodeficiency virus. J Infect Dis. 2000;182:708–15.
46. Smith SM, Mefford M, Klase Z, Sodora D, Alexander N, Hess D, et al. Topical estrogen protects against SIV vaginal transmission without evidence of systemic effect. AIDS. 2004;18:1637–43.
47. Parr MB, Kepple L, McDermott M, Drew MD, Bozzola JJ, Parr EL. A mouse model for studies of mucosal immunity to vaginal infection by herpes simplex virus type 2. Lab Invest. 1994;70:369–80.
48. Mauck C. Overview of why hormones may be an issue. J Acquir Immune Defic Syndr. 2005;38:S11–12.
49. Bahamondes L, Trevisan M, Andrade L, Marchi NM, Castro S, Dı'az J, et al. The effect upon the human vaginal histology of the long-term use of the injectable

contraceptive Depo-Provera. Contraception. 2000;62:23–7.
50. Miller L, Patton DL, Meier A, Thwin SS, Hooton TM, Eschenbach DA. Depomedroxyprogesterone-induced hypoestrogenism and changes in vaginal flora and epithelium. Obstet Gynecol. 2000;96:431–9.
51. Roberts JN, Buck CB, Thompson CD, et al. Genital transmission of HPV in a mouse model is potentiated by nonoxynol-9 and inhibited by carrageenan. Nat Med. 2007;13:857–61.
52. Joyce JG, Tung JS, Przysiecki CT, Cook JC, Lehman ED, Sands JA. The L1 major capsid protein of human papillomavirus type 11 recombinant virus-like particles interacts with heparin and cell-surface glycosaminoglycans on human keratinocytes. J Biol Chem. 1999;274:5810–22.
53. Donders GG. Definition and classification of abnormal vaginal flora. Best Pract Res Clin Obstet Gynaecol. 2007;21:355–73.
54. Gillet E, Meys JF, Verstraelen H, Bosire C, De Sutter P, Temmerman M, et al. Bacterial vaginosis is associated with uterine cervical human papillomavirus infection: a meta-analysis. BMC Infect Dis. 2011;11:10.
55. Dunne EF, Unger ER, Sternberg M, McQuillan G, Swan DC, Patel SS, et al. Prevalence of HPV infection among females in the United States. JAMA. 2007;297(8):813–19.
56. Kjaer SK, Breugelmans G, Munk C, Junge J, Watson M, Iftner T. Population-based prevalence, type- and age-specific distribution of HPV in women before introduction of an HPV-vaccination program in Denmark. Int J Cancer. 2008;123(8):1864–70. doi:10.1002/ijc.23712.
57. de Sanjosé S, Diaz M, Castellsagué X, Clifford G, Bruni L, Muñoz N, et al. Worldwide prevalence and genotype distribution of cervical human papillomavirus DNA in women with normal cytology: a meta-analysis. Lancet Infect Dis. 2007;7(7):453–9.
58. Castle PE, Rodríguez AC, Burk RD, Herrero R, Wacholder S, Alfaro M, et al. Short term persistence of human papillomavirus and risk of cervical precancer and cancer: population based cohort study. BMJ. 2009;339:b2569.
59. Doorbar J. The papillomavirus life cycle. J Clin Virol. 2005;32 suppl 1:S7–15.
60. Woodman CB, Collins SI, Young LS. The natural history of cervical HPV infection: unresolved issues. Nat Rev Cancer. 2007;7:11–22.
61. Stanley M. Pathology and epidemiology of HPV infection in females. Gynecol Oncol. 2010;117(2 suppl):S5–10.
62. Münger K, Phelps WC, Bubb V, Howley PM, Schlegel R. The E6 and E7 genes of the human papillomavirus type 16 together are necessary and sufficient for transformation of primary human keratinocytes. J Virol. 1989;63:4417.
63. Ghittoni R, Accardi R, Hasan U, Gheit T, Sylla B, Tommasino M. The biological properties of E6 and E7 oncoproteins from human papillomaviruses. Virus Genes. 2010;40:1–13.
64. Piersma SJ. Immunosuppressive tumor microenvironment in cervical cancer patients. Cancer Microenviron. 2011;4(3):361–75.
65. Dunn GP, Bruce AT, Ikeda H, Old LJ, Schreiber RD. Cancer immunoediting: from immunosurveillance to tumor escape. Nat Immunol. 2002;3:991–8.
66. Walboomers JM, Jacobs MV, Manos MM, Bosch FX, Kummer JA, Shah KV, et al. Human papillomavirus is a necessary cause of invasive cervical cancer worldwide. J Pathol. 1999;189:12–9.
67. IARC Working Group on the Evaluation of Carcinogenic Risks to Humans. Biological agents. Volume 100 B. A review of human carcinogens. IARC Monogr Eval Carcinog Risks Hum. 2012;100(Pt B):1–441.
68. International Collaboration of Epidemiological Studies of Cervical Cancer, Appleby P, Beral V, Berrington de González A, Colin D, Franceschi S, et al. Cervical cancer and hormonal contraceptives: collaborative reanalysis of individual data for16,573 women with cervical cancer and 35,509 women without cervical cancer from 24 epidemiological studies. Lancet. 2007;370(9599):1609–21.
69. International Collaboration of Epidemiological Studies of Cervical Cancer, Appleby P, Beral V, Berrington de González A, Colin D, Franceschi S, et al. Carcinoma of the cervix and tobacco smoking: collaborative reanalysis of individual data on 13,541 women with carcinoma of the cervix and 23,017 women without carcinoma of the cervix from 23 epidemiological studies. Int J Cancer. 2006;118: 1481–95.
70. Arnheim Dahlström L, Andersson K, Luostarinen T, Thoresen S, Ögmundsdottír H, Tryggvadottír L, et al. Prospective seroepidemiologic study of human papillomavirus and other risk factors in cervical cancer. Cancer Epidemiol Biomarkers Prev. 2011;20:2541–50.
71. Linhares IM, Summers PR, Larsen B, Giraldo PC, Witkin SS. Contemporary perspectives on vaginal pH and lactobacilli. Am J Obstet Gynecol. 2011;204:120. e1–5.
72. Atashili J, Poole C, Ndumbe PM, Adimora AA, Smith JS. Bacterial vaginosis and HIV acquisition: a meta-analysis of published studies. AIDS. 2008;22(12): 1493–501.
73. Wiesenfeld HC, Hillier SL, Krohn MA, Landers DV, Sweet RL. Bacterial vaginosis is a strong predictor of Neisseria gonorrhoeae and Chlamydia trachomatis infection. Clin Infect Dis. 2003;36:663–8.
74. Allsworth JE, Lewis VA, Peipert JF. Viral sexually transmitted infections and bacterial vaginosis: 2001-2004 National Health and Nutrition Examination Survey data. Sex Transm Dis. 2008;35:791–6.
75. Cauci S. Vaginal immunity in bacterial vaginosis. Curr Infect Dis Rep. 2004;6:450–6.
76. Figueiredo PG, Sarian LO, Tambascia JK, Simões JA, Rabelo-Santos SH, Discacciati MG, et al. Increased detection of clue cells in smears from cervical intraepithelial lesions with reduced expression of COX-2. Diagn Cytopathol. 2008;36:705–9.

77. Burghardt E, Ostor AG. Site and origin of squamous cervical cancer: a histomorphologic study. Obstet Gynecol. 1983;62:117–27.
78. Schiffman M, Herrero R, Desalle R, Hildesheim A, Wacholder S, Rodriguez AC, et al. The carcinogenicity of human papillomavirus types reflects viral evolution. Virology. 2005;337:76–84.
79. Guan P, Howell-Jones R, Li N, Bruni L, de Sanjosé S, Franceschi S, et al. Human papillomavirus types in 115,789 HPV-positive women: a meta-analysis from cervical infection to cancer. Int J Cancer. 2012;131: 2349–59.
80. Li N, Franceschi S, Howell-Jones R, Snijders PJ, Clifford GM. Human papillomavirus type distribution in 30,848 invasive cervical cancers worldwide: variation by geographical region, histological type and year of publication. Int J Cancer. 2011;128: 927–35.
81. Martens JE, Smedts FM, Ploeger D, Helmerhorst TJ, Ramaekers FC, Arends JW, et al. Distribution pattern and marker profile show two subpopulations of reserve cells in the endocervical canal. Int J Gynecol Pathol. 2009;28(4):381–8.
82. Burd EM. Human papillomavirus and cervical cancer. Clin Microbiol Rev. 2003 Jan;16(1):1–17.
83. Kjær SK, Frederiksen K, Munk C, Iftner T. Long-term absolute risk of cervical intraepithelial neoplasia grade 3 or worse following human papillomavirus infection: role of persistence. J Natl Cancer Inst. 2010;102(19):1478–88. doi:10.1093/jnci/djq356. Epub 2010 Sep 14.
84. Forman D, de Martel C, Lacey CJ, Soerjomataram I, Lortet-Tieulent J, Bruni L, et al. Global burden of human papillomavirus and related diseases. Vaccine. 2012;30 Suppl 5:F12–23. doi:10.1016/j.vaccine.2012.07.055.
85. Muñoz N, Castellsagué X, de Conzález AB, Gissmann L. Chapter 1: HPV in the etiology of human cancer. Vaccine. 2006;24 Suppl 3:S3/1–10.
86. Doorbar J, Foo C, Coleman N, Medcalf L, Hartley O, Prospero T, et al. Characterization of events during the late stages of HPV16 infection in vivo using high-affinity synthetic Fabs to E4. Virology. 1997;238(1): 40–52.

Antibodies Against Citrullinated Proteins and Filaggrin

34

Katleen F.E. Van Steendam and Dieter L. Deforce

Contents

Rheumatoid arthritis (RA) is an autoimmune disease characterized by chronic inflammation of the joints. In autoimmune diseases, the immune system loses the ability to distinguish between self and nonself, so self-antigens are recognized as nonself and a T- and/or B-cell immune response is initialized. The best known dysregulation of the adaptive immune response in autoimmune diseases is the production of autoantibodies. In RA, the presence of autoantibodies has already been reported: autoantibodies against fibrinogen, vimentin, filaggrin, etc. In this chapter, we will focus on the autoreactivity in RA against filaggrin and the role of these antibodies in the disease.

34.1 Autoreactivity Against Filaggrin

Filaggrin, a cytokeratin filament of the epithelial cells, is involved in the organization of cytoskeletal structures in epithelial cells. Reactivity to filaggrin was first detected by Nienhuis and Mandena [1] and Young et al. [2]. They detected the antiperinuclear factor (APF) and the anti-keratin antibodies (AKA), respectively. APF antibodies bind to pro-filaggrin in the perinuclear granules in epithelial cells of the human buccal mucosa [3]. The antigenic substrate of AKA, on the other hand, was rat esophagus, but AKA also recognize filaggrin in the human epidermis [4]. Interestingly, both antibodies showed a high specificity for RA (87–100 % and 73–100 %,

K.F.E. Van Steendam, PhD, PharmD
D.L. Deforce, PhD, PharmD (✉)
Laboratory for Pharmaceutical Biotechnology, University Ghent, Harelbekestraat 72, Ghent, East-Flanders 9000, Belgium
e-mail: dieter.deforce@ugent.be; katleen.vansteendam@ugent.be

J.P. Thyssen, H. Maibach (eds.), *Filaggrin*,
DOI 10.1007/978-3-642-54379-1_34,

respectively) [3]. The widespread use of these tests, however, was limited because of the problems in standardizing the natural substrates and since APF-positive buccal mucosa cells can only be found in approximately 5 % of the donors [5, 6]. Later on, APF and AKA antibodies were classified as anti-filaggrin antibodies (AFA) since they both recognize filaggrin [3]. More precisely, AFA are reactive with the acidic/neutral isoform of filaggrin, leading to the hypothesis that citrulline residues might be present on the epitopes [7]. Indeed, in the late 1990s, it was shown that these antibodies targeted citrullinated epitopes of (pro) filaggrin.

Moreover, citrullination was essential for the autoantigenicity of filaggrin [5, 7]. Citrullination of albumin revealed no reactivity with sera from RA patients, indicating that the amino acid sequence was important as well [7]. AFA were also correlated with disease activity and severity, and their presence was found before the clinical symptoms of RA [8].

34.2 Citrullination

Citrullination or deimination is a posttranslational modification in which the arginine residue in a protein is converted to a citrulline residue. This modification is effected by peptidylarginine deiminase (PAD) enzymes, which need a high calcium concentration (10^{-2} mmol/L) (Fig. 34.1) [9, 10].

The PAD family consists of five members (PAD1, PAD2, PAD3, PAD4, PAD6) marked by a different tissue distribution, with PAD2 being the most widely expressed PAD enzyme (Table 34.1) [9, 11, 12].

During citrullination, the charge of the protein changes: a positively charged arginine residue is replaced by a neutral citrulline. This change in charge can lead to different intra- and intermolecular interactions, potentially resulting in an altered protein folding [9, 14]. This change in three-dimensional structure of the protein/peptide can alter the proteolytic processing and the MHC class II presentation, which can lead to the exposure of new antigens to the immune system [9, 15–17].

34.3 Citrullinated Proteins

Citrullinated proteins are found in healthy subjects in the outer epidermis (skin, nails, and hair) and in the central nervous system (CNS) [11, 18, 19]. Citrullination of keratin and filaggrin in the skin is necessary for the cornification of the epidermis [11, 20]. In the CNS, citrullination of myelin basic protein is involved in the assembly and the structural stabilization of the myelin sheaths [21]. Citrullination of histones is involved in transcriptional regulation [22] and in the decondensation of chromatin when forming neutrophil extracellular traps (NETs) [23].

Table 34.1 Expression of PAD in various tissues [9–13]

	Tissue/organ	Localization in the cell
PAD1	Epidermis, uterus	Cytosol
PAD2	Brain, uterus, salivary gland, macrophage, lymphocyte, spleen, pancreas, skeletal muscle, bone marrow, skin, synovial membrane, synovial fluid	Cytosol
PAD3	Hair follicle	Cytosol
PAD4	Eosinophil, neutrophil, monocyte, macrophage, bone marrow, synovial membrane, synovial fluid	Cytosol/ nucleus
PAD6	Oocyte, ovary, early embryo	Cytosol

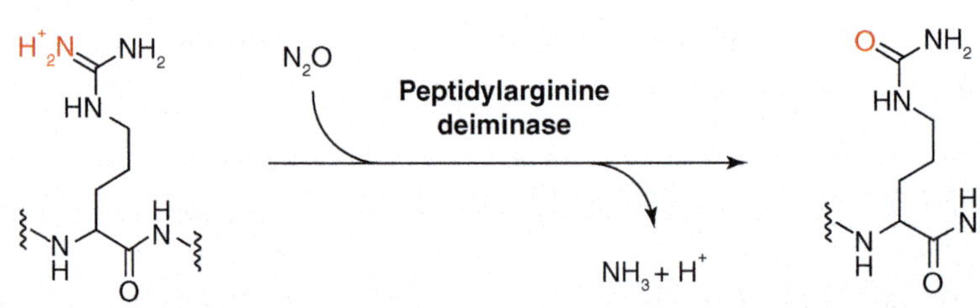

Fig. 34.1 Deimination of peptidylarginine to peptidylcitrulline by the calcium-dependent enzyme peptidylarginine deiminase (PAD)

Citrullinated proteins are also found in several diseases, such as cancer, Alzheimer's, and RA [24]. During the last two decades, citrullinated proteins have gained a lot of interest in the field of RA since ACPA were specifically found in this disease.

34.4 Antibodies Against Citrullinated Proteins (ACPA)

Nowadays, determination of the ACPA titer is commonly used for the diagnosis of RA. The diagnostic test for ACPA is based on the ELISA technique with synthetic cyclic citrullinated proteins/peptides as coated antigens (anti-CCP ELISA). The peptides are made cyclic in order to optimally expose the citrulline residue to antibodies. Anti-CCP antibodies were originally detected by the use of citrullinated cyclic filaggrin peptides (later called the anti-CCP1 ELISA) since antibodies against citrullinated filaggrin showed a high specificity for RA [7]. A few years later, the anti-CCP1 ELISA was replaced by anti-CCP2 ELISA, which consisted not only of filaggrin but a pool of cyclic citrullinated peptides. Comparative studies between anti-CCP1 ELISA and anti-CCP2 ELISA revealed that the anti-CCP2 ELISA had a higher sensitivity (68 vs. 53 %) compared to the anti-CCP1 ELISA and a comparable specificity (95–96 %) [25, 26]. This resulted in a significant higher area under the curve (AUC) in ROC analyses for the anti-CCP2 ELISA compared to anti-CCP1 ELISA. The anti-CCP2 test was also superior in identifying patients with an increased rate of joint damage progression [25]. Therefore, the anti-CCP1 ELISA is inferior to the current anti-CCP2 ELISA, indicating that a pool of citrullinated peptides has a better diagnostic and prognostic value compared to an ELISA based on citrullinated filaggrin alone.

The anti-CCP2 ELISA has a good sensitivity (around 70 %) and a high specificity (98 %), making the anti-CCP titer a very useful diagnostic marker (reviewed in [27–33]).

The role of ACPA in the pathogenesis of RA is emphasized by the presence of anti-CCP antibodies before the onset of the disease [34–36]. This predictive value is of major importance in RA since treatment at the early stages of disease can delay or even avoid joint destruction.

Additionally, it has been shown that ACPA-positive patients have a more severe and erosive disease course compared to ACPA-negative patients [35–38]. The presence of ACPA is also associated with human leukocyte antigen (HLA) DRB1 alleles, a gene that is associated with RA and the severity of the disease [39]. Moreover, ACPA reactivity divides RA patients into two major subsets, which should be treated as separate entities when studying the molecular pathophysiology of RA [15].

Since the discovery of ACPA, a lot of research has been conducted on the identity of these citrullinated proteins involved in the pathogenesis of RA. The best-studied citrullinated proteins in the context of RA are filaggrin, fibrinogen, vimentin, fibronectin, collagen type II (CII), and α-enolase [8, 29, 40–54].

34.5 Is (Pro)Filaggrin the Trigger for ACPA Production?

An important difference between (pro)filaggrin and other proteins that are considered as possible triggers for the ACPA production (CII, α-enolase, fibrinogen, and vimentin) is their presence in the joint. (Pro)filaggrin has not been found in the joints of RA patients, whereas CII, α-enolase, fibrin(ogen), and vimentin have been detected in large amounts [41, 50, 52, 55, 56]. Since citrulline-reactive autoantibodies are synthesized locally by plasma cells in the pannus, the trigger for ACPA production must be a (self-)antigen present in the inflamed joint [57, 58]. Therefore, one could presume that citrullinated (pro)filaggrin is not the autoantigen that drives the ACPA response [57]. Citrullinated CII, α-enolase, fibrin(ogen), and vimentin, on the other hand, are detected in the joints of RA patients [41, 55] and are, therefore, considered as more relevant candidates to trigger the ACPA production.

Although filaggrin is not present in the joint, Masson-Bessiere et al. [8] found that anti-filaggrin autoantibodies (AFA) could bind proteins from the joint, more precisely citrullinated α- and β-fibrin chains. Moreover, citrullinated fibrin is targeted by a larger proportion of RA sera compared to (pro)filaggrin [8]. AFA are, therefore, postulated as antibodies reactive to citrullinated epitopes from a cross-reactive protein [5, 8]. This was confirmed by Baeten et al. [59], who found by means of double immunofluorescence that AFA reactivity colocalized with anti-citrulline reactivity in the synovium but not with monoclonal AFA, which recognized both filaggrin and (pro)filaggrin. This cross-reactivity of AFA with other citrullinated proteins was explained by Schellekens et al. [5], who suggested that citrulline is essential for the antigenic properties of proteins recognized by RA-specific antibodies. Additionally, it was found that not citrullinated proteins in the synovium as such, but rather the intracellular citrullinated proteins from the synovium, were specific for RA, leading to the assumption that citrullinated filaggrin is not the most specific antigen in RA [59, 60].

Van de Stadt et al. also compared the reactivity of ACPA-positive patients before RA diagnosis [61]. They found that the first positive sample was most often directed against a citrullinated fibrinogen peptide, followed by a citrullinated vimentin peptide, a citrullinated alpha-enolase peptide, and then CCP1, indicating that the antigens of the CCP1 test, citrullinated filaggrin, are not the trigger for ACPA production.

Besides its localization and antibody reactivity, researchers who wanted to determine the trigger for ACPA production also focused on the capacity of citrullinated proteins to induce T-cell proliferation, since there is a strong association between ACPA titers and HLA-DRB1 alleles in RA and since ACPA are class-switched antibodies [62]. The proliferative response of peripheral blood mononuclear cells to both filaggrin and citrullinated filaggrin was rarely observed and did not significantly differ between RA patients and healthy controls [63], again indicating that (citrullinated) (pro)filaggrin is not the trigger for RA. However, as mentioned above, citrullinated filaggrin was once the only antigen used to detect ACPA, but more and more reports indicate that this is due to cross-reactivity, based on the citrulline residues rather than on the protein backbone.

References

1. Nienhuis RL, Mandema E. A new serum factor in patients with rheumatoid arthritis; the antiperinuclear factor. Ann Rheum Dis. 1964;23:302–5.
2. Young BJ, Mallya RK, Leslie RD, Clark CJ, Hamblin TJ. Anti-keratin antibodies in rheumatoid arthritis. Br Med J. 1979;2(6182):97–9.
3. Sebbag M, Simon M, Vincent C, Masson-Bessière C, Girbal E, Durieux J, et al. The antiperinuclear factor and the so-called antikeratin antibodies are the same rheumatoid arthritis-specific autoantibodies. J Clin Invest. 1995;95(6):2672–9.
4. Simon M, Girbal E, Sebbag M, Gomes-Daudrix V, Vincent C, Salama G, et al. The cytokeratin filament-aggregating protein filaggrin is the target of the so-called "antikeratin antibodies," autoantibodies specific for rheumatoid arthritis. J Clin Invest. 1993;92(3):1387–93.
5. Schellekens G, de Jong B, van den Hoogen F, van de Putte L, van Venrooij W. Citrulline is an essential constituent of antigenic determinants recognized by rheumatoid arthritis-specific autoantibodies. J Clin Invest. 1998;101(1):273–81.
6. Vander Cruyssen B, Cantaert T, Nogueira L, Clavel C, De Rycke L, Dendoven A, et al. Diagnostic value of anti-human citrullinated fibrinogen ELISA and comparison with four other anti-citrullinated protein assays. Arthritis Res Ther. 2006;8(4):R122.
7. Girbal-Neuhauser E, Durieux JJ, Arnaud M, Dalbon P, Sebbag M, Vincent C, et al. The epitopes targeted by the rheumatoid arthritis-associated antifilaggrin autoantibodies are posttranslationally generated on various sites of (pro)filaggrin by deimination of arginine residues. J Immunol. 1999;162(1):585–94.
8. Masson-Bessiere C, Sebbag M, Girbal-Neuhauser E, Nogueira L, Vincent C, Senshu T, et al. The major synovial targets of the rheumatoid arthritis-specific antifilaggrin autoantibodies are deiminated forms of the alpha- and beta-chains of fibrin. J Immunol. 2001;166(6):4177–84.
9. Vossenaar ER, Zendman AJ, van Venrooij WJ, Pruijn GJ. PAD, a growing family of citrullinating enzymes: genes, features and involvement in disease. Bioessays. 2003;25(11):1106–18.
10. Vossenaar ER, Radstake TR, van der Heijden A, van Mansum MA, Dieteren C, de Rooij DJ, et al. Expression and activity of citrullinating peptidylarginine deiminase enzymes in monocytes and macrophages. Ann Rheum Dis. 2004;63(4):373–81.

11. Gyorgy B, Toth E, Tarcsa E, Falus A, Buzas EI. Citrullination: a posttranslational modification in health and disease. Int J Biochem Cell Biol. 2006;38(10):1662–77.
12. Suzuki A, Yamada R, Yamamoto K. Citrullination by peptidylarginine deiminase in rheumatoid arthritis. Ann N Y Acad Sci. 2007;1108:323–39.
13. Hojo-Nakashima I, Sato R, Nakashima K, Hagiwara T, Yamada M. Dynamic expression of peptidylarginine deiminase 2 in human monocytic leukaemia THP-1 cells during macrophage differentiation. J Biochem. 2009;146(4):471–9.
14. Tarcsa E, Marekov LN, Mei G, Melino G, Lee SC, Steinert PM. Protein unfolding by peptidylarginine deiminase. Substrate specificity and structural relationships of the natural substrates trichohyalin and filaggrin. J Biol Chem. 1996;271(48):30709–16.
15. Klareskog L, Ronnelid J, Lundberg K, Padyukov L, Alfredsson L. Immunity to citrullinated proteins in rheumatoid arthritis. Annu Rev Immunol. 2008;26:651–75.
16. van Venrooij WJ, Pruijn GJ. Citrullination: a small change for a protein with great consequences for rheumatoid arthritis. Arthritis Res. 2000;2(4):249–51.
17. Wegner N, Lundberg K, Kinloch A, Fisher B, Malmstrom V, Feldmann M, et al. Autoimmunity to specific citrullinated proteins gives the first clues to the etiology of rheumatoid arthritis. Immunol Rev. 2010;233(1):34–54.
18. Nicholas AP, King JL, Sambandam T, Echols JD, Gupta KB, McInnis C, et al. Immunohistochemical localization of citrullinated proteins in adult rat brain. J Comp Neurol. 2003;459(3):251–66.
19. Senshu T, Akiyama K, Kan S, Asaga H, Ishigami A, Manabe M. Detection of deiminated proteins in rat skin: probing with a monospecific antibody after modification of citrulline residues. J Invest Dermatol. 1995;105(2):163–9.
20. Ishida-Yamamoto A, Senshu T, Eady R, Takahashi H, Shimizu H, Akiyama M, et al. Sequential reorganization of cornified cell keratin filaments involving filaggrin-mediated compaction and keratin 1 deimination. J Invest Dermatol. 2002;118(2):282–7.
21. Harauz G, Musse AA. A tale of two citrullines – structural and functional aspects of myelin basic protein deimination in health and disease. Neurochem Res. 2007;32(2):137–58.
22. Wang Y, Wysocka J, Sayegh J, Lee YH, Perlin JR, Leonelli L, et al. Human PAD4 regulates histone arginine methylation levels via demethylimination. Science. 2004;306(5694):279–83.
23. Wang Y, Li M, Stadler S, Correll S, Li P, Wang D, et al. Histone hypercitrullination mediates chromatin decondensation and neutrophil extracellular trap formation. J Cell Biol. 2009;184(2):205–13.
24. De Ceuleneer M, Van Steendam K, Dhaenens M, Deforce D. In vivo relevance of citrullinated proteins and the challenges in their detection. Proteomics. 2012;12(6):752–60.
25. van Gaalen FA, Visser H, Huizinga TW. A comparison of the diagnostic accuracy and prognostic value of the first and second anti-cyclic citrullinated peptides (CCP1 and CCP2) autoantibody tests for rheumatoid arthritis. Ann Rheum Dis. 2005;64(10):1510–12.
26. Avouac J, Gossec L, Dougados M. Diagnostic and predictive value of anti-cyclic citrullinated protein antibodies in rheumatoid arthritis: a systematic literature review. Ann Rheum Dis. 2006;65(7):845–51.
27. De Rycke L, Peene I, Hoffman IE, Kruithof E, Union A, Meheus L, et al. Rheumatoid factor and anticitrullinated protein antibodies in rheumatoid arthritis: diagnostic value, associations with radiological progression rate, and extra-articular manifestations. Ann Rheum Dis. 2004;63(12):1587–93.
28. van Boekel MA, Vossenaar ER, van den Hoogen FH, van Venrooij WJ. Autoantibody systems in rheumatoid arthritis: specificity, sensitivity and diagnostic value. Arthritis Res. 2002;4(2):87–93.
29. Van Steendam K, Tilleman K, Deforce D. The relevance of citrullinated vimentin in the production of antibodies against citrullinated proteins and the pathogenesis of rheumatoid arthritis. Rheumatology (Oxford). 2011;50(5):830–7.
30. Dejaco C, Klotz W, Larcher H, Duftner C, Schirmer M, Herold M. Diagnostic value of antibodies against a modified citrullinated vimentin in rheumatoid arthritis. Arthritis Res Ther. 2006;8(4):R119.
31. Soos L, Szekanecz Z, Szabo Z, Fekete A, Zeher M, Horvath IF, et al. Clinical evaluation of anti-mutated citrullinated vimentin by ELISA in rheumatoid arthritis. J Rheumatol. 2007;34(8):1658–63.
32. Sghiri R, Bouajina E, Bargaoui D, Harzallah L, Fredj HB, Sammoud S, et al. Value of anti-mutated citrullinated vimentin antibodies in diagnosing rheumatoid arthritis. Rheumatol Int. 2008;29(1):59–62.
33. Rodriguez-Mahou M, Lopez-Longo FJ, Sanchez-Ramon S, Estecha A, Garcia-Segovia A, Rodriguez-Molina JJ, et al. Association of anti-cyclic citrullinated peptide and anti-Sa/citrullinated vimentin autoantibodies in rheumatoid arthritis. Arthritis Rheum. 2006;55(4):657–61.
34. van Gaalen FA, Linn-Rasker SP, van Venrooij WJ, de Jong BA, Breedveld FC, Verweij CL, et al. Autoantibodies to cyclic citrullinated peptides predict progression to rheumatoid arthritis in patients with undifferentiated arthritis: a prospective cohort study. Arthritis Rheum. 2004;50(3):709–15.
35. Kroot EJ, de Jong BA, van Leeuwen MA, Swinkels H, van den Hoogen FH, van't Hof M, et al. The prognostic value of anti-cyclic citrullinated peptide antibody in patients with recent-onset rheumatoid arthritis. Arthritis Rheum. 2000;43(8):1831–5.
36. Nielen MM, van Schaardenburg D, Reesink HW, van de Stadt RJ, van der Horst-Bruinsma IE, de Koning MH, et al. Specific autoantibodies precede the symptoms of rheumatoid arthritis: a study of serial

measurements in blood donors. Arthritis Rheum. 2004;50(2):380–6.

37. Forslind K, Ahlmen M, Eberhardt K, Hafstrom I, Svensson B. Prediction of radiological outcome in early rheumatoid arthritis in clinical practice: role of antibodies to citrullinated peptides (anti-CCP). Ann Rheum Dis. 2004;63(9):1090–5.
38. Ronnelid J, Wick MC, Lampa J, Lindblad S, Nordmark B, Klareskog L, et al. Longitudinal analysis of citrullinated protein/peptide antibodies (anti-CP) during 5 year follow up in early rheumatoid arthritis: anti-CP status predicts worse disease activity and greater radiological progression. Ann Rheum Dis. 2005;64(12):1744–9.
39. Weyand CM, Hicok KC, Conn DL, Goronzy JJ. The influence of HLA-DRB1 genes on disease severity in rheumatoid arthritis. Ann Intern Med. 1992;117(10):801–6.
40. Rowley MJ, Nandakumar KS, Holmdahl R. The role of collagen antibodies in mediating arthritis. Mod Rheumatol. 2008;18(5):429–41.
41. Matsuo K, Xiang Y, Nakamura H, Masuko K, Yudoh K, Noyori K, et al. Identification of novel citrullinated autoantigens of synovium in rheumatoid arthritis using a proteomic approach. Arthritis Res Ther. 2006;8(6):R175.
42. Tilleman K, Van Steendam K, Cantaert T, De Keyser F, Elewaut D, Deforce D. Synovial detection and autoantibody reactivity of processed citrullinated isoforms of vimentin in inflammatory arthritides. Rheumatology (Oxford). 2008;47(5):597–604.
43. Vossenaar ER, Smeets TJ, Kraan MC, Raats JM, van Venrooij WJ, Tak PP. The presence of citrullinated proteins is not specific for rheumatoid synovial tissue. Arthritis Rheum. 2004;50(11):3485–94.
44. Chapuy-Regaud S, Sebbag M, Baeten D, Clavel C, Foulquier C, De Keyser F, et al. Fibrin deimination in synovial tissue is not specific for rheumatoid arthritis but commonly occurs during synovitides. J Immunol. 2005;174(8):5057–64.
45. Takizawa Y, Suzuki A, Sawada T, Ohsaka M, Inoue T, Yamada R, et al. Citrullinated fibrinogen detected as a soluble citrullinated autoantigen in rheumatoid arthritis synovial fluids. Ann Rheum Dis. 2006;65(8):1013–20.
46. Tabushi Y, Nakanishi T, Takeuchi T, Nakajima M, Ueda K, Kotani T, et al. Detection of citrullinated proteins in synovial fluids derived from patients with rheumatoid arthritis by proteomics-based analysis. Ann Clin Biochem. 2008;45(Pt 4):413–17.
47. Bang H, Egerer K, Gauliard A, Luthke K, Rudolph PE, Fredenhagen G, et al. Mutation and citrullination modifies vimentin to a novel autoantigen for rheumatoid arthritis. Arthritis Rheum. 2007;56(8):2503–11.
48. Lundberg K, Nijenhuis S, Vossenaar E, Palmblad K, van Venrooij W, Klareskog L, et al. Citrullinated proteins have increased immunogenicity and arthritogenicity and their presence in arthritic joints correlates with disease severity. Arthritis Res Ther. 2005;7(3):R458–67.
49. Burkhardt H, Sehnert B, Bockermann R, Engstrom A, Kalden JR, Holmdahl R. Humoral immune response to citrullinated collagen type II determinants in early rheumatoid arthritis. Eur J Immunol. 2005;35(5):1643–52.
50. Yoshida M, Tsuji M, Kurosaka D, Kurosaka D, Yasuda J, Ito Y, et al. Autoimmunity to citrullinated type II collagen in rheumatoid arthritis. Mod Rheumatol. 2006;16(5):276–81.
51. Uysal H, Bockermann R, Nandakumar KS, Sehnert B, Bajtner E, Engstrom A, et al. Structure and pathogenicity of antibodies specific for citrullinated collagen type II in experimental arthritis. J Exp Med. 2009;206(2):449–62.
52. Kinloch A, Tatzer V, Wait R, Peston D, Lundberg K, Donatien P, et al. Identification of citrullinated alpha-enolase as a candidate autoantigen in rheumatoid arthritis. Arthritis Res Ther. 2005;7(6):R1421–9.
53. Lundberg K, Kinloch A, Fisher BA, Wegner N, Wait R, Charles P, et al. Antibodies to citrullinated alpha-enolase peptide 1 are specific for rheumatoid arthritis and cross-react with bacterial enolase. Arthritis Rheum. 2008;58(10):3009–19.
54. Steiner G. Auto-antibodies and autoreactive T-cells in rheumatoid arthritis: pathogenetic players and diagnostic tools. Clin Rev Allergy Immunol. 2007;32(1):23–36.
55. Tilleman K, Van Steendam K, Cantaert T, De Keyser F, Elewaut D, Deforce D. Synovial detection and autoantibody reactivity of processed citrullinated isoforms of vimentin in inflammatory arthritides. Rheumatology. 2008;47(5):597–604.
56. Kinloch A, Lundberg K, Wait R, Wegner N, Lim NH, Zendman AJ, et al. Synovial fluid is a site of citrullination of autoantigens in inflammatory arthritis. Arthritis Rheum. 2008;58(8):2287–95.
57. Masson-Bessiere C, Sebbag M, Durieux J, Nogueira L, Vincent C, Girbal-Neuhauser E, et al. In the rheumatoid pannus, anti-filaggrin autoantibodies are produced by local plasma cells and constitute a higher proportion of IgG than in synovial fluid and serum. Clin Exp Immunol. 2000;119(3):544–52.
58. Reparon-Schuijt CC, van Esch WJ, van Kooten C, Schellekens GA, de Jong BA, van Venrooij WJ, et al. Secretion of anti-citrulline-containing peptide antibody by B lymphocytes in rheumatoid arthritis. Arthritis Rheum. 2001;44(1):41–7.
59. Baeten D, Peene I, Union A, Meheus L, Sebbag M, Serre G, et al. Specific presence of intracellular citrullinated proteins in rheumatoid arthritis synovium: relevance to antifilaggrin autoantibodies. Arthritis Rheum. 2001;44(10):2255–62.

60. De Rycke L, Nicholas A, Cantaert T, Kruithof E, Echols J, Vandekerckhove B, et al. Synovial intracellular citrullinated proteins colocalizing with peptidyl arginine deiminase as pathophysiologically relevant antigenic determinants of rheumatoid arthritis-specific humoral autoimmunity. Arthritis Rheum. 2005;52(8):2323–30.
61. van de Stadt LA, de Koning MH, van de Stadt RJ, Wolbink G, Dijkmans BA, Hamann D, et al. Development of the anti citrullinated peptide antibody repertoire prior to the onset of rheumatoid arthritis. Arthritis Rheum. 2011;63(11):3226–33.
62. Cantaert T, Brouard S, Thurlings RM, Pallier A, Salinas GF, Braud C, et al. Alterations of the synovial T cell repertoire in anti-citrullinated protein antibody-positive rheumatoid arthritis. Arthritis Rheum. 2009; 60(7):1944–56.
63. Fritsch R, Eselbock D, Skriner K, Jahn-Schmid B, Scheinecker C, Bohle B, et al. Characterization of autoreactive T cells to the autoantigens heterogeneous nuclear ribonucleoprotein A2 (RA33) and filaggrin in patients with rheumatoid arthritis. J Immunol. 2002;169(2):1068–76.

35 Filaggrin and the Oral Mucosa

Anne Marie Lynge Pedersen and Jesper Reibel

Contents

A.M.L. Pedersen, PhD, DDS (✉)
Section of Oral Medicine, Clinical Oral Physiology, Oral Pathology and Anatomy,
Department of Odontology, Faculty of Health and Medical Sciences, University of Copenhagen, Norre Alle 20, Copenhagen N 2200, Denmark
e-mail: amlp@sund.ku.dk

J. Reibel, PhD
Department of Odontology, Faculty of Health and Medical Sciences, University of Copenhagen, Norre Alle 20, Copenhagen N 2200, Denmark
e-mail: jrei@sund.ku.dk

35.1 Organization of the Oral Mucosa

All areas of the oral mucosa are lined with stratified squamous epithelium; however, marked differences in epithelial differentiation patterns are seen from one area to another. Traditionally, the oral mucosa is divided into lining, masticatory, and specialized types, reflecting the functions in different areas [1]. Lining mucosa, supported by lamina propria and submucosa with underlying striated muscle, is nonkeratinized and characterized by being flexible and distensible in order to accommodate the functions related to speech and manipulating food in the chewing process. Masticatory mucosa, supported by a lamina propria firmly attached to the underlying bone, is keratinized and has a rigid surface in order to participate in macerating food and to resist the mechanical friction related to mastication. Specialized mucosa is represented by the dorsal surface of the tongue containing papillae with taste buds and papillae participating in the chewing and swallowing process. The latter papillae are keratinized, whereas most other areas are nonkeratinized.

In contrast to skin, hair follicles and sweat glands are not present in the oral mucosa. Sebaceous glands, however, are present in many individuals in the labial and buccal mucosa sometimes presenting clinically as yellow spots (Fordyce's spots). Their presence is not easily explainable, as lubrication of the oral mucosa is taken care of by the salivary glands, which are

J.P. Thyssen, H. Maibach (eds.), *Filaggrin*,
DOI 10.1007/978-3-642-54379-1_35, © Springer-Verlag Berlin Heidelberg 2014

responsible for the moist character of the mucosa and furthermore have an important protective function as a first line of defense.

35.2 Structure of Oral Epithelia

The oral epithelium is thick compared to the epidermis in most areas of the skin. Interfollicular epidermis in the iliac crest region is 120/35 μm thick in epithelial ridges/connective tissue papillae regions. The corresponding figures for buccal mucosa are 580/150 μm and for palatal mucosa, 310/120 μm [2]. Nonkeratinized mucosa contains a mitotically active basal layer that advances through a spinous cell layer to a superficial layer where cells flatten and desquamate at the surface [2] (Fig. 35.1). Buccal and labial epithelia are thick with an undulated epithelial connective tissue interface formed by epithelial rete ridges and connective tissue papillae, whereas the epithelium of the floor of the mouth is comparably thin, with a rather straight epithelium connective tissue interface. Keratinized palatal and gingival epithelium differ in keratinization type, as palatal epithelium usually is orthokeratinized (Fig. 35.2) and gingival epithelium is parakeratinized (Fig. 35.3), although both keratinization types can be seen in both regions. Thus, the general stratification

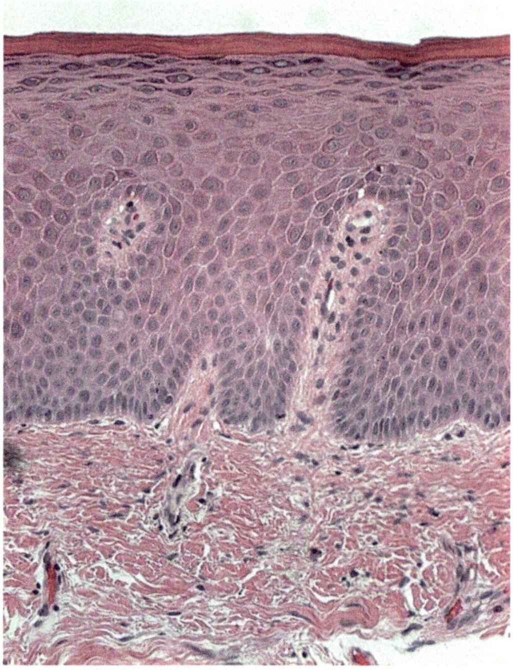

Fig. 35.2 Normal orthokeratinized human palatal epithelium

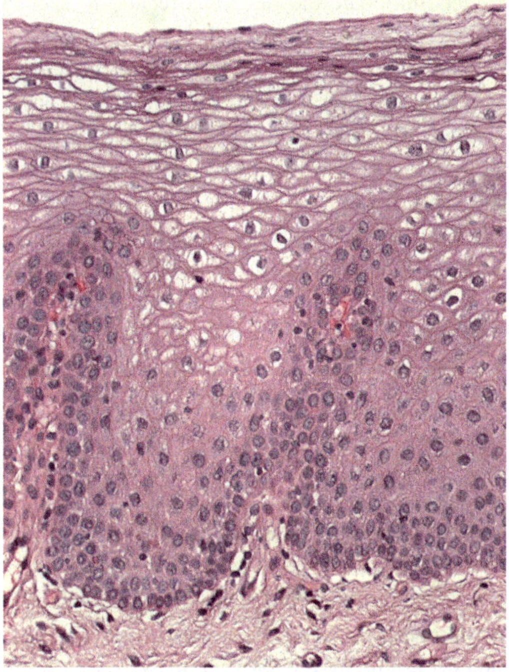

Fig. 35.1 Normal nonkeratinized human buccal epithelium

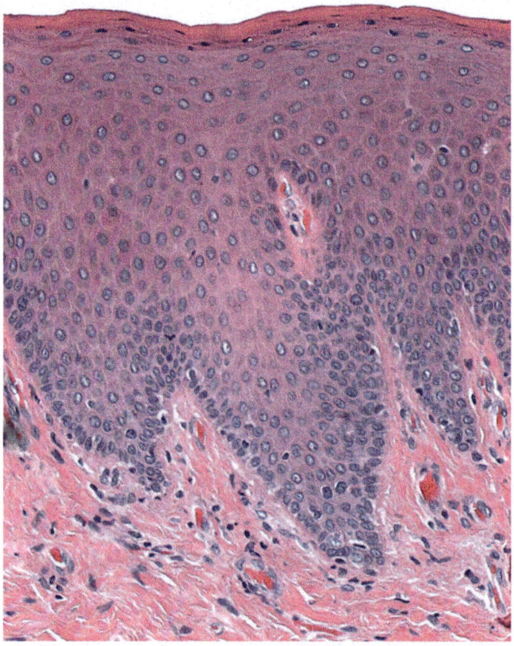

Fig. 35.3 Normal parakeratinized human gingival epithelium

of palatal epithelium includes a basal cell layer, a spinous cell layer, a granular cell layer, and a stratum corneum of orthokeratin, whereas the gingival epithelium lacks a distinct granular cell layer and the stratum corneum consists of parakeratin with pycnotic nuclei. Both epithelia are rather thick, with well-formed epithelial rete ridges in the gingiva with a slender and variable appearance.

35.3 The Permeability of the Oral Mucosa

The oral epithelia are relatively leaky, being more than ten times more permeable to water compared to the epidermis [3]. The difference in permeability varies considerably between various regions in the oral cavity. The keratinized palate and gingiva have the lowest permeability, followed by the buccal mucosa, whereas the area with highest permeability is the sublingual mucosa/floor of mouth. In areas of damaged mucosa, the barrier may be impaired, leading to an increased permeability [4]. The turnover time for the buccal epithelium has been estimated at 5–8 days [5], which is most likely representative for the oral mucosa as a whole.

The keratinized epithelia contain neutral lipids like ceramides and acylceramides, which have been associated with the barrier function. These epithelia are relatively impermeable to water. In contrast, nonkeratinized epithelia, such as the floor of the mouth and the buccal epithelia, only contain small amounts of ceramides and no acylceramides. Furthermore, the permeability barrier in the oral mucosa may also be a result of the presence of membrane-coating granules [6]. These are made when cells undergo differentiation, and at the apical cell surfaces, they fuse with the plasma membrane, and their contents are released into the intercellular spaces at the upper third of the epithelium [7]. This barrier comprises about 200 μm of the superficial layer. The membrane-coating granules of keratinized epithelium are composed of lamellar lipid stacks, while the nonkeratinized epithelium contains non-lamellar membrane-coating granules. The lipids of the membrane-coating granules in the keratinized epithelia include ceramides, glucosylceramides, and sphingomyelin, whereas the lipids in nonkeratinized epithelia comprise cholesterol esters, cholesterol, and glycosphingolipids [7]. The difference in lipid composition may contribute to explain why water permeability is higher in the nonkeratinized epithelia and the fact that larger molecules like toxins and enzymes more easily penetrate keratinized than nonkeratinized epithelia [7].

The surface layer of the oral epithelia is, under normal conditions, bathed in saliva. One of the functions of saliva is to dilute and remove surface material, food substances, bacteria, etc., from the oral cavity but also to form a barrier layer of the oral mucosa [8]. In this regard, salivary mucins like MG1 and MG2 play an essential role as they contribute to cell-cell adhesion, lubrication of the oral mucosa, and limitation of the microbial attachment to the mucosal surface [9]. More than 70 % of the mucins are secreted by the minor salivary glands.

Saliva plays a major protective role for the hard and soft tissues in the oral cavity. It lubricates the tissues and protects them from abrasion by rough materials and from chemicals. Saliva comprises 1 % organic and inorganic substances and 90 % water. The salivary pH varies from 5.5 to 7, depending on the flow rate. At high flow rates, the sodium chloride and bicarbonate concentrations increase, leading to an increase in the pH. Under normal conditions, the daily saliva volume is between 0.5 and 1 L.

Salivary gland hypofunction often leads to oral mucosal problems like erosions and ulcers and symptoms including sensation of oral dryness as well as itching and burning most likely due to increased permeability of the oral mucosa [8].

35.4 Molecular Differentiation in Oral Epithelium

35.4.1 Keratins and Cornified Envelope Proteins

The expression of the major structural proteins in the oral epithelium, the keratins, varies in lining and masticatory areas [10, 11]. In all areas, like in the epidermis, the basal cells possess keratin 5 and 14, and in palatal and gingival epithelia, the suprabasal cells contain keratin 1 and 10, resembling the pattern in epidermis. Nonkeratinized lining epithelia, however, contain keratin 4 and 13 in suprabasal layers. Keratin 19 is sometimes present in nonkeratinized epithelia, and keratin 16 sometimes can be isolated from normal palatal epithelium [10]. In diseased states, normally nonkeratinized epithelia can become keratinized, e.g., leukoplakia, and a shift in keratin expression is seen as suprabasal cells cease expressing keratin 4 and 13 and now express keratin 1 and 10 like normal palatal epithelium [12]. Thus, the terminal differential pattern can change profoundly in oral epithelia.

Suprabasally, the spinous cells synthesize proteins engaging in the later process of thickening of the cell membrane by cross-linking of proteins by transglutaminases at the interior aspect of the cell membranes (cornified envelopes). In oral epithelia, involucrin and transglutaminase can be detected suprabasally in keratinized as well as nonkeratinized epithelia [11, 13]. In the latter epithelia, the expression of involucrin and transglutaminase begins at a lower stage in the spinous cell layer than in keratinized epithelium [11]. Thus, although it is discussed if cornified envelopes are seen in nonkeratinized epithelium [2], the proteins and the enzyme necessary for their formation are present.

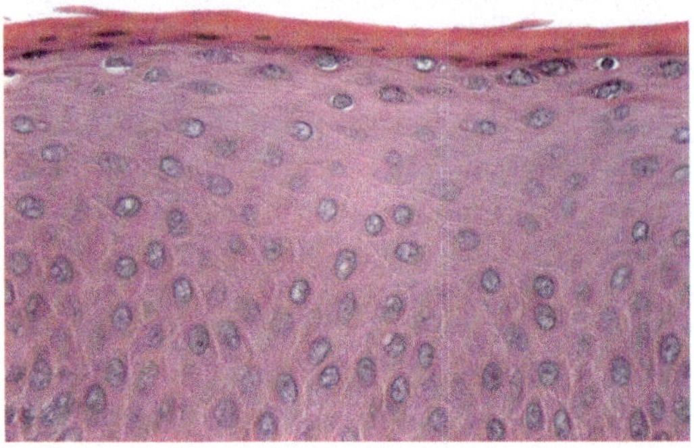

Fig. 35.4 Superficial part of normal parakeratinized human gingival epithelium showing scattered keratohyaline granules in few cells immediately beneath the cornified layer

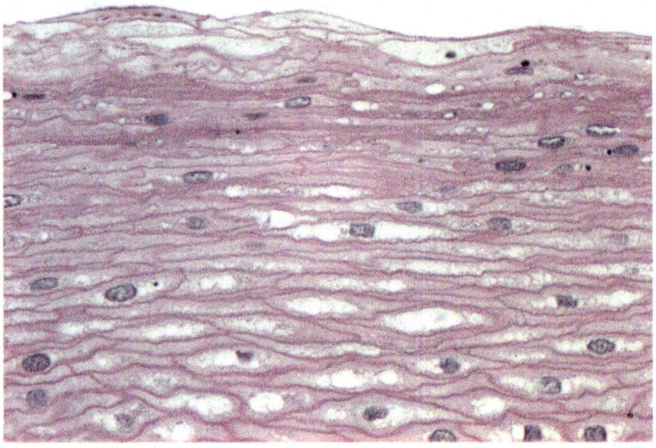

Fig. 35.5 Superficial part of normal nonkeratinized human buccal epithelium showing very few scattered keratohyaline granules

35.5 Filaggrin

In the oral mucosal epithelium, distinct keratohyaline granules are present in the granular cell layer of orthokeratinized epithelium (see Fig. 35.2) and in a more diffuse pattern in parakeratinized epithelium (Fig. 35.4). In nonkeratinized epithelium, few scattered keratohyaline granules can often be found in some specimens (Fig. 35.5) [2, 14]. However, some studies do not report on keratohyaline granules in nonkeratinized oral epithelium [15].

Immunohistochemical and protein analyses of oral epithelia are scarce. Antibodies against profilaggrin/filaggrin have shown the presence of these proteins in orthokeratinized and parakeratinized specimens (Fig. 35.6) [11, 15–18]. However, in nonkeratinized specimens, the presence of profilaggrin/filaggrin is variably reported, as some studies show the presence of these proteins in a weak and scattered pattern (Fig. 35.7) [11, 15, 16, 19], whereas others did not detect their presence [20]. The reported presence in nonkeratinized oral epithelia may be caused by a tendency to hyperkeratinization caused by physical stimuli [15], as also suggested in an electron microscopic study [14]. This is in line with a study showing by RT-PCR that oral keratinocytes derived from smokers overexpressed filaggrin compared to those from nonsmokers [21] and with a study showing immunoreactivity in organotypic cultures exposed to khat but not in control organotypic samples [22].

The supposed functions of filaggrin, other than participating in the keratinization process, are in line with the presence in nonkeratinized epithelia; however, the weak and variable expression suggests that epithelial homeostasis can be

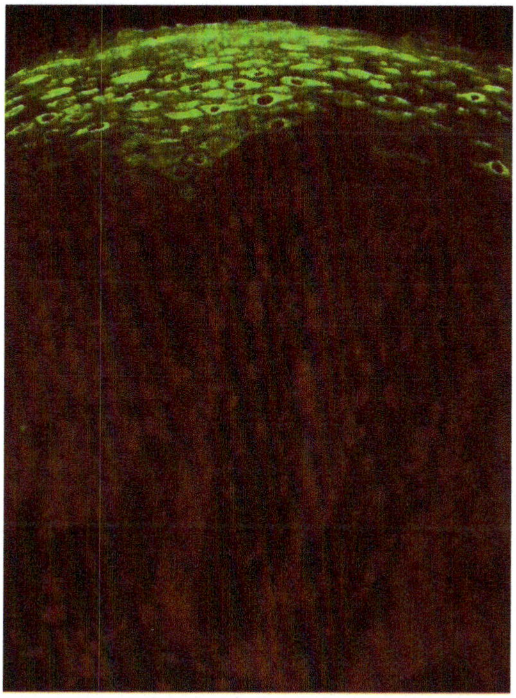

Fig. 35.6 Normal orthokeratinized human palatal epithelium stained with antibodies against profilaggrin as earlier described [11]. Dense staining of granular cell layer

Fig. 35.7 Normal nonkeratinized human buccal epithelium stained with antibodies against profilaggrin as described earlier [11]. Scattered positive granules in the superficial part of the epithelium

achieved without the presence of profilaggrin/filaggrin. In diseased states characterized by hyperkeratosis (keratinization of the normally nonkeratinized buccal mucosa) such as hyperkeratinized irritational hyperplasias ("fibromas"), leukoplakia and lichen planus, keratin expression is induced in a pattern comparable to the normal expression in palatal epithelium [11, 17, 20].

35.6 A Role for Filaggrin in Oral Contact Allergic Lesions?

Preliminary results of recent studies indicate that filaggrin may play a role in relation to the development of contact allergy (Pedersen and Reibel, unpublished data). Contact allergic lesions in the oral cavity are often caused by dental materials such as amalgam, nickel, chromium, cobalt, or composite but also by additives used in dental hygiene products as well as in food substances. The clinical manifestations display large diversity, ranging from localized lichenoid lesions, diffuse erythematous changes like stomatitis, swellings, and irritation of labial mucosa (cheilitis) and gingiva, perioral dermatitis, vesicles, aphtae, and ulcers [23]. The symptoms may be an itching, burning, and/or stinging sensation in the oral mucosa, oral dryness, and taste disturbances. Itching, especially in the palate, usually occurs in the oral allergy syndrome, which is related to type I allergy. The symptoms, the clinical manifestations as well as the histopathological and immunopathological manifestations, in oral contact allergy may be difficult to differentiate from those seen in oral mucosal diseases like oral lichen planus [23], pemphigus vulgaris, pemphigoid, discoid lupus erythematosus, and oral candidiasis. Diagnosis and treatment of these reactions, therefore, display a significant challenge in the clinic.

The importance of filaggrin in relation to maintain a normal barrier function in the epithelium has been substantiated in recent studies on genes coding for profilaggrin. Filaggrin gene (*FLG*) mutations may result in development of a number of cutaneous diseases like atopic dermatitis [24]. Similarly, *FLG* mutations may also play a role in development of oral lesions and oral contact allergy. The integrity of the oral mucosa is mainly maintained by the structural components of the epithelium, including the keratin filaments. They comprise a coherent skeleton in the epithelium that via binding to cell contacts terminally forms stratum corneum by interaction with filaggrin, which functions as a matrix [25]. Filaggrin and its metabolites bind nickel, and it is possible that it also binds other metals and may also be of importance to the penetration and accumulation of amalgam and consequently influence the risk of allergy to amalgam. Along this line, the results of a recent study on mice (tails) suggest that the stratum corneum barrier can be weakened against penetration of particular small metal ions, e.g., Cr (chromium) (III), by filaggrin deficiency [26].

References

1. Squier CA, Kremer MJ. Biology of oral mucosa and esophagus. J Natl Cancer Inst Monogr. 2001;29:7–15.
2. Schroeder HE. Differentiation of human stratified epithelia. Basel: S. Karger; 1981.
3. Lesch CA, Squier CA, Cruchley A, Williams DM, Speight P. The permeability of human oral mucosa and skin to water. J Dent Res. 1989;68(9):1345–9.
4. Riber E, Kaaber S. Barrier properties of inflamed denture-loaded palatal mucosa to water. Scand J Dent Res. 1978;86(5):386–91.
5. Hill MW. Cell renewal in oral epithelia. In: Meyer J, Squier CA, Gerson SJ, editors. The structure and function of oral mucosa. Oxford: Pergamon Press; 1984. p. 53–81.
6. Wertz PW, Squier CA. Cellular and molecular basis of barrier function in oral epithelium. Crit Rev Ther Drug Carrier Syst. 1991;8(3):237–69.
7. Squier CA. The permeability of oral mucosa. Crit Rev Oral Biol Med. 1991;2(1):13–32.
8. Pedersen AM, Bardow A, Jensen SB, Nauntofte B. Saliva and gastrointestinal functions of taste, mastication, swallowing and digestion. Oral Dis. 2002;8(3): 117–29.
9. Tabak LA, Levine MJ, Mandel ID, Ellison SA. Role of salivary mucins in the protection of the oral cavity. J Oral Pathol. 1982;11(1):1–17.
10. Clausen H, Vedtofte P, Moe D, Dabelsteen E, Sun T-T. Differentiation-dependent expression of keratins in human oral epithelia. J Invest Dermatol. 1986;86: 249–54.
11. Reibel J, Clausen H, Dale BA, et al. Immunohistochemical analysis of stratum corneum components in oral squamous epithelia. Differentiation. 1989;41:237–44.
12. Reibel J, Clausen H, Dabelsteen E. Staining patterns of human pre-malignant oral epithelium and squamous cell carcinomas by monoclonal anti-keratin antibodies. Acta Pathol Microbiol Immunol Scand A. 1985;93:323–30.
13. Banks-Schlegel S, Green H. Involucrin synthesis and tissue assembly by keratinocytes in natural and cultured human epithelia. J Cell Biol. 1981;90(3):732–7.
14. Silverman S. Ultrastructure studies of oral mucosa. I. Comparison of normal and hyperkeratotic human buccal epithelium. J Dent Res. 1967;46(6):1433–43.
15. Smith SA, Dale BA. Immunologic localization of filaggrin in human oral epithelia and correlation with keratinization. J Invest Dermatol. 1986;86(2):168–72.
16. Itoiz ME, Conti CJ, Lanfranchi HE, Mamrack M, Klein-Szanto AJ. Immunohistochemical detection of filaggrin in preneoplastic and neoplastic lesions of the human oral mucosa. Am J Pathol. 1985;119(3):456–61.
17. Vigneswaran N, Peters KP, Hornstein OP, Haneke E. Comparison of cytokeratin, filaggrin and involucrin profiles in oral leukoplakias and squamous carcinomas. J Oral Pathol Med. 1989;18(7):377–90.
18. De Benedetto A, Qualia CM, Baroody FM, Beck LA. Filaggrin expression in oral, nasal, and esophageal mucosa. J Invest Dermatol. 2008;128(6):1594–7.
19. Dale BA, Gown AM, Fleckman P, Kimball JR, Resing KA. Characterization of two monoclonal antibodies to human epidermal keratohyalin: reactivity with filaggrin and related proteins. J Invest Dermatol. 1987;88(3):306–13.
20. Boisnic S, Ouhayoun JP, Branchet MC, Frances C, Béranger JY, Le Charpentier Y, Szpirglas H. Alteration of cytokeratin expression in oral lichen planus. Oral Surg Oral Med Oral Pathol Oral Radiol Endod. 1995;79(2):207–15.
21. Renò F, Rocchetti V, Migliario M, Rizzi M, Cannas M. Chronic exposure to cigarette smoke increases matrix metalloproteinases and Filaggrin mRNA expression in oral keratinocytes: role of nicotine stimulation. Oral Oncol. 2011;47(9):827–30.
22. Lukandu OM, Neppelberg E, Vintermyr OK, Johannessen AC, Costea DE. Khat alters the phenotype of in vitro-reconstructed human oral mucosa. J Dent Res. 2010;89(3):270–5.
23. Koch P, Bahmer FA. Oral lesions and symptoms related to metals used in dental restorations: a clinical, allergological, and histologic study. J Am Acad Dermatol. 1999;41(3 Pt 1):422–30.
24. Palmer CNA, Irvine AD, Terron-Kwiatkowski A, Zhao Y, Liao H, Lee SP, et al. Common loss-of-function variants of the epidermal barrier protein filaggrin are major predisposing factor for atopic dermatitis. Nat Genet. 2006;38:441–6.
25. Presland RB, Dale BA. Epithelial structural proteins of the skin and oral cavity: function in health and disease. Crit Rev Oral Biol Med. 2000;11: 383–408.
26. Kubo A, Ishizaki I, Kubo A, Kawasaki H, Nagao K, Ohashi Y, Amagai M. The stratum corneum comprises three layers with distinct metal-ion barrier properties. Sci Rep. 2013;3:1731. doi:10.1038/srep01731.

Part VII

Management

Therapeutics and Other Interventions

36

Berit Christina Carlsen

Contents

B.C. Carlsen, MD, PhD
Department of Dermato-Allergology, National Allergy Research Centre, Gentofte Hospital, University of Copenhagen,
Niels Andersens Vej 65, Hellerup 2900, Denmark
e-mail: bccarlsen@dadlnet.dk

36.1 Introduction

The improved understanding of the filaggrin pathway and the secondary and tertiary consequences of dysfunction on skin barrier homeostasis have revealed several potential targets for future novel treatments. Superior to them all is enhancement or replacement of the missing components pro-filaggrin, filaggrin, and natural moisturizing factors (NMFs). A large spectrum of cutaneous and non-cutaneous diseases is associated with the filaggrin gene (*FLG*) mutations, which hold potential for wide applicability for any new drug targeting filaggrin or other parts of the filaggrin cascade. There are reasons to be optimistic that such drugs can be used both to control disease and also for primary prevention. Non-pharmacological intervention should be directed against minimizing or avoiding exposure to environmental dangers known to deteriorate the skin barrier or cause secondary disease, which may have detrimental effects on already vulnerable skin.

This chapter focuses on targets for future novel treatments, promising new candidate drugs and their relation to the individual components of the filaggrin pathway. A discussion on current available treatments and their ability to restore the skin barrier is also provided as well as the applicability of filaggrin pathway-targeting therapeutics in relation to disease spectrum, primary and secondary intervention, and personalized treatment strategies. Finally, an overview of other important non-pharmacological interventions that form the basis for counseling strategies is given.

J.P. Thyssen, H. Maibach (eds.), *Filaggrin*,
DOI 10.1007/978-3-642-54379-1_36, © Springer-Verlag Berlin Heidelberg 2014

36.2 Therapeutics: Restoring the Skin Barrier

It is important to understand the properties and functions of the various components in the filaggrin pathway and the secondary and tertiary consequences of dysfunction, in order to fully appreciate the effects of current treatments and potential targets for future novel treatments relating to filaggrin. The filaggrin pathway is briefly summarized below. Properties and functions of the individual components are summarized in Table 36.1 and consequences of dysfunction summarized in Table 36.2.

Processing of pro-filaggrin via filaggrin into NMFs constitutes the filaggrin pathway. Pro-filaggrin is a large polyprotein consisting of an N-terminal domain with calcium-binding and nucleus-homing properties, 10–12 filaggrin repeats with keratin-binding properties, and a C-terminal domain essential for further processing of pro-filaggrin [1]. Pro-filaggrin is involved in enucleation of keratinocytes important for the cornification of the epidermis [2] and may play a role in proper lamellar body loading and secretion important for a well-organized extracellular lipid lamellar bilayer and proper functioning antimicrobial defense. Lamellar bodies contain lipid precursors, lipid-processing antibodies, and antimicrobial peptides [3].

Filaggrin monomers are produced by dephosphorylation and proteolysis of the pro-filaggrin molecule. Filaggrin acts as a keratin filament aggregator necessary for the intracellular cytoskeleton formation. The cytoskeleton collapses the cells of the epidermis into the well-known flattened appearance of the stratum corneum [2] and acts as a scaffold for the extracellular lamellar bilayers [3].

Subsequent proteolysis and deiminition of the filaggrin monomers produce free amino acids and

Table 36.1 Properties and function of the individual components in the filaggrin pathway

Pro-filaggrin	Filaggrin	Natural moisturizing factor
Cornification of the epidermis by enucleation of keratinocytes and calcium-binding properties	Cytoskeleton formation by aggregation of keratin filaments	Maintenance of skin hydration by osmolytic retention of water
Proper lamellar body loading and secretion important for lamellar bilayer formation, skin cohesion, and antimicrobial defense	Disruption of the nucleus and nuclear envelope	Maintenance of proper antimicrobial activity by acidification of the skin surface and secondary to proper lipid metabolism
	Scaffold for extracellular lipid lamellar bilayers	Maintenance of proper skin lipid metabolism via skin pH-regulatory effects on enzyme activity
		Maintenance of proper skin cohesion via skin pH-regulatory effects on enzyme activity
		Immunomodulating effects via skin pH-regulatory effects on enzyme activity
		Direct photo-protective functions by trans/cis conversion of urocanic acid

Table 36.2 Abnormalities observed in *FLG*-deficient dysfunctional skin

Skin architecture	Extracellular lipid phase skin lipid profile	Skin cohesion
Disorganization of keratin filaments and the intracellular cytoskeleton	Abnormal extracellular lipid lamellar bilayer organization	Reduced corneodesmosome density
Reduced levels of filaggrin	Abnormal extracellular lipid lamellar bilayer distribution	Reduced tight junction formation
Reduced levels of natural moisturizing factors	Abnormal extracellular lamellar bilayer maturation	
Paucity of keratohyalin granules	Abnormal skin lipid composition	
Paucity of cell layers of stratum granulosum	Impaired loading of intracellular lamellar bodies	
Parakeratosis and hyperkeratosis	Inhomogeneous secretion of intracellular lamellar bodies	

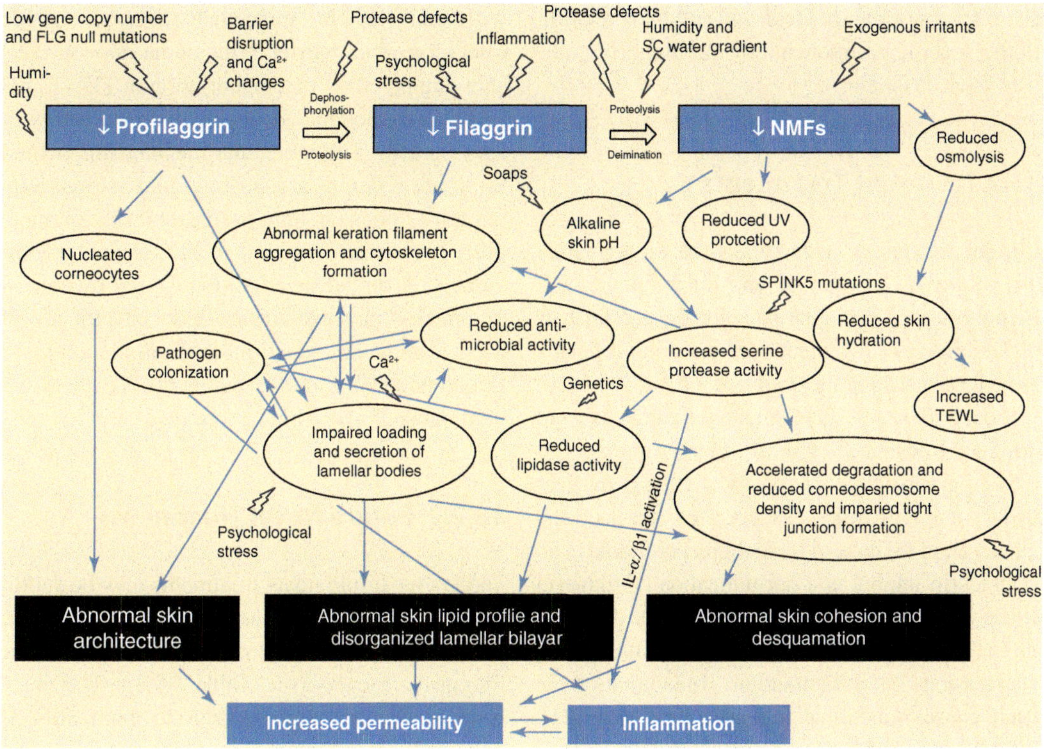

Fig. 36.1 The dysfunctional filaggrin cascade and intercorrelations. *Pro-filaggrin* pro-filaggrin, *NMF* natural moisturizing factor, *SC* stratum corneum, *TEWL* transepidermal water loss

derivatives hereof, the so-called NMFs [4]. NMFs maintain skin hydration by osmolytic effects and stabilize skin pH important for proper antimicrobial barrier function, skin cohesion, immunomodulation, and skin lipid metabolism via pH-regulatory effects of enzyme activity and protect the epidermis against UV-induced DNA damage via cis/trans conversion of urocanic acid (UCA) [5–9].

When the filaggrin pathway is upset, disturbances in skin architecture [5, 10–12], skin cohesion [12], and lipid processing and extracellular lipid phase [5, 7, 12–14] are noticed (see Table 36.2). These secondary events are highly intercorrelated, illustrated in Fig. 36.1. Ultimately, the skin barrier is impaired, displaying disturbances in biophysical properties, e.g., alkaline pH [10, 15, 16], reduced skin hydration [15], increased permeability [14, 16], and increased transepidermal water loss (TEWL) both at baseline [11, 15, 16] and aggravated under compromised conditions [13, 15]. Dry skin [17], inflammation [14, 18], and, ultimately, clinical disease are the tertiary consequences.

Table 36.3 Potential targets for future novel treatments

Level of intervention	Targets
Primary	Enhancement of pro-filaggrin processing
	Enhancement or Replacement of filaggrin
	Replacement of natural moisturizing factors
Secondary	Restoring extracellular skin lipid profile
	Restoring skin pH levels
	Restoring skin hydration
Tertiary	Controlling inflammation

Evidently, the optimal treatment scope in filaggrin-deficient individuals is to restore the skin barrier. This may theoretically be achieved by targeting the missing components, targeting the secondary abnormalities, or addressing the tertiary inflammation. Potential targets and level of intervention are illustrated in Table 36.3. How this relates to and is achieved by current treatments or possibly can be achieved by future novel treatments is discussed below, along with promising new candidate drugs and their relation to the specific components of the filaggrin pathway mentioned above.

36.2.1 Current Treatments

Current treatments within the field of dermatitis and ichthyosis consist of anti-inflammatory and immunosuppressive pharmaceuticals, modulators of keratinocyte differentiation, keratolytic agents as well as basic treatment with emollients with and without humectants. None of these treatments were designed to enhance or substitute missing parts of the filaggrin pathway or specifically target the secondary abnormalities except for humectants. However, in the case of topical glucocorticoids and calcineurin inhibitors a normalization of filaggrin levels have been noticed upon treatment [19, 20] probably by means of a secondary effect related to suppression of inflammation. Inflammation in itself causes functional filaggrin deficiency [21]. Intensified treatment with current available anti-inflammatory drugs would be one therapeutic strategy in order to try to restore the skin barrier and control skin disease in filaggrin-deficient individuals where inflammation is present.

There is, however, is evidence of reduced effect of the current available treatments in individuals with *FLG* mutations compared to wild-type individuals, [22]. In particular, atopic dermatitis (AD) patients carrying the R501X mutation were two times more likely to use topical steroids and the least responsive to therapy [22]. Additional prospective studies assessing treatment outcome in *FLG* mutation carriers are ongoing (http://ClinicalTrials.gov/show/NCT01689805 and http://ClinicalTrials.gov/show/NCT01569906). Topical glucocorticoids and calcineurin inhibitors impair epidermal differentiation by a reduced expression of epidermal differentiation proteins [20, 23], by decreased epidermal lipid synthesis, by decreased density and secretion of intracellular lamellar bodies [23, 24], and by decreased density of corneodesmosomes and tight junction proteins [25]. Keratolytic agents and retinoids (modulators of skin differentiation) worsen skin permeability objectified by increased levels of TEWL [26, 27]. Finally, non-physiological emollients with and without humectants have a variable effect on skin barrier with some moisturizers showing no effect or a deteoriation in TEWL, pH and skin capacitance in healthy skin and ichthyosis skin [28]. The molecular mechanisms behind these observations are not fully understood. Skin integrity, cohesion, lipid profile, and skin permeability are already affected in FLG-deficient individuals. A further deterioration caused by current treatments may explain the reduced efficacy of these drugs in filaggrin-deficient individuals. Development of new treatments is therefore warranted.

36.2.2 Future Novel Treatments

Targets for future novel treatments may be subdivided into three areas based on whether they target primary, secondary, or tertiary events in the filaggrin cascade (see Table 36.3). The most obvious target – and superior to them all – is enhancement or replacement of the missing parts: pro-filaggrin, filaggrin, and NMFs. This may potentially correct all of the underlying secondary defects caused by filaggrin deficiency. Another approach would be to correct the secondary imbalances, e.g., skin lipid organization and composition, skin pH, and skin hydration, and the consequential inflammation in a more specific manner than achieved by the current available drugs. Treatments targeting the secondary or tertiary events may be beneficial in individuals, where filaggrin deficiency is secondary to inflammation or abnormal processing of pro-filaggrin and filaggrin.

Studies on candidate drugs are still preliminary. Very few studies have been performed in humans and, in these cases, primarily in healthy skin. Studies performed in diseased skin are, in most cases, based on individuals with AD without assessing their *FLG* mutation carrier status. The majority of studies are, however, in vitro and animal studies. Data from monolayer keratinocyte in vitro studies cannot be extrapolated to human conditions, and data from animal studies may be rodent specific. Data from AD individuals and data generated from studies in healthy skin cannot necessarily be extrapolated to filaggrin-deficient skin. Further research into candidate drugs, testing their efficacy and safety in filaggrin-deficient skin, is highly warranted.

36.2.3 Filaggrin Enhancement and Replacement

Data from a filaggrin copy number variation study indicate that even a small increase in filaggrin level by around 5–10 % substantially reduces the risk of AD [29]. To support this, genotypic configuration, filaggrin protein levels, and phenotype expression are positively correlated in a dose-dependent manner [12, 16, 30]. There seems to be a lot to gain, if increased levels of filaggrin can be achieved.

Ongoing research into identifying molecules with filaggrin-enhancing properties is currently being performed. Candidates may be peroxisome proliferators and antimicrobial peptides. *FLG* mutation heterozygous individuals, but not homozygous individuals, may benefit from such therapeutics. Heterozygous individuals possess one wild-type allele that potentially can be upregulated to produce more filaggrin to compensate for the defective allele. However, some heterozygous individuals already express normal levels of filaggrin despite having one mutated allele [21]. It remains to be determined whether a filaggrin overexpression can be reached in heterozygous individuals.

Peroxisome proliferator-activated receptors (PPARs) are nuclear hormone receptors activated by endogenous lipids. Topical peroxisome proliferator-activated receptor-β/δ agonists (GW1514), PPAR-α agonists (clofibrate, docosahexaenoic acid, and WY-14,643), and PPAR-γ agonists (ciglitazone and troglitazone) all stimulate filaggrin production and keratinocyte differentiation in vitro and in mice [31–33]. PPAR-β/δ agonists also induce keratinocyte differentiation in animal models with abnormal epidermis, suggesting that they may be beneficial in diseases with dysfunctional differentiation. In addition to increasing filaggrin levels, PPAR-β/δ agonists exhibit an anti-inflammatory effect exceeding the effect of topical glucocorticoids, whereas PPAR-γ agonists reduce inflammation with an effect comparable to clobetasol [33]. An increase in epidermal lipids may potentially also be reached by topical application of PPAR-α (WY14643), PPAR-δ (GW1514), and PPAR-γ (ciglitazone) agonists [34]. On the downside, PPAR agonists exhibit antiproliferative properties and induce apoptosis in varying degrees [31–33]. Thiazolidinediones (PPAR-γ agonists) are already used in diabetes treatment. An efficacy study of docosahexaenoic acid, a PPAR-α agonist, on skin hydration and skin barrier in humans is currently performed (http://ClinicalTrials.gov/show/NCT01305057).

Studies on other candidate drugs with upregulatory effects on filaggrin expression have been published. For each candidate drug, only single studies have yet assessed their efficacy by means of either in vitro or animal models. These studies need to be replicated. The candidates include sericin-rich diets (animal study) [35]; apigenin, an active constituent in chrysanthemum (in vitro study) [36]; interferon-γ (in vitro study) [21]; hyaluronic acid; and bacterial antimicrobial peptide plantaricin A (in vitro study) [37]. Urea, a well-known humectant, also possesses filaggrin upregulatory effects and may prove more versatile than previously assumed [38].

All known *FLG* mutations are null mutations resulting in a truncated pro-filaggrin molecule [39] that cannot be further processed. Mutant mRNA is degraded by the nonsense-mediated decay (NMD) pathway [40]. If degradation of mutant mRNA can be stopped, drugs able to read through the nonsense mutations could potentially restore protein levels. Both *FLG* mutation homozygous and heterozygous individuals may benefit from such treatment. Read-through could be achieved either by skipping of the nonsense-mutation-containing exon during RNA splicing or by incorporation of amino acids at the mutation site. Certain kinds of drugs are known to let cells read through nonsense mutations. These include aminoglycoside members gentamicin, G418 (Geneticin), amikacin, PTC124 (Ataluren), the antibiotic negamycin [41, 42], and the recently reported drug amlexanox [43]. Read-through drugs are currently studied in in vitro studies and in clinical trials for other genetic diseases (e.g., cystic fibrosis, Duchenne muscular dystrophy, coagulation factor deficiencies) and for cancer therapy [41, 42, 43, 45] with promising results.

Several factors need to be considered before such drugs can go into clinical trials. First of all, aminoglycosides have a narrow therapeutic index. Second, the efficacy of the same read-through-inducing molecule varies, depending on the stop codon [41, 46] present and on the nucleotide context upstream and downstream from the stop codon [41, 46], creating potential for large interindividual variability in efficacy. Finally, limited amounts of mutant mRNA may decrease read-through efficiency. Candidates should possess both NMD inhibitory and read-through-inducing properties. The development of read-through drugs for the treatment of ichthyosis vulgaris and atopic conditions is patented.

36.2.4 Replacement of NMFs

The pool of NMFs consists of filaggrin degradation products (e.g., free amino acids, pyrrolidone carboxylic acid, and UCA) but also of non-filaggrin-derived inorganic salts, lactate, and urea [4]. NMFs regulate skin hydration, and filaggrin-deficient skin shows both reduced levels of NMFs and reduced skin hydration [11]. Replacement by topical application constitutes the second potential target for future novel treatments.

Two studies have studied the effects of topically applied NMFs on xerosis. Unfortunately, *FLG* mutation carrier status was not determined in the trial subjects in either one of the studies. In the first study [47], two topical formulations were examined containing the active ingredients glyceryl glucoside, NMFs, and ceramide, and, therefore, not NMFs alone. They differed in the vehicle ingredients. Improvement in skin hydration assessed by corneometry and in TEWL was noted when the topical formulations with the active ingredients were compared to control skin, but not when compared to a vehicle formulation supplemented with urea, sodium lactate, and lactic acid. Urea and lactate are known humectants utilized in currently available emollients. It remains to be determined whether the addition of filaggrin-derived NMFs to topical formulations shows an additional beneficial moisturizing effect compared to urea- and lactate-containing formulations. In the second study, application of pyrrolidone acid on human skin did not show any effect on skin hydration even with a fivefold increase in concentration [48]. Topically applied UCA has only been assessed in relation to photo-protection and immunosuppression [49, 50].

36.2.5 Restoring Skin Lipid Profile

The extracellular lipid phase in normal skin is comprised of three key lipids: ceramides, cholesterol, and fatty acids arranged in lamellar bilayers [8]. Only very few studies have studied the extracellular lipid phase in filaggrin-deficient skin with somewhat divergent results. The overall quantities of lamellar bilayers seem normal; however, the bilayers are inhomogeneous, distributed with areas of non-lamellar, electron-dense material and areas of incompletely processed lamellar material dispersed between areas with mature lamellar bilayers [12]. Increased levels of cholesterol, decreased levels of triglycerides, normal levels of free fatty acids, and normal total levels of ceramides but reduced amounts of the ceramide subtype CER(EOH) were noted in one study [13]. Another study did not find any difference in CER(EOH) or cholesterol levels [11]. It has repeatedly been shown both in humans and in mice that the increased skin permeability in filaggrin-deficient skin occurs via a paracellular route [12, 14]. Therapeutics restoring the extracellular lipid phase to normal or near-normal conditions may prove very important. Emollients containing physiological lipids (ceramides, cholesterol, and free fatty acids) are possible candidates.

Current emollients with nonphysiological lipids do not restore the extracellular lipid phase. They work via simple occlusion without any influence on endogenous lipid synthesis. They are not intracellularly processed but disperse into the extracellular phase of the stratum corneum, creating a non-lamellar hydrophobic phase splitting lamellae into clefts with bulks of hydrophobic amorphous material [51]. Physiological lipids, on the other hand, penetrate both intact and disturbed stratum corneum and are taken up by the nucleated cell layers of the epidermis, incorporated into lamellar bodies, and secreted along endogenous lipids into the extracellular space, restoring the lamellar bilayers [51–53].

Several emollients containing physiological lipids have been assessed in animals and humans with AD, where they perform superiorly to nonphysiological lipids on TEWL measurements [51] and perform equally well as moderate-potent TCS and topical calcineurin inhibitors on disease severity scores and biophysical parameters [54, 55]. They have not been assessed in *FLG* mutation carriers.

Several factors need to be addressed concerning physiological lipids. There seems to be a narrow range of molar ratios wherein physiological lipids improve skin barrier [53]; furthermore, the chosen mixture of lipids (e.g., identity of lipids [53], absolute quantities, and type of skin assessed [52, 53, 56]) is not unimportant. Application of only ceramides, fatty acids, or cholesterol impedes barrier repair [52], whereas equimolar ratios of ceramides, fatty acids, and cholesterol in the same mix allow normal skin recovery [52, 52]. It is crucial to understand the lipid changes in filaggrin-deficient skin and to test various mixtures of physiological lipids (type of lipids, relative and absolute quantities) in filaggrin-deficient individuals in future research.

36.2.6 Superacidification: Restoring Skin pH

Skin pH is increased in filaggrin-deficient skin [10, 15, 16]. Neutral to alkaline pH increases serine protease activity that accelerates degradation of corneodesmosomes and deactivates lipid-processing enzymes [57]. Increased serine protease activity also directly stimulates the release of pro-inflammatory cytokines IL-1α and β [8, 10] and reduces antimicrobial activity [9]. Acidification of the skin may counteract the secondary abnormalities caused by elevated skin pH in filaggrin-deficient skin. Superacidification of murine skin improves barrier homeostasis [58] without induction of inflammation. On the contrary, treating healthy humans with creams of different pH values (respectively, 4.0 and 7.5) did not result in any pH-related differences in TEWL, blood flow, or skin capacitance [59]. The divergent results may be the result of use of different acids with varying effect on skin barrier, varying bioavailability, and lack of compatibility between the models studied. Whether or not acidification of the skin leads to barrier improvement in filaggrin-deficient human skin has not been investigated and remains to be determined.

36.2.7 Cytokine and Inflammation Control

Several cytokines downregulate both mRNA and protein levels of filaggrin along other keratinocyte differentiation markers in in vitro studies. These include the Th2 cytokines IL-4 and IL-13 [21], IL-25 produced by eosinophils, basophils and dendritic cells [60], and IL-22 produced by Th17 and Th22 cells [61]. Cytokine-antagonizing drugs rather than a general anti-inflammatory effect exhibited by current treatments may be a target for future treatments.

Dupilumab is an IL-4 receptor (IL-4R)α mutein that blocks the IL-4Rα subunit involved in the signaling pathways of both IL-4 and IL-13. Data from a phase 1B study in AD was recently released (dupilumab, Sanofi/Regeneron) and showed promising improvement in global assessment and severity scores after 4 weeks of treatment [62]. Unfortunately, no data on *FLG* mutation carrier status in the AD test subjects were presented. Another IL-4Rα mutein, Pitrakinra by Bayer, is also currently investigated. Preliminary results from a phase II clinical trial in patients with AD showed reduced levels of skin inflammation [63]. A third IL-4R mutein, Aeroderm, did not, however, improve disease severity scores compared with placebo but reduced the level of eczema exacerbation [64]. Previous testing of anti-IL-13 did not achieve any effect on filaggrin [65].

Other possible targets may be IL-17 and the IL-1 receptor (IL-1R). Inflammation in filaggrin-deficient mice is dominated by Th17 cells [18], and the pro-inflammatory cytokines IL-1α and β and IL-1R are increased in both humans with *FLG* null mutations and in a murine model [10]. An interleukin-1 receptor antagonist, anakinra, is currently used in rheumatoid arthritis but has also been used off-label in AD [66]. Whether or not IL-1 blockade will have beneficial or deleterious effects on the skin barrier in filaggrin-deficient skin remains to be determined.

36.2.8 Additional Targets

Inhibition of serine proteases is the final possible target for future novel treatments. Serine protease activity is upregulated secondarily to an increased pH in filaggrin-deficient skin. One study treated patients with Netherton syndrome for 3 weeks with topically applied protease inhibitors. No effect compared with placebo could be documented [67]. Only five subjects were included.

36.3 Applicability of FLG Pathway-Directed Therapeutics

36.3.1 Disease Spectrum

All diseases with affected filaggrin levels may potentially benefit from filaggrin-enhancing therapeutics. Four different mechanisms result in lack of filaggrin: null mutations in the pro-FLG gene [38], functional deficiency driven by inflammation [21], reduced levels due to low intragenic copy number [27], or by means of genetic or inflammation-driven variation in enzymes processing pro-filaggrin or filaggrin [61, 68, 69].

FLG mutations are associated with a wide variety of relatively common cutaneous and non-cutaneous diseases and skin morphological features. A full list is given in Table 36.4. This large spectrum of diseases, combined with their relatively high prevalences and the frequent occurrence of *FLG* mutations in the general population, holds potential for wide applicability for any new drug targeting the filaggrin pathway.

Diseases with defects in other proteins besides filaggrin involved in the terminal differentiation or defects in lipid-processing enzymes may also benefit from filaggrin pathway-directed therapeutics. Such diseases result in secondary changes very similar to changes seen in filaggrin-deficient skin and present with ichthyotic phenotypes [81, 82].

36.3.2 Efficacy and Individualized Medicine

Expectations of efficacy of any new filaggrin-directed therapeutics should be attuned to the level of causality between the *FLG* mutations and the individual diseases, as this varies (see Table 36.4). *FLG* mutations are both causal, risk factors and modifiers of disease. The greatest effect may be warranted for diseases in which filaggrin deficiency is causal if filaggrin levels can be restored completely.

Diseases in which *FLG* mutations act as cofactor may also require other interventions

Table 36.4 Diseases associated with *FLG* null mutations

Level of causality	Diseases		
	Cutaneous diseases	Non-cutaneous diseases	Skin morphological features
Causal	Ichthyosis vulgaris [28]		
Cofactor	Atopic dermatitis (AD) [68]	Asthma in combination with AD [69]	Keratosis pilaris [28, 70]
	Nickel allergy [71]	Allergic rhinitis [69]	Palmar hyperlinearity [28, 70]
	Eczema herpeticum [72]	Aeroallergen sensitization [69]	Dry skin [17]
	Staphylococcal skin infection [70]	Peanut allergy [73]	Fissured skin [74]
	Hand eczema [75, 76]		
	Irritant contact dermatitis [75, 77]		
	Allergic contact dermatitis in combination with AD [78]		

besides filaggrin-directed therapeutics in order to firmly control disease. AD is the model disease. The underlying pathogenesis is a complex interplay between a dysfunctional skin barrier and abnormalities relating to inflammation and the immune system [83]. Both environmental and genetic predisposing factors are involved in the etiology of disease [83]. Only 15 % of individuals with mild to moderate AD and 30–50 % with moderate to severe disease have *FLG* mutations [84, 85], leaving 50 % or more still experiencing disease despite possessing a wild-type genotype. *FLG* mutations are the single most important genetic factor for the development of this disease; however, the etiological fraction only amounts to 12–15 % [84], and 33 other polymorphisms in genes associated with either the epidermal skin barrier or the immune system have been linked to AD [86]. It is obvious that case selection is vital for upmost effect of any filaggrin-directed therapeutics in diseases with a multifactorial etiology.

36.3.3 Primary Intervention

An impaired skin barrier can be detected as early as 3 months of age in humans carrying *FLG* mutations, prior to any clinical inflammation [87]. If the skin barrier can be restored at this point in time, there are reasons to be optimistic that early intervention can prevent disease. Preventive studies utilizing petrolatum-based emollient therapy in preterm infants and in children predisposed to atopic disease point toward a protective effect with reductions in the incidence of clinical disease [87, 89]. None of these studies, however, assessed the protective effect of emollient therapy in *FLG* mutation carriers. Even non-cutaneous allergic diseases involved in the atopic march may be prevented by primary intervention; the suspected mechanism is discussed in detail below.

Prospective studies on preventive effects of therapy in filaggrin-deficient individuals are warranted and on their way. These include the UK and US BEEP study treating children with sunflower oil, Aquaphor ointment, or Cetaphil cream (www.beepstudy.org and http://ClinicalTrials.gov/show/NCT01142999); a Finnish study treating infants with Lipikar Balm AP (http://ClinicalTrials.gov/show/NCT01577628); and a study examining the effect of EpiCeram, an FDA-approved emollient containing the physiological lipids ceramides, cholesterol, and free fatty acids in a 3:1:1 molar ratio in an acidic formulation (pH 5.0) [90]. The initial studies on EpiCeram in newborns proved its safety [90].

36.4 Other Interventions

An intact skin barrier provides excellent protection against penetration of exogenous agents. Conversely, when it is impaired, the increased permeability increases the risk of allergen and microbial penetration and secondary disease. Additionally, several factors are known to deteriorate the skin barrier, which may have detrimental effects on already vulnerable skin. These dangers represent important targets for non-pharmacological intervention in patients inherently predisposed to a defective skin barrier and are discussed in detail below. An overview is given in Table 36.5.

36.4.1 Allergen Avoidance

FLG mutation carriers have an increased risk of allergic diseases (e.g., type I sensitizations [aeroallergens and food allergies], allergic rhinitis, and asthma in combination with AD) [71, 84]. *FLG* mutation carriers also have an increased risk of nickel allergy and of type IV sensitizations to other chemicals besides nickel but only in individuals suffering from AD [73, 80]. Filaggrin is not expressed in nasal, bronchial, and likely only modestly in esophageal mucosa [91, 92]. The link between *FLG* mutations and the non-cutaneous allergic diseases is believed to be sensitization via the skin because of enhanced cutaneous allergen penetration, enhanced cutaneous inflammatory responses, and development of systemic inflammatory responses upon topical application of protein allergens [14, 18]. The hypothesis is supported

Table 36.5 Non-pharmacological interventions in filaggrin-deficient individuals: basis for counseling strategies

Non-pharmacological interventions	Examples
Avoidance of excessive exposure to protein allergens	Avoiding topical application of food allergens Avoiding cat ownership Reducing house dust mite levels
Avoidance of excessive topical exposure to chemical haptens	Avoid use of nickel-containing jewelry and other personal items Attention toward proper choice of consumer products and moisturizers Use of preventive measures if excessively exposed occupationally
Avoidance of excessive topical exposure to irritants	Use of acidic cleansers rather than neutral or alkaline cleaners and soaps Avoid frequent hand washing and frequent use of gloves Use of preventive measures both at home and occupationally if excessively exposed
Avoidance of other skin barrier stressors	Minimize psychological stress Use protective measures against cold, dry weather Avoid air-conditioning
UV protection	Use of sunscreens Use of physical UV barriers Rational behavior when exposed to sun
Avoidance of tobacco smoking	

by the findings of a dose-dependent increased risk of peanut allergy in relation to cutaneous peanut exposure either by application of peanut oil to inflamed skin [93] or non-oral exposure to peanut-containing foods [94]. *FLG* mutations are strongly associated with peanut allergy [75]. Enhanced cutaneous allergen penetration and enhanced cutaneous inflammatory responses upon topical application of chemical haptens have also been noted and explain the association with type IV sensitizations [14, 18]. Reducing excessive exposure to allergens and haptens known to cause type I and type IV sensitizations should be emphasized in any counseling.

36.4.2 Irritant Avoidance

Filaggrin-deficient murine skin shows decreased thresholds for development of irritant contact dermatitis (ICD) [14], and *FLG* mutations are associated with both acute and chronic ICD [77, 79]. Mechanical and chemical barrier disruption reduces mRNA levels of filaggrin [3], and soaking of the skin and exposure to solvents increases TEWL probably as a result of extraction of NMF components, since NMF application can reverse these findings [95]. Skin surface pH is increased with regular use of conventional soaps and decreases again after the change to an acidic cleanser [8]. Most conventional soaps are alkaline.

Avoidance of irritants that can deteriorate an already inherently impaired skin barrier, both professionally and at home, should be emphasized in counseling of filaggrin-deficient individuals and considered in pre-employment counseling.

36.4.3 Avoidance of Other Skin Barrier Stressors

Psychological stress may impair the skin barrier by means of inhibited lipid synthesis and secretion, decreased density of corneodesmosomes, as well as decreased production of filaggrin and other terminal differentiation-related proteins [96]. Cockroach and house dust mite allergens can aggravate permeability homeostasis and delay barrier recovery in murine skin via increased protease activity [97]. Finally, exposure to reduced environmental humidity accelerates TEWL [98]. Protection against psychological stress; cold, dry weather; air-conditioning; and house dust mites may be another focus for intervention.

36.4.4 UV Protection

Lack of filaggrin increases UV sensitivity [5, 6]. UCA has direct photo-protective properties [5, 6] and is reduced in filaggrin-deficient skin [10, 11]. An increased risk of nonmelanoma skin cancer has been documented in AD [99] but has not been assessed in relation to *FLG* mutation carrier status. The reports on increased UV sensitivity should encourage individuals with filaggrin mutations to rational UV protective behavior and encourage clinicians to limit the use of UV treatments in FLG mutation carriers, if possible. The findings should also be addressed in counseling settings.

36.4.5 Intervening with Gene-Environment Interactions

Cat ownership and tobacco smoking constitute an additional increased risk of, respectively, AD and asthma in *FLG* mutation carriers [100, 101]. The positive association between cat ownership in combination with *FLG* mutations and AD was not associated with increased sensitivity rates. The effect may be mediated by other mechanisms. Regarding tobacco smoking, this risk succeeded the risk of asthma among tobacco smokers in general. Avoidance of cat ownership seems to be another way to reduce the risk of development of AD in *FLG* mutation carriers, and tobacco smoking should cease in order to eliminate the additional effect on the risk of asthma. Data on gene-environment interactions relating to *FLG* mutations are sparse. Additional relevant gene-environment interactions may be revealed in the future.

36.4.6 Surveillance as Non-pharmacological Intervention

Awareness and surveillance of primary and secondary diseases are the final non-pharmacological interventional strategies with the potential for early intervention and even prevention.

Patients with filaggrin deficiency have an approximately seven times increased risk of bacterial skin infection with the pathogen *Staphylococcus aureus* [72] and ten times increased risk of eczema herpeticum caused by the viral pathogen herpes simplex [74] compared to wild-type carriers.

Some subgroups of *FLG* mutation carriers seem to require special attention. Whereas one-third of all individuals with *FLG* mutations will develop asthma [100]; three-fourths of all individuals with *FLG* mutations *and* food sensitivity will develop asthma; but *all* individuals with *FLG* mutations, dermatitis, and food sensitivity within the first 3 years of life will develop asthma [102]. The risk of ICD is increased by AD and by *FLG* mutations independently by a factor of 2, whereas the combination of AD and *FLG* mutations causes a synergistic effect on the risk of ICD (factor of ~5) [79].

Conclusion

Determination of *FLG* mutation carrier status is fundamental in order to identify this vulnerable population and improve surveillance of primary and secondary diseases related to filaggrin deficiency with the potential for early intervention and even prevention. Knowledge of *FLG* mutation carrier status also forms the basis for proper targeting of pharmacological and non-pharmacological interventions, personalized strategies, and counseling.

Patients genetically predisposed to a defective skin barrier should be warned against the increased risk of cutaneous and non-cutaneous diseases. Counseling should be directed at minimizing or avoiding exposure to agents known to deteriorate the skin barrier (e.g., irritants, low humidity, and psychological stress). Excessive topical exposure to agents essential for development of allergic disease should also be avoided, and food derivate should not be applied to the skin. These dangers should also be taken into account in preemployment counseling of filaggrin-deficient individuals. Avoidance of unnecessary UV exposure should also be emphasized. Finally, advice to avoid factors that aggravate an already increased inherent risk of AD and asthma (cat ownership and tobacco smoking) should be given.

The discoveries of *FLG* mutations and their pronounced impact on common skin diseases have inspired researchers and led to a quickly expanding understanding of the filaggrin pathway and skin barrier homeostasis. These scientific breakthroughs have revealed several potential targets for future therapy, and intensive research into promising candidate drugs is currently being conducted. These are reasons for optimism for better disease control in the future. Furthermore, primary prevention seems plausible.

References

1. Presland RB, Haydock PV, Fleckman P, Nirunsuksiri W, Dale BA. Characterization of the human epidermal proFLG gene. J Biol Chem. 1992;267:23772–81.
2. Dale BA, Presland RB, Lewis SP, Underwood RA, Fleckman P. Transient expression of epidermal FLG in cultured cells causes collapse of intermediate filament networks with alteration of cell shape and nuclear integrity. J Invest Dermatol. 1997;108:179–87.
3. Elias PM, Wakefield JS. Therapeutic implications of a barrier-based pathogenesis of AD. Clin Rev Allergy Immunol. 2011;41(3):282–95.
4. Rawlings AV, Harding CR. Moisturization and skin barrier function. Dermatol Ther. 2004;17:43–8.
5. Mildner M, Jin J, Eckhart L, Kezic S, Gruber F, Barresi C, et al. Knockdown of FLG impairs diffusion barrier function and increases UV sensitivity in a human skin model. J Invest Dermatol. 2010;130:2286–94.
6. Barresi C, Stremnitzer C, Mlitz V, Kezic S, Kammeyer A, Ghannadan M, et al. Increased sensitivity of histidinemic mice to UVB radiation suggests a crucial role of endogenous UCA in photoprotection. J Invest Dermatol. 2011;131:188–94.
7. Hachem J-P, Man M-Q, Crumrine D, Uchida Y, Brown BE, Rogiers V, et al. Sustained serine proteases activity by prolonged increase in pH leads to degradation of lipid processing enzymes and profound alterations of barrier function and stratum corneum integrity. J Invest Dermatol. 2005;125:510–20.
8. Elias PM, Hatano Y, Williams ML. Basis for the barrier abnormality in AD: outside-inside-outside pathogenic mechanisms. J Allergy Clin Immunol. 2008;121:1337–43.
9. Korting HC, Braun-Falco O. The effect of detergents on skin pH and its consequences. Clin Dermatol. 1996;14:23–7.
10. Kezic S, O'Regan GM, Lutter R, Jakasa I, Koster ES, Saunders S, et al. FLG loss-of-function mutations are associated with enhanced expression of IL-1 cytokines in the stratum corneum of patients with AD and in a murine model of FLG deficiency. J Allergy Clin Immunol. 2012;129:1031–9.
11. Mlitz V, Latreille J, Gardinier S, Jdid R, Drouault Y, Hafnagl P, et al. Impact of FLG mutations on Raman spectra and biophysical properties of the stratum corneum in mild to moderate AD. J Eur Acad Dermatol Venereol. 2012;26:983–90.
12. Gruber R, Elias PM, Crumrine D, Lin T-K, Brandner JM, Hachem J-P, et al. FLG genotype in ichthyosis vulgaris predicts abnormalities in epidermal structure and function. Am J Pathol. 2011;178:2252–63.
13. Angelova-Fischer I, Mannheimer A-C, Hinder A, Ruether A, Franke A, Neubert RHH, et al. Distinct barrier integrity phenotypes in FLG-related atopic eczema following sequential tape stripping and lipid profiling. Exp Dermatol. 2011;20:351–6.
14. Scharschmidt TC, Man M-Q, Hatano Y, Crumrine D, Gunathilake R, Sundberg JP, et al. FLG deficiency confers a paracellular barrier abnormality that reduces inflammatory thresholds to irritants and haptens. J Allergy Clin Immunol. 2009;124:496–506.
15. Jungersted JM, Scheer H, Mempel M, Baurecht H, Cifuentes L, Høgh JK, et al. Stratum corneum lipids, skin barrier function and FLG mutations in patients with atopic eczema. Allergy. 2010;65:911–18.
16. Winge MCG, Hoppe T, Berne B, Vahlquist A, Nordenskjöld M, Bradley M, et al. FLG genotype determines functional and molecular alterations in skin of patients with AD and ichthyosis vulgaris. PLoS ONE. 2011;6:e28254.
17. Sergeant A, Campbell LE, Hull PR, Porter M, Palmer CNA, Smith FJD, et al. Heterozygous null alleles in FLG contribute to clinical dry skin in young adults and the elderly. J Invest Dermatol. 2009;129:1042–5.
18. Oyoshi MK, Murphy GF, Geha RS. FLG-deficient mice exhibit T_H17-dominated skin inflammation and permissiveness to epicutaneous sensitization with protein antigen. J Allergy Clin Immunol. 2009;124:485–93.
19. Jensen J-M, Pfeiffer S, Witt M, Bräutigam M, Neumann C, Weichenthal M, et al. Different effects of pimecrolimus and betamethasone on the skin barrier in patients with AD. J Allergy Clin Immunol. 2009;123:1124–33.
20. Jensen J-M, Scherer A, Wanke C, Bräutigam M, Bongiovanni S, Letzkus M, et al. Gene expression is differently affected by pimecrolimus and betamethasone in lesional skin of AD. Allergy. 2011;67:413–23.
21. Howell MD, Kim BE, Gao P, Grant AV, Boguniewicz M, DeBenedetto A, et al. Cytokine modulation of AD FLG skin expression. J Allergy Clin Immunol. 2007;120:150–5.
22. Margolis DJ, Apter AJ, Gupta J, Hoffstad O, Papadopoulos M, Campbell LE, et al. The persistence of AD and FLG (FLG) mutations in a US longitudinal cohort. J Allergy Clin Immunol. 2012;130:912–17.
23. Kao JS, Fluhr JW, Man M-Q, Fowler AJ, Hachem J-P, Crumrine D, et al. Short-term glucocorticoid treatment compromises both permeability barrier homeostasis and stratum corneum integrity: inhibition of

epidermal lipid synthesis accounts for functional abnormalities. J Invest Dermatol. 2003;120:456–64.

24. Kim M, Jung M, Hong S-P, Jeon H, Kim M-J, Cho M-Y, et al. Topical calcineurin inhibitors compromise stratum corneum integrity, epidermal permeability and antimicrobial barrier function. Exp Dermatol. 2009;19:501–10.
25. Lee SE, Choi Y, Kim S-E, Noh EB, Kim S-C. Differential effects of topical corticosteroid and calcineurin inhibitor on the epidermal tight junction. Exp Dermatol. 2013;22:54–80.
26. Gånemo A, Virtanen M, Vahlquist A. Improved topical treatment of lamellar ichthyosis: a double-blind study of four different cream formulations. Br J Dermatol 1999;141:1027–32.
27. Effendy I, Kwangsukstith C, Lee JY, Maibach HI. Functional changes in human stratum corneum induced by topical glycolic acid: comparison with all-trans retinoic acid. Acta Derm Venereol 1995;75: 455–8.
28. Hoppe T, Winge MCG, Bradley M, Nordenskjöld M, Vahlquist A, Berne B et al. X-linked recessive ichthyosis: an impaired barrier function evokes limited gene responses before and after moisturizing treatments. Br J Dermatol 2012;167:514–22.
29. Brown SJ, Kroboth K, Sandilands A, Campbell LE, Pohler E, Kezic S, et al. Intragenic copy number variation within FLG contributes to the risk of AD with a dose-dependent effect. J Invest Dermatol. 2012;132:98–104.
30. Smith FJ, Irvine AD, Terron-Kwiatkowski A, Sandilands A, Campbell LE, Zhao Y, et al. Loss-of-function mutations in the gene encoding FLG cause ichthyosis vulgaris. Nat Genet. 2006;38:337–42.
31. Zhang C, Gurevich I, Aneskievich BJ. Organotypic modeling of human keratinocyte response to peroxisome proliferators. Cells Tissues Organs. 2012;196:431–41.
32. Schmuth M, Haqq CM, Cairns WJ, Holder JC, Dorsam S, Chang S, et al. Peroxisome proliferator-activated receptor (PPAR)-β/δ stimulates differentiation and lipid accumulation in keratinocytes. J Invest Dermatol. 2004;122:971–83.
33. Mao-Qiang M, Fowler AJ, Schmuth M, Lau P, Chang S, Brown BE, et al. Peroxisome-proliferator-activated receptor (PPAR)-γ activation stimulates keratinocyte differentiation. J Invest Dermatol. 2004;123:305–12.
34. Mao-Chiang M, Choi E-H, Schmuth M, Crumrine D, Uchida Y, Elias PM, et al. Basis for improved permeability barrier homeostasis induced by PPAR and LXR activators: liposensors stimulate synthesis, lamellar body secretion, and post-secretory lipid processing. J Invest Dermatol. 2006;126:386–92.
35. Kim H, Lim YJ, Park JH, Cho Y. Dietary silk protein, sericin, improves epidermal hydration with increased levels of FLGs and free amino acids in NC/Nga mice. Br J Nutr. 2012;108:1726–35.
36. Hou M, Sun R, Hupe M, Kim PL, Park K, Crumrine D, et al. Topical apigenin improves epidermal permeability barrier homeostasis in normal murine skin by divergent mechanisms. Exp Dermatol. 2013;22:210–15.
37. Marzani B, Pinto D, Minervini F, Calasso M, Cagno RD, Giuliani G, et al. The antimicrobial peptide pheromone Plantaricin A increases antioxidant defences of human keratinocytes and modulates the expression of FLG, involucrin, β-defensin 2 and tumor necrosis factor-α genes. Exp Dermatol. 2012;21:665–71.
38. Grether-Beck S, Felsner I, Brenden H, Kohne Z, Majora M, Marini A, et al. Urea uptake enhances function and antimicrobial defense in humans by regulating epidermal gene expression. J Invest Dermatol. 2012;132:1561–72.
39. Chen H, Common JEA, Haines RL, Balakrishnan A, Brown SJ, Goh CSM, et al. Wide spectrum of FLG-null mutations in AD highlights differences between Singaporean Chinese and European populations. Br J Dermatol. 2011;165:106–14.
40. Mühlemann O, Eberle AB, Stalder L, Zamudio Orozco R. Recognition and elimination of nonsense mRNA. Biochim Biophys Acta. 2008;1779:538–49.
41. Floquet C, Deforges J, Rousset J-P, Bidou L. Rescue of non-sense mutated p53 tumor suppressor gene by aminoglycosides. Nucleic Acids Res. 2011;39:3350–62.
42. Pinotti M, Rizzotto L, Pinton P, Ferraresi P, Chuansumrit A, Charoenkwan P, et al. Intracellular readthrough of nonsense mutations by aminoglycosides in coagulation factor VII. J Thromb Haemost. 2006;4:1308–14.
43. Gonzalez-Hilarion S, Beghyn T, Jia J, Debreuck N, Berte G, Mamchaoui K, et al. Rescue of nonsense mutations by amlexanox in human cells. Orphanet J Rare Dis. 2012;7:58.
44. Sermet-Gaudelus I, Renouil M, Fajac A, Bidou L, Parbaille B, Pierrot S, et al. In vitro prediction of stop-codon suppression by intravenous gentamicin in patients with cystic fibrosis: a pilot study. BMC Med. 2007;5:5.
45. Kayali R, Ku JM, Khitrov G, Jung ME, Prikhodko O, Bertoni C. Read-through compound 13 restores dystrophin expression and improves muscle function in the mdx mouse model for Duchenne muscular dystrophy. Hum Mol Genet. 2012;21:4007–20.
46. Floquet C, Hatin I, Rousset J-P, Bidou L. Statistical analysis of readthrough levels for nonsense mutations in mammalian cells reveals a major determinant of response to gentamicin. PLoS Genet. 2012;8:e1002608.
47. Weber TM, Kausch M, Rippke F, Schoelermann AM, Filbry AW. Treatment of xerosis with a topical formulation containing glyceryl glucoside, NMFs, and ceramide. J Clin Aesthet Dermatol. 2012;5:29–39.
48. Sugawara T, Kikuchi K, Tagami H, Aiba S, Sakai S. Decreased lactate and potassium levels in NMF from the stratum corneum of mild AD patients are involved with the reduced hydration state. J Dermatol Sci. 2012;66:154–9.
49. de Fine Olivarius F, Wulf HC, Crosby J, Norval M. The sunscreening effect of UCA. Photodermatol Photoimmunol Photomed. 1996;12:95–9.
50. Jansen CT, Pylkkanen L, Laihia J, et al. Improvement of epidermal barrier function and erythema by cis-UCA emulsion cream in adult patients with mild to

moderate AD. J Invest Dermatol. 2010;130 Suppl 2:S67.
51. Mao-Qiang M, Brown BE, Wu S, Feingold KR, Elias PM. Exogenous non-physiological vs. physiological lipids: divergent mechanisms for correction of permeability barrier dysfunction. Arch Dermatol. 1995;131: 809–16.
52. Mao-Qiang M, Feingold KR, Elias PM. Exogenous lipids influence permeability barrier recovery in acetone-treated murine skin. Arch Dermatol. 1993; 129:728–38.
53. Mao-Qiang M, Feingold KR, Thornfeldt CR, Elias PM. Optimization of physiological lipid mixtures for barrier repair. J Invest Dermatol. 1996;106:1096–101.
54. Sugarman JL, Parish LJ. A topical physiologic lipid-based, barrier repair formulation (Epiceram™ cream) is highly-effective monotherapy for moderate-to-severe pediatric AD: a multicenter, investigator-blinded trial comparing a barrier repair formulation, Epiceram™, to fluticasone propionate (Cutivate®) cream. J Invest Dermatol. 2008;128:S54.
55. Simpson EL, Berry T, Tofte S, Hanifin J, Eichenfield L. Epiceram for the treatment of mild to moderate AD – a pilot study. J Invest Dermatol. 2008;128:S77.
56. Zettersten EM, Ghadially R, Feingold KR, Crumrine D, Elias PM. Optimal ratios of topical stratum corneum lipids improve barrier recovery in chronologically aged skin. J Am Acad Dermatol. 1997; 37:403–8.
57. Hachem JA, Fowler AJ, Behne M, Fluhr JW, Feingold R, Elias PM. Increased stratum corneum pH promotes activation and release of primary cytokines from the strum corneum attributable to activation of serine proteases. J Invest Dermatol. 2002;119:258.
58. Hachem J-P, Roelandt T, Schürer N, Pu X, Fluhr J, Giddelo C, et al. Acute acidification of stratum corneum membrane domains using polyhydroxyl acids improves lipid processing and inhibits degradation of corneodesmosomes. J Invest Dermatol. 2010;130:500–10.
59. Buraczewska I, Lodén M. Treatment of surfactant-damaged skin in humans with creams of different pH values. Pharmacology. 2005;73:1–7.
60. Hvid M, Vestergaard C, Kemp K, Christensen GB, Deleuran B, Deleuran M. IL-25 in AD: a possible link between inflammation and skin barrier dysfunction? J Invest Dermatol. 2011;131:150–7.
61. Gutowska-Owsiak D, Schaupp AL, Salimi M, Taylor S, Ogg GS. Interleukin-22 downregulates FLG expression and affects expression of proFLG processing enzymes. Br J Dermatol. 2011;165:492–8.
62. Sanofi and Regeneron report positive proof-of-concept data for Dupilumab, an IL-4R alpha antibody, in AD. Press Release March 2nd 2013. http://investor.regeneron.com/releasedetail.cfm?releaseid=744703.
63. Antoniu SA. Pitrakinra, a dual IL-4/IL-13 antagonist for the potential treatment of asthma and eczema. Curr Opin Investig Drugs. 2010;11:1286–94.
64. Groves RW, Wilbraham D, Fuller R, Longphre M. Inhibition of IL-4 and IL-13 with an IL-4 mutein (Aeroderm) protects against flares in atopic eczema. J Invest Dermatol. 2007;127:S54.
65. Sehra S, Yao Y, Howell MD, Nguyen ET, Kansas GS, Leung DYM, et al. IL-4 regulates skin homeostasis and the predisposition toward allergic skin inflammation. J Immunol. 2010;184:3186–90.
66. Pazyar N, Felly A, Yaghoobi R. An overview of interleukin-1 receptor antagonist, anakinra, in the treatment of cutaneous diseases. Curr Clin Pharmacol. 2012;7:271–5.
67. Mazereeuw-Hautier J, Cope J, Ong C, Green A, Hovnanian A, Harper JI. Topical recombinant alpha-1-antitrypsin: a potential treatment for Netherton syndrome? Arch Dermatol. 2006;142:396–8.
68. Hoste E, Kemperman P, Devos M, Denecker G, Kezic S, Yau N, et al. Caspase-14 is required for FLG degradation to NMFs in the skin. J Invest Dermatol. 2011;131:2233–41.
69. Tan SP, Abdul-Ghaffar S, Weller RB, Brown SB. Protease-antiprotease imbalance may be linked to potential defects in proFLG proteolysis in AD. Br J Dermatol. 2012;166:1137–40.
70. Palmer CNA, Irvine AD, Terron-Kwiatkowski A, Zhao Y, Liao H, Lee SP, et al. Common loss-of-function variants of the epidermal barrier protein FLG are a major predisposing factor for AD. Nat Genet. 2006;38:441–6.
71. Rodriguez E, Baurecht H, Herberich E, Wagenpfeil S, Brown SJ, Cordell HJ, et al. Meta-analysis of FLG polymorphisms in eczema and asthma: robust risk factors in atopic disease. J Allergy Clin Immunol. 2009;123:1361–70.
72. Cai SCS, Chen H, Koh W-P, Common JEA, van Bever HP, McLean WHI, et al. FLG mutations are associated with recurrent skin infection in Singaporean Chinese patients with AD. Br J Dermatol. 2012;166: 200–3.
73. Novak N, Baurecht H, Schäfer T, Rodriguez E, Wagenpfeil S, Klopp N, et al. Loss-of-function mutations in the FLG gene and allergic contact sensitization to nickel. J Invest Dermatol. 2008;128:1430–5.
74. Gao P-S, Rafaels NM, Hand T, Murray T, Boguniewicz M, Hata T, et al. FLG mutations that confer risk of AD confer greater risk for eczema herpeticum. J Allergy Clin Immunol. 2009;124:507–13.
75. Brown SJ, Asai Y, Cordell HJ, Campbell LE, Zhao Y, Liao H, et al. Loss-of-function variants in the FLG gene are a significant risk factor for peanut allergy. J Allergy Clin Immunol. 2011;127:661–7.
76. Thyssen JP, Ross-Hansen K, Johansen JD, Zachariae C, Carlsen BC, Linneberg A, et al. FLG loss-of-function mutation R501X and 2282del4 carrier status is associated with fissured skin on the hands: results from a cross-sectional population study. Br J Dermatol. 2012;166:46–53.
77. De Jongh CM, Khrenova L, Verberk MM, Calkoen F, van Dijk FJ, Voss H, et al. Loss-of-function polymorphisms in the FLG gene are associated with an increased susceptibility to chronic ICD: a case-control study. Br J Dermatol. 2008;159:621–7.

78. Thyssen JP, Carlsen BC, Menné T, Linneberg A, Nielsen NH, Meldgaard M, et al. FLG null mutations increase the risk and persistence of hand eczema in subjects with AD: results from a general population study. Br J Dermatol. 2010;163:115–20.
79. Visser MJ, Landeck L, Campbell LE, McLean WHI, Weidinger S, Calkoen F, et al. Impact of AD and loss-of-function mutations in the FLG gene on the development of occupational ICD. Br J Dermatol. 2013;168:326–32.
80. Thyssen JP, Linneberg A, Ross-Hansen K, Carlsen BC, Meldgaard M, Szecsi PB, et al. FLG mutations are strongly associated with contact sensitization in individuals with dermatitis. Contact Dermatitis. 2013;68(5):273–6.
81. Sevilla LM, Nachat R, Groot KR, Klement JF, Uitto J, Dijan P, et al. Mice deficient in involucrin, envoplakin, and periplakin have a defective epidermal barrier. J Cell Biol. 2007;179:1599–612.
82. Holleran WM, Ginns EI, Menon GK, Grundmann J-U, Fartasch M, McKinney CE, et al. Consequences of β-glucocerebrosidase deficiency in epidermis. J Clin Invest. 1994;93:1756–64.
83. Williams HC. Epidemiology of AD. Clin Exp Dermatol. 2000;25:522–9.
84. Marenholz I, Nickel R, Rüschendorf F, Schulz F, Esparza-Gordillo J, Kerscher T, et al. FLG loss-of-function mutations predispose to phenotypes involved in the atopic march. J Allergy Clin Immunol. 2006;118:866–71.
85. Sandilands A, Terron-Kwiatkowski A, Hull PR, O'Regan GM, Clayton TH, et al. Comprehensive analysis of the gene encoding FLG uncovers prevalent and rare mutations in ichthyosis vulgaris and atopic eczema. Nat Genet. 2007;39:650–4.
86. Brown SJ, McLean WH. Eczema genetics: current state of knowledge and future goals. J Invest Dermatol. 2009;129:866–71.
87. Flohr C, England K, Radulovic S, McLean WHI, Campbell LE, Barker J, et al. FLG loss-of-function mutations are associated with early-onset eczema, eczema severity and TEWL at 3 months of age. Br J Dermatol. 2010;163:1333–6.
88. Simpson EL, Berry TM, Brown PA, Hanifin JM. A pilot study of emollient therapy for the primary prevention of AD. J Am Acad Dermatol. 2010;63: 587–93.
89. Soll RF, Edwards WH. Topical ointment for preventing infection in preterm infants. Cochrane Database Syst Rev. 2000;2, CD001150.
90. Lowe AJ, Tang MLK, Dharmage SC, Varigos G, Forster D, Gurrin LC, et al. A phase I study of daily treatment with a ceramide-dominant triple lipid mixture commencing in neonates. BMC Dermatol. 2012;12:3.
91. Ying S, Meng Q, Corrigan CJ, Lee TH. Lack of FLG expression in the human bronchial mucosa. J Allergy Clin Immunol. 2006;118:1386–8.
92. De Benedetto A, Qualia CM, Baroody FM, Beck LA. FLG expression in oral, nasal and esophageal mucosa. J Invest Dermatol. 2008;128:1594–7.
93. Lack G, Fox D, Northstone K, Golding J. Factors associated with the development of peanut allergy in childhood. N Engl J Med. 2003;348:977–85.
94. Fox AT, Sasieni P, Toit GD, Syed H, Lack G. Household peanut consumption as a risk factor for the development of peanut allergy. J Allergy Clin Immunol. 2009;123:417–23.
95. Visscher MO, Tolia GT, Wickett RR, Hoath SB. Effect of soaking and NMF on stratum corneum water-handling properties. J Cosmet Sci. 2003;54:289–300.
96. Choi E-H, Brown BE, Crumrine D, Chang S, Man M-Q, Elias PM, et al. Mechanisms by which psychologic stress alters cutaneous permeability barrier homeostasis and stratum corneum integrity. J Invest Dermatol. 2005;124:587–95.
97. Jeong SK, Kim HJ, Youm J-K, Ahn SK, Choi EH, Sohn MH, et al. Mite and cockroach allergens activate protease-activated receptor 2 and delay epidermal permeability barrier recovery. J Invest Dermatol. 2008;128:1930–9.
98. Scott IR, Harding CR. FLG breakdown to water binding compounds during development of the rat stratum corneum is controlled by the water activity of the environment. Dev Biol. 1986;115:84–92.
99. Jensen AO, Svaerke C, Kormendine FD, et al. AD and risk of skin cancer: a Danish nationwide cohort study (1977–2006). Am J Clin Dermatol. 2012;13:29–36.
100. Bisgaard H, Simpson A, Palmer CNA, Bønnelykke K, McLean I, Mukhopadhyay S, et al. Gene-environment interaction in the onset of eczema in infancy: FLG loss-of-function mutations enhanced by neonatal cat exposure. PLoS One. 2008;5:e131.
101. Berg ND, Husemoen LLN, Thuesen BH, Hersoug L-G, Elberling J, Thyssen JP, et al. Interaction between FLG null mutations and tobacco smoking in relation to asthma. J Allergy Clin Immunol. 2012;129:374–80.
102. Marenholz I, Kerscher T, Bauerfeind A, Esparza-Gordillo J, Nickel R, Keil T, et al. An interaction between FLG mutations and early food sensitization improves the prediction of childhood asthma. J Allergy Clin Immunol. 2009;123:911–16.

The manufacturer's authorised representative in the EU is Springer Nature Customer Service Centre GmbH, Europaplatz 3, 69115 Heidelberg, Germany. If you have any concerns regarding our products, please contact ProductSafety@springernature.com

Printed and bound by CPI Group (UK) Ltd, Croydon, CR0 4YY
22/07/2026
02174186-0001